AF804786

122

Advances in Experimental Medicine and Biology

Volume 1030

Editorial Board
IRUN R. COHEN, *The Weizmann Institute of Science, Rehovot, Israel*
ABEL LAJTHA, *N.S. Kline Institute for Psychiatric Research, Orangeburg, NY, USA*
JOHN D. LAMBRIS, *University of Pennsylvania, Philadelphia, PA, USA*
RODOLFO PAOLETTI, *University of Milan, Milan, Italy*
NIMA REZAEI, *Tehran University of Medical Sciences Children's Medical Center, Children's Medical Center Hospital, Tehran, Iran*

More information about this series at http://www.springer.com/series/5584

Anwar Sunna • Andrew Care
Peter L. Bergquist
Editors

Peptides and Peptide-based Biomaterials and their Biomedical Applications

Springer

Editors
Anwar Sunna
Department of Chemistry and
Biomolecular Sciences
Macquarie University
North Ryde, NSW, Australia

Biomolecular Discovery and Design
Research Centre
Macquarie University
North Ryde, NSW, Australia

Andrew Care
Department of Chemistry and
Biomolecular Sciences
Macquarie University
North Ryde, NSW, Australia

ARC Centre of Excellence for
Nanoscale BioPhotonics (CNBP)
Macquarie University
North Ryde, NSW, Australia

Peter L. Bergquist
Department of Chemistry and
Biomolecular Sciences
Macquarie University
North Ryde, NSW, Australia

Biomolecular Discovery and Design
Research Centre
Macquarie University
North Ryde, NSW, Australia

Department of Molecular Medicine
& Pathology
Medical School, University of Auckland
Auckland, New Zealand

ISSN 0065-2598 ISSN 2214-8019 (electronic)
Advances in Experimental Medicine and Biology
ISBN 978-3-319-66094-3 ISBN 978-3-319-66095-0 (eBook)
https://doi.org/10.1007/978-3-319-66095-0

Library of Congress Control Number: 2017954955

© Springer International Publishing AG 2017
This work is subject to copyright. All rights are reserved by the Publisher, whether the whole or part of the material is concerned, specifically the rights of translation, reprinting, reuse of illustrations, recitation, broadcasting, reproduction on microfilms or in any other physical way, and transmission or information storage and retrieval, electronic adaptation, computer software, or by similar or dissimilar methodology now known or hereafter developed.
The use of general descriptive names, registered names, trademarks, service marks, etc. in this publication does not imply, even in the absence of a specific statement, that such names are exempt from the relevant protective laws and regulations and therefore free for general use.
The publisher, the authors and the editors are safe to assume that the advice and information in this book are believed to be true and accurate at the date of publication. Neither the publisher nor the authors or the editors give a warranty, express or implied, with respect to the material contained herein or for any errors or omissions that may have been made. The publisher remains neutral with regard to jurisdictional claims in published maps and institutional affiliations.

Printed on acid-free paper

This Springer imprint is published by Springer Nature
The registered company is Springer International Publishing AG
The registered company address is: Gewerbestrasse 11, 6330 Cham, Switzerland

Preface

The chapters in this volume demonstrate that peptides in various forms have been used increasingly over the past decade as molecular building blocks in nanobiotechnology. Biologically, their isolation has been facilitated by the creation of combinatorial libraries based on expression by bacteriophage (usually M13) and bacterial display systems (usually by *Escherichia coli*). Many of these peptides show selectivity and bind with high affinity to the surfaces of a diverse range of solid materials, e.g. metals, metal oxides, metal compounds, magnetic materials, semiconductors, carbon materials, polymers and minerals. These solid-binding peptides can direct the assembly and functionalisation of materials and have the ability to mediate the synthesis and construction of nanoparticles and complex nanostructures. The field is being influenced by the design and construction of solid-state synthetic peptides and the possibility of altering folding patterns, for example, by the incorporation of 'unnatural' β-amino acids that are without some of the constraints of natural α-amino acids.

Self-assembling biomolecules to create nanoscale-ordered templates have emerged as an important area in nanotechnology. Peptides are particularly attractive as molecular building blocks because their structure, folding and stability have been studied in detail already along with measuring equipment developed for the study of surface chemistry in other systems. Self-assembling peptides can adopt diverse 3-D architectures such as vesicles, micelles, monolayers, bilayers, fibres, tubes, ribbons and tapes. Importantly, short peptides can be produced easily by standard chemical synthetic methods, thus avoiding the overall complexities of synthesising large proteins. Their biocompatibility makes them ideal candidates for the stabilisation of enzymes, as used in biosensors, for example.

In this volume, we have collected chapters from researchers investigating both basic and applied questions that are of contemporary interest in the development of the rules governing small peptides that have shown promise in applications from biosensors to anticancer agents capable of penetrating cellular membranes of mammals. The importance of biomolecular self-assembly on solid materials and molecular recognition plays a critical role in biological interactions and has become recognised increasingly as essential in many biomedical situations.

The chapter by Shiba sets the scene by describing the overall requirements for peptides to act as surface functionalisation agents to create 'programmable biosurfaces' with the implication that the surfaces are endowed with specific biological functions that are active and derived from either natural or artificial peptide sequences. They are identified by those that bind selectively to a given target. Cell-penetrating peptides that bind to cell membranes including ones that can recognise membrane vesicles with highly curved surfaces that may play a role in devices that are designed for the differentiation of subclasses of extracellular vesicles are of interest. The chapter has a commentary on the four methods of immobilising the peptides on surfaces, and the author concludes that the main advantages of the utilisation of peptides for surface functionalisation are their definability and programmability.

Care et al. describe the isolation and application of peptides isolated from combinatorial libraries involving M13 that are able to bind to metal and mineral surfaces with emphasis on inexpensive matrices derived from silica. These solid-binding peptides provide simple and flexible bioconjugation methods that avoid conventional chemical techniques that may cause biomolecules to attach to surfaces with altered conformations and random orientations that cause a loss of activity. The authors have utilised a fusion protein, Linker-Protein G (LPG), as an anchoring point for the orientated immobilisation and functionalisation of nanomaterials with antibodies in a range of applications such as capture, detection and imaging and for the functionalisation of lanthanide-doped upconversion nanocrystals (UCNCs) which have considerable potential as probes and delivery vehicles for imaging, diagnostics and therapeutics. The chapter describes examples such as the controlled synthesis of nanomaterials and nanostructures, formation of hybrid biomaterials, immobilisation of functional proteins and improved nanomaterial biocompatibility.

A critical review is provided by Walsh of the fundamental knowledge gaps that need to be resolved before unambiguous molecular simulations can be proposed for the binding of peptides to surfaces rather than a compendium of solutions based on current data in the literature. Much current information is defective in that physiologically relevant conditions have not been part of the modelling and two specific surfaces are used to illustrate the extent of the lack of pertinent data and what studies need to be done to fill the gap. However, the successful implementation in current medical applications is largely on a trial-and-error basis, and molecular simulation approaches can complement experimental characterisation techniques and provide relevant details at the atomic scale.

Seoudi and Mechler discuss the design principles relating to self-assembled nanomaterials based on peptides that are centred on bottom-up nanofabrication through small molecule precursors using standard thermodynamic principles so that energy minimisation of smaller molecules drives the formation of larger, well-ordered structures. They examine the role of non-covalent interactions in the self-assembly of peptides and discuss the bio-inspired nature of the components for construction of more complex formats and the role that unnatural β-peptides could play in bottom-up self-assembly.

The Williams group chapter provides a comprehensive account of bioprinting and biofabrication processes that have been evoked by the potential offered by regenerative medicine with possibilities offered for organ repair or replacement, drug delivery and tissue engineering in general. A major challenge is the development of a suitable matrix that can carry out the functions of the extracellular matrix to provide the appropriate behavioural cues to cells as well as physical support. Techniques are being developed to print blends of biomaterials including living cells in suspension as hydrogels (termed 'bio-inks') that protect cells during the printing process. They record that the procedures outlined have found use in both organ engineering and non-organ tissue engineering applications.

The chapter by Horsley and co-workers deals with the application of peptides as bio-inspired molecular electronic materials and the need to understand the unique electronic properties of single peptides in biology. They describe several factors known to influence electron transfer in peptides and provide a case study illustrating the function of peptides in electronics and the need to understand what controls the mechanism of charge transfer as a key requirement to realise the application of peptides as molecular electronic materials.

How cartilaginous tissue offers special problems relating to regeneration after injury as it does not contain vascular or nervous elements is described in the chapter by Hastar et al. Consequently, scaffolds formed by biomaterials are promising tools for cartilage regeneration. The extracellular matrix is responsible for the lubrication and articulation functions of cartilage and scaffolds that mimic the extracellular matrix can serve as a temporary replacement for cartilage tissue. Modified polymers can recruit mesenchymal stem cells to the damaged area with the aid of specific peptides as well as supplying the environmental conditions necessary for cartilage regeneration. These scaffolds must be biocompatible and biodegradable to eliminate immunogenic responses and exhibit the mechanical properties of newly-formed cartilaginous tissue for new tissue support and provision of signals for cellular recruitment and differentiation.

Ulapane et al. address the role of peptides in drug delivery to cancer cells, primarily to lower side effects, but especially their selective ability in binding to the appropriate receptors targeted. Some active peptides have been derived from endogenous molecules such as endorphins and oxytoxin, while others have been derived from the active regions of much larger proteins, for example, the Arg-Gly-Asp (RDG) sequence derived from a number of extracellular matrix proteins that is recognised by various receptors on the cell surface. They remark on the use of cell-penetrating peptides (CPPs) as possible peptide-drug conjugates for therapeutic cancer treatment and discuss several complex examples. An interesting discussion follows on the possibility of using peptides to enhance the paracellular permeation of molecules the size of drugs.

Peptide lipidation is a promising strategy to improve pharmacokinetic and pharmacodynamic profiles of peptide-based drugs. Self-adjuvanting peptide-based vaccines commonly utilise the powerful TLR2 agonist Pam_nCys lipid to stimulate adjuvant activity. The chemical synthesis of lipidated peptides can be challenging and time-consuming. Accordingly, efficient synthetic chemical routes to access homogeneous lipid-tagged peptides are of considerable interest. The Brimble group describes the occurrence of natural examples of peptide lipidation found with microorganisms as candidates for the generation of novel compounds with improved pharmacokinetic and pharmacodynamic characteristics. In particular, there is a focus on synthetic approaches allowing the incorporation of Pam_nCys-based Toll-like receptor 2 lipidated ligands into peptides with the possibility of generating self-adjuvanting vaccine constructions. Protein lipidation is not uncommon in nature, and naturally derived peptides have been optimised synthetically in the quest for effective peptide-based drug candidates, and there is a comprehensive evaluation of the Pam_n Cys ligand as an adjuvant for peptide-based vaccines.

Animal venoms are a valuable source of novel therapeutic peptides. The chapter by Daniel and Clark on cone snails and their potent and fast-acting paralytic venoms describes the discovery and application of the short bioactive peptide molecules responsible for the rapid inhibition of neuromuscular currents in animal tissues. *Conus* venoms comprise an astoundingly diverse cocktail of peptide toxins (conopeptides) that are potent and highly selective modulators of important neurophysiological ion channels, G-protein-coupled receptors (GPCRs) and membrane transporters. Various synthetic approaches that have been used to engineer conotoxin analogues with improved structural and pharmacological properties are outlined.

Cell-penetrating peptides (CPPs) have received significant attention due to their inherent ability to cross plasma membranes or to facilitate the cellular entry of drug molecules, macromolecules and nanoparticles into cells. However, despite their versatility for delivery of drug cargoes, their non-selectivity and lack of efficiency have led to research promoting rational design using CPPs as a building block for nanostructure formation by way of self-assembly providing functionality for intracellular delivery. The chapter on the uptake mechanisms for cell-penetrating peptides (CPPs) by Gestin et al. comments on their extensive use for delivering cargoes unable to cross the cell membrane and describes numerous cell-penetrating peptides offering a wide library of structures. The uptake mechanisms of CPPs and their respective cargoes are poorly understood although they have been employed extensively to transport cargo molecules. The authors discuss the roles of compounds such as heparan sulphate proteoglycans and neuropilin-1 in the uptake of CPPs and their cargoes and the association of peptide motifs with the energy-independent pathway and several variations on the process of endocytosis.

Shi et al. describe strategies for the design of self-assembling CCP hydrophobic conjugates and their manipulation for enhanced intracellular delivery. They review the design of self-assembling CPPs as conjugates, first by adding appropriate hydrophobic segments such as lipid tails to the CPP sequence (which is hydrophilic). A second method uses the formation of hydrogen bonding and π-π stacking to drive the assembly process. They discuss how CPP drug conjugates can be formed readily since the majority of CPPs are hydrophobic while most therapeutic drugs are hydrophilic and the two components are conjugatable by way of a specific linkage that is designed to be released on its cleavage.

The general inability of proteins to penetrate mammalian cells or specifically target tumour cells decreases their value as potential therapeutic agents for a number of diseases, but recently, cell-penetrating peptides (CPPs) have been shown to bring about the delivery of therapeutic proteins or peptides into living cells. Feni and Nuendorf outline the several conjugation procedures used for linking the various 'cargoes' – small molecules, peptides, proteins and drugs – by covalent binding with CPPs, most frequently by a disulphide linkage. They cite the example of the CPP R8 (octaarginine) conjugated to the anticancer drug taxol via disulphide linkage. Covalent linkage methods raise the concern that the introduction of the bond between the CPP and the active cargo may cause alterations in biological activity. As a result, non-covalent complexes are employed even for the delivery of small molecule drugs.

Our intention in editing this volume is to create an awareness of the potential of peptides and to trace how and where the research has emerged and to outline the opportunities we see to develop novel and validated tools for specific biomedical therapeutic scenarios. We invite interested researchers to develop and expand applications of these versatile biomolecules.

North Ryde, NSW, Australia

Anwar Sunna
Andrew Care
Peter L. Bergquist

Contents

About the Editors

Anwar Sunna is an associate professor in the Department of Chemistry and Biomolecular Sciences at Macquarie University (MQ), Sydney, Australia. He obtained a PhD from the Hamburg University of Technology in Germany. He was manager of the Environmental Biotechnology Cooperative Research Centre at MQ and later was the recipient of the prestigious Vice-Chancellor's Innovation Fellowship. His recent research has been on the interaction between biomolecules and inorganic compounds including new synthetic peptide linkers with applications in the functionalisation of nanoparticles, bioimaging and cancer therapy. Anwar is a member of the MQ Biomolecular Discovery and Design Research Centre, MQ Biosecurity Futures Research Centre, Australian Research Council (ARC) Training Centre for Molecular Technology in the Food Industry and the ARC Centre of Excellence for Nanoscale BioPhotonics. He is also one of the directors of Synthetic Biology Australasia.

Andrew Care is a research fellow in the ARC Centre of Excellence for Nanoscale BioPhotonics, a transdisciplinary research centre that aims to develop innovative nanotechnologies to investigate complex living systems. He obtained his PhD from Macquarie University (MQ) in Sydney, Australia. Andrew is also a member of both the MQ Biomolecular Discovery and Design Research Centre and the MQ BioFocus Research Centre. His current research is focused on the development of engineered biomolecules, including solid-binding peptides, to control the self-assembly and biofunctionalisation of materials and biomolecules in a range of biomedical and biotechnological applications.

Peter L. Bergquist is emeritus professor in the Biomolecular Discovery and Design Research Centre at Macquarie University and emeritus professor in the Department of Molecular Medicine and Pathology at the University of Auckland Medical School. He has a PhD and DSc from the University of Auckland in New Zealand and has been a postdoctoral fellow and research fellow at Harvard Medical School and Yale and Oxford Universities and a visiting fellow at the New York University School of Medicine. His recent publications reflect research performed with the Environmental Biotechnology Cooperative Research Centre in Australia, and he developed research expertise in gas phase catalysis and biofunctionalisation of inorganic matrices containing silica. He has been on the editorial boards of several significant journals such as *Applied and Environmental Microbiology* and the *Journal of Bacteriology*.

1 Programmable Bio-surfaces for Biomedical Applications

Kiyotaka Shiba

Abstract

A peptide can be used as a functional building block to construct artificial systems when it has sufficient transplantability and functional independence in terms of its assigned function. Recent advances in *in vitro* evolution systems have been increasing the list of peptides that specifically bind to certain targets, such as proteins and cells. By properly displaying these peptides on solid surfaces, we can endow the inorganic materials with various biological functions, which will contribute to the development of diagnosis and therapeutic medical devices. Here, the methods for the peptide-based surface functionalization are reviewed by focusing on sources of peptides as well as methods of immobilization.

Keywords

Peptide aptamer • In vitro evolution • Artificial peptide • Motif programing • Material surface • Diagnostic • Exosome

1.1 Introduction

The functionalization of surfaces of materials employs three major approaches: physical, chemical, and biological. In the biological functionalization of material surfaces, various biomolecules, including proteins, peptides, DNA, RNA, sugar, and lipids, can be used as material units for exerting certain biological functions. Among them, in this review, peptides are focused on, and current state-of-the-art peptide-based bio-surfaces are introduced.

"Programmable bio-surfaces" represent the material surfaces that are endowed with certain biological functions by peptide immobilization. "Programmable" implies the elasticity of peptide-based functionalization, *i.e.*, we can not only select the functional peptides from natural sources, but also even create arti-

K. Shiba (✉)
Division of Protein Engineering, The Cancer Institute of Japanese Foundation for Cancer Research, Ariake, Koto, Tokyo 135-8550, Japan
e-mail: kshiba@jfcr.or.jp

© Springer International Publishing AG 2017

A. Sunna et al. (eds.), *Peptides and Peptide-based Biomaterials and their Biomedical Applications*, Advances in Experimental Medicine and Biology 1030, https://doi.org/10.1007/978-3-319-66095-0_1

ficial peptides using an *in vitro* evolution system. If a peptide has sufficient transplantability and functional independence in terms of its assigned function, we can recuperate the function on the surface of the materials (Shiba 2010b).

Figure 1.1 shows four main technical elements that must be considered to make bio-surfaces using peptides. "Peptides" represent the source of peptides to be used. They may be selected from natural sources and can be artificially created by appropriate methodologies. The "Function" is exerted by an immobilized peptide. The primary function of peptides is specific binding. All biological activities rely on this specific binding between biomolecules.

When we make bio-surfaces with peptides, we must first consider which function of peptides is needed to realize certain functionalities. If needed, we can create novel peptides having the necessary affinity. Then, an appropriate method of "Immobilization" is required, by which peptides are anchored on the surfaces of materials. And, "Material" is the one, on which peptides are immobilized. Below, sources of peptides and methods of immobilization will be descried in detail, referring to peptide functions and materials to be surface modified.

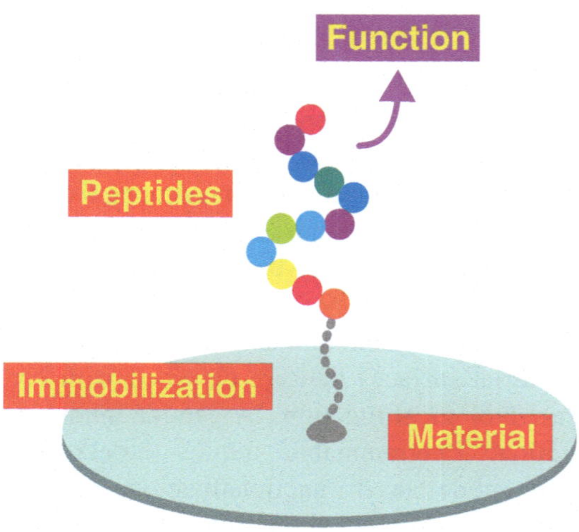

Fig. 1.1 Four technical elements that are discussed for programmable bio-surfaces

1.2 Sources of Peptide

Here, the sources of peptides that are immobilized on material surfaces are discussed. Although the difference between peptides and proteins is elusive, in this review, a peptide is defined as an amino acid polymer whose length is within the range of chemically synthesizable ones (current peptide synthesis technologies allow for the robust synthesis of peptides having 50 residues or so). In this definition, a peptide is not necessary chemically synthesized. Short polypeptides that are prepared from fragmentation and fractionation from natural proteins are also regarded as peptides. This definition of a peptide also includes those that have non-natural amino acids and those with chemical modifications. As shown in Fig. 1.2, the sources of peptides for bio-surfaces can be summarized into two major categories: natural and artificial, depending on whether their sequences are extracted from natural polypeptides or artificially created by researchers.

1.2.1 Natural Peptide

Organisms produce various biologically functional peptides (Kastin 2006). Examples include peptidic hormones, bactericidal peptides (Mahlapuu et al. 2016), anti-freezing peptides, mineralization peptides (Weiner and Hood 1975; Mann 2001), venom peptide (Vetter et al. 2011), gene regulating peptides (Lauressergues et al. 2015) and communication peptide of virus (Erez et al. 2017). These natural peptides are produced from the translation of short open reading frames on the genome, from post translational processing from larger precursors, or from non-ribosomal peptide synthesis (Weissman 2015). As an example of the usage of natural peptides in bio-surfaces, Yoshinari et al. have used histatin 5, an antimicrobial peptide secreted from the human salivary gland (Oppenheim et al. 1988), for the purpose of functionalizing the surface of titanium implants by fusing the peptide to an artificial peptide having an affinity to titanium (see Sect. 1.2.3) (Yoshinari et al. 2010).

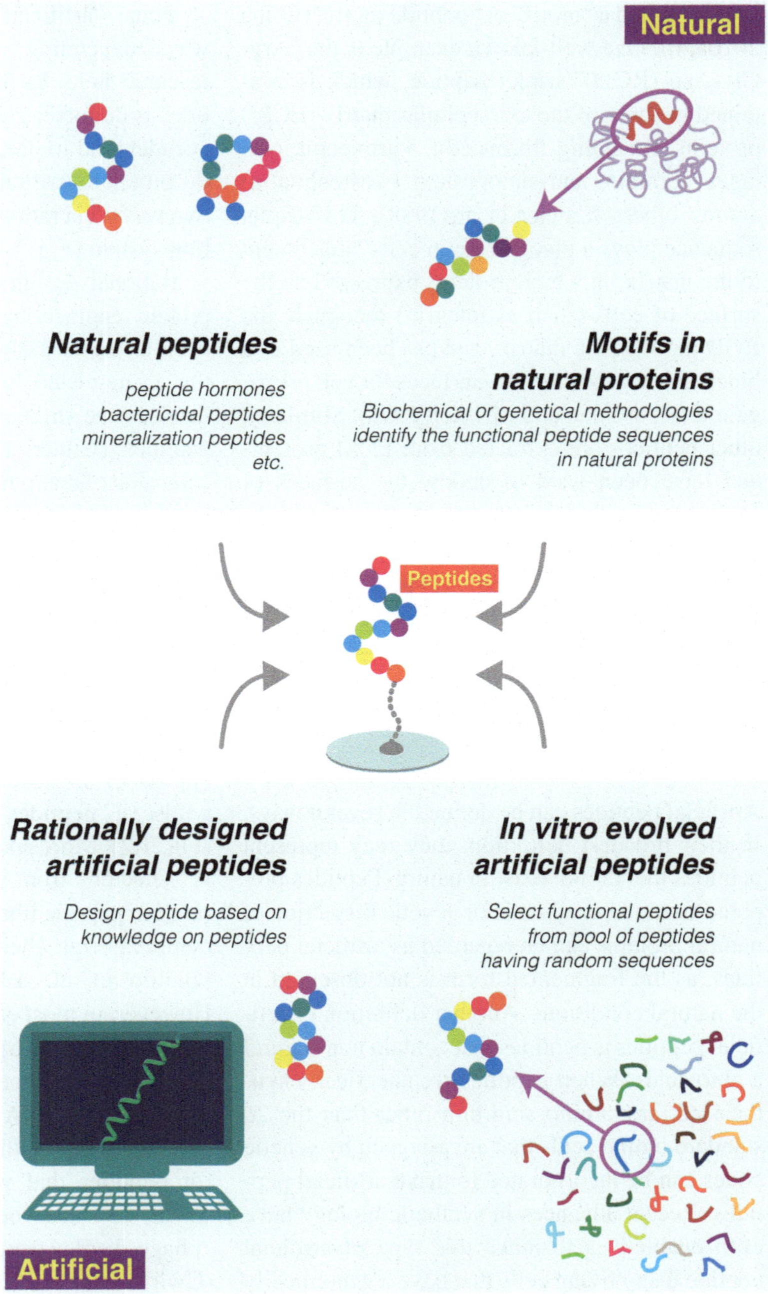

Fig. 1.2 Sources of peptides used to make bio-surfaces

1.2.2 Peptide Derived from Natural Protein

In some cases, the function of a protein, or a part of the function of a multifunctional protein, can be attributed to a short stretch of a peptide sequence contained in the protein sequence (Fig. 1.2). Biochemical or genetical methodologies are used for the identification of the peptide sequence. The identified peptide often recapitulates the function of its parental protein by itself (as a peptide); furthermore, the function can be transferred to other molecules including not only polypeptides, but also various non-peptidic molecules. This kind of short peptide sequence is usually called a "motif" or "peptide motif" (Shiba 2010a, b). One well-known example is the "Arg-Gly-Asp (RGD)" triplet peptide, which is contained in some of the extracellular matrix (ECM) proteins, including fibronectin, vitronectin, collagen, laminin, and sialoprotein I (osteopontin) among others (Lawler et al. 1988). This triplet sequence plays a pivotal role in cells' attachment to the matrix, in which proteins expressed on the surface of cells (such as integrin) recognize the RGD motif in the matrix, and has been used as a block unit to fabricate bio-surfaces for cell propagation (Houseman and Mrksich 1998). Similarly, other peptides are extracted from ECM proteins and have been used to endow the surfaces of materials with cell attachment activities (Melkoumian et al. 2010; Klim et al. 2010). In some cases, these natural protein-derived peptides were exploited to peptidic pharmaceuticals (Lima e Silva et al. 2017).

1.2.3 Artificial Peptide

Artificial peptides can be defined in several ways. In their broadest definition, they may represent peptides that do not exist in nature. Peptides prepared from the enzymatic or genetic dissection of natural proteins can be regarded as artificial peptides, as the fragmented form is not observed in the natural conditions. Another definition of artificial peptides is peptides that contain non-natural amino acids. When peptides are chemically synthesized, any amino acid unit other than the 20 standard amino acids that are assigned by genetic codes can be incorporated to make artificial peptides. Recent advances in synthetic biology have even enabled us to make this type of artificial peptide using living cells that have a genomically re-coded translation system (Young et al. 2011; Lajoie et al. 2013). In addition, various chemical modifications are post-translationally introduced into peptides to enhance peptides' activity (Ng et al. 2012). Alternatively, these modifications can be incorporated using modified amino acids as the building blocks during peptide synthesis.

Thus, "artificial" can be defined in several ways depending on the frameworks of the research field. In this review, an "artificial peptide" is defined as a peptide whose sequences are not identical to natural peptides or parts of natural proteins. Artificial peptides can be created in two ways: via rational design or an *in vitro* evolution system (Fig. 1.2).

Rational design is an approach employed in protein engineering, in which researchers have been trying to design artificial peptides based on the accumulated knowledge on the relationship between the structure and function of proteins or peptides (Ulmer 1983). The recent admirable extension of computing capabilities has enabled us to freely design almost any peptide that forms specific three-dimensional conformations (Bhardwaj et al. 2016). Of course, in the designing process, people avoid any resemblances in their sequences between designed peptides and natural proteins.

The other approach for the creation of artificial peptides is to use an *in vitro* evolution system (Szostak 1992; Shiba 1998). In this approach, an artificial peptide is created from pools of peptides having random sequences (Fig. 1.2). More specifically, an artificial peptide is "selected" from a naïve library of peptides. A random peptide library can be prepared from the combinatorial chemical synthesis of peptides (Gallop et al. 1994; Gordon et al. 1994). However, in most cases, the pool of random peptides is prepared by the translation of near-random DNA sequences, as the synthesis of random sequences of DNA is much easier than that of peptides. One of the *in vitro* evolution systems for peptides that was established in the early 1990s and has been widely employed is a "phage-display peptide library system" (Fig. 1.3) (Cwirla et al. 1990; Scott and Smith 1990; Devlin et al. 1990). In this system, near-random sequences of DNA (they are not random, but are skewed to reduce the appearance of translation termination codons) are inserted into the phage (bacterial virus) genome so inserted random sequences can be translated as fusion proteins to the phage's coat proteins. The formula of the M13 (fd) phage and its gp3 coat protein is fre-

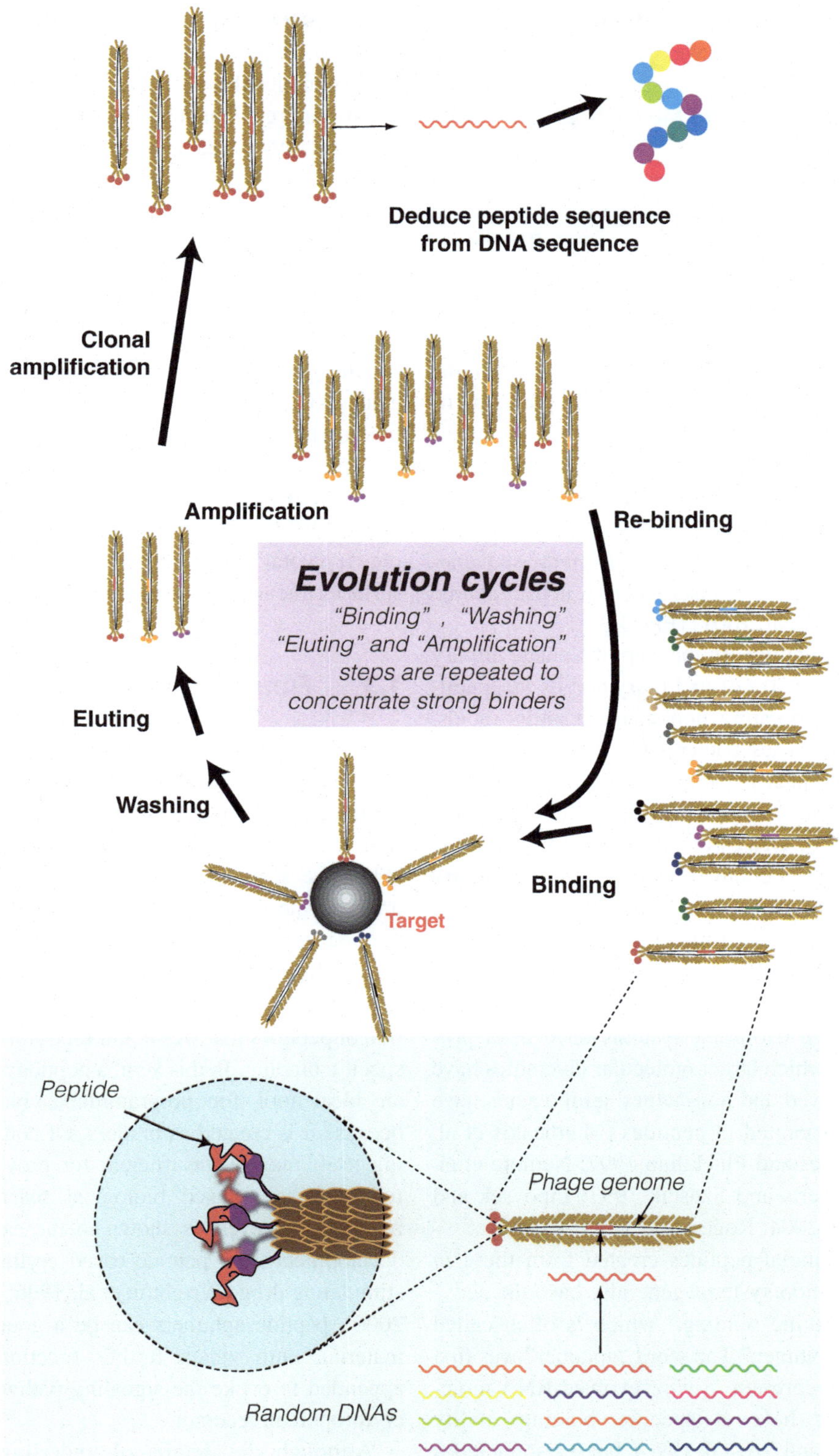

Fig. 1.3 Scheme of the phage-display peptide library system

quently used, but other combinations of phages and their proteins have been also established (Sternberg and Hoess 1995; Mikawa et al. 1996; Kozlovska et al. 1996; Mottershead et al. 1997; Castagnoli et al. 2001; Danner and Belasco 2001), or microorganisms and their proteins (Agterberg et al. 1990; Georgiou et al. 1996; Lu et al. 1995; Boder and Wittrup 1997; Samuelson et al. 2002). Typically, 7- or 12-mer peptides are displayed on the surface of a phage. Each phage displays few molecules of peptides having an identical sequence that is translated from the phage genome packaged within the phage particles. Therefore, after repeating evolution cycles ("binding", "washing", "eluting" and "amplification") to concentrate strong binders, the sequence of peptides that are displayed on the binders can be easily deduced from the sequencing of the portion of inserted DNA in the genome (Barbas III et al. 2001) (Fig. 1.3).

The limitation of the peptide phage library system is the size of its library and its incapability of incorporating non-standard amino acids. The system uses bacteria (*Escherichia coli*) to propagate phages, and the size of the library (i.e., molecular diversity of peptide) is limited by the efficiency of DNA's incorporation into bacterial cells. A typical commercial peptide phage library has a diversity of approximately 10^9. In addition, it is generally very difficult to use a non-standard amino acid as a building block in this system. To overcome this limitation, various types of cell-free *in vitro* evolution systems have been proposed, by which larger molecular diversities have been achieved and non-natural amino acids have been incorporated in peptides (Mattheakis et al. 1994; Hanes and Pluckthun 1997; Nemoto et al. 1997; Roberts and Szostak 1997; Lipovsek and Pluckthun 2004; Rogers and Suga 2015).

The artificial peptides created from these *in vitro* evolution systems generally have the activity of "specific binding," which is often called "peptide aptamer." The word "aptamer" was first coined to represent *in vitro* evolved RNA molecules that have a specific binding ability (Ellington and Szostak 1990), but it also includes DNA (Ellington and Szostak 1992) and peptide aptamers (Colas et al. 1996). To date, tremendous amounts of peptide aptamers have been created (He et al. 2016), the targets of which include not only biomolecules such as proteinous receptors, but also inorganic materials such as oxidized iron (Brown 1992), semiconductor (Whaley et al. 2000), titanium (Sano and Shiba 2003), carbon nanotubes (Wang et al. 2003; Kase et al. 2004), conducting polymer (Sanghvi et al. 2005), among others (Shiba 2010a; Seker and Demir 2011; He et al. 2016; Thota and Perry 2016). The aptamers that recognize non-biological molecules are often called "genetically engineered peptides for inorganics (GEPI)" (Sarikaya et al. 2003), "solid binding peptides (SBPs)" (Baneyx and Schwartz 2007; Care et al. 2015), or "material binding peptides (MBPs)" (Hattori et al. 2008; Seker and Demir 2011). These peptide aptamers should play a pivotal role in making programmable bio-surfaces due to their programmability.

1.3 Functions of Peptides

Figure 1.4 summarizes the functions of peptides that are immobilized on bio-surfaces. Natural peptides are often assigned their biological functions such as hormone and antibacterial activities. When we seek the molecular mechanisms that underlie these biological functions of peptides, the first step of the complexed biological reaction should be the specific interaction between the peptide and its target. Therefore, the major function of peptides that we should focus on should be specific binding. In this vein, a peptide aptamer is an ideal tool for programmable bio-surfaces because it is created as binders. Of course, binding itself may be insufficient for provoking the following complexed biological reaction after binding. However, as shown in the case of the development of a peptide-based erythropoiesis-stimulating drug (Wrighton et al. 1996; Fan et al. 2006), peptide aptamers can be a good starting material, onto which further function can be appended to evoke the signaling pathway of the erythropoietin receptor.

Although the details of underling mechanisms have not yet been well understood, some bactericidal peptides seem to directly bind to

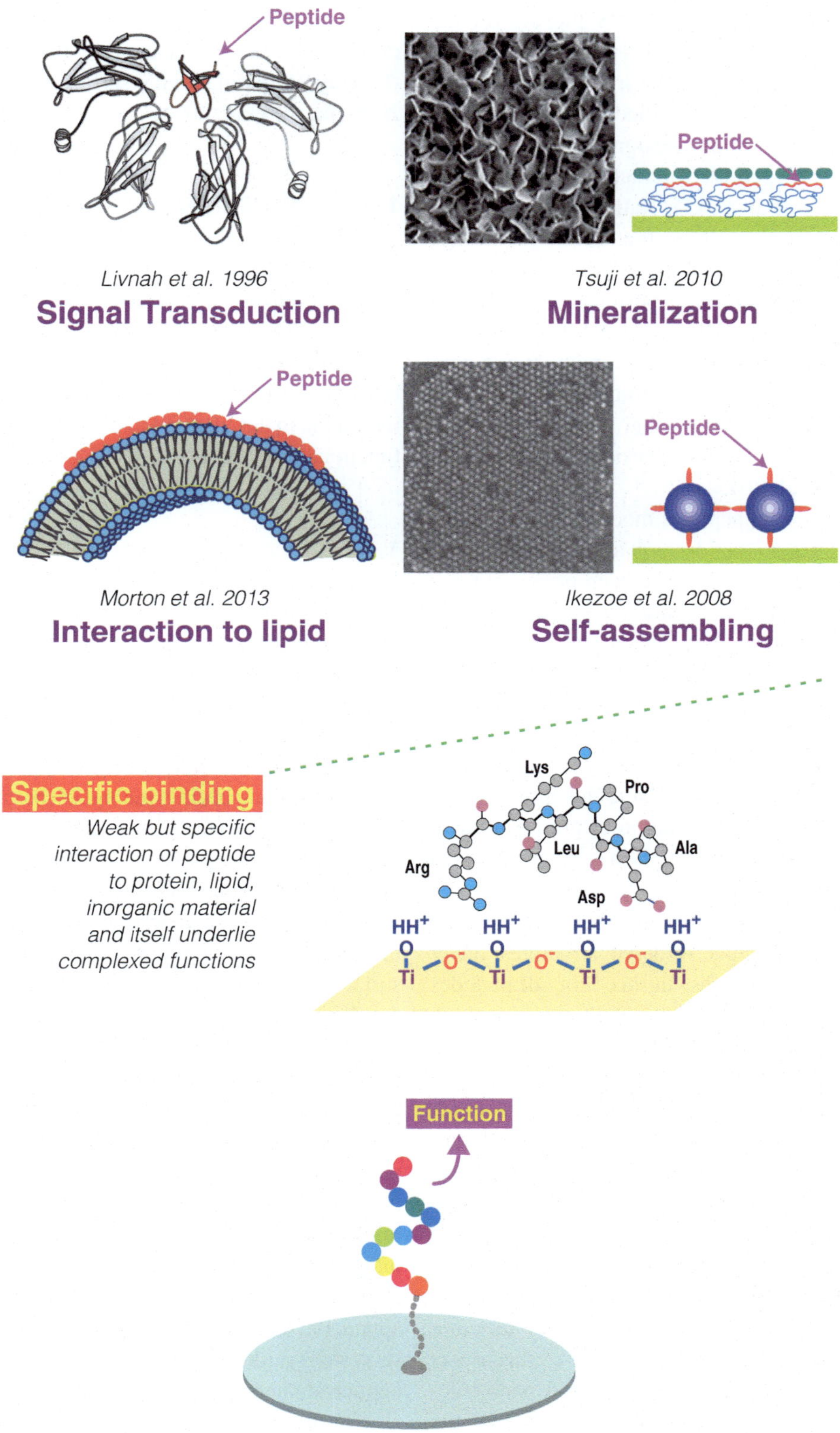

Fig. 1.4 Biological functions of peptides

lipid membranes, followed by the destruction of the membrane integrity (Hong and Su 2011; Mahlapuu et al. 2016). Several cell-penetrating peptides have been identified from natural proteins, or by rational design, which are also examples of peptide-interacting lipid bilayers (Kauffman et al. 2015). Rational design based on the natural proteins that sense membrane curvature has created an artificial peptide that recognizes membrane vesicles with highly curved surfaces (Saludes et al. 2012; Morton et al. 2013). These membrane-recognizing peptides are expected to play important roles in the bio-surface of devices that differentiate subclasses of extracellular vesicles or exosomes (Morton et al. 2013).

Anti-freezing peptides modulate the crystallization of water (Baardsnes et al. 2001). Similarly, many natural peptides have been proposed to be involved in biomineralization, *i.e.*, the controlled crystallization of inorganic materials by organisms (Weiner and Hood 1975; Mann 2001). Interestingly, artificial peptides that have been created as binders to inorganic material often possesses the modulating activity for mineralization (Sano et al. 2005b). Using this bifunctionality of peptides, the formation of a mineral layer on the surface of materials has been proposed (see below).

Most peptides have the capacity to self-assemble into higher-ordered structures under appropriate conditions. In this vein, all peptides should be bifunctional or multi-functional. A group led by S. I. Stupp has been exploring this self-assembling capacity of peptides to make various types of medical devices (Aida et al. 2012). When the self-assembly of material-binding peptides occurs on the surface of a material, it should be very a complex process because the peptide is involved in both peptide-peptide interaction and in peptide-surface interaction. Under the proper conditions, the peptide can contribute to the emergence of an orderly structure on the surface. So et al. have explored this aspect of peptides to fabricate the ordered nanostructures on the surface of graphite by modifying the sequences of graphite-binding peptides (So et al. 2012). When nano-caged protein particles are ornamented with a carbon-binding peptide, the protein particle can make a two-dimension crystal array on the material's surface (Matsui et al. 2007; Ikezoe et al. 2008).

1.4 Immobilization Methods

In this section, the methods by which peptides are immobilized on the surfaces of materials are discussed (Fig. 1.5). There is no universal method that works for any materials, and different immobilization strategies should be considered depending on the materials used and the purposes of the functionalized surfaces.

1.4.1 Physical Adsorption

Nearly all of surfaces of inorganic materials tend to strongly absorb biomolecules on them when they are exposed to biological fluids, such as blood or saliva. This is explained by the high free energy of material surfaces. Therefore, considerable efforts have been devoted to suppressing this apparently non-specific biding of biomolecules on material surfaces. At the same time, this interaction could be employed as the simplest and the least expensive approach to peptide immobilization. The underlying mechanisms of the physical interaction between biomolecules and material surfaces are complex (Israelachvili 2001). Van der Waals as well as electrostatic interactions should play pivotal roles, but absorbing reactions proceed slowly in multiple steps, resulting in the evolving character of the bio-surface (a typical example is the slow changing of the biomolecular corona that is formed on the surface of nanoparticles in body fluids). In this vein, the physical immobilization of peptides on the surfaces of inorganic materials contain some uncontrollable aspects. In a sense, the interaction of peptide aptamers (see Sect. 1.4.4) with their target substrates is regarded as physical adsorption, although peptide aptamers have subtle target specificity.

The streptavidin-biotin system is often used to immobilize peptides or other biomolecules on materials' surfaces, wherein the streptavidin is

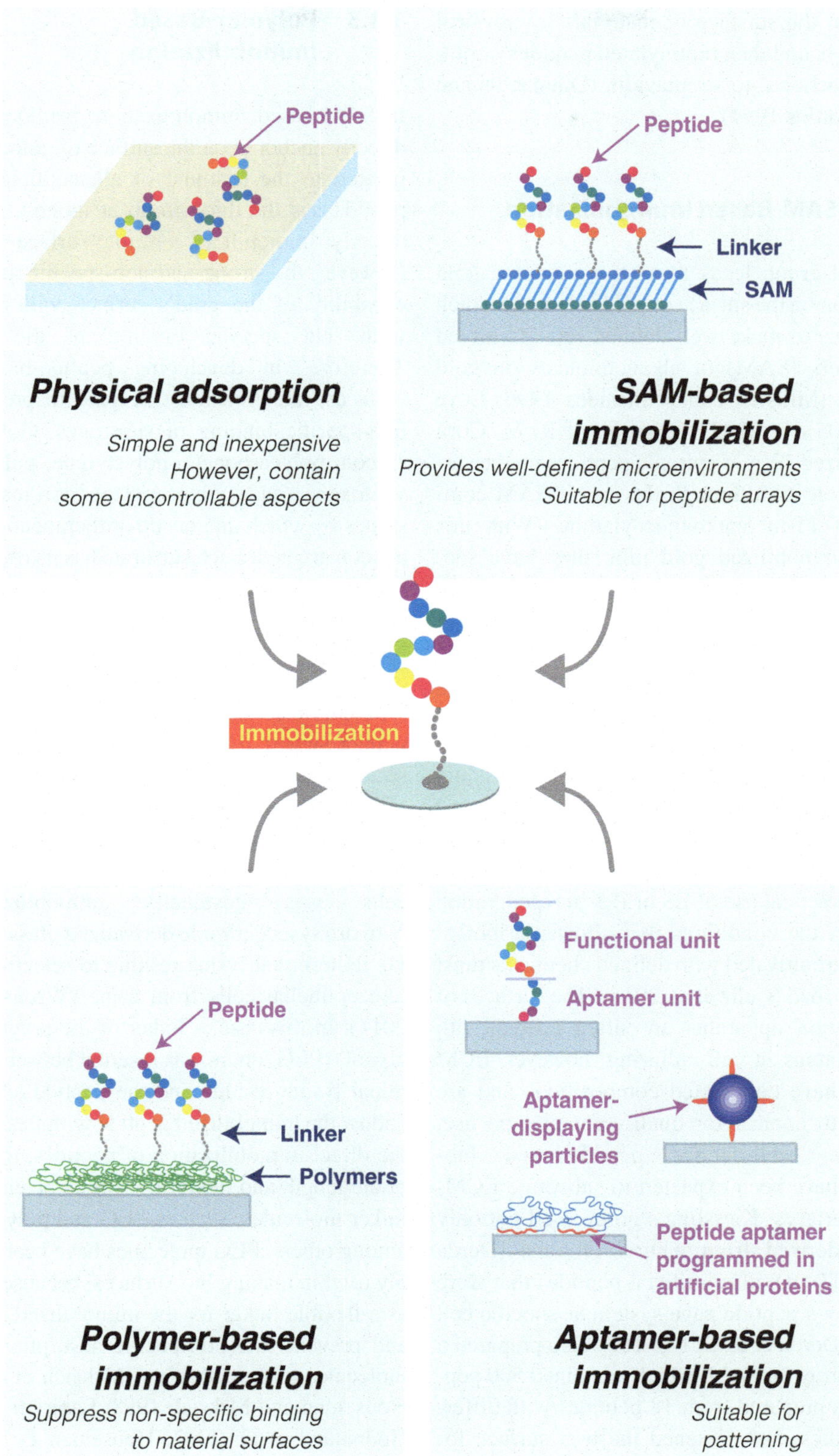

Fig. 1.5 Four major methods for immobilizing peptides on inorganic materials

coated on the sur-face of materials by physical adsorption, and then biotinylated peptides or others are anchored to streptavidin (Diamandis and Christopoulos 1991).

1.4.2 SAM-Based Immobilization

The thiol group leads to a strong sulphur-gold interaction on the surface of gold substrate, which enabled us to make well-defined self-assembled monolayers (SAM) of alkanethiolates on gold substrate (Mrksich and Whitesides 1996; Love et al. 2005). In 2002, the group of R. M. Corn immobilized 12-mer epitopic peptides on the surface of gold film through the use of SAM comprised of 11-mercaptoundecylamine. With this epitope-immobilized gold film, they have succeeded in observing the binding of antibody to the epitopic peptide via surface plasmon resonance imaging (Wegner et al. 2002). This is an example in which a bio-surface was applied to develop a chip-type sensor by quantifying the interaction between the immobilized peptide and its target molecule. The group of L. L. Kiessling has been developing various types of SAM-based bio-surfaces for providing well-defined microenvironments to grow embryonic stem cells *in vitro*. For the practical use of ES or iPS in regeneration medicine, the conditions that allow pluripotent cells to be cultivated with defined chemicals must be established (Celiz et al. 2014). The surfaces of a cultivation apparatus are often coated with ECM proteins in cell culturing; however, ECM proteins have complexed compositions and are difficult to control for quality for clinical use. Therefore, well-defined peptide-based bio-surfaces have been expected to substitute ECM-coated surfaces. Kiessling's group used not only peptide derived from natural proteins (Derda et al. 2007), but also artificial peptides that were created by a peptide page system as specific cell binders (Derda et al. 2010). They also prepared a peptide array on SAM, which contained 500 peptide spots prepared from 18 peptides with different densities, and screened the best surface for supporting the long-term growth of ES cells (Klim et al. 2010).

1.4.3 Polymer-Based Immobilization

In SAM-based immobilization, peptides are not directly anchored on the surface of gold, but conjugated to the terminal of alkanethiolates. If a peptide has the thiol group at its end, it can be directly immobilized on the surface of gold. However, the strong and non-specific adsorption capability of the gold's surface will mask the weak but specific binding of the peptide. Therefore, in developing peptide-based bio-surfaces, attention must be paid to prevent this non-specific binding of substrates. Gold, glass, silicon, polycarbonate, polystyrene, polydimethylsiloxane, and polypropylene are major substrates by which diagnostic or therapeutic devices are constructed. Each substrate has its own unique surface characteristics, and various surface processing may affect its adsorption property. Similar, each peptide has its own specificity and affinity, which will be affected by its microenvironment. In certain conditions of a substrate and a peptide, where the non-specific affinity of a substrate is sufficiently weak to create the specific affinity of the peptide, simple immobilization will work as programmable bio-surfaces. For example, a peptidic aptamer that binds to urothelial cells was covalently immobilized on *N*-hydroxysuccinimide derivatized glass substrate via its terminal lysine residue to selectively capture epithelial cells from urine (Wronska et al. 2014). In this case, a linker of the poly(ethylene glycol) (PEG) chain was inserted between the terminal lysine residue and the peptide aptamer to endow the immobilized peptide with flexibility. In the direct immobilization of peptides, the appropriate length and flexibility should be ensured by linker molecules, such as PEG and polypeptides, among others. PEG molecules have been preferably used in making bio-surfaces, because it works as a flexible linker for the immobilized molecule and prevents the nonspecific adsorption of biomolecules to the substrate (Mrksich et al. 1995; Houseman and Mrksich 1998; Lahiri et al. 1999; Hodneland et al. 2002; Houseman et al. 2003; Meyers and Grinstaff 2012; Hassert and Beck-Sickinger 2013).

SAM provides chemically-well defined and controllable surfaces not only on gold, but also on silicon wafers and glass slides, among others (Mrksich and Whitesides 1996; Love et al. 2005). For these reasons, SAM has played a pivotal role in developing bio-surfaces. Recently, in addition to SAM, synthetic polymers have come under the spotlight as foundation materials for peptide immobilization. Synthetic polymers have already been used as surface modification materials for biomedical materials to support cell growth in culture dishes (Li et al. 2006; Villa-Diaz et al. 2010; Mei et al. 2010) or to suppress the non-specific binding of biomolecules on material surfaces (Watanabe and Ishihara 2008; Xu et al. 2010; Meyers and Grinstaff 2012). Further functionalization of these synthetic polymers has been attempted to conjugate peptides to polymers. For instance, research groups from Corning Inc. and Geron Corporation have conjugated peptides that correspond to parts of natural ECM proteins (such as bone sialoprotein and to acrylate polymer to develop a culture flask that allows the long-term culture of human embryonic stem cells (Melkoumian et al. 2010).

A methacrylate, 2-methacryloyloxyethyl phosphorylcholine (MPC), has a phosphoryl-choline zwitterionic group and was designed as a membrane-mimicking polymer unit in 1978 (Kadoma et al. 1978). Hydrophilic zwitterionic molecules are believed to suppress the non-specific biding of biomolecules (Holmlin et al. 2001), which has been demonstrated by the surface modification of various materials by MPC-derivative polymers (Sibarani et al. 2007; Watanabe and Ishihara 2008; Xu et al. 2010). In these cases, MPC was co-polymerized with a hydrophobic polymer unit, such as *n*-butyl methacrylate (BMA), with various ratios to make MPC-based coating agents, in which the BMA unit is responsible for the physical interaction with substrates. To further functionalize MPC polymers, an enzyme or antibody has been conjugated via a doped reactive polymer unit (Sakai-Kato et al. 2004; Kim et al. 2012). Similarly, a peptide aptamer to (Shiba et al. 2012) one of the epithelial cell makers (Litvinov et al. 1994), has been conjugated to MPC-based polymer, to make the coating agent, EpiVeta, which endows the surfaces of materials with the affinity to the EpCAM molecule (Shiba et al. 2013).

1.4.4 Aptamer-Based Immobilization

Targets of peptide aptamers are not limited to proteins, cells, or organs. Artificial peptides that have an affinity to inorganic materials, such as gold (Brown 1997), semiconductors (Whaley et al. 2000), titanium (Sano and Shiba 2003), carbon nanomaterials (Wang et al. 2003; Kase et al. 2004), and polymer films (Sanghvi et al. 2005; Kumada 2014), have been created from a phage-display peptide library system or other *in vitro* evolution systems. These material-binding peptides have also been employed to make bio-surfaces. For example, a peptide aptamer that has an affinity to polypyrrole was conjugated either to laminin-derived peptide or PEG and has been used to endow the surface of polypyrrole with an affinity to neural cells or with stealth properties (Nickels and Schmidt 2012).

Hexapeptidic minTBP-1 is the minimal sequence responsible for the titanium binding of 12-mer TBP-1, which was isolated from a phage-display peptide library system (Sano and Shiba 2003). This hexapeptide has been explored to functionalize the surfaces of materials in various ways. As mentioned in Sect. 1.2.1, the peptide was fused to natural histatin 5 peptide to endow the surface of titanium with antimicrobial activity, aiming to functionalize the titanium implant (Yoshinari et al. 2010). The peptide was also used to immobilize cytokines on the titanium (Kashiwagi et al. 2009), the goal of which was to enhance the osseointegration of titanium. As a property of peptide aptamers, the interaction between peptides and their targets is reversible. This is because in the phage-display peptide library system, binders are usually detached under acidic conditions to collect. Thus, peptide aptamers inherently bind to targets in a reversible manner. This reversible binding of aptamers is advantageous and disadvantageous for making bio-surfaces. The irreversible immobilization of signaling molecules on solid surfaces prevent the interlunation of the molecule within cells, which is often required to evoke the signaling pathway for some biological activities (Hartung et al. 2006). However, the strength of this reversible interaction is not sufficient to anchor large molecules on the surface. Although the strength can be enhanced by multivalent usages of the peptide, a stronger interaction seemed to be required to functionalize titanium implant *in vivo* (Yuasa et al. 2014).

Various combinations of minTBP-1 with other peptidic motifs have been explored by the MolCraft system, by which multiple peptide motifs are programmed to make artificial proteins (Shiba 2004). When minTBP-1 was combined with the RGD motif, it served as a synthetic matric protein on a titanium plate (Kokubun et al. 2008). When minTBP-1 was shuffled with two mineralization-related peptide motifs derived from dentin matrix protein 1, the resultant proteins mediated the nucleation of octacalcium phosphate on titanium substrates (Tsuji et al. 2010).

Nanostructured cargos including natural and artificial ones have also been utilized as tethers of minTBP-1. For natural cargo, the gene for minTBP-1 was fused with the gene for a subunit of ferritin (Sano et al. 2005a). Because ferritin particles are formed from the self-assembly of 24 subunits, 24 copies of minTBP-1 peptides are displayed on 12 nm of ferritin molecules. As introduced above, minTBP-1 is a bifunctional aptamer, which has both the binding activity to Ti and the mineralization activity for titania and silica (Sano et al. 2005b). When this recombinant ferritin was incubated with the substrate that has the nanoscaled patterns of titanium film on a platinum base, the minTBP-1 displaying ferritin bound to the titanium-covered regions. Because only some of the 24 peptides were engaged in this specific binding (note that 24 peptides are nearly evenly distributed on spherical surface), other peptides on ferritin were free to access chemicals in solution. By taking advantage of the bifunctionality of minTBP-1, a thin film of titania or silica was formed on the nanoscaled patterns (Fig. 1.6). In addition, inner nanospaces of ferritin can serve as a carrier for inorganic nanodots, which enable us to fabricate a complex nanostructure on inorganic substrates (Sano et al. 2006, 2007; Sano and Shiba 2008). Similarly, the 3.5 nm nano-cage formed from the self-assembly of 12 metal ions and 24 chemical ligands was conjugated with minTBP-1 (Sato et al. 2015).

In addition to TBP-1, several other titanium-binding peptide aptamers have been isolated, some of which have been used to make bio-surfaces, such as PEG-coated (Khoo et al. 2009) and RGD motif-displayed surfaces (Meyers et al. 2007).

1.5 Utilization of Programmable Bio-surfaces

Figure 1.7 illustrates some examples in which programmable bio-surfaces are expected to play active roles. The area of regenerative medicine (including tissue engineering and cell-based therapy) could be the one of most promising areas for programmable bio-surfaces. In this field, well-defined culturing conditions that allow the prolonged propagation of pluripotent cells without unexpected differentiation need to be established (Baker 2011; Celiz et al. 2014; Martins et al. 2016). Alternatively, certain surface conditions will be required to differentiate stem cells into an appropriate direction in culture dishes. Some examples of applications in regenerative medicine were introduced in previous sections, in which protein-derived peptides or peptide aptamers are immobilized on substrate using appropriate immobilization methods (Celiz et al. 2014; Derda et al. 2007, 2010; Hudalla et al. 2011; Li et al. 2011; Singh et al. 2014). Such *ex vivo* culturing of cells is performed in flasks made of polystyrene, among others, or in microfluidics devices made of glass, silicon, and polydimethylsiloxane, among others. Microfluidics devices will play an important role in continuously processing cells. Microfluidics or chip-type devices are also explored to establish cell-based or exosome-based diagnostic devices (Shiba et al. 2013; Pu et al. 2017). For the development of diagnostic devices, bio-surfaces on semiconductive or metal materials will sensitively detect biomolecules, which will be translated into

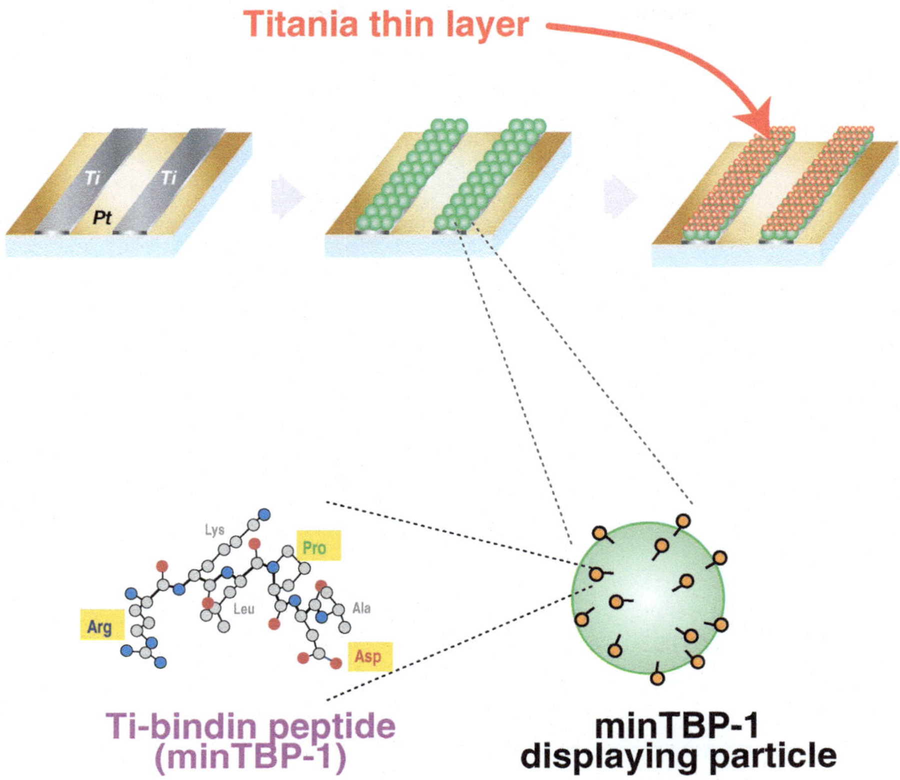

Fig. 1.6 An example of applications of titanium-binding aptamers in bio-surfaces

electronic signals (Pavan and Berti 2012; Cui et al. 2012). The fabrication of elaborate patterns is possible using conventional lithographic methodology, which can be translated into nano-patterned bio-surfaces (Sarikaya et al. 2003; Sano and Shiba 2008; Tamerler and Sarikaya 2009; Sato et al. 2015).

Programmed bio-surfaces are also applied to the materials that are used *in vivo*. Titanium has already been widely used in dental, orthopedic, and cardiovascular fields, and further functionalization with biological methods has been challenged (Brunette et al. 2001; Hassert and Beck-Sickinger 2013; Panayotov et al. 2015). In addition, mesh (Shao et al. 2012, 2015; Li et al. 2013) or hydrogel (Lin and Anseth 2009; Gungormus et al. 2010; Hamilton et al. 2013) made of synthetic polymers or other materials can be endowed with biological functions with peptides and has been aimed for use with implantable artificial organs or regenerative medical devices. Alternatively, fibrils with affinity have been fabricated from peptides (Yolamanova et al. 2013). When nanoparticles are ornamented with various peptides, they will acquire organ specificity, which will enhance the performance of nanoparticles in diagnostic imaging or photodynamic therapy (Cutler et al. 2013; Lee 2013; Gautam et al. 2014; Chen et al. 2014). Further, artificial peptides by themselves (Wang et al. 2003; Grigoryan et al. 2011; Li et al. 2015), or peptide-PEG conjugates (Matsumura et al. 2007; Matsumura et al. 2009), have been used to endow hydrophobic carbon nano-materials with hydrophilicity, preventing the aggregation of the particles in aqueous conditions.

1.6 Conclusions

The main advantages of the utilization of peptides for surface functionalization are their definability and programmability. Peptides can be chemically synthesized in large quantities and

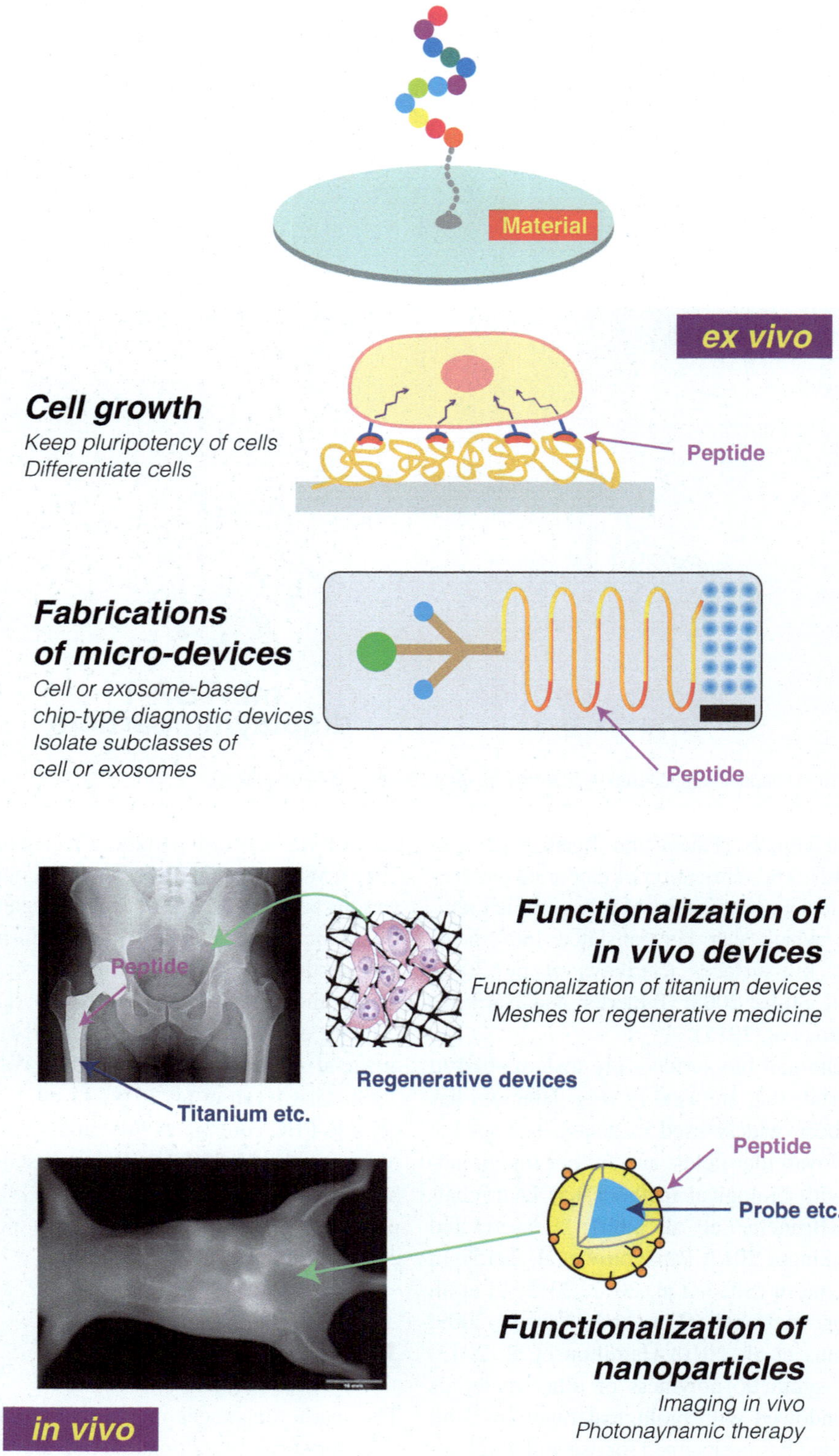

Fig. 1.7 Fields in which bio-surfaces play active roles

their quality control is easier than biological macromolecules such as proteins or polysaccharides. Furthermore, because they are chemical compounds, methods for conjugations with foreign molecules as well as modifications have been well established. Furthermore, novel peptides having the capacity for specific binding can be readily created using established *in vitro* revolution systems. Considering the fact that all biological activities are based on the specific molecular recognition between biomolecules, this programmability of artificial peptides is a particularly compelling property for developing functional surfaces. Although peptides have several disadvantages in terms of their cost-effectiveness (Celiz et al. 2014) and capriciousness of exerting their assigned function (Shiba 2010b), they have unparalleled functionality as a biomaterial carrying genetic information. Thus, peptides will play a pivotal role in developing diagnostic and therapeutic medical devices.

References

Agterberg M, Adriaanse H, van Bruggen A, Karperien M, Tommassen J (1990) Outer-membrane phoe protein of *Escherichia coli* k-12 as an exposure vector: possibilities and limitations. Gene 88:37–45

Aida T, Meijer EW, Stupp SI (2012) Functional supramolecular polymers. Science 335(6070):813–817. doi:10.1126/science.1205962

Baardsnes J, Jelokhani-Niaraki M, Kondejewski LH, Kuiper MJ, Kay CM, Hodges RS, Davies PL (2001) Antifreeze protein from shorthorn sculpin: identification of the ice-binding surface. Protein Sci 10(12):2566–2576

Baker M (2011) Stem cells in culture: defining the substrate. Nat Methods 8(4):293–297. doi:10.1038/nmeth0411-293

Baneyx F, Schwartz DT (2007) Selection and analysis of solid-binding peptides. Curr Opin Biotechnol 18(4):312–317. doi:10.1016/j.copbio.2007.04.008

Barbas CF III, Burton DR, Scott JK, Silverman GJ (2001) Phage display : a laboratory manual. Cold Spring Harbor laboratory press. Cold Spring Harbor, New York

Bhardwaj G, Mulligan VK, Bahl CD, Gilmore JM, Harvey PJ, Cheneval O, Buchko GW, Pulavarti SV, Kaas Q, Eletsky A, Huang PS, Johnsen WA, Greisen PJ, Rocklin GJ, Song Y, Linsky TW, Watkins A, Rettie SA, Xu X, Carter LP, Bonneau R, Olson JM, Coutsias E, Correnti CE, Szyperski T, Craik DJ, Baker D (2016) Accurate de novo design of hyperstable constrained peptides. Nature 538(7625):329–335. doi:10.1038/nature19791

Boder ET, Wittrup KD (1997) Yeast surface display for screening combinatorial polypeptide libraries. Nat Biotechnol 15(6):553–557

Brown S (1992) Engineered iron oxide-adhesion mutants of the escherichia coli phage l receptor. Proc Natl Acad Sci U S A 89(18):8651–8655

Brown S (1997) Metal-recognition by repeating polypeptides. Nat Biotechnol 15(3):269–272

Brunette DM, Tengvall P, Textor M, Thomsen P (2001) Titanium in medicin. Engineering materials. Springer Verlag Berlin, Heidelberg/Berlin

Care A, Bergquist PL, Sunna A (2015) Solid-binding peptides: smart tools for nanobiotechnology. Trends Biotechnol 33(5):259–268. doi:10.1016/j.tibtech.2015.02.005

Castagnoli L, Zucconi A, Quondam M, Rossi M, Vaccaro P, Panni S, Paoluzi S, Santonico E, Dente L, Cesareni G (2001) Alternative bacteriophage display systems. Comb Chem High Throughput Screen 4(2):121–133

Celiz AD, Smith JG, Langer R, Anderson DG, Winkler DA, Barrett DA, Davies MC, Young LE, Denning C, Alexander MR (2014) Materials for stem cell factories of the future. Nat Mater 13(6):570–579. doi:10.1038/nmat3972

Chen ZY, Wang YX, Lin Y, Zhang JS, Yang F, Zhou QL, Liao YY (2014) Advance of molecular imaging technology and targeted imaging agent in imaging and therapy. Biomed Res Int 2014:1. doi:10.1155/2014/819324

Colas P, Cohen B, Jessen T, Grishina I, Mccoy J, Brent R (1996) Genetic selection of peptide aptamers that recognize and inhibit cyclin-dependent kinase 2. Nature 380(6574):548–550

Cui Y, Kim SN, Naik RR, McAlpine MC (2012) Biomimetic peptide nanosensors. Acc Chem Res 45:696. doi:10.1021/ar2002057

Cutler CS, Chanda N, Shukla R, Sisay N, Cantorias M, Zambre A, McLaughlin M, Kelsey J, Upenandran A, Robertson D, Deutscher S, Kannan R, Katti K (2013) Nanoparticles and phage display selected peptides for imaging and therapy of cancer. Recent Results Cancer Res 194:133–147. doi:10.1007/978-3-642-27994-2_8

Cwirla SE, Peters EA, Barrett RW, Dower WJ (1990) Peptides on phage – a vast library of peptides for identifying ligands. Proc Natl Acad Sci U S A 87(16):6378–6382

Danner S, Belasco JG (2001) T7 phage display: a novel genetic selection system for cloning rna-binding proteins from cdna libraries. Proc Natl Acad Sci U S A 98(23):12954–12959

Derda R, Li L, Orner BP, Lewis RL, Thomson JA, Kiessling LL (2007) Defined substrates for human embryonic stem cell growth identified from surface arrays. ACS Chem Biol 2(5):347–355. doi:10.1021/cb700032u

Derda R, Musah S, Orner BP, Klim JR, Li L, Kiessling LL (2010) High-throughput discovery of synthetic surfaces that support proliferation of pluripotent cells. J Am Chem Soc 132(4):1289–1295. doi:10.1021/ja906089g

Devlin JJ, Panganiban LC, Devlin PE (1990) Random peptide libraries: a source of specific protein binding molecules. Science 249(4967):404–406

Diamandis EP, Christopoulos TK (1991) The biotin-(strept)avidin system: principles and applications in biotechnology. Clin Chem 37(5):625–636

Ellington AD, Szostak JW (1990) *In vitro* selection of rna molecules that bind specific ligands. Nature 346:818–822

Ellington AD, Szostak JW (1992) Selection in vitro of single-stranded DNA molecules that fold into specific ligand-binding structures. Nature 355(6363):850–852

Erez Z, Steinberger-Levy I, Shamir M, Doron S, Stokar-Avihail A, Peleg Y, Melamed S, Leavitt A, Savidor A, Albeck S, Amitai G, Sorek R (2017) Communication between viruses guides lysis-lysogeny decisions. Nature 541(7638):488–493. doi:10.1038/nature21049

Fan Q, Leuther KK, Holmes CP, Fong KL, Zhang J, Velkovska S, Chen MJ, Mortensen RB, Leu K, Green JM, Schatz PJ, Woodburn KW (2006) Preclinical evaluation of hematide, a novel erythropoiesis stimulating agent, for the treatment of anemia. Exp Hematol 34(10):1303–1311. doi:10.1016/j.exphem.2006.05.012

Gallop MA, Barrett RW, Dower WJ, Fodor SPA, Gordon EM (1994) Applications of combinatorial technologies to drug discovery 1. Background and peptide combinatorial libraries. J Med Chem 37(9):1233–1251

Gautam A, Kapoor P, Chaudhary K, Kumar R, Open Source Drug Discovery C, Raghava GP (2014) Tumor homing peptides as molecular probes for cancer therapeutics, diagnostics and theranostics. Curr Med Chem 21(21):2367–2391

Georgiou G, Stephens DL, Stathopoulos C, Poetschke HL, Mendenhall J, Earhart CF (1996) Display of beta-lactamase on the Escherichia coli surface: outer membrane phenotypes conferred by lpp′-OmpA′-beta-lactamase fusions. Protein Eng 9(2):239–247

Gordon EM, Barrett RW, Dower WJ, Fodor SPA, Gallop MA (1994) Applications of combinatorial technologies to drug discovery .2. Combinatorial organic synthesis, library screening strategies, and future directions. J Med Chem 37(10):1385–1401

Grigoryan G, Kim YH, Acharya R, Axelrod K, Jain RM, Willis L, Drndic M, Kikkawa JM, DeGrado WF (2011) Computational design of virus-like protein assemblies on carbon nanotube surfaces. Science 332(6033):1071–1076. doi:10.1126/science.1198841

Gungormus M, Branco M, Fong H, Schneider JP, Tamerler C, Sarikaya M (2010) Self assembled bi-functional peptide hydrogels with biomineralization-directing peptides. Biomaterials 31(28):7266–7274. doi:10.1016/j.biomaterials.2010.06.010

Hamilton PT, Jansen MS, Ganesan S, Benson RE, Hyde-Deruyscher R, Beyer WF, Gile JC, Nair SA, Hodges JA, Gron H (2013) Improved bone morphogenetic protein-2 retention in an injectable collagen matrix using bifunctional peptides. PLoS One 8(8):e70715. doi:10.1371/journal.pone.0070715

Hanes J, Pluckthun A (1997) In vitro selection and evolution of functional proteins by using ribosome display. Proc Natl Acad Sci U S A 94(10):4937–4942

Hartung A, Bitton-Worms K, Rechtman MM, Wenzel V, Boergermann JH, Hassel S, Henis YI, Knaus P (2006) Different routes of bone morphogenic protein (bmp) receptor endocytosis influence bmp signaling. Mol Cell Biol 26(20):7791–7805. doi:10.1128/MCB.00022-06

Hassert R, Beck-Sickinger AG (2013) Tuning peptide affinity for biofunctionalized surfaces. Eur J Pharm Biopharm 85(1):69–77. doi:10.1016/j.ejpb.2013.02.006

Hattori T, Umetsu M, Nakanishi T, Tsumoto K, Ohara S, Abe H, Naito M, Asano R, Adschiri T, Kumagai I (2008) Grafting of material-binding function into antibodies functionalization by peptide grafting. Biochem Biophys Res Commun 365(4):751–757. doi:10.1016/j.bbrc.2007.11.062

He B, Chai G, Duan Y, Yan Z, Qiu L, Zhang H, Liu Z, He Q, Han K, Ru B, Guo FB, Ding H, Lin H, Wang X, Rao N, Zhou P, Huang J (2016) Bdb: biopanning data bank. Nucleic Acids Res 44(D1):D1127–D1132. doi:10.1093/nar/gkv1100

Hodneland CD, Lee YS, Min DH, Mrksich M (2002) Selective immobilization of proteins to self-assembled monolayers presenting active site-directed capture ligands. Proc Natl Acad Sci U S A 99(8):5048–5052. doi:10.1073/pnas.072685299

Holmlin RE, Chen X, Chapman RG, Takayama S, Whitesides GM (2001) Zwitterionic sams that resist nonspecific adsorption of protein from aqueous buffer. Langmuir 17(9):2841–2850. doi:10.1021/la0015258

Hong M, Su Y (2011) Structure and dynamics of cationic membrane peptides and proteins: insights from solid-state nmr. Protein Sci 20(4):641–655. doi:10.1002/pro.600

Houseman BT, Mrksich M (1998) Efficient solid-phase synthesis of peptide-substituted alkanethiols for the preparation of substrates that support the adhesion of cells. J Org Chem 63(21):7552–7555

Houseman BT, Gawalt ES, Mrksich M (2003) Maleimide-functionalized self-assembled monolayers for the preparation of peptide and carbohydrate biochips. Langmuir 19(5):1522–1531. doi:10.1021/la0262304

Hudalla GA, Koepsel JT, Murphy WL (2011) Surfaces that sequester serum-borne heparin amplify growth factor activity. Adv Mater 23(45):5415–5418. doi:10.1002/adma.201103046

Ikezoe Y, Kumashiro Y, Tamada K, Matsui T, Yamashita I, Shiba K, Hara M (2008) Growth of giant two-dimensional crystal of protein molecules from a three-phase contact line. Langmuir 24(22):12836–12841. doi:10.1021/la802104f

Israelachvili JN (2001) Intermolecular and surface forces. Academic, San Deigo

Kadoma Y, Nakabayashi E, Masuhara J, Yamauchi E (1978) Synthesis and hemolysis test of the polymer containing phosphorylcholine groups. Kobunshi Ronbunshu 35(7):423–427

Kase D, Kulp JL III, Yudasaka M, Evans JS, Iijima S, Shiba K (2004) Affinity selection of peptide phage libraries against single-wall carbon nanohorns identifies a peptide aptamer with conformational variability. Langmuir 20(20):8939–8941

Kashiwagi K, Tsuji T, Shiba K (2009) Directional bmp-2 for functionalization of titanium surfaces. Biomaterials 30(6):1166–1175. doi:10.1016/j.biomaterials.2008.10.040

Kastin AJ (2006) Handbook of biologically active peptides. Academic

Kauffman WB, Fuselier T, He J, Wimley WC (2015) Mechanism matters: a taxonomy of cell penetrating peptides. Trends Biochem Sci 40(12):749–764. doi:10.1016/j.tibs.2015.10.004

Khoo X, Hamilton P, O'Toole GA, Snyder BD, Kenan DJ, Grinstaff MW (2009) Directed assembly of pegylated-peptide coatings for infection-resistant titanium metal. J Am Chem Soc 131(31):10992–10997. doi:10.1021/ja9020827

Kim G, Yoo CE, Kim M, Kang HJ, Park D, Lee M, Huh N (2012) Noble polymeric surface conjugated with zwitterionic moieties and antibodies for the isolation of exosomes from human serum. Bioconjug Chem 23(10):2114–2120. doi:10.1021/bc300339b

Klim JR, Li L, Wrighton PJ, Piekarczyk MS, Kiessling LL (2010) A defined glycosaminoglycan-binding substratum for human pluripotent stem cells. Nat Methods 7(12):989–994. doi:10.1038/nmeth.1532

Kokubun K, Kashiwagi K, Yoshinari M, Inoue T, Shiba K (2008) Motif-programmed artificial extracellular matrix. Biomacromolecules 9(11):3098–3105. doi:10.1021/bm800638z

Kozlovska TM, Cielens I, Vasiljeva I, Strelnikova A, Kazaks A, Dislers A, Dreilina D, Ose V, Gusars I, Pumpens P (1996) RNA phage Q beta coat protein as a carrier for foreign epitopes. Intervirology 39(1-2):9–15

Kumada Y (2014) Site-specific immobilization of recombinant antibody fragments through material-binding peptides for the sensitive detection of antigens in enzyme immunoassays. Biochim Biophys Acta 1844(11):1960–1969. doi:10.1016/j.bbapap.2014.07.007

Lahiri J, Isaacs L, Tien J, Whitesides GM (1999) A strategy for the generation of surfaces presenting ligands for studies of binding based on an active ester as a common reactive intermediate: a surface plasmon resonance study. Anal Chem 71(4):777–790

Lajoie MJ, Rovner AJ, Goodman DB, Aerni HR, Haimovich AD, Kuznetsov G, Mercer JA, Wang HH, Carr PA, Mosberg JA, Rohland N, Schultz PG, Jacobson JM, Rinehart J, Church GM, Isaacs FJ (2013) Genomically recoded organisms expand biological functions. Science 342(6156):357–360. doi:10.1126/science.1241459

Lauressergues D, Couzigou JM, Clemente HS, Martinez Y, Dunand C, Becard G, Combier JP (2015) Primary transcripts of micrornas encode regulatory peptides. Nature 520(7545):90–93. doi:10.1038/nature14346

Lawler J, Weinstein R, Hynes RO (1988) Cell attachment to thrombospondin: the role of arg-gly-asp, calcium, and integrin receptors. J Cell Biol 107(6):2351–2361. doi:10.1083/jcb.107.6.2351

Lee JH (2013) Conjugation approaches for construction of aptamer-modified nanoparticles for application in imaging. Curr Top Med Chem 13(4):504–512

Li YJ, Chung EH, Rodriguez RT, Firpo MT, Healy KE (2006) Hydrogels as artificial matrices for human embryonic stem cell self-renewal. J Biomed Mater Res A 79A(1):1–5. doi:10.1002/jbm.a.30732

Li L, Klim JR, Derda R, Courtney AH, Kiessling LL (2011) Spatial control of cell fate using synthetic surfaces to potentiate tgf-beta signaling. Proc Natl Acad Sci U S A 108(29):11745–11750. doi:10.1073/pnas.1101454108

Li Q, Wang Z, Zhang S, Zheng W, Zhao Q, Zhang J, Wang L, Wang S, Kong D (2013) Functionalization of the surface of electrospun poly(epsilon-caprolactone) mats using zwitterionic poly(carboxybetaine methacrylate) and cell-specific peptide for endothelial progenitor cells capture. Mater Sci Eng C Mater Biol Appl 33(3):1646–1653. doi:10.1016/j.msec.2012.12.074

Li Z, Kameda T, Isoshima T, Kobatake E, Tanaka T, Ito Y, Kawamoto M (2015) Solubilization of single-walled carbon nanotubes using a peptide aptamer in water below the critical micelle concentration. Langmuir 31(11):3482–3488. doi:10.1021/la504777b

Lima e Silva R, Kanan Y, Mirando AC, Kim J, Shmueli RB, Lorenc VE, Fortmann SD, Sciamanna J, Pandey NB, Green JJ, Popel AS, Campochiaro PA (2017) Tyrosine kinase blocking collagen iv–derived peptide suppresses ocular neovascularization and vascular leakage. Sci Transl Med 9(373):eaai8030. doi:10.1126/scitranslmed.aai8030

Lin CC, Anseth KS (2009) Controlling affinity binding with peptide-functionalized poly(ethylene glycol) hydrogels. Adv Funct Mater 19(14):2325. doi:10.1002/adfm.200900107

Lipovsek D, Pluckthun A (2004) In-vitro protein evolution by ribosome display and mRNA display. J Immunol Methods 290(1–2):51–67

Litvinov SV, Velders MP, Bakker HA, Fleuren GJ, Warnaar SO (1994) Ep-CAM: a human epithelial antigen is a homophilic cell-cell adhesion molecule. J Cell Biol 125(2):437–446

Love JC, Estroff LA, Kriebel JK, Nuzzo RG, Whitesides GM (2005) Self-assembled monolayers of thiolates on metals as a form of nanotechnology. Chem Rev 105(4):1103–1170. doi:10.1021/cr0300789

Lu Z, Murray KS, Cleave VV, LaVallie ER, Stahl ML, McCoy JM (1995) Expression of thioredoxin random peptide libraries on the *Escherichia coli* cell surface as functional fusions to flagellin: a system designed for exploring protein-protein interactions. Biotechnology 13:366–372

Mahlapuu M, Hakansson J, Ringstad L, Bjorn C (2016) Antimicrobial peptides: an emerging category of therapeutic agents. Front Cell Infect Microbiol 6:194. doi:10.3389/fcimb.2016.00194

Mann S (2001) Biomineralization-principles and concepts in bioinorganic materials chemistry. Oxford chemistry masters. Oxford University Press, Oxford

Martins IM, Reis RL, Azevedo HS (2016) Phage display technology in biomaterials engineering: progress and opportunities for applications in regenerative medicine. ACS Chem Biol 11(11):2962–2980. doi:10.1021/acschembio.5b00717

Matsui T, Matsukawa N, Iwahori K, Sano K, Shiba K, Yamashita I (2007) Realizing a two-dimensional ordered array of ferritin molecules directly on a solid surface utilizing carbonaceous material affinity peptides. Langmuir 23(4):1615–1618

Matsumura S, Ajima K, Yudasaka M, Iijima S, Shiba K (2007) Dispersion of cisplatin-loaded carbon nanohorns with a conjugate comprised of an artificial peptide aptamer and polyethylene glycol. Mol Pharm 4(5):723–729

Matsumura S, Sato S, Yudasaka M, Tomida A, Tsuruo T, Iijima S, Shiba K (2009) Prevention of carbon nanohorn agglomeration using a conjugate composed of comb-shaped polyethylene glycol and a peptide aptamer. Mol Pharm 6(2):441–447. doi:10.1021/mp800141v

Mattheakis LC, Bhatt RR, Dower WJ (1994) An *in vitro* polysome display system for identifying ligands from very large peptide libraries. Proc Natl Acad Sci U S A 91:9022–9026

Mei Y, Saha K, Bogatyrev SR, Yang J, Hook AL, Kalcioglu ZI, Cho SW, Mitalipova M, Pyzocha N, Rojas F, Van Vliet KJ, Davies MC, Alexander MR, Langer R, Jaenisch R, Anderson DG (2010) Combinatorial development of biomaterials for clonal growth of human pluripotent stem cells. Nat Mater 9(9):768–778. doi:10.1038/nmat2812

Melkoumian Z, Weber JL, Weber DM, Fadeev AG, Zhou Y, Dolley-Sonneville P, Yang J, Qiu L, Priest CA, Shogbon C, Martin AW, Nelson J, West P, Beltzer JP, Pal S, Brandenberger R (2010) Synthetic peptide-acrylate surfaces for long-term self-renewal and cardiomyocyte differentiation of human embryonic stem cells. Nat Biotechnol 28(6):606–610. doi:10.1038/nbt.1629

Meyers SR, Grinstaff MW (2012) Biocompatible and bioactive surface modifications for prolonged in vivo efficacy. Chem Rev 112(3):1615–1632. doi:10.1021/cr2000916

Meyers SR, Hamilton PT, Walsh EB, Kenan DJ, Grinstaff MW (2007) Endothelialization of titanium surfaces. Adv Mater 19(18):2492–2498

Mikawa YG, Maruyama IN, Brenner S (1996) Surface display of proteins on bacteriophage lambda heads. J Mol Biol 262(1):21–30

Morton LA, Yang H, Saludes JP, Fiorini Z, Beninson L, Chapman ER, Fleshner M, Xue D, Yin H (2013) Marcks-ed peptide as a curvature and lipid sensor. ACS Chem Biol 8(1):218–225. doi:10.1021/cb300429e

Mottershead D, van der Linden I, von Bonsdorff CH, Keinanen K, OkerBlom C (1997) Baculoviral display of the green fluorescent protein and rubella virus envelope proteins. Biochem Biophys Res Commun 238(3):717–722

Mrksich M, Whitesides GM (1996) Using self-assembled monolayers to understand the interactions of man-made surfaces with proteins and cells. Annu Rev Biophys Biomol Struct 25:55–78. doi:10.1146/annurev.bb.25.060196.000415

Mrksich M, Grunwell JR, Whitesides GM (1995) Biospecific adsorption of carbonic anhydrase to self-assembled monolayers of alkanethiolates that present benzenesulfonamide groups on gold. JACS 117(48):12009–12010. doi:10.1021/ja00153a029

Nemoto N, Miyamoto-Sato E, Husimi Y, Yanagawa H (1997) In vitro virus: bonding of mrna bearing puromycin at the 3′-terminal end to the C-terminal end of its encoded protein on the ribosome in vitro. FEBS Lett 414(2):405–408

Ng S, Jafari MR, Derda R (2012) Bacteriophages and viruses as a support for organic synthesis and combinatorial chemistry. ACS Chem Biol 7(1):123–138. doi:10.1021/cb200342h

Nickels JD, Schmidt CE (2012) Surface modification of the conducting polymer, polypyrrole, via affinity peptide. J Biomed Mater Res A 101A(5):1464–1471. doi:10.1002/jbm.a.34435

Oppenheim FG, Xu T, McMillian FM, Levitz SM, Diamond RD, Offner GD, Troxler RF (1988) Histatins, a novel family of histidine-rich proteins in human parotid secretion. Isolation, characterization, primary structure, and fungistatic effects on candida albicans. J Biol Chem 263(16):7472–7477

Panayotov IV, Vladimirov BS, Dutilleul PC, Levallois B, Cuisinier F (2015) Strategies for immobilization of bioactive organic molecules on titanium implant surfaces - a review. Folia Med (Plovdiv) 57(1):11–18. doi:10.1515/folmed-2015-0014

Pavan S, Berti F (2012) Short peptides as biosensor transducers. Anal Bioanal Chem 402(10):3055–3070. doi:10.1007/s00216-011-5589-8

Pu K, Li C, Zhang N, Wang H, Shen W, Zhu Y (2017) Epithelial cell adhesion molecule independent capture of non-small lung carcinoma cells with peptide modified microfluidic chip. Biosens Bioelectron 89(Pt 2):927–931. doi:10.1016/j.bios.2016.09.092

Roberts RW, Szostak JW (1997) Rna-peptide fusions for the in vitro selection of peptides and proteins. Proc Natl Acad Sci U S A 94(23):12297–12302

Rogers JM, Suga H (2015) Discovering functional, non-proteinogenic amino acid containing, peptides using genetic code reprogramming. Org Biomol Chem 13(36):9353–9363. doi:10.1039/c5ob01336d

Sakai-Kato K, Kato M, Ishihara K, Toyo'oka T (2004) An enzyme-immobilization method for integration of biofunctions on a microchip using a water-soluble amphiphilic phospholipid polymer having a reacting group. Lab Chip 4(1):4–6. doi:10.1039/b310932a

Saludes JP, Morton LA, Ghosh N, Beninson L, Chapman ER, Fleshner M, Yin H (2012) Detection of highly curved membrane surfaces using a cyclic peptide derived from synaptotagmin-I. ACS Chem Biol 7(10):1629–1635. doi:10.1021/cb3002705

Samuelson P, Gunneriusson E, Nygren PA, Stahl S (2002) Display of proteins on bacteria. J Biotechnol 96(2):129–154

Sanghvi AB, Miller KP, Belcher AM, Schmidt CE (2005) Biomaterials functionalization using a novel peptide that selectively binds to a conducting polymer. Nat Mater 4(6):496–502

Sano K, Shiba K (2003) A hexapeptide motif that electrostatically binds to the surface of titanium. J Am Chem Soc 125(47):14234–14235

Sano K, Shiba K (2008) *In aqua* manufacturing of a three-dimensional nanostructure using a peptide aptamer. MRS Bull 33(05):524–529

Sano K, Ajima K, Iwahori K, Yudasaka M, Iijima S, Yamashita I, Shiba K (2005a) Endowing a ferritin-like cage protein with high affinity and selectivity for certain inorganic materials. Small 1(8–9):826–832

Sano K, Sasaki H, Shiba K (2005b) Specificity and biomineralization activities of Ti-binding peptide-1 (TBP-1). Langmuir 21(7):3090–3095

Sano K, Sasaki H, Shiba K (2006) Utilization of the pleiotropy of a peptidic aptamer to fabricate heterogeneous nanodot-containing multilayer nanostructures. J Am Chem Soc 128(5):1717–1722

Sano K, Yoshii S, Yamashita I, Shiba K (2007) In aqua structuralization of a three-dimensional configuration using biomolecules. Nano Lett 7(10):3200–3202

Sarikaya M, Tamerler C, Jen AK, Schulten K, Baneyx F (2003) Molecular biomimetics: nanotechnology through biology. Nat Mater 2(9):577–585. doi:10.1038/nmat964

Sato S, Ikemi M, Kikuchi T, Matsumura S, Shiba K, Fujita M (2015) Bridging adhesion of a protein onto an inorganic surface using self-assembled dual-functionalized spheres. J Am Chem Soc 137(40):12890–12896. doi:10.1021/jacs.5b06184

Scott JK, Smith GP (1990) Searching for peptide ligands with an epitope library. Science 249(4967):386–390

Seker UO, Demir HV (2011) Material binding peptides for nanotechnology. Molecules 16(12):1426–1451. doi:10.3390/molecules16021426

Shao Z, Zhang X, Pi Y, Wang X, Jia Z, Zhu J, Dai L, Chen W, Yin L, Chen H, Zhou C, Ao Y (2012) Polycaprolactone electrospun mesh conjugated with an msc affinity peptide for msc homing in vivo. Biomaterials 33(12):3375–3387. doi:10.1016/j.biomaterials.2012.01.033

Shao Z, Zhang X, Pi Y, Yin L, Li L, Chen H, Zhou C, Ao Y (2015) Surface modification on polycaprolactone electrospun mesh and human decalcified bone scaffold with synovium-derived mesenchymal stem cells-affinity peptide for tissue engineering. J Biomed Mater Res A 103(1):318–329. doi:10.1002/jbm.a.35177

Shiba K (1998) *In vitro* constructive approaches to the origin of coding sequences. J Biochem Mol Biol 31(3):209–220

Shiba K (2004) MolCraft: a hierarchical approach to the synthesis of artificial proteins. J Mol Catal B Enzym 28(4-6):145–153

Shiba K (2010a) Exploitation of peptide motif sequences and their use in nanobiotechnology. Curr Opin Biotechnol 21(4):412–425. doi:10.1016/j.copbio.2010.07.008

Shiba K (2010b) Natural and artificial peptide motifs: their origins and the application of motif-programming. Chem Soc Rev 39(1):117–126. doi:10.1039/b719081f

Shiba K, Kokubun K, Suga K (2012) Peptides that bind to epithelial cell adhesion molecule. WO2014042209 A1

Shiba K, Hibino K, Yoshida M (2013) Conjugate composed of EpCAM-binding peptide aptamer and phosphorylcholine polymer copolymer. WO2014168230 A1

Sibarani J, Takai M, Ishihara K (2007) Surface modification on microfluidic devices with 2-methacryloyloxyethyl phosphorylcholine polymers for reducing unfavorable protein adsorption. Colloids Surf B Biointerfaces 54(1):88–93. doi:10.1016/j.colsurfb.2006.09.024

Singh A, Corvelli M, Unterman SA, Wepasnick KA, McDonnell P, Elisseeff JH (2014) Enhanced lubrication on tissue and biomaterial surfaces through peptide-mediated binding of hyaluronic acid. Nat Mater 13(10):988–995. doi:10.1038/nmat4048

So CR, Hayamizu Y, Yazici H, Gresswell C, Khatayevich D, Tamerler C, Sarikaya M (2012) Controlling self-assembly of engineered peptides on graphite by rational mutation. ACS Nano 6(2):1648–1656. doi:10.1021/nn204631x

Sternberg N, Hoess RH (1995) Display of peptides and proteins on the surface of bacteriophage λ. Proc Natl Acad Sci U S A 92(5):1609–1613

Szostak JW (1992) *In vitro* genetics. Trends Biochem Sci 17(3):89–93

Tamerler C, Sarikaya M (2009) Molecular biomimetics: nanotechnology and bionanotechnology using genetically engineered peptides. Philos Trans A Math Phys Eng Sci 367(1894):1705–1726. doi:10.1098/rsta.2009.0018

Thota V, Perry CC (2016) A review on recent patents and applications of inorganic material binding peptides. Recent Pat Nanotechnol 9(999):1

Tsuji T, Oaki Y, Yoshinari M, Kato T, Shiba K (2010) Motif-programmed artificial proteins mediated nucleation of octacalcium phosphate on titanium substrates. Chem Commun 46(36):6675–6677. doi:10.1039/c0cc01512a

Ulmer KM (1983) Protein engineering. Science 219(4585):666–671

Vetter I, Davis JL, Rash LD, Anangi R, Mobli M, Alewood PF, Lewis RJ, King GF (2011) Venomics: a new paradigm for natural products-based drug discovery. Amino Acids 40(1):15–28. doi:10.1007/s00726-010-0516-4

Villa-Diaz LG, Nandivada H, Ding J, Nogueira-de-Souza NC, Krebsbach PH, O'Shea KS, Lahann J, Smith GD (2010) Synthetic polymer coatings for long-term growth of human embryonic stem cells. Nat Biotechnol 28(6):581–583. doi:10.1038/nbt.1631

Wang S, Humphreys ES, Chung SY, Delduco DF, Lustig SR, Wang H, Parker KN, Rizzo NW, Subramoney S, Chiang YM, Jagota A (2003) Peptides with selective affinity for carbon nanotubes. Nat Mater 2(3):196–200

Watanabe J, Ishihara K (2008) Establishing ultimate biointerfaces covered with phosphorylcholine groups. Colloids Surf B Biointerfaces 65(2):155–165. doi:10.1016/j.colsurfb.2008.04.006

Wegner GJ, Lee HJ, Corn RM (2002) Characterization and optimization of peptide arrays for the study of epitope-antibody interactions using surface plasmon resonance imaging. Anal Chem 74(20):5161–5168

Weiner S, Hood L (1975) Soluble protein of the organic matrix of mollusk shells: a potential template for shell formation. Science 190(4218):987–989

Weissman KJ (2015) The structural biology of biosynthetic megaenzymes. Nat Chem Biol 11(9):660–670. doi:10.1038/nchembio.1883

Whaley SR, English DS, Hu EL, Barbara PF, Belcher AM (2000) Selection of peptides with semiconductor binding specificity for directed nanocrystal assembly. Nature 405:665–668

Wrighton NC, Farrell FX, Chang R, Kashyap AK, Barbone FP, Mulcahy LS, Johnson DL, Barrett RW, Jolliffe LK, Dower WJ (1996) Small peptides as potent mimetics of the protein hormone erythropoietin. Science 273(5274):458–463

Wronska DB, Krajewska M, Lygina N, Morrison JC, Juzumiene D, Culp WD, Nair SA, Darby M, Hofmann CM (2014) Peptide-conjugated glass slides for selective capture and purification of diagnostic cells: applications in urine cytology. BioTechniques 57(2):63–71. doi:10.2144/000114195

Xu Y, Takai M, Ishihara K (2010) Phospholipid polymer biointerfaces for lab-on-a-chip devices. Ann Biomed Eng 38(6):1938–1953. doi:10.1007/s10439-010-0025-3

Yolamanova M, Meier C, Shaytan AK, Vas V, Bertoncini CW, Arnold F, Zirafi O, Usmani SM, Muller JA, Sauter D, Goffinet C, Palesch D, Walther P, Roan NR, Geiger H, Lunov O, Simmet T, Bohne J, Schrezenmeier H, Schwarz K, Standker L, Forssmann WG, Salvatella X, Khalatur PG, Khokhlov AR, Knowles TP, Weil T, Kirchhoff F, Munch J (2013) Peptide nanofibrils boost retroviral gene transfer and provide a rapid means for concentrating viruses. Nat Nanotechnol 8(2):130–136. doi:10.1038/nnano.2012.248

Yoshinari M, Kato T, Matsuzaka K, Hayakawa T, Shiba K (2010) Prevention of biofilm formation on titanium surfaces modified with conjugated molecules comprised of antimicrobial and titanium-binding peptides. Biofouling 26(1):103–110. doi:10.1080/08927010903216572

Young TS, Young DD, Ahmad I, Louis JM, Benkovic SJ, Schultz PG (2011) Evolution of cyclic peptide protease inhibitors. Proc Natl Acad Sci U S A 108(27):11052–11056. doi:10.1073/pnas.1108045108

Yuasa K, Kokubu E, Kokubun K, Matsuzaka K, Shiba K, Kashiwagi K, Inoue T (2014) An artificial fusion protein between bone morphogenetic protein 2 and titanium-binding peptide is functional in vivo. J Biomed Mater Res A 102(4):1180–1186. doi:10.1002/jbm.a.34765

Solid-Binding Peptides in Biomedicine

2

Andrew Care, Peter L. Bergquist, and Anwar Sunna

Abstract

Some peptides are able to bind to inorganic materials such as silica and gold. Over the past decade, Solid-binding peptides (SBPs) have been used increasingly as molecular building blocks in nanobiotechnology. These peptides show selectivity and bind with high affinity to a diverse range of inorganic surfaces e.g. metals, metal oxides, metal compounds, magnetic materials, semiconductors, carbon materials, polymers and minerals. They can be used in applications such as protein purification and synthesis, assembly and the functionalization of nanomaterials. They offer simple and versatile bioconjugation methods that can increase biocompatibility and also direct the immobilization and orientation of nanoscale entities onto solid supports without impeding their functionality. SBPs have been employed in numerous nanobiotechnological applications such as the controlled synthesis of nanomaterials and nanostructures, formation of hybrid biomaterials, immobilization of functional proteins and improved nanomaterial biocompatibility. With advances in nanotechnology, a multitude of novel nanomaterials have been designed and synthesized for diagnostic and therapeutic applications. New approaches have been developed recently to exert

A. Care
Department of Chemistry and Biomolecular Sciences, Macquarie University, North Ryde, NSW, Australia

ARC Centre of Excellence for Nanoscale BioPhotonics (CNBP), Macquarie University, North Ryde, NSW, Australia

P.L. Bergquist
Department of Chemistry and Biomolecular Sciences, Macquarie University, North Ryde, NSW, Australia

Biomolecular Discovery and Design Research Centre, Macquarie University, North Ryde, NSW, Australia

Department of Molecular Medicine & Pathology, Medical School, University of Auckland, Auckland, New Zealand

A. Sunna (✉)
Department of Chemistry and Biomolecular Sciences, Macquarie University, North Ryde, NSW, Australia

Biomolecular Discovery and Design Research Centre, Macquarie University, North Ryde, NSW, Australia
e-mail: anwar.sunna@mq.edu.au

© Springer International Publishing AG 2017

A. Sunna et al. (eds.), *Peptides and Peptide-based Biomaterials and their Biomedical Applications*, Advances in Experimental Medicine and Biology 1030, https://doi.org/10.1007/978-3-319-66095-0_2

a greater control over bioconjugation and eventually, over the optimal and functional display of biomolecules on the surfaces of many types of solid materials. In this chapter we describe SBPs and highlight some selected examples of their potential applications in biomedicine.

Keywords

Bioconjugation • Biomaterials • Functionalization • Biomedicine • Solid-binding peptides

2.1 Introduction

Solid-binding peptides (SBPs) exhibit selectivity and high binding affinity towards the surfaces of a wide variety of solid materials e.g. metals, carbon-based materials, minerals and polymers (Care et al. 2015). They can be used in simple applications such as protein purification after being incorporated into a structural gene to form a fusion protein, they can be involved in more sophisticated reactions such as contributing to, or directing, the assembly and functionalization of inorganic materials and may have the ability to regulate the synthesis of nanoparticles. Many nanomaterials suffer from low solubility and poor biocompatibility, presenting potential safety concerns for *in vivo* applications. In addition, conventional bioconjugation techniques such as the EDC (1-ethyl-3-[3-dimethylaminopropyl]carbodiiamide hydrochloride and the NHS (N-hydroxysuccinamide) esters and maleimide reactions used to functionalize nanomaterials are often laborious and inefficient, as well as possibly interfering with the recognition of the immobilized protein towards its specific receptor or molecular partner. New approaches have been developed recently to exert greater control over the bioconjugation, and eventually, over the optimal and functional display of biomolecules on the surfaces of many types of solid materials (Avvakumova et al. 2014). The initial origin of the study of peptides that could recognize surfaces appears to be related to observations made in late 1988 where two groups reported the use of gene fusion products that could recognize and bind to nickel chelates as a method for purifying heterologously-expressed recombinant protein in *E. coli* (Hochuli et al. 1988). This approach was developed subsequently into a general method for recombinant protein purification, along with other affinity tags as competitors but was not exploited at the time as a general constructive process to generate organic-inorganic interactions and fusions.

These distinctive peptides have the unique ability to act as 'molecular linkers' that direct the facile and oriented immobilization of biomolecules onto solid surfaces to provide biological functionality (Care et al. 2014a), or as 'material synthesizers' that initiate and regulate the synthesis of both simple and complex materials e.g. nanoparticles (Naik et al. 2002b) (see Table 2.1). Thus, SBPs show increasing potential as biotechnological tools for a range of biomedical applications, including surgical implantation, drug delivery, vaccine formulation, and biosensing devices.

2.1.1 SBP Binding and Synthesis Mechanisms

The binding constants observed between SBPs and their corresponding solids are often in the nanomolar to low micromolar range. This high affinity is the sum of multiple non-covalent interactions that occur under ambient conditions, including electrostatic interactions, van der Waals forces, hydrophobic interactions, and hydrogen bonds (Tang et al. 2013). Furthermore, some SBPs have been shown to mediate the synthesis of solid materials with high precision. In this situation, SBPs lower the surface energies of their target solids upon binding, which drives thermodynamic forces that promote the nucleation and growth of solid materials (Naik et al. 2002b). Nevertheless,

Table 2.1 Selected SBPs and their applications

Target solid	Sequence	pI[a]	Charge[b]	Applications	References[c]
Metals					
Gold	MHGKTQATSGTIQS	8.52	+1	Biomolecule immobilization; biocompatibility; multi-material fabrication	Brown et al. (2000)
	WALRRSIRRQSY	12.00	+4	Biomolecule immobilization; biocompatibility	Hnilova et al. (2008)
	WAGAKRLVLRRE	11.71	+3	Biocompatibility	Hnilova et al. (2008)
	LKAHLPPSRLPS	11.00	+2	Material-specific antibody generation	Nam et al. (2006)
	VSGSSPDS	3.80	0	Biomaterial production; multi-material fabrication	Huang et al. (2005)
Palladium	TSNAVHPTLRHL	9.47	+1	Nanostructure synthesis; controlled catalysis	Pacardo et al. (2009)
Platinum	PTSTGQA	5.96	0	SBP binding studies	Sarikaya et al. (2003)
	TLTTLTN	5.19	0	Nanostructure synthesis	Li and Huang (2010)
	SSFPQPN	5.24	0	Nanostructure synthesis	Li and Huang (2010)
	CSQSVTSTKSC	8.06	+1	Biocompatibility	Sarikaya et al. (2003)
Silver	AYSSGAPPMPPF	5.57	0	Biomaterial production	Naik et al. (2002b)
	NPSSLFRYLPSD	5.84	0	Biomaterial production	Naik et al. (2002b)
Titanium	RPRENRGRERGL	11.82	+3	Biocompatibility	Khatayevich et al. (2010)
	RKLPDA	8.75	+1	Multi-material fabrication; nanostructure synthesis	Sano and Shiba (2003)
Metal oxides					
Iron oxide	RRTVKHHVN	12.01	+3	Biomolecule immobilization	Brown (1997)
Upconversion nanoparticles	ACTARSPWICG	8.11	+1	Biomolecule immobilization; biocompatibility	Zhang et al. (2012)
Silica	MSPHPHPRHHHT	9.59	+1	Biomolecule immobilization; nanostructure synthesis	Naik et al. (2002a)
	SSKKSGSYSGSKGSKRRIL	11.22	+6	Multi-material fabrication	Kröger et al. (1999)
	HPPMNASHPHMH	7.10	0	Multi-material fabrication	Eteshola et al. (2005)
	RKLPDA	8.75	+1	Multi-material fabrication; nanostructure synthesis	Sano and Shiba (2003)
Quartz	PPPWLPYMPPWS	5.95	0	Biomolecule immobilization; biocompatibility; biomaterial production; multi-material fabrication	Oren et al. (2007)
Zeolites	VKTQATSREEPPRLPSKHRPG	10.90	+3	Biomolecule immobilization	Nygaard et al. (2002)
Zinc oxide	EAHVMHKVAPRP	8.86	+1	Material-specific antibody generation	Umetsu et al. (2005)
	RPHRK	12.01	+3	Biomolecule immobilization	Thai et al. (2004)

(continued)

Table 2.1 (continued)

Target solid	Sequence	pI[a]	Charge[b]	Applications	References[c]
Minerals					
Calcium phosphate	KDVVVGVPGGQD	4.21	−1	Biomolecule immobilization; biomaterial production	Chiu et al. (2012)
Hydroxyapatite	NPYHPTIPQSVH	6.92	0	Biomaterial production	Chung et al. (2011)
Carbon materials					
Graphene	EPLQLKM	6.10	0	Biomolecule immobilization; biomaterial production	Cui et al. (2010)
Single-walled carbon nanotubes	HSSYWYAFNNKT	8.50		Biomolecule immobilization, biomaterial production; multi-material fabrication	Pender et al. (2005)
	DYFSSPYYEQLF	3.67	−2	Biocompatibility	Kase et al. (2004)
	DSPHTELP	4.35	−2	Biomaterial production	Dang et al. (2011)
Semiconductors					
Cadmium sulphide	CTYSRKHKC	9.39	+3	Multi-material fabrication; nanostructure synthesis	Flynn et al. (2003)
Gallium arsenide	AQNPSDNNTHTH	5.97	0	SBP binding studies	Whaley et al.)2000)
Zinc sulphide	CGPAGDSSGVDSRSVGPC	4.21	−1	Biomolecule immobilization; biomaterial production	Zhou et al. (2010)
	CNNPMHQNC	6.72	0	Biomaterial production	Mao et al. (2003)
	LRRSSEAHNSIV	9.61	+1	Multi-material fabrication; nanostructure synthesis	Mao et al. (2003)

Adapted from Care et al. (2015). Copyright (2015), with permission from Elsevier

[a]Isoelectric points were calculated using Compute pI/Mw tool (http://web.expasy.org/compute_pi/)

[b]Calculated by subtracting the number of basic residues (R and K) from the number acidic residues (D and E)

[c]Reference from which the SBP sequences were identified originally

due to the complex and dynamic nanoenvironment at the peptide-material interface, many of the mechanisms involved in SBP recognition, selectivity and affinity remain poorly understood. At present, the various properties of peptides (e.g. amino acid composition and sequence, peptide structure and physicochemistry) in combination with the surfaces of solid materials (e.g. crystallinity, charge and topology) and the surrounding solution (e.g. water) are known to influence binding (Corni et al. 2013; Goede et al. 2004; Nel et al. 2009; Tang et al. 2013).

2.1.2 Isolation of SBPs

A number of methods have been developed and used to obtain SBPs that bind selectively to solid materials. These include isolation from biological sources, rational design using computational biology followed by chemical synthesis (Oren et al. 2007), or recombinant techniques such as directed mutagenesis and multimerization (Brown 1997; Kim et al. 2010; Seker et al. 2008). However, SBPs are most commonly isolated via combinatorial display technologies, particularly phage display and cell-surface display. These methodologies are simple, robust and allow extensive libraries of peptides to be screened for solid binding functionality. Following selection and enrichment, whole phage particles (Huang et al. 2005; Whaley et al. 2000) or bacterial cells (Park et al. 2009) that display the SBP on their surface can be utilized as biological carriers of solid materials (Ghosh et al. 2014). The peptide itself may also be produced by solid-phase peptide synthesis and then chemically conjugated to other entities (Nochomovitz et al. 2010); or the peptide as a domain within a fusion protein can be used to direct the selective attachment of biomolecules to a solid surface (Ko et al. 2009).

2.2 Applications in Biomedicine

We outline below selected examples of the potential for applications of SBPs in some areas of biomedicine.

2.2.1 Vaccine Development

Some nanoparticles (NPs) have adjuvant properties and therefore can be exploited for antigen delivery in vaccine development. However, conventional NP functionalization reactions often reduce the stability and specificity of antigens, preventing the desired host immune responses.

Ha et al. (2016) recently employed a ZnO-binding peptide (ZBP) for the facile functionalization of ZnO nanoparticles (ZNPs) with the bacterial antigen, ScaA of *Orientia tsutsugamushi*, which causes the infectious disease scrub typhus. Mice vaccinated with the ZBP-ScaA functionalized ZNPs exhibited strong anti-ScaA antibody responses and gained protective immunity against lethal challenges of *O. tsutsugamushi*. Thus, a functional vaccine delivery technology was developed by using the binding properties of an SBP.

SBPs have also been used to functionalize NPs with tumor-specific antigens for antigen delivery in tumor immunotherapy. For example, a ZBP was reported to facilitate the attachment of carcinoembryonic antigen (CEA) to optically-active ZnO-coated magnetic NPs (Cho et al. 2011). The ZBP-CEA functionalized NPs were recognized and taken up by dendritic cells (DCs). Mice that were immunized with the modified DCs displayed T-cell responses specific to CEA, suppressed tumor growth, and showed longer survival times when compared to control mice. Furthermore, ZBP-CEA remained bound to the NPs in culture media for 4 h, before slowly disassociating over the following 2 days. It was proposed that this gradual and continuous release of antigens inside DCs enhanced their capacity to induce antigen-specific immune responses.

Zhou et al. (2014) exploited both the material synthesizing and molecular linking properties of SBPs to develop a novel method for the self-assembly of vaccine formulations. They demonstrated that when a calcium phosphate (CaP)-binding peptide capable of driving the synthesis of CaP NPs was genetically fused to an antigen, the resulting fusion protein could be shown to mediate the single-step synthesis of antigen-coated CaP NPs. Mice vaccinated with these functional NPs demonstrated greater antigen-specific and longer-lasting T cell responses when compared to the fusion protein alone.

2.2.2 Bioimaging

Materials that emit near-infrared (NIR) light are appropriate for the development of non-invasive *in vivo* fluorescent biomedical imaging applications because of its high tissue penetration and reduced light scattering in comparison to visible light, which eliminates problematic background autofluorescence from biological tissues (Yi et al. 2014). Single-walled carbon nanotubes (SWNTs) demonstrate fluorescent emission in the second-window near-infrared (NIR-II: 900–1400 nm). M13 bacteriophage has been coated with SWNTs by displaying a SWNT-binding peptide along the p8 major coat protein of M13 (Dang et al. 2011; Ghosh et al. 2014). These optically-active M13-SWNT complexes also were engineered to present targeting ligands on the p3 minor coat protein to facilitate targeted fluorescent imaging. For example, p3 was genetically modified to display a targeting peptide that binds Secreted Protein Acidic and Rich in Cysteines proteins (SPARC) that are over-expressed in certain cancer types, including ovarian cancer (Ghosh et al. 2014). The resulting SWNT-SPARC-M13 nanoprobe was injected intravenously into a mouse model bearing SPARC-expressing ovarian tumors. Whole-body NIR imaging indicated high tumor-to-background uptake and showed better signal-to-noise imaging performance when compared with conventional visible fluorescent and near-infrared (NIR-I) dyes for defining tumors. Subsequent surgical removal of submillimeter tumors was improved significantly with the image guidance provided by the nanoprobe. Thus, the exploitation of SBPs to construct unique biomaterials, such as SWNT-SPARC-M13 nanoprobe, further demonstrates their potential as tools for biomedicine.

2.2.3 Drug Delivery

In many conventional cancer therapies, anticancer drugs diffuse and become distributed throughout the body upon systematic administration, causing adverse side-effects in healthy cells and tissues and a reduction in clinical efficacy (Bae and Park 2011). Drugs must be targeted selectively, delivered, and released controllably at their primary site-of-action (e.g. tumors) to avoid these complications and provide an effective and localized treatment for cancer.

Emerging nanotechnologies show great promise as drug delivery systems for cancer treatment. In particular, nanoparticles have the ability to enhance drug targeting and pharmacokinetics and to lower drug toxicity (Singh and Lillard 2009). Unfortunately, very few synthetic NPs have progressed to become approved clinical treatments. This result is due to significant limitations such as their physicochemical heterogeneity, problematic functionalization, instability in physiological solutions, poor tumor penetration and toxicity in biological systems (Bertrand et al. 2014).

For example, lanthanide-doped upconversion nanoparticles (UCNPs) are luminescent probes for bioimaging, diagnostics and therapeutics (Chen et al. 2014). However, the bioconjugation techniques for functionalizing UCNPs with antibodies (e.g. amine-reactive crosslinking) tend to be laborious and inefficient. In most cases, antibodies attach to UCNPs with altered conformations and random orientations that cause a reduction or complete loss of their antigen-binding activity. Furthermore, UCNPs have a propensity to aggregate after functionalization, which presents potential safety concerns for *in vivo* applications. The conventional techniques for antibody immobilization (e.g., adsorption and covalent binding) often lead to random antibody orientations that decrease the density and availability of antigen-binding sites. As an alternative, antibody-binding proteins (ABPs, such as Protein A and G from bacteria) strongly bind to the Fc region of antibodies, ensuring their correct orientation (Akerström and Björck 1989). Therefore,

SBPs have been fused genetically to ABPs (de Juan-Franco et al. 2013; Sunna et al. 2013b), circumventing the need to specifically modify antibodies and to increase versatility. The SBP selectively directs the immobilization onto materials while the ABP binds antibodies so that they retain their biological functionality. This strategy has been shown to enhance significantly the orientation and density of antibodies attached to material surfaces when compared to traditional covalent methods (de Juan-Franco et al. 2013). We have developed a platform bioconjugation technology based upon the novel fusion protein, Linker–Protein G (LPG). LPG is composed of two functionally-unique domains: (a) a silica-specific SBP (referred to as the linker) which binds strongly to silica surfaces; and (b) *Streptococcus* Protein G', which binds to antibodies (Care et al. 2014a, b; Sunna et al. 2013a, b). LPG has been shown to provide an anchorage point for the functional immobilization of antibodies onto the outer surfaces of silica-coated nanoparticles, including UCNPs, within minutes and without the need for any complex chemical reactions or harsh physical treatments. This strategy has been used successfully in various practical applications, including cell capture, detection, and imaging (Care et al. 2014a, b; Lu et al. 2014; Sayyadi et al. 2016; Sunna et al. 2013a, b).

We reported recently the practical use of LPG in the targeted photodynamic therapy (PDT) of cancer. PDT is a promising cancer treatment due to its outstanding selectivity and minimal invasiveness (Chen et al. 2014; Lucky et al. 2015). To destroy solid tumors, PDT relies on the photochemical reactions between excitation light and photosensitizers to produce cytotoxic reactive oxygen species (ROS) from cellular oxygen that induce cell death (Lucky et al. 2015). UCNPs can be excited with near-infrared (NIR) light (~980 nm) to emit bright visible light, allowing their emission to penetrate through biological tissues for both deep non-invasive imaging and therapy. Therefore, UCNPs can improve the localization and penetration depth of PDT by transforming deep-penetrating NIR light to visible light to activate photosensitizers via Förster resonance energy transfer (FRET) (Liang et al. 2016, 2017). We showed that LPG was able to mediate the effective attachment of tumor-specific antibodies to the surface of UCNPs with photosensitizer-doped silica coatings (Si-UCNPs) (Fig. 2.1a, b) (Liang et al. 2016). These functionalized Si-UCNPs bound selectively to human colon adenocarcinoma HT-29 cells and were internalized subsequently by the live cells (Fig. 2.1c). Under NIR excitation, the internalized Si-UCNPs emitted visible light, which activated the photosensitizer within the silica shell, resulting in the conversion of intracellular molecular oxygen into cytotoxic ROS (i.e. singlet oxygen $^1O_2$) (Fig. 2.1d) that induced phototoxicity to suppress the growth of the targeted cancer cells *in vitro*. Specifically, the viability of colon cancer cells exposed to 100 µg/mL of Si-UCNPs was reduced by ~40% after 10 min of NIR irradiation (Fig. 2.1e).

Overall, this approach allows the integration of a diverse selection of silica-coated nanoparticles, photosensitizing agents (or other anticancer therapeutics) and virtually any antibody that binds cancer cells. Therefore, LPG provides a platform technology for antibody – NP coupling in targeted cancer therapy and imaging.

2.2.4 LPG – Cell Capture

LPG also shows potential for use in applications that require cell capture e.g. microfluidic diagnostics. In one example, LPG mediated the attachment of different cell-specific antibodies to silica-coated magnetic microparticles. Each of the biofunctionalized particles were shown to mediate the selective binding and recovery of different target cell types (e.g., human stem cells, *Legionella*, *Cryptosporidium* and *Giardia*) from solution, enabling their rapid and simple visualization and identification (Fig. 2.2) (Care et al. 2014a).

2.2.5 Biocompatibility

Nanomaterials (e.g. nanoparticles) show great promise in biomedicine. However, concerns over the biological safety of many of these materials has prevented their clinical use. This reaction is

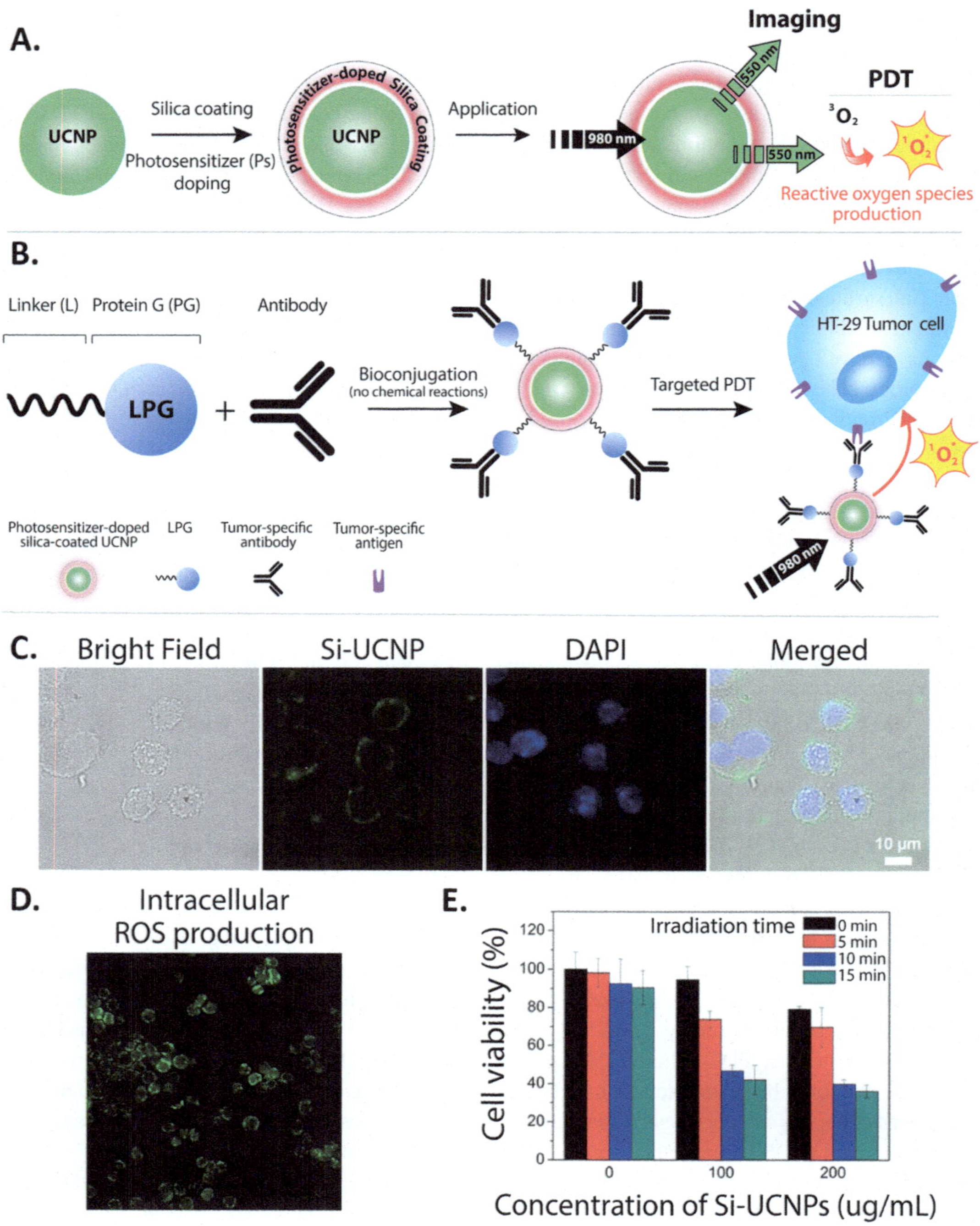

Fig. 2.1 (**a**) Schematic illustration of the fabrication of UCNPs with photosensitizer-doped silica coatings (Si-UCNPs) and its functionality and (**b**) LPG-mediated bioconjugation of Si-UCNPs with tumor-specific antibodies and their application in targeted PDT. (**c**) Fluorescent images of human colon adenocarcinoma HT-29 cells after incubation with Si-UCNPs + LPG + tumor-specific antibody. (**d**) Intracellular ROS generated (green fluorescence) in HT-29 cells treated with NIR-excited bio-functionalized UCNPs particles (**e**) Cell viability of HT-29 cells treated with various concentrations (0, 100, and 200 μg/mL) of Si-UCNPs and then exposed to NIR-excitation for different durations (0, 5, 10, and 15 min) and quantified by standard MTT assay. Each value represents the mean ± standard deviation of five replicates (Adapted with permission from Liang et al. (2016). Copyright (2016) American Chemical Society)

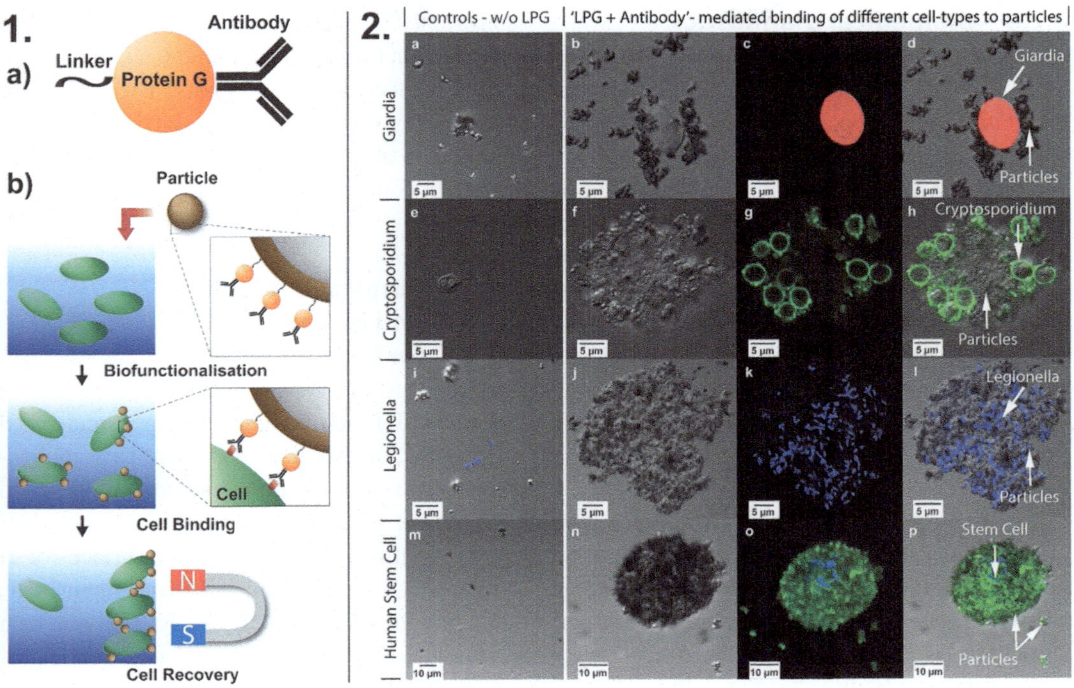

Fig. 2.2 (**1**) Illustration showing the antibody-mediated recovery of cells using biofunctionalized particles. (*a*) "LPG-Antibody complex". (*b*) Silica-coated magnetic particles functionalized with "LPG-Antibody complexes" are added to a solution containing target cell types. The particles bind to the targeted cells and are then recovered from solution using magnetic separation. (**2**) "LPG-Antibody complex"-mediated binding of different cell types to silica-coated magnetic particles. (*a*) *Giardia* cysts (red) after binding assay with anti-*Giardia* antibody but without LPG. (*b–d*) *Giardia* cysts (red) after binding assay with LPG + anti-*Giardia* antibody. (*e*) *Cryptosporidium* oocysts after binding assay with anti-*Cryptosporidium* antibody (labeled with green fluorescein (FITC)) but without LPG. (*f–h*) *Cryptosporidium* oocysts after binding assay with LPG + anti-*Cryptosporidium*-FITC. (**i**) *Legionella pneumophila* cells after binding assay with a cell-specific monoclonal antibody but without LPG. Cells visualized using 0.1% Evans blue stain. (*j–l*) *Legionella pneumophila* cells after binding assay with LPG and a cell-specific monoclonal antibody. Cells visualized using 0.1% Evans blue stain. (*m*) Human adipose-derived mesenchymal stem cells after binding assay with cell-specific CD90-FITC antibody but without LPG. Cellular DNA fluorescently dyed with Hoechst stain 33342 (blue). (*n–p*) Human adipose-derived mesenchymal stem cells after binding assay with LPG + CD90-FITC. Cellular DNA fluorescently dyed with Hoechst stain 33342 (blue) (Adapted with permission from Care et al. 2014a. Copyright (2014) Springer)

in part due to their inconsistent surface properties (e.g. charge, size, structure and functionalization), which causes instability in physiological solutions, decreased biocompatibility and toxicity *in vivo*. Recently, the surfaces of nanomaterials have been modified with proteins or peptides to help improve the safety of their application within biological systems.

SBPs can form biocompatible coatings on nanomaterial surfaces under biological conditions without inducing any immunogenic or cytotoxic responses. For example, a peptide (RE-1) with binding affinity towards lanthanide-based upconversion nanoparticles (UCNPs) was shown to form stable coatings on UCNP surfaces (Zhang et al. 2012). This action prevented UCNP aggregation and reduced unwanted interactions with cells, both *in vitro* and *in vivo*, and thus significantly improved the safety of UCNPs. In other work, SBPs were conjugated to polyethylene glycol (PEG), enabling the directed assembly of PEG coatings on solid surfaces (i.e. platinum and gold) that rendered them bio-inert (Khatayevich et al. 2010). This approach also could be applied to nanoparticle surfaces to improve their biocompatibility for biomedical applications.

2.2.6 Biomedical Diagnostics

Graphene has unique electrical properties that lends itself to chemical and biological sensing applications. However, the covalent functionalization of graphene can easily degrade its delicate structure leading to a loss in physical and electronic properties (Mann and Dichtel 2013). As an alternative, graphene-binding peptides have mediated successfully the facile non-covalent immobilization of biomolecules on graphene-based biosensors. For example, a graphene-binding peptide has been fused to an antimicrobial peptide isolated from frog skin, which selectively binds various pathogenic bacteria (e.g. *Helicobacter pylori*) (Mannoor et al. 2012). This bifunctional peptide was used to functionalize a graphene-based wireless biosensor without degrading its electronic sensing properties. The graphene was printed onto soluble silk, allowing the biosensor to be transferred easily to biological surfaces such as bone and skin. As a proof-of-concept, the biosensor was placed onto a tooth to enable the real-time monitoring of respiration and the effective detection of *H. pylori* in saliva (observed lower detection limit ~100 cells) (Fig. 2.3) (Mannoor et al. 2012). This wireless biosensor also was attached to the outside of an intravenous (IV) bag to test its usage in hospital sanitation and biohazard monitoring. As a result, contamination of the IV bag by an antibiotic-resistant strain of *Staphylococcus aureus*, a pathogen responsible for many post-surgical wound infections, was monitored by the sensor which was shown to have a detection limit of 1 bacterium μl^{-1} (Mannoor et al. 2012).

2.2.7 Anti-microbial Therapeutics

Silver nanoparticles have well-documented anti-bacterial properties and are present in various pharmaceuticals e.g. wound dressings and topical creams. Recently, a silver-binding peptide capable of precipitating silver from aqueous silver nitrate was genetically engineered onto the interior surface of a spherical protein cage (Giessen and Silver 2016). This action enabled the size-constrained synthesis of silver nanoparticles (Ag NPs) within the cage under ambient and environmentally sustainable conditions. The antimicrobial activity of the protein cage-coated Ag NPs was tested and shown to be greater than silver nitrate or conventional citrate-capped Ag NPs.

2.2.8 Surgical Implants and Biomaterials for Biomedicine

Titanium (Ti) and titanium alloys are widely used to manufacture surgical implants. This application is due to their high biocompatibility and exceptional physical and chemical properties of high strength and low weight. The surfaces of Ti implants can be modified to enhance their interactions with biological tissues (e.g. bone) with functional molecules. However, some of the processes used to attach these molecules to titanium involve harsh chemicals (e.g. organic solvents), which can reduce the implants safe osseointegration significantly (de Oliveira et al. 2007).

Alternatively, SBPs have been used to functionalize titanium-based materials under biologically-friendly conditions and without any complex pre-treatments. For example, the adhesion and proliferation of osteoblast and fibroblast cells was improved on implant-grade Ti using a bifunctional peptide that contained the cell attachment motif RGDS (Arg-Gly-Asp-Ser) and a Ti-binding peptide (Yazici et al. 2013). In addition, a fusion protein consisting of the human bone morphogenetic protein 2 (BMP-2) and multiple repeats of a Ti-binding peptide was shown to self-assemble on Ti-coated implants, and promote the production of bone *in vivo* (rats) (Yuasa et al. 2014).

Pathogenic bacterial biofilms are the major cause of surgical implant-associated infection, causing inflammation in the surrounding tissue, loss of the supporting bone and failure of the implanted device. The ability to control and reduce bacterial colonization and biofilm formation on implant surfaces is therefore highly desirable. In a sophisticated example using a hybrid peptide that was bifunctional with respect to a titanium-binding peptide, possessed a suitable

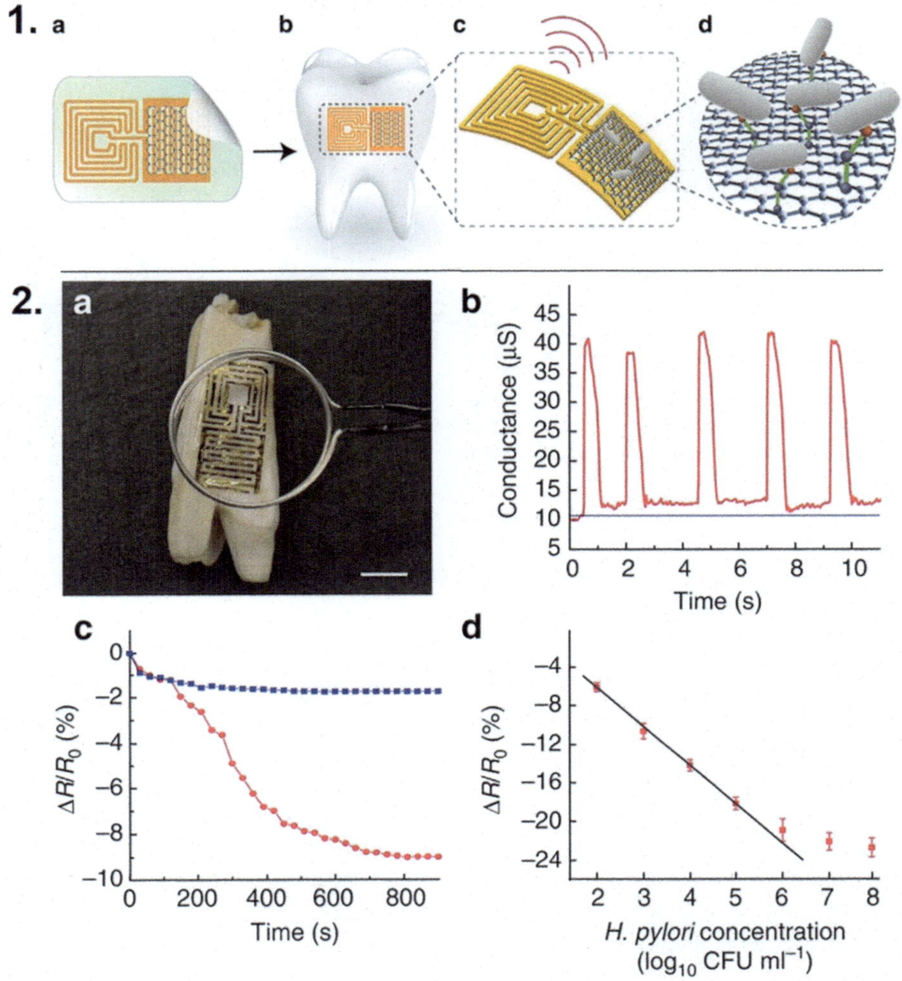

Fig. 2.3 (**1**) Biotransferable graphene wireless nanosensor. (*a*) Graphene is printed onto bioresorbable silk and contacts are formed containing a wireless coil. (*b*) Biotransfer of the nanosensing architecture onto the surface of a tooth. (*c*) magnified schematic of the sensing element, illustrating wireless readout. (*d*) Binding of pathogenic bacteria by peptides self-assembled on the graphene nanotransducer. (**2**) Tooth sensor monitoring of breath and saliva. (*a*) optical image of the graphene wireless sensor biotransferred onto the surface of a tooth. Scale bar is 1 cm. (*b*) Electrical conductance versus time upon exposure of the sensor to pulses of exhaled breath (red line). Baseline conductance is shown as *blue line*. (*c*) Percentage change in graphene resistance versus time following exposure to ~100 *H. pylori cells* in human saliva (*red line*). The response to 'blank' saliva solution is shown as *blue line*. (*d*) Percentage change in graphene resistance versus concentration of *H. pylori* cells. Error bars show s.d. (N = 3) (Adapted by permission from Macmillan Publishers Ltd: Nat Commun (Mannoor et al. 2012, 3:763), copyright (2012))

spacer and incorporated an antimicrobial peptide able to bind to the metal component of a transplant, Yazici et al. (2016) showed that bacterial colonization with a variety of pathogens such as *Streptococcus mutans*, *Staphylococcus epidermis* and *Escherichia coli* was reduced as compared to titanium surfaces without a coating of the hybrid peptide (Fig. 2.4).

Ti-binding peptides have been used to facilitate the immobilization of antimicrobial peptides onto Ti surfaces to help prevent biofilm formation (Yoshinari et al. 2010). In another example,

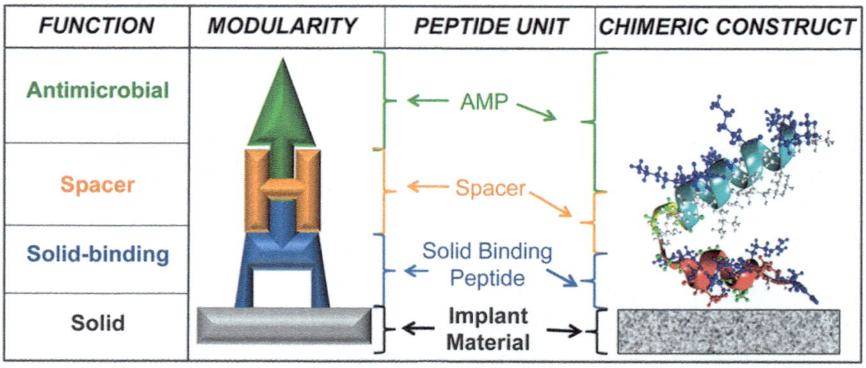

Fig. 2.4 Illustration depicting the design of a bifunctional peptide for imparting antimicrobial properties to a solid surface. The resulting bifunctional peptide is composed of a SBP and an antimicrobial peptide (Adapted with permission from Yazici et al. (2016). Copyright (2016) American Chemical Society)

Yoshinari et al. (2011) applied Ti-binding peptides to immobilize an antimicrobial peptide onto a titanium surface, which resulted in a significant reduction in the attachment of the oral pathogen *Porphyromonas gingivalis*. Hence, SBPs can be utilized to modify implant surfaces to enhance their functionality, efficacy, and biological safety.

SBPs have the capacity to mediate the construction of composite materials for biomedical purposes. A method for coating the surfaces of titanium implants with hydroxyapatite using SBPs was developed recently (Kelly et al. 2015). They used SBPs that bind to either Ti or hydroxyapatite and were fused to each other and showed that they acted as a linking molecule between a Ti surface and a subsequent hydroxyapatite coating. Human osteoblast cells deposit a collagenous matrix that is mineralized to produce new bone tissue. They were able to proliferate and also deposit collagen on the resulting hydroxyapatite-coated surfaces. Overall, this SBP-based coating process is faster, cheaper and less complicated than current standard commercial methods used in the manufacture of implants.

In nature, many proteins and peptides are association with the formation of hard tissues (such as bone and teeth). SBPs that bind mineral constituents of bone, e.g., apatite and hydroxyapatite (HA) show great potential for orthopaedics and dentistry and were isolated and characterized by Gungormus et al. (2008). For example, HA-binding peptides have been incorporated into green fluorescent protein and used in conjunction with fluorescent microscopy to monitor the real-time mineralization of synthetically formed hydroxyapatite and to visualize mineralized regions on biological tissues such as human teeth (Yuca et al. 2011). In another example, a bifunctional peptide composed of a broad-spectrum antimicrobial peptide and a HA-binding peptide have been developed to inhibit biofilm formation on tooth enamel analogues by the oral pathogenic bacterium *Streptococcus mutans*. It was also shown to have cytocompatibility with human cells and functional stability in human saliva, thus demonstrating an alternative strategy for oral pathogen control (Huang et al. 2016).

SBPs that bind to bone have been used to mediate the formation of biomaterials. For example, an apatite-binding peptide (apatite is another major component of bone) was conjugated to a peptide that binds to human bone marrow stromal cells (hBMSCs) (Ramaraju et al. 2014). The resulting bifunctional peptide was capable of functionalizing the surface of apatite and mediating the attachment of hBMSCs, significantly enhancing cell spreading, proliferation, and their osteogenic differentiation *in vitro*. Thus, the potential of SBPs in cell-based therapies for skeletal regeneration could be demonstrated.

2.3 Concluding Remarks and Future Perspectives

The development of suitable bioconjugation strategies best suited for functionalization of an ever-increasing number of new nanomaterials with diverse surface chemistries and physical properties continues to lag behind their discovery and synthesis. SBPs can be selected readily for virtually any nanomaterial via combinatorial display technologies and represent a biological route towards bioconjugation strategies specific for individual nanomaterials, resulting in the provision of improved solubility, colloidal stability and biocompatibility. This approach would overcome many of the technical issues that prevent their translation from the laboratory bench to biomedical applications. However, in some cases, the stability and functionality of SBPs are compromised within living systems, resulting in dissociation from their target substrates via exchange with free host proteins or complete degradation by host proteases (Care et al. 2014b). A more application-orientated approach would include the isolation of SBPs using combinatorial display libraries that bind nanomaterials during the biopanning procedure while in the presence of the appropriate biological fluid (e.g., whole blood, serum, urine) that they will come into contact with during their application *in vivo* or *ex vivo*. Such an action would go a long way to ensuring that the stability and binding function of the isolated SBPs were retained when exposed to complex biological environments, consequently improving their feasibility in biomedical applications. Another important point to consider is that combinatorial display technologies are not without their disadvantages. Peptide libraries display some compositional, positional, and expression bias that may lead to the under-representation of sequences with desirable properties (Umlauf et al. 2014). For example, general biopanning protocols performed against solid materials demonstrate intrinsic bias towards positively-charged sequences that bind via electrostatic interactions. Sequences that bind via non-electrostatic interactions are discriminated against and as a consequence, many strongly binding SBPs are not isolated and remain unexplored (Puddu and Perry 2012).

SBPs have a demonstrated potential in biomedical applications as shown by the selected examples reviewed here of fundamental research on their role in the prevention of infection in implant surgery but more extensive *in vivo* studies are required on the safety and efficiency of specific hybrid materials intended for transplant protection. A more immediate application appears to be their possible uses in cell capture and in diagnosis, especially of infectious microorganisms at low concentrations. The fundamental research reported offers the possibility of rapid, enabling identification outside of clinical laboratories, for example, by the adaptation and use of smart-phone technologies.

The isolation or design of SBPs appears to have advanced rapidly without a parallel increase in the extent of understanding of the mechanisms by which they function. One way to expand knowledge in this area would be set up a comprehensive database as has been done for other systems, some of which are related to research on SBPs. They could be introduced into a molecular toolbox for the rational design of stable basic building blocks (individual immobilized proteins of various sorts) with predefined functions in the same way as inorganic and organic compounds can act as nanoscale building blocks that can be combined to assemble functional nanomaterials. Such a molecular toolbox would have three main interchangeable building blocks: inorganic matrices, matrix-specific SBPs, biomedically-relevant proteins (e.g. antibodies, antimicrobial peptides and/or specific microbial probes). Researchers would have the flexibility to 'pick and mix' multiple building blocks for the assembly of bifunctionalized materials that could be designed for a particular biomedical purpose. Such a molecular toolbox would rely on continued improvements in large-scale data-sharing initiatives (databases) that would allow researchers access to detailed information on different building blocks (e.g. materials and biomolecules). Several databases that are relevant to this approach are currently available. These sources include the newly released 'Nano' database from Springer-Nature (https://nano.nature.com/), which is highly relevant to this approach. Nano provides indexed and structured information on a wide range of nanomaterials and nanodevices. Databases comprising various bioactive peptides (e.g., anticancer peptides (Tyagi et al. 2015) and antimicrobial peptides (Gogoladze et al. 2014) and detailed information on their properties also are available. However, only a limited number of SBPs have been incorporated into such databases (e.g., MimoDB 2.0, Huang et al. 2012), thus there is currently no specialized database for SBPs. For this reason, we propose the development and construction of a comprehensive SBP database that provides complete information on the properties and functions of SBPs that are relevant to biomedical applications (e.g., binding affinity and selectivity). A repository of such data would facilitate the widespread use of SBPs in conjunction with other building blocks to construct novel and unique functional materials for biomedicine.

References

Akerström B, Björck L (1989) Protein L: an immunoglobulin light chain-binding bacterial protein. Characterization of binding and physicochemical properties. J Biol Chem 264:19740–19746

Avvakumova S, Colombo M, Tortora P, Prosperi D (2014) Biotechnological approaches toward nanoparticle biofunctionalization. Trends Biotechnol 32:11–20

Bae YH, Park K (2011) Targeted drug delivery to tumors: myths, reality and possibility. J Control Release 153:198–205

Bertrand N, Wu J, Xu X, Kamaly N, Farokhzad OC (2014) Cancer nanotechnology: the impact of passive and active targeting in the era of modern cancer biology. Adv Drug Deliv Rev 66:2–25

Brown S (1997) Metal-recognition by repeating polypeptides. Nat Biotech 15:269–272

Brown S, Sarikaya M, Johnson E (2000) A genetic analysis of crystal growth. J Mol Biol 299:725–735

Care A, Chi F, Bergquist P, Sunna A (2014a) Biofunctionalization of silica-coated magnetic particles mediated by a peptide. J Nanopart Res 16:1–9

Care A, Nevalainen H, Bergquist P, Sunna A (2014b) Effect of *Trichoderma reesei* proteinases on the affinity of an inorganic-binding peptide. Appl Biochem Biotechnol 173:2225–2240

Care A, Bergquist PL, Sunna A (2015) Solid-binding peptides: smart tools for nanobiotechnology. Trends Biotechnol 33:259–268

Chen G, Qiu H, Prasad PN, Chen X (2014) Upconversion nanoparticles: design, nanochemistry, and applications in theranostics. Chem Rev 114:5161–5214

Chiu D, Zhou W, Kitayaporn S, Schwartz DT, Murali-Krishna K, Kavanagh TJ, Baneyx F (2012) Biomineralization and size control of stable calcium phosphate core–protein shell nanoparticles: potential for vaccine applications. Bioconjug Chem 23:610–617

Cho N-H, Cheong T-C, Min JH, Wu JH, Lee SJ, Kim D, Yang J-S, Kim S, Kim YK, Seong S-Y (2011) A multifunctional core-shell nanoparticle for dendritic cell-based cancer immunotherapy. Nat Nano 6:675–682

Chung W-J, Kwon K-Y, Song J, Lee S-W (2011) Evolutionary screening of collagen-like peptides that nucleate hydroxyapatite crystals. Langmuir 27:7620–7628

Corni S, Hnilova M, Tamerler C, Sarikaya M (2013) Conformational behavior of genetically-engineered dodecapeptides as a determinant of binding affinity for gold. J Phys Chem C Nanomater Interfaces 117:16990–17003

Cui Y, Kim SN, Jones SE, Wissler LL, Naik RR, McAlpine MC (2010) Chemical functionalization of graphene enabled by phage displayed peptides. Nano Lett 10:4559–4565

Dang X, Yi H, Ham M-H, Qi J, Yun DS, Ladewski R, Strano MS, Hammond PT, Belcher AM (2011) Virus-templated self-assembled single-walled carbon nanotubes for highly efficient electron collection in photovoltaic devices. Nat Nano 6:377–384

de Juan-Franco E, Caruz A, Pedrajas J, Lechuga LM (2013) Site-directed antibody immobilization using a protein A–gold binding domain fusion protein for enhanced SPR immunosensing. Analyst 138:2023–2031

de Oliveira PT, Zalzal SF, Beloti MM, Rosa AL, Nanci A (2007) Enhancement of in vitro osteogenesis on titanium by chemically produced nanotopography. J Biomed Mater Res A 80:554–564

Eteshola E, Brillson LJ, Lee SC (2005) Selection and characteristics of peptides that bind thermally grown silicon dioxide films. Biomol Eng 22:201–204

Flynn CE, Mao C, Hayhurst A, Williams JL, Georgiou G, Iverson B, Belcher AM (2003) Synthesis and organization of nanoscale II-VI semiconductor materials using evolved peptide specificity and viral capsid assembly. J Mater Chem 13:2414–2421

Ghosh D, Bagley AF, Na YJ, Birrer MJ, Bhatia SN, Belcher AM (2014) Deep, noninvasive imaging and surgical guidance of submillimeter tumors using targeted M13-stabilized single-walled carbon nanotubes. Proc Natl Acad Sci U S A 111:13948–13953

Giessen TW, Silver PA (2016) Converting a natural protein compartment into a nanofactory for the size-constrained synthesis of antimicrobial silver nanoparticles. ACS Synth Biol 5:1497–1504

Goede K, Busch P, Grundmann M (2004) Binding specificity of a peptide on semiconductor surfaces. Nano Lett 4:2115–2120

Gogoladze G, Grigolava M, Vishnepolsky B, Chubinidze M, Duroux P, Lefranc M-P, Pirtskhalava M (2014)

DBAASP: database of antimicrobial activity and structure of peptides. FEMS Microbiol Lett 357:63–68
Gungormus M, Fong H, Kim IW, Evans JS, Tamerler C, Sarikaya M (2008) Regulation of in vitro calcium phosphate mineralization by combinatorially selected hydroxyapatite-binding peptides. Biomacromolecules 9:966–973
Ha N-Y, Shin HM, Sharma P, Cho HA, Min C-K, H-i K, Yen NTH, Kang J-S, Kim I-S, Choi M-S, Kim YK, Cho N-H (2016) Generation of protective immunity against *Orientia tsutsugamushi* infection by immunization with a zinc oxide nanoparticle combined with ScaA antigen. J Nanobiotechnol 14:76
Hnilova M, Oren EE, Seker UOS, Wilson BR, Collino S, Evans JS, Tamerler C, Sarikaya M (2008) Effect of molecular conformations on the adsorption behavior of gold-binding peptides. Langmuir 24:12440–12445
Hochuli E, Bannwarth W, Döbeli H, Gentz R, Stüber D (1988) Genetic approach to facilitate purification of recombinant proteins with a novel metal chelate adsorbent. Nat Biotechnol 6:1321–1325
Huang Y, Chiang C-Y, Lee SK, Gao Y, Hu EL, Yoreo JD, Belcher AM (2005) Programmable assembly of nanoarchitectures using genetically engineered viruses. Nano Lett 5:1429–1434
Huang J, Ru B, Zhu P, Nie F, Yang J, Wang X, Dai P, Lin H, Guo F-B, Rao N (2012) MimoDB 2.0: a mimotope database and beyond. Nucleic Acids Res 40:D271–D277
Huang Z-B, Shi X, Mao J, S-q G (2016) Design of a hydroxyapatite-binding antimicrobial peptide with improved retention and antibacterial efficacy for oral pathogen control. Sci Rep 6:38410
Kase D, Kulp JL, Yudasaka M, Evans JS, Iijima S, Shiba K (2004) Affinity selection of peptide phage libraries against single-wall carbon nanohorns identifies a peptide aptamer with conformational variability. Langmuir 20:8939–8941
Kelly M, Williams R, Aojula A, O'Neill J, Trzińscka Z, Grover L, Scott RA, Peacock AF, Logan A, Stamboulis A (2015) Peptide aptamers: novel coatings for orthopaedic implants. Mater Sci Eng C Mater Biol Appl 54:84–93
Khatayevich D, Gungormus M, Yazici H, So C, Cetinel S, Ma H, Jen A, Tamerler C, Sarikaya M (2010) Biofunctionalization of materials for implants using engineered peptides. Acta Biomater 6:4634–4641
Kim J, Rheem Y, Yoo B, Chong Y, Bozhilov KN, Kim D, Sadowsky MJ, Hur H-G, Myung NV (2010) Peptide-mediated shape- and size-tunable synthesis of gold nanostructures. Acta Biomater 6:2681–2689
Ko S, Park TJ, Kim H-S, Kim J-H, Cho Y-J (2009) Directed self-assembly of gold binding polypeptide-protein A fusion proteins for development of gold nanoparticle-based SPR immunosensors. Biosens Bioelectron 24:2592–2597
Kröger N, Deutzmann R, Sumper M (1999) Polycationic peptides from diatom biosilica that direct silica nanosphere formation. Science 286:1129–1132
Li Y, Huang Y (2010) Morphology-controlled synthesis of platinum nanocrystals with specific peptides. Adv Mater Res 22:1921–1925
Liang L, Care A, Zhang R, Lu Y, Packer NH, Sunna A, Qian Y, Zvyagin AV (2016) Facile assembly of functional upconversion nanoparticles for targeted cancer imaging and photodynamic therapy. ACS Appl Mater Interfaces 8:11945–11953
Liang L, Lu Y, Zhang R, Care A, Ortega TA, Deyev SM, Qian Y, Zvyagin AV (2017) Deep-penetrating photodynamic therapy with KillerRed mediated by upconversion nanoparticles. Acta Biomater 51:461–470
Lu Y, Zhao J, Zhang R, Liu Y, Liu D, Goldys EM, Yang X, Xi P, Sunna A, Lu J, Shi Y, Leif RC, Huo Y, Shen J, Piper JA, Robinson JP, Jin D (2014) Tunable lifetime multiplexing using luminescent nanocrystals. Nat Photon 8:32–36
Lucky SS, Soo KC, Zhang Y (2015) Nanoparticles in photodynamic therapy. Chem Rev 115:1990–2042
Mann JA, Dichtel WR (2013) Noncovalent functionalization of graphene by molecular and polymeric adsorbates. J Phys Chem Lett 4:2649–2657
Mannoor MS, Tao H, Clayton JD, Sengupta A, Kaplan DL, Naik RR, Verma N, Omenetto FG, McAlpine MC (2012) Graphene-based wireless bacteria detection on tooth enamel. Nat Commun 3:763
Mao C, Flynn CE, Hayhurst A, Sweeney R, Qi J, Georgiou G, Iverson B, Belcher AM (2003) Viral assembly of oriented quantum dot nanowires. Proc Natl Acad Sci U S A 100:6946–6951
Naik RR, Brott LL, Clarson SJ, Stone MO (2002a) Silica-precipitating peptides isolated from a combinatorial phage display peptide library. J Nanosci Nanotechnol 2:95–100
Naik RR, Stringer SJ, Agarwal G, Jones SE, Stone MO (2002b) Biomimetic synthesis and patterning of silver nanoparticles. Nat Mater 1:169–172
Nam KT, Kim D-W, Yoo PJ, Chiang C-Y, Meethong N, Hammond PT, Chiang Y-M, Belcher AM (2006) Virus-enabled synthesis and assembly of nanowires for lithium ion battery electrodes. Science 312:885–888
Nel AE, Madler L, Velegol D, Xia T, Hoek EMV, Somasundaran P, Klaessig F, Castranova V, Thompson M (2009) Understanding biophysicochemical interactions at the nano-bio interface. Nat Mater 8:543–557
Nochomovitz R, Amit M, Matmor M, Ashkenasy N (2010) Bioassisted multi-nanoparticle patterning using single-layer peptide templates. Nanotechnology 21:145305
Nygaard S, Wendelbo R, Brown S (2002) Surface-specific zeolite-binding proteins. Adv Mater Res 14:1853–1856
Oren EE, Tamerler C, Sahin D, Hnilova M, Seker UOS, Sarikaya M, Samudrala R (2007) A novel knowledge-based approach to design inorganic-binding peptides. Bioinformatics 23:2816–2822
Pacardo DB, Sethi M, Jones SE, Naik RR, Knecht MR (2009) Biomimetic synthesis of Pd nanocatalysts for the Stille coupling reaction. ACS Nano 3:1288–1296

Park TJ, Zheng S, Kang YJ, Lee SY (2009) Development of a whole-cell biosensor by cell surface display of a gold-binding polypeptide on the gold surface. FEMS Microbiol Lett 293:141–147

Pender MJ, Sowards LA, Hartgerink JD, Stone MO, Naik RR (2005) Peptide-mediated formation of single-wall carbon nanotube composites. Nano Lett 6:40–44

Puddu V, Perry CC (2012) Peptide adsorption on silica nanoparticles: evidence of hydrophobic interactions. ACS Nano 6:6356–6363

Ramaraju H, Miller SJ, Kohn DH (2014) Dual-functioning phage-derived peptides encourage human bone marrow cell-specific attachment to mineralized biomaterials. Connect Tissue Res 55:160–163

Sano K, Shiba K (2003) A hexapeptide motif that electrostatically binds to the surface of titanium. J Am Chem Soc 125:14234–14235

Sarikaya M, Tamerler C, Jen AKY, Schulten K, Baneyx F (2003) Molecular biomimetics: nanotechnology through biology. Nat Mater 2:577–585

Sayyadi N, Care A, Connally RE, Try AC, Bergquist PL, Sunna A (2016) A novel universal detection agent for time-gated luminescence bioimaging. Sci Rep 6:27564

Seker UOS, Wilson B, Sahin D, Tamerler C, Sarikaya M (2008) Quantitative affinity of genetically engineered repeating polypeptides to inorganic surfaces. Biomacromolecules 10:250–257

Singh R, Lillard JW (2009) Nanoparticle-based targeted drug delivery. Exp Mol Pathol 86:215–223

Sunna A, Chi F, Bergquist PL (2013a) Efficient capture of pathogens with a zeolite matrix. Parasitol Res 112:2441–2452

Sunna A, Chi F, Bergquist PL (2013b) A linker peptide with high affinity towards silica-containing materials. New Biotechnol 30:485–492

Tang Z, Palafox-Hernandez JP, Law W-C, Hughes ZE, Swihart MT, Prasad PN, Knecht MR, Walsh TR (2013) Biomolecular recognition principles for bionanocombinatorics: an integrated approach to elucidate enthalpic and entropic factors. ACS Nano 7:9632–9646

Thai CK, Dai H, Sastry MSR, Sarikaya M, Schwartz DT, Baneyx F (2004) Identification and characterization of Cu2O- and ZnO-binding polypeptides by *Escherichia coli* cell surface display: toward an understanding of metal oxide binding. Biotechnol Bioeng 87:129–137

Tyagi A, Tuknait A, Anand P, Gupta S, Sharma M, Mathur D, Joshi A, Singh S, Gautam A, Raghava GPS (2015) CancerPPD: a database of anticancer peptides and proteins. Nucleic Acids Res 43:D837–D843

Umetsu M, Mizuta M, Tsumoto K, Ohara S, Takami S, Watanabe H, Kumagai I, Adschiri T (2005) Bioassisted room-temperature immobilization and mineralization of zinc oxide—the structural ordering of ZnO nanoparticles into a flower-type morphology. Adv Mater Res 17:2571–2575

Umlauf BJ, McGuire MJ, Brown KC (2014) Introduction of plasmid encoding for rare tRNAs reduces amplification bias in phage display biopanning. BioTechniques 58:81

Whaley SR, English DS, Hu EL, Barbara PF, Belcher AM (2000) Selection of peptides with semiconductor binding specificity for directed nanocrystal assembly. Nature 405:665–668

Yazici H, Fong H, Wilson B, Oren EE, Amos FA, Zhang H, Evans JS, Snead ML, Sarikaya M, Tamerler C (2013) Biological response on a titanium implant-grade surface functionalized with modular peptides. Acta Biomater 9:5341–5352

Yazici H, O'Neill MB, Kacar T, Wilson BR, Oren EE, Sarikaya M, Tamerler C (2016) Engineered chimeric peptides as antimicrobial surface coating agents toward infection-free implants. ACS Appl Mater Interfaces 8:5070–5081

Yi X, Wang F, Qin W, Yang X, Yuan J (2014) Near-infrared fluorescent probes in cancer imaging and therapy: an emerging field. Int J Nanomedicine 9:1347

Yoshinari M, Kato T, Matsuzaka K, Hayakawa T, Shiba K (2010) Prevention of biofilm formation on titanium surfaces modified with conjugated molecules comprised of antimicrobial and titanium-binding peptides. Biofouling 26:103–110

Yoshinari M, Matsuzaka K, Inoue T (2011) Surface modification by cold-plasma technique for dental implants—bio-functionalization with binding pharmaceuticals. Jpn Dent Sci Rev 47:89–101

Yuasa K, Kokubu E, Kokubun K, Matsuzaka K, Shiba K, Kashiwagi K, Inoue T (2014) An artificial fusion protein between bone morphogenetic protein 2 and titanium-binding peptide is functional in vivo. J Biomed Mater Res A 102:1180–1186

Yuca E, Karatas AY, Seker UOS, Gungormus M, Dinler-Doganay G, Sarikaya M, Tamerler C (2011) In vitro labeling of hydroxyapatite minerals by an engineered protein. Biotechnol Bioeng 108:1021–1030

Zhang Y, Zheng F, Yang T, Zhou W, Liu Y, Man N, Zhang L, Jin N, Dou Q, Zhang Y, Li Z, Wen L-P (2012) Tuning the autophagy-inducing activity of lanthanide-based nanocrystals through specific surface-coating peptides. Nat Mater 11:817–826

Zhou W, Schwartz DT, Baneyx F (2010) Single-pot biofabrication of zinc sulfide immuno-quantum dots. J Am Chem Soc 132:4731–4738

Zhou W, Moguche AO, Chiu D, Murali-Krishna K, Baneyx F (2014) Just-in-time vaccines: biomineralized calcium phosphate core-immunogen shell nanoparticles induce long-lasting $CD8^+$ T cell responses in mice. Nanomedicine 10:571–578

Molecular Modelling of Peptide-Based Materials for Biomedical Applications

3

Tiffany R. Walsh

Abstract

The molecular-level interactions between peptides and medically-relevant biomaterials, including nanoparticles, have the potential to advance technologies aimed at improving performance for medical applications including tissue implants and regenerative medicine. Peptides can possess materials-selective non-covalent adsorption properties, which in this instance can be exploited to enhance the biocompatibility and possible multi-functionality of medical implant materials. However, at present, their successful implementation in medical applications is largely on a trial-and-error basis, in part because a deep comprehension of general structure/function relationships at these interfaces is currently lacking. Molecular simulation approaches can complement experimental characterisation techniques and provide a wealth of relevant details at the atomic scale. In this Chapter, progress and prospects for advancing peptide-mediated medical implant surface treatments via molecular simulation is summarised for two of the most widely-found medical implant interfaces, titania and hydroxyapatite.

Keywords

Peptides • Implant materials • Titania • Hydroxyapatite • Molecular simulation

3.1 Introduction

Peptide-based coatings are finding wider use in medical implant materials, based on exploitation of the phenomenon of non-covalent peptide-materials recognition (Care et al. 2015). Building on this approach, a more generalized peptide-based strategy for tuning the medical bio-interface

T.R. Walsh (✉)
Institute for Frontier Materials, Deakin University, Geelong, VIC 3216, Australia
e-mail: tiffany.walsh@deakin.edu.au

© Springer International Publishing AG 2017
A. Sunna et al. (eds.), *Peptides and Peptide-based Biomaterials and their Biomedical Applications*, Advances in Experimental Medicine and Biology 1030, https://doi.org/10.1007/978-3-319-66095-0_3

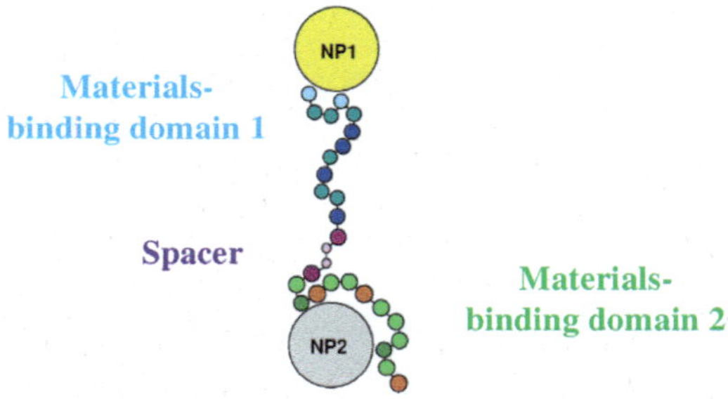

Fig. 3.1 Schematic of a "chimeric" peptide, which in this case mediates binding between two nanoparticles (NP1 and NP2) of different chemical compositions. *Small coloured spheres* represent peptide residues, with *blue spheres* indicating the NP1-binding domain, and *green spheres* indicating the NP2-binding domain. The *small purple spheres* indicate non-binding spacer residues

can be accomplished by constructing more complex multi-domain peptides – *i.e.* a sequence comprising a materials-binding domain, conjugated to an additional functional domain (which itself may or may not be a peptide), as illustrated in Fig. 3.1 In its simplest form, this "chimeric-peptide" approach was initially used as a passive coating strategy to prevent biofouling of implant surfaces, for example by conjugating the materials-binding peptide to an oligomeric polyethylene glycol (PEG) chain (Khatayevich et al. 2010; Khoo et al. 2009). More recently, advances in this approach have evolved to a more pro-active coating strategy, where the functional domain can carry out an active biological function, rather than merely passively resist biomolecule adsorption in general. Examples include the use of integrin sequences to elicit the biomolecular cues that underpin cell attachment and proliferation for use in bio-compatible implant materials (*i.e.* to promote osseointegration), (Khatayevich et al. 2010; Meyers et al. 2007) along with the use of antifungal and/or antibacterial sequences to reduce bacterial adhesion in orthopaedic implants (Yucesoy et al. 2015).

However, the central underlying assumptions associated with the "chimeric peptide" strategy are that: (1) the two domains in the chimeric molecule do not interact with each other, and (2) that the functional domain also does not preferentially bind to the target substrate – *i.e.* that the functional domain will be displayed to the external host medium, located away from the implant surface, in order to carry out its key function. One strategy to ensure this outcome is to use a central "spacer" motif (Fig. 3.1). While the chimeric approach is very promising, in this field it is also widely known and appreciated that structural determination of these molecules in the adsorbed state under aqueous conditions is an extremely challenging task. Very little is known regarding the structure of these chimeric peptides at the bio-interface; at present, even details of the molecular-level structure of these molecules in solution (in the absence of the surface) are sparse (Tamerler and co-workers (Yucesoy et al. 2015) provide a recent example). This lack of fundamental structural knowledge prevents the systematic improvement and refinement of the "chimeric peptide" strategy, to ensure that the two key assumptions listed above are actually satisfied. The use of molecular simulations has helped to bridge this gap in our understanding, and provides a platform for making systematic adaptations to these peptide sequences.

In this Chapter, we focus on how molecular simulations have contributed to our current understanding of how peptide-surface adsorption can be manipulated, for two key materials that are widely used in medical implants, namely hydroxyapatite and titania. While molecular simulations of "chimeric" peptide systems adsorbed at such solid surfaces are currently a rarity, the progress made in computational modelling of the abiotic/biotic interface, as described herein, paves the way for meeting these needs in the future.

3.2 Hydroxyapatite-Based Materials

Hydroxyapatite (HAp) is a crystalline mineral with the chemical formula $Ca_{10}(PO_4)_6(OH)_2$, as illustrated in Fig. 3.2. Biogenic HAp, or variations thereof, can be found throughout Nature, *e.g.* in the bones and teeth of vertebrates, and in eggshells. The mechanical properties, biocompatibility and accommodating nature of HAp (with respect to dopants) make this material a

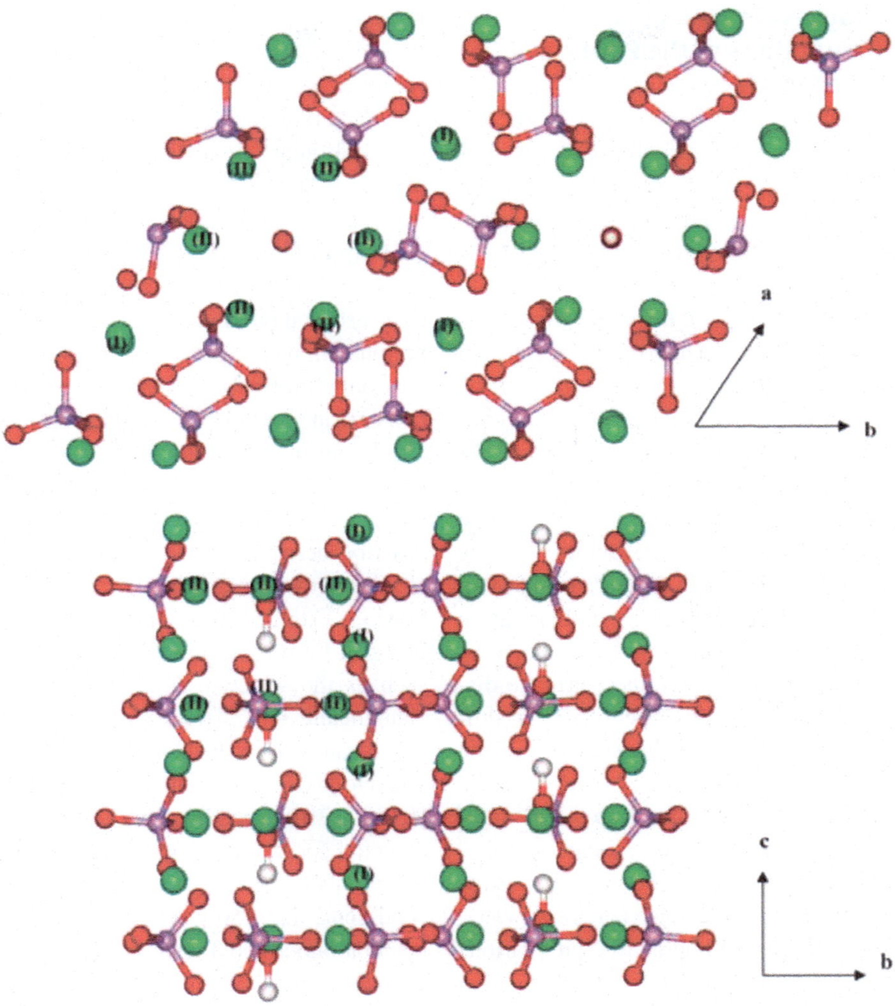

Fig. 3.2 Plan and side views of the hydroxyapatite structure, showing the hydroxy groups in hexagonal channels surrounded by Ca ions in the Ca(II) positions, where the Ca(I) positions are within the rest of the lattice (O ¼ *red*, Ca ¼ *green*, P ¼ *pink*, H ¼ *white*, Ca(I) and Ca(II) indicated on some positions) (Reproduced from de Leeuw (2010). Reproduced with permission from the copyright owner)

versatile and desirable candidate for myriad medical uses. Examples of the use of HAp in biomedical applications are numerous and range from the obvious, such as in coatings for joint implants (Suchanek and Yoshimura 1998) or for tissue regeneration (Zhou and Lee 2011), through to lesser-known examples in drug delivery (Syamchand and Sony 2015), *in vivo* imaging (Syamchand and Sony 2015), and ophthalmic surgery (Baino and Vitale-Brovarone 2014). Biofunctionalisation of HAp surfaces is a key element of many of these applications, and therefore, a critical and in-depth understanding of the structures and properties of biomolecules such as peptides, when they are adsorbed at the substrate interface, is needed. This need is particularly acute for tissue engineering, where it is essential that the implanted scaffold can approximately reproduce the environment of the extracellular matrix (ECM).

3.2.1 Progress in Molecular Simulations of the Peptide-HAp Interface

Compared with other medically-relevant inorganic substrates such as Au or titania, reports of the use of molecular simulations as applied to peptide adsorption are relatively scarce for HAp surfaces (herein we limit discussions to non-collagenous peptides only). For molecular simulations of the peptide-HAp interface to be relevant to physiological conditions, consideration of the presence of liquid water is an essential requirement. Many, but not all, of the simulation studies described herein have modelled the interface using a molecular-level description of liquid water. Another aspect for consideration is the choice of crystallographic orientation of the HAp surface, and also the idealised crystalline nature of these HAp surface models, particularly at physiologically relevant pH values. Prior to discussing the current progress of peptide-HAp simulation efforts, these considerations will be described in detail.

A number of electronic structure theory studies, principally based on density functional theory (DFT) have explored the structures and energetics of small molecules (such as water, citrate and amino acids) interacting with various terminations of the HAp surface (Astala and Stottt 2008; Corno et al. 2009; Filgueiras et al. 2006; Lou et al. 2012; Rimola et al. 2008), including the (001), $(1\bar{1}0)$, and (100) orientations. While valuable, these DFT studies were done under anhydrous conditions, which limits the applicability of these finds to biologically-relevant (*i.e.* aqueous) conditions, given that it is currently thought that the structuring of the interfacial solvent can strongly influence the adsorption preferences and structures of molecules adsorbed at the aqueous interface (Schneider and Colombi Ciacchi 2012; Skelton et al. 2009). Due to practical limitations, these DFT studies were also limited to consideration of idealised crystalline HAp surfaces, where the surface was constructed as a perfect truncate of the bulk crystal. Despite its high relevance, consideration of the presence of disorder in the surface structure of HAp has been the subject of relatively few studies (de Leeuw 2010; Lee et al. 2000; Slepko and Demkov 2013; Wu et al. 2016).

In addition, a crucial omission in almost all of the molecular simulation studies of *aqueous* peptide-HAp interfaces to date is the failure to capture the protonation state of phosphate groups on the HAp surface. Along with the lack of disorder or surface reconstruction in these idealised bulk-truncate surface models, the lack of surface-phosphate protonation means that these simulations are relevant only to extremely basic (*i.e.* pH $\geq$14) conditions (Lin and Heinz 2016), which is not physiologically relevant. The reason for this omission is presumably due to the lack of interatomic potentials (herein referred to as force-fields) to describe these protonation states and how they may change with solution pH. Instead, most of these previous studies used force-fields that were intended for use on bulk HAp crystalline systems, such as the widely-used force-field reported by Hauptmann *et al.* (Hauptmann et al. 2003). Significantly, to address some of these limitations inherent to currently-available force-fields, Lin and Heinz (2016) recently published an inter-atomic potential for the aqueous HAp interface that can describe both acidic (pH $\approx$ 5) and basic (pH $\approx$ 10) conditions, that also takes surface phosphate protonation into account under these conditions (with surface coverage of hydrogen phosphate and dihydrogen phosphate, respectively).

The use of molecular simulations to predict the structures of peptides adsorbed at the aqueous HAp interface is a relatively under-developed area. On a related note, the study of the interface between HAp surfaces and liquid water has also been somewhat under-studied (Chen et al. 2007; de Leeuw 2010; Lin and Heinz 2016; Zahn and Hochrein 2003; Zhao et al. 2014). Not unexpectedly, all of these studies reported the presence of several strongly-structured solvent layers closest to the HAp surface (Fig. 3.3), for some orientations of the HAp surface. However, only the most recent of these studies, reported by Lin and Heinz (2016), used a structural model that did not correspond with extremely basic conditions. These authors considered the (100), (010), (020) and (101) interfaces at solvent pH values of 14, 10

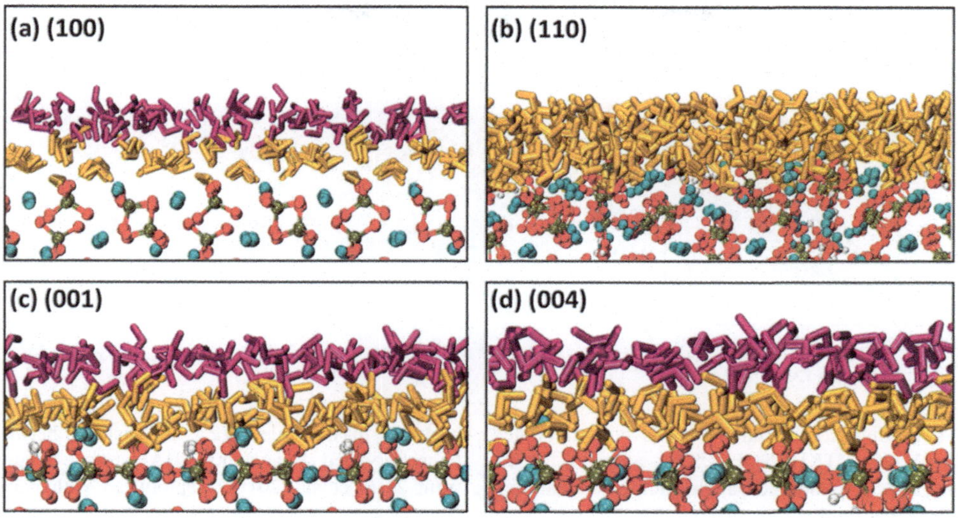

Fig. 3.3 Water adsorption structures for the first two hydration layers on four HAp surfaces: (**a**) (100), (**b**) (110), (**c**) (001), and (**d**) (004). Water molecules in the first and second layers are shown by *yellow* and *purple* licorice representation, respectively. Only the first hydration layer is displayed on the (110) face. Water molecules in the bulk aqueous solution are not shown for clarity. Figure 3.1. HAp colour designation: *cyan*, calcium; *brown*, phosphorus; *red*, oxygen; *white*, hydrogen (Reproduced from Zhao et al. (2014). Reproduced with permission from the copyright owner)

and 5, and found that not only was the solvent structuring different for these four different crystallographic orientations, but also that this structuring varied with pH for each of these surfaces.

As an intermediate step on the path to investigating aqueous peptide-HAp interfaces, a limited number of studies have reported the adsorption of small molecules such as citrate and amino acids (including phosphorylated amino acids) at the aqueous HAp interface (Filgueiras et al. 2006; Pan et al. 2007; Xu et al. 2014), with a strong emphasis on negatively-charged adsorbates. Both Pan et al. (2007) and Xu et al. (2014) attempted to estimate the adsorption free energy of such molecules at the HAp interface, using non-equilibrium molecular dynamics simulations and umbrella sampling, respectively. These studies indicated that solvent-mediated adsorption states could be supported, akin to those reported for the aqueous titania interface (Schneider and Ciacchi 2012; Skelton et al. 2009). Molecules possessing negatively-charged species, such as PO_4^{3-} and COO^- were found to bind strongly, and also exhibit facet-specific differences in binding strength (Pan et al. 2007; Xu et al. 2014), while uncharged adsorbates such as glycine were found to support weaker interfacial binding.

Despite the fact that several experimental studies (using selection techniques such as phage display) have reported HAp-binding sequences (Gungormus et al. 2008; Roy et al. 2008; Segvich et al. 2009), very few interfacial simulation studies have focused on these HAp-binding peptides. The exceptions to this are the studies of Zhao et al. (2016), who investigated phage-display peptides (see Fig. 3.4) reported by Segvich et al. (2009), and Lin and Heinz (2016) who investigated a phage-display sequence reported by Roy et al. (2008). The study reported by Lin and Heinz impressively explored pH effects of peptide adsorption at the aqueous HAp (001) interface (considered at pH values of 5 and 10). Their findings suggested that charged species bound strongly to the (001) interface at pH = 10 (under these conditions the surface phosphates were singly-protonated, as HPO_4^{2-}). In contrast, these authors found that at pH = 5 (where the surface phosphates were doubly-protonated, as $H_2PO_4^-$), polar and hydrophobic species made greater contact at the aqueous interface. However, these

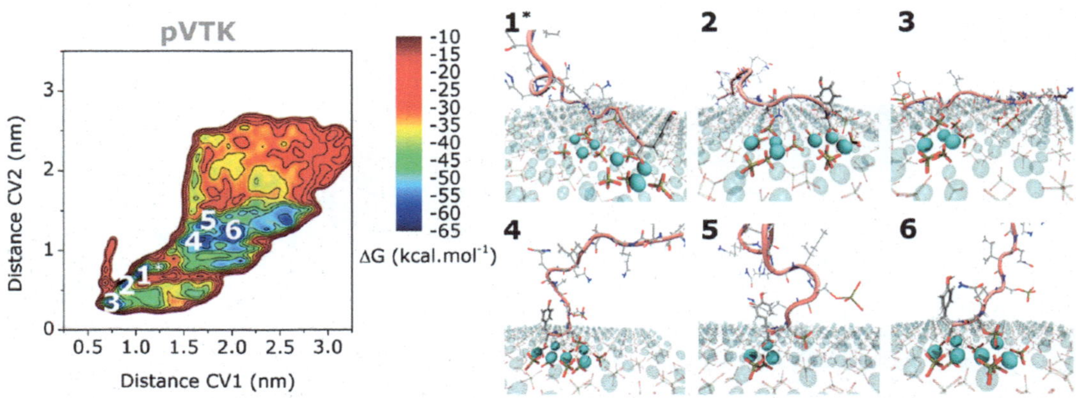

Fig. 3.4 Binding configurations of the phosphorylated peptide pVTK (VTKHLNQIS$_p$QS$_p$Y) adsorbed at the aqueous (100) HAp interface, at the additional local minima. The global minimum is marked here by an asterisk. CV1 is the distance between the surface and the peptide centre of mass, while CV2 is the distance between one phosphorylated serine and the HAP surface (Reproduced from Zhao et al. (2016). Reproduced with permission from the copyright owner)

authors did not use advanced conformational sampling approaches in their simulations, which is a known limitation in the field of peptide-interface simulation in general. In contrast, the simulation study reported by Zhao et al. (2016) focused on the aqueous HAp (100) interface, and used advanced conformational sampling, including both steered molecular dynamics simulations and parallel-tempering metadynamics simulations (Bussi et al. 2006), and based their investigation on a phage-display-derived HAp-binding sequence (Segvich et al. 2009). However, these authors made use of a surface model generated via the idealised truncate of bulk HAp, and did not consider protonation of the surface phosphates. Therefore, these simulations correspond only with extremely basic solution conditions and are not relevant to the physiological conditions of the experiments. Consistent with the findings of Lin and Heinz (2016) at pH = 10, the negatively- and positively-charged residues (such as Lys an Glu) were found to support the highest degree of contact with the HAp surface.

In terms of molecular simulations of the aqueous peptide-HAp interface, many of the previous studies have focused on the interfacial structures and properties of peptides derived from osteopontin (Lai et al. 2014, 2017) adsorbed at the interface. These peptides were first reported by Addison et al. (2010), and comprise two 18-residue sequences, taken from residues 198–295 and 115–132 of the osteopontin protein (also referred to as bone sialoprotein 1, or BSP-1), and two phosphorylated variants of the latter sequence (which carried a substantial overall negative charge). Osteopontin is found in the extracellular matrix in bone, and is thought to act as a natural adhesive in this complex high-performance composite material. These authors focused on using molecular dynamics simulations of the aqueous interface of HAp (100), and used non-equilibrium simulations to laterally pull the peptides across the HAp surface. In common with many other previously-published simulation studies of the aqueous HAp interface, these authors used the ideal bulk-truncate of HAp as their surface model, and did not capture phosphate protonation. This means the outcomes of these studies correspond with a HAp surface at a solution pH $\geq$ 14. Consistent with other studies conducted under basic conditions, these authors found that the negatively- and positively-charged residues in the osteopontin-based peptides dictated the interactions with the HAp surface. Unfortunately, as the work of Lin and Heinz (2016) suggests, these observations may not necessarily translate to milder pH conditions. Moreover, the large size of these peptides (18 residues) demands substantial efforts to convincingly sample the conformational ensemble. However, these authors did not combine their steered molecular dynamics simulations with advanced conformational sampling, the neglect of which may lead to biased results.

The final class of peptides that have been the subject of interfacial HAp molecular simulations are phosphorylated short peptides (Huq et al. 2000; Villarreal-Ramirez et al. 2014) associated with body fluids such as saliva, *e.g.* dentin binding, or with non-phosphorylated collagen fragments (Almora-Barrios and de Leeuw 2010). Huq et al. (2000) pioneered peptide-HAp simulations, reporting the first all-atom molecular modelling for this class of interfaces. Based on the Hauptmann force-field, these authors considered the (100) and (010) interfaces, in contact with a phosphorylated penta-peptide SSSEE. The surface models used in this instance were the unprotonated ideal surfaces, and unsurprisingly, as with many studies that followed this, the highly charged sequence was found to adsorb strongly (in this case to both interfaces). It should be noted that, due to the vintage of this modelling study, these authors did not report results of molecular simulations, but instead performed docking calculations, which did not include consideration of the presence of liquid water. In a separate study, Almora-Barrios and de Leeuw (2010) used standard molecular dynamics simulations and found that an uncharged collagen-relevant tri-peptide Hyp-Pro-Gly appeared to interact more strongly with the ($1\bar{1}0$) aqueous interface compared with the (001) interface.

More recently, Villarreal-Ramirez et al. (2014) reported molecular simulations of ten different peptides derived from human dentin phosphoprotein (DPP), in both regular and phosphorylated forms, with an overall charge ranging from -2*e* to -22*e*, in contact with the aqueous HAp (100) interface. These peptides were around 15 residues or more in length; however, these peptides were modelled using only standard molecular dynamics simulations of relatively short duration (20 ns). It is currently accepted that peptides of this size cannot be reliably modelled without recourse to advanced conformational sampling strategies. Moreover, these authors used the common structural model of HAp corresponding with a pH $\geq$14, and therefore supported a conclusion similar to many previous studies, identifying the highly-charged phosphorylated sequences as binding the strongest. However, as remarked earlier, these findings have not been obtained at physiologically-relevant conditions, due to the limitations of the surface model.

3.2.2 Outlook for Molecular Simulations of the Peptide-HAp Interface

The chemical composition and atomic-scale surface structure of HAp are key aspects that need further improvement and refinement in future peptide-HAp simulation studies. Relevant traits regarding the state of the HAp surface include: the degree of crystallinity of the surface, the charge density of the surface under aqueous conditions and over a range of pH values, not to mention the protonation state of the surface phosphate units. To date, no published molecular dynamics simulation study has presented a surface structural model of HAp under physiological (pH $\approx$ 7) conditions. The closest that previous studies have come to meeting this requirement can be found in the study reported by Lin and Heinz. While these authors were able to explore pH-dependent peptide binding, their model permits the study of acidic or basic conditions only. Nevertheless, this model is preferable to the widely-used ideal truncate of the bulk solid, which corresponds with extremely basic conditions. Construction of a force-field and corresponding structural model that is relevant to a pH $\approx$ 7 should be an urgent priority in this area.

The use of advanced conformational sampling is also an essential requirement for future molecular simulations of peptide-HAp interfaces. This is especially relevant to chimeric peptides, where one domain comprises the HAp-binding domain, while the second domain may be used for a different function (such as an integrin, or an antifungal peptide sequence), and therefore may comprise relatively long (and complex) peptides. The molecular simulation of such chimeric peptides in contact with HAp have not yet been reported, despite their recent applications in experimental studies, and the growing need to identify the structural modes of adsorption of these sequences to ensure both domains in the chimeric peptide can optimally carry out their intended function.

3.3 Titania-Based Materials

The titanium surface is another key substrate relevant to the bio-engineering of tissue implant surfaces. Under physiological conditions, titanium naturally oxidizes to form a titania layer on the surface, and when in contact with solution at neutral pH, this surface features a negative surface charge density. Titania is used in myriad biomedical implant applications, due to its high strength relative to its light weight, its corrosion resistance, and its lack of bio-toxicity. The use of titania-binding peptides as a coating agent for the titania surface, without recourse to covalent attachment to the substrate, remains a prime strategy to tune and enhance the bio-compatibility of titania implant materials. Moreover, materials-binding peptides in and of themselves can confer benefits in addition to material-specific recognition; *e.g.* Shiba and co-workers found that their titania-binding minTi-1 hexapeptide (Sano and Shiba 2003) could prevent the formation of biofilms on titania substrates (Yoshinari et al. 2010).

3.3.1 Progress in Molecular Simulations of the Peptide-Titania Interface

The vast majority of molecular simulations of the aqueous peptide-titania interface have been reported for crystalline substrates, particularly the crystalline rutile TiO_2 (110) interface. An abundance (relative to many other liquid/solid interfaces) of atomic-scale details of the rutile TiO_2 (110) surface in contact with liquid water are available. Experimental data along these lines includes truncation X-ray standing wave and crystal truncation rod measurements on single-crystal substrates (Zhang et al. 2004), which has enabled the adaptation and refinement of interfacial force-fields to describe the aqueous rutile (110) interface (Predota et al. 2004). Moreover, the experimental results have provided valuable benchmarks for validating the data produced by molecular simulations using such force-fields (Predota et al. 2004; Skelton and Walsh 2007) for rutile (110). To date, credible structural models and force-fields for other crystalline orientations of the rutile surface, in addition to different polymorphs of titania (*e.g.* anatase) are much less well developed.

Use of the crystalline rutile TiO_2 (110) interface, however, presents a complication in terms of making close links with experimental findings, since most peptide-titania characterization experiments have been reported for the native, thermally-oxidized titanium substrate, which might be best approximated *via* a structural model of amorphous titania. The work of Schneider and Colombi Ciacchi (2011) is impressive for addressing this challenge; in this work, the authors constructed and used an amorphous titania substrate, in addition to a corresponding peptide-surface force-field, in their simulations. Notably, Roddick-Lanzilotta et al. (1998) used *in situ* spectroscopy of lysine adsorbed at the aqueous titania interface to compare adsorption for amorphous titania surfaces *versus* crystalline anatase surfaces. These authors found no discernible differences in adsorption in this instance. This finding indicates that crystalline surfaces might provide suitable structural surrogates for amorphous substrates, for the purposes of molecular simulation. Another aspect of the peptide-TiO_2 interfacial simulations that has only recently received attention is the use of the Reax-FF (van Duin et al. 2001) force-field to model deprotonation events at the aqueous titania interface in the context of peptide adsorption (Li et al. 2012; Monti et al. 2012).

The use of simulations to predict the adsorption of small biologically-relevant molecules, such as amino acids (or their analogues), can provide clear fundamental benchmarks for interpreting more complex data arising from measurements or simulations of peptide-surface adsorption. While there is a dearth of experimental data corresponding to such measurements, several molecular simulation studies have reported adsorption free energies at aqueous titania interfaces for amino acids (or related molecules) on the rutile (110) (Monti and Walsh 2010; Sultan et al. 2014) surface. These simulation data were reported for both charge-neutral and hydroxylated, negatively-charged surfaces, and under both of these conditions identified that the analogues of charged amino acids, in particular Arg, Glu and Asp, exhibited the strongest adsorption at the titania interface, while hydrophobic and aromatic residues (such as Leu and Trp) were repelled by the surface. These findings are consistent with experimental observations (Paszti and Guczi 2009; Roddick-Lanzilotta et al. 1998; Roddick-Lanzilotta and McQuillan 1999; (Paszti and Guczi 2009; Roddick-Lanzilotta et al. 1998; Roddick-Lanzilotta and McQuillan 2000), and have also been recently reported for the calculation of adsorption free energies of amino acids on titania nanoparticles in solution (Fig. 3.5), predicted *via* molecular simulation (Liu et al. 2016). However, in this instance, the nanoparticle surface was not hydroxylated and the nanoparticle itself did not carry an overall charge; both of these aspects should be considered in future. In contrast, umbrella sampling molecular simulations of amino acid analogues at the uncharged aqueous rutile (100) interface (Brandt and Lyubartsev 2015) indicated that polar and aromatic residues supported the strongest adsorption, which is not consistent with available experimental data.

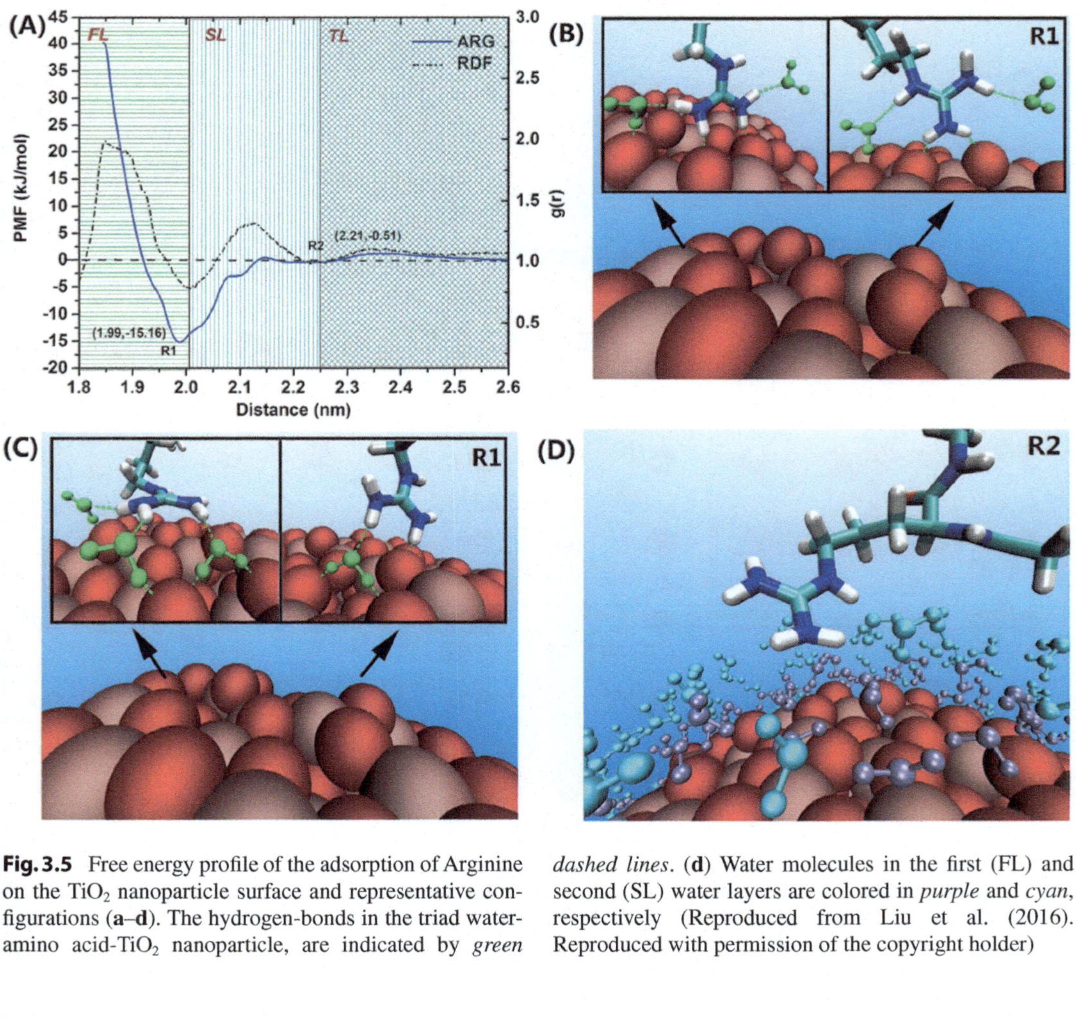

Fig. 3.5 Free energy profile of the adsorption of Arginine on the TiO_2 nanoparticle surface and representative configurations (**a–d**). The hydrogen-bonds in the triad water-amino acid-TiO_2 nanoparticle, are indicated by *green dashed lines*. (**d**) Water molecules in the first (FL) and second (SL) water layers are colored in *purple* and *cyan*, respectively (Reproduced from Liu et al. (2016). Reproduced with permission of the copyright holder)

Much of the previously-published molecular simulation work concerned with aqueous peptide-titania interfaces involves short peptides with a propensity to aggregate (or form filaments). Monti and co-workers reported some of the earliest simulations of short peptides (di- and tri-peptides) adsorbed at the aqueous crystalline rutile TiO_2 (110) interface, including peptide fragments derived from collagen (Fig. 3.6) (Carravetta and Monti 2006; Iucci et al. 2007; Monti et al. 2008; Monti et al. 2007; Zheng et al. 2016). Given the findings reported for amino acids (and their analogues) adsorbed at the aqueous titania interface, it is not surprising that charged residues were reported as being of the most strongly adsorbing. Of particular note is the abundance of simulation studies concerning the adsorption of the tri-peptide Arg-Gly-Asp (RGD) (Chen et al. 2010; Wu et al. 2011, 2012, 2013). RGD is a ubiquitous adhesive motif in proteins, and has been previously used in experiments based on "chimeric" peptide constructs for implant materials, and therefore, a deeper understanding of how the RGD motif interacts with these interfaces is a key requirement. These studies have indicated how the charged Arg and Asp residues are the chief mediators of the peptide-surface contact, although it appears that the presence of cations such as Na^+ and Ca^{2+} in solution at the interface can modulate this contact.

In terms of larger peptides, molecular simulation studies in this area have focused chiefly on peptide sequences identified from phage display experiments. The minTi-1 titania-binding peptide (RKLPDA), first experimentally isolated by Sano and Shiba (2003), has been the subject of molecular simulation studies (Schneider and Colombi Ciacchi 2012; Skelton et al. 2009),

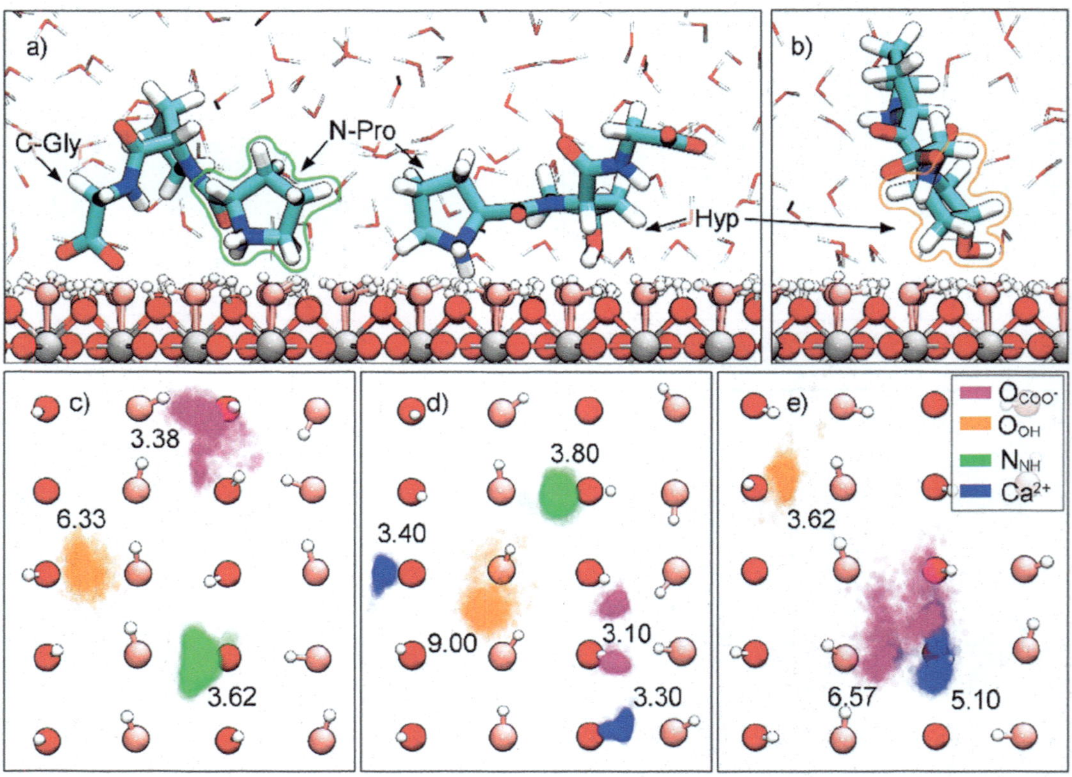

Fig. 3.6. The adsorption conformations of the Pro-Hyp-Gly tripeptide on the negatively charged surface (**a**) and (**b**), and the distribution map (**c**) and (**d**) of absorbed functional groups on the TiO_2 surface. The values in (**c**)–(**e**) represent the z-positions (peak value) of the functional groups or calcium ions with respect to the surface Ti atoms (Reproduced from Zheng et al. (2016). Reproduced with permission from the copyright owner)

which identified that the presence charged side-chains (in this case the Arg, Lys and Asp residues) ensured close interfacial contact. These studies also indicated that the presence of the strongly structured interfacial solvent layers nearest to the titania surface exerted a strong influence on peptide adsorption. In particular, these studies suggested that stable surface-adsorbed peptide configurations could be mediated *via* the first interfacial water layer, in addition to being mediated *via* direct surface contact. Moreover, a recent simulation study (Sultan et al. 2016) of two different titania-binding peptides identified from phage display experiments (Puddu et al. 2013) adsorbed at the aqueous rutile (110) interface similarly showed that both solvent-mediated and direct-contact modes of adsorption were possible (Fig. 3.7). Interfacial solvent structuring was also found to play an influential role in peptide adsorption, as suggested by the molecular simulations of Friedrichs et al. (2013), who considered the rutile (100) interface, and investigated two different peptides identified from a different set of phage-display selection experiments.

3.3.2 Outlook for Molecular Simulations of the Peptide-Titania Interface

In summary, the current body work in the area of peptide-titania simulations provides a strong foundation for enabling developments with the potential to advance effective surface-coating strategies for medical implant materials. Most of the existing peptide-titania simulation data has been obtained for short peptides, and moreover has been focused on exploring the characteristics of single-peptide adsorption. The prospects for extending this knowledge base into more complex scenarios is promising. The influence on peptide-surface binding due to the presence of more than one surface-adsorbed peptide (*i.e.* multi-peptide effects) at the aqueous rutile (110)

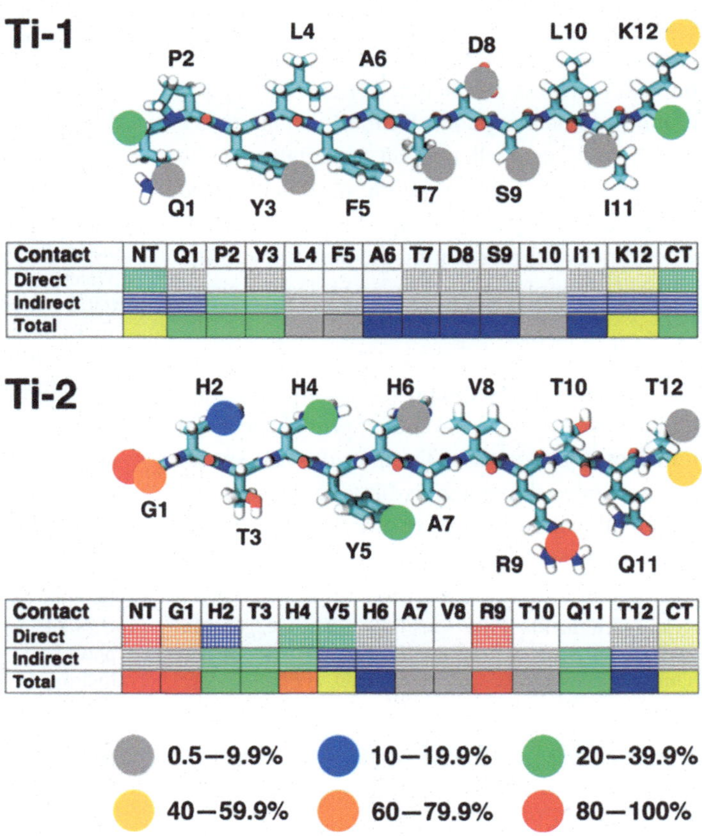

Fig. 3.7. Degree of residue – surface contact for peptides Ti-1 and Ti-2, generated from REST molecular dynamics simulation data. *Coloured dots* superimposed on the molecules indicate the degree of residue – surface direct contact. Tables use the same colour scheme to indicate the amounts of direct and indirect (solvent mediated) surface contact in addition to the total contact (direct + indirect) (Reproduced from Sultan et al. (2016). Reproduced with permission from the copyright owner)

interface is a clear direction that could be readily studied *via* molecular simulation. On a related note, the adsorption of peptides on the surface of titania nanoparticles in solution is a similarly under-explored area. However, a pre-requisite to realizing this goal is first to construct credible structural models (that incorporate physically-reasonable surface charge density and surface hydroxylation states) of such titania nanoparticles.

3.4 Outlook and Future Prospects

Molecular simulations of the aqueous peptide-solid interface can provide a valuable and fundamental structural basis which can directly inform and guide the rational design of more complex molecules such as the "chimeric", two-domain peptides. Such peptides show enormous promise in providing reliable, non-covalent surface-coating strategies for medical implants. Recent experimental evidence suggests that the effectiveness of hydroxyapatite as a coating for implant materials can be tuned *via* the adsorption of multi-domain peptides. Similarly, titanium is a widely-used medical implant material, and the non-covalent functionalisation of the oxidized substrate, namely titania, can also provide a reliable strategy for modulating the biocompatibility of titanium implants. To date, no molecular simulations have yet been reported regarding the adsorption of "chimeric" peptides at these aqueous interfaces of HAp and titania. In particular, molecular simulations could provide critical input into the design of the inter-domain spacer that separates the materials-binding domain and the functional domain, to ensure that these molecules can provide optimal performance *in vivo*.

Acknowledgments TRW gratefully acknowledges past funding and support from the EPSRC, AFOSR, Deakin University, **veski**, ARC and the AOARD.

References

Addison WN, Masica DL, Gray JJ, McKee MD (2010) Phosphorylation-dependent inhibition of mineralization by osteopontin ASARM peptides is regulated by PHEX cleavage. J Bone Min Res 25:695–705

Almora-Barrios N, de Leeuw NH (2010) Modelling the interaction of a Hyp-pro-Gly peptide with hydroxyapatite surfaces in aqueous environment. Cryst Eng Comm 12:960–967

Astala R, Stott MJ (2008) First-principles study of hydroxyapatite surfaces and water adsorption. Phys Rev B 78:075427

Baino F, Vitale-Brovarone C (2014) Bioceramics in ophthalmology. Acta Biomater 10:3372–3397

Brandt EG, Lyubartsev AP (2015) Molecular dynamics simulations of adsorption of amino acid side chain analogues and a titanium binding peptide on the TiO_2 (100) surface. J Phys Chem C 119:18126–18139

Bussi G, Gervasio FL, Laio A, Parrinello M (2006) Free-energy landscape for beta hairpin folding from combined parallel tempering and metadynamics. J Amer Chem Soc 128:13435–13441

Care A, Bergquist PL, Sunna A (2015) Solid-binding peptides: smart tools for nanobiotechnology. Trends Biotechnol 33:259–268

Carravetta V, Monti S (2006) Peptide-TiO_2 surface interaction in solution by ab initio and molecular dynamics simulations. J Phys Chem B 110:6160–6169

Chen X, Wang Q, Shen JW, Pan HH, Wu T (2007) Adsorption of leucine-rich amelogenin protein on hydroxyapatite (001) surface through -COO- claws. J Phys Chem C 111:1284–1290

Chen MJ, Wu CY, Song DP, Li K (2010) RGD tripeptide onto perfect and grooved rutile surfaces in aqueous solution: adsorption behaviors and dynamics. Phys Chem Chem Phys 12:406–415

Corno M, Busco C, Bolis V, Tosoni S, Ugliengo P (2009) Water adsorption on the stoichiometric (001) and (010) surfaces of hydroxyapatite: a periodic B3LYP study. Langmuir 25:2188–2198

de Leeuw NH (2010) Computer simulations of structures and properties of the biomaterial hydroxyapatite. J Mat Chem 20:5376–5389

Filgueiras MRT, Mkhonto D, de Leeuw NH (2006) Computer simulations of the adsorption of citric acid at hydroxyapatite surfaces. J Cryst Growth 294:60–68

Friedrichs W, Koppen S, Langel W (2013) Titanium binding dodecapeptides and the impact of water structure. Surf Sci 617:42–52

Gungormus M, Fong H, Kim IW, Evans JS, Tamerler C, Sarikaya M (2008) Regulation of in vitro calcium phosphate mineralization by combinatorially selected hydroxyapatite-binding peptides. Biomacromolecules 9:966–973

Hauptmann S, Dufner H, Brickmann J, Kast SM, Berry RS (2003) Potential energy function for apatites. Phys Chem Chem Phys 5:635–639

Huq NL, Cross KJ, Reynolds EC (2000) Molecular modelling of a multiphosphorylated sequence motif bound to hydroxyapatite surfaces. J Mol Model 6:35–47

Iucci G, Battocchio C, Dettin M, Gambaretto R, Di Bello C, Borgatti F, Carravetta V, Monti S, Polzonetti G (2007) Peptides adsorption on TiO_2 and au: molecular organization investigated by NEXAFS, XPS and IR. Surf Sci 601:3843–3849

Khatayevich D, Gungormus M, Yazici H, So C, Cetinel S, Ma H, Jen A, Tamerler C, Sarikaya M (2010) Biofunctionalization of materials for implants using engineered peptides. Acta Biomater 6:4634–4641

Khoo XJ, Hamilton P, O'Toole GA, Snyder BD, Kenan DJ, Grinstaff MW (2009) Directed assembly of PEGylated-peptide coatings for infection-resistant titanium metal. J Amer Chem Soc 131:10992–10997

Lai ZB, Wang MC, Yan C, Oloyede A (2014) Molecular dynamics simulation of mechanical behavior of osteopontin-hydroxyapatite interfaces. J Mech Behav Biomed Mater 36:12–20

Lai ZB, Bai RX, Yan C (2017) Effect of nano-scale constraint on the mechanical behaviour of osteopontin-hydroxyapatite interfaces. Comput Mater Sci 126:59–65

Lee WT, Dove MT, Salje EKH (2000) Surface relaxations in hydroxyapatite. J Phys Condens Matter 12:9829–9841

Li C, Monti S, Carravetta V (2012) Journey toward the surface: how glycine adsorbs on titania in water solution. J Phys Chem C 116:18318–18326

Lin TJ, Heinz H (2016) Accurate force field parameters and pH resolved surface models for hydroxyapatite to understand structure, mechanics, hydration, and biological interfaces. J Phys Chem C 120:4975–4992

Liu ST, Meng XY, Perez-Aguilar JM, Zhou RH (2016) An in silico study of TiO_2 nanoparticles interaction with twenty standard amino acids in aqueous solution. Sci Rep 6:37761

Lou ZY, Zeng Q, Chu X, Yang F, He DW, Yang ML, Xiang ML, Zhang XD, Fan HS (2012) First-principles study of the adsorption of lysine on hydroxyapatite (100) surface. Appl Surf Sci 258:4911–4916

Meyers SR, Hamilton PT, Walsh EB, Kenan DJ, Grinstaff MW (2007) Endothelialization of titanium surfaces. Adv Mater 19:2492–2498

Monti S, Walsh TR (2010) Free energy calculations of the adsorption of amino acid analogues at the aqueous titania interface. J Phys Chem C 114:22197–22206

Monti S, Carravetta V, Zhang WH, Yang JL (2007) Effects due to interadsorbate interactions on the dipeptide/TiO_2 surface binding mechanism investigated by molecular dynamics simulations. J Phys Chem C 111:7765–7771

Monti S, Carravetta V, Battocchio C, Iucci G, Polzonetti G (2008) Peptide/TiO_2 surface interaction: a theoretical and experimental study on the structure of adsorbed ALA-GLU and ALA-LYS. Langmuir 24:3205–3214

Monti S, van Duin ACT, Kim SY, Barone V (2012) Exploration of the conformational and reactive dynamics of glycine and diglycine on TiO_2: computational investigations in the gas phase and in solution. J Phys Chem C 116:5141–5150

Pan HH, Tao JH, Xu XR, Tang RK (2007) Adsorption processes of Gly and Glu amino acids on hydroxyapatite surfaces at the atomic level. Langmuir 23:8972–8981

Paszti Z, Guczi L (2009) Amino acid adsorption on hydrophilic TiO_2: a sum frequency generation vibrational spectroscopy study. Vib Spectrosc 50:48–56

Predota M, Bandura AV, Cummings PT, Kubicki JD, Wesolowski DJ, Chialvo AA, Machesky ML (2004) Electric double layer at the rutile (110) surface. 1. Structure of surfaces and interfacial water from molecular dynamics by use of ab initio potentials. J Phys Chem B 108:12049–12060

Puddu V, Slocik JM, Naik RR, Perry CC (2013) Titania binding peptides as templates in the biomimetic synthesis of stable titania nanosols: insight into the role of buffers in peptide-mediated mineralization. Langmuir 29:9464–9472

Rimola A, Corno M, Zicovich-Wilson CM, Ugliengo P (2008) Ab initio modeling of protein/biomaterial interactions: glycine adsorption at hydroxyapatite surface. J Amer Chem Soc 130:16181–16183

Roddick-Lanzilotta A, McQuillan AJ (1999) An in situ infrared spectroscopic investigation of lysine peptide and polylysine adsorption to TiO_2 from aqueous solutions. J Colloid Interface Sci 217:194–202

Roddick-Lanzilotta AD, McQuillan AJ (2000) An in situ infrared spectroscopic study of glutamic acid and of aspartic acid adsorbed on TiO_2: implications for the biocompatibility of titanium. J Colloid Interface Sci 227:48–54

Roddick-Lanzilotta AD, Connor PA, McQuillan AJ (1998) An in situ infrared spectroscopic study of the adsorption of lysine to TiO_2 from an aqueous solution. Langmuir 14:6479–6484

Roy MD, Stanley SK, Amis EJ, Becker ML (2008) Identification of a highly specific hydroxyapatite-binding peptide using phage display. Adv Mater 20:1830–1836

Sano KI, Shiba K (2003) A hexapeptide motif that electrostatically binds to the surface of titanium. J Amer Chem Soc 125:14234–14235

Schneider J, Colombi Ciacchi L (2011) A classical potential to model the adsorption of biological molecules on oxidized titanium surfaces. J Chem Theory Comput 7:473–484

Schneider J, Colombi Ciacchi L (2012) Specific material recognition by small peptides mediated by the interfacial solvent structure. J Amer Chem Soc 134:2407–2413

Segvich SJ, Smith HC, Kohn DH (2009) The adsorption of preferential binding peptides to apatite-based materials. Biomaterials 30:1287–1298

Skelton AA, Walsh TR (2007) Interaction of liquid water with the rutile TiO_2 (110) surface. Mol Simulat 33:379–389

Skelton AA, Liang TN, Walsh TR (2009) Interplay of sequence, conformation, and binding at the

peptide-titania interface as mediated by water. ACS Appl Mater Interfaces 1:1482–1491
Slepko A, Demkov AA (2013) First principles study of hydroxyapatite surface. J Chem Phys 139:044714
Suchanek W, Yoshimura M (1998) Processing and properties of hydroxyapatite-based biomaterials for use as hard tissue replacement implants. J Mat Res 13:94–117
Sultan AM, Hughes ZE, Walsh TR (2014) Binding affinities of amino acid analogues at the charged aqueous titania interface: implications for titania-binding peptides. Langmuir 30:13321–13329
Sultan AM, Westcott ZC, Hughes ZE, Palafox-Hernandez JP, Giesa T, Puddu V, Buehler MJ, Perry CC, Walsh TR (2016) Aqueous peptide-TiO_2 interfaces: Isoenergetic binding via either entropically or enthalpically driven mechanisms. ACS Appl Mater Interfaces 8:18620–18630
Syamchand SS, Sony G (2015) Multifunctional hydroxyapatite nanoparticles for drug delivery and multimodal molecular imaging. Microchim Acta 182:1567–1589
van Duin ACT, Dasgupta S, Lorant F, Goddard WA (2001) ReaxFF: a reactive force field for hydrocarbons. J Phys Chem A 105:9396–9409
Villarreal-Ramirez E, Garduno-Juarez R, Gericke A, Boskey A (2014) The role of phosphorylation in dentin phosphoprotein peptide absorption to hydroxyapatite surfaces: a molecular dynamics study. Connect Tissue Res 55:134–137
Wu CY, Skelton AA, Chen MJ, Vlcek L, Cummings PT (2011) Modeling the interaction between integrin-binding peptide (RGD) and rutile surface: the effect of Na^+ on peptide adsorption. J Phys Chem C 115:22375–22386
Wu CY, Skelton AA, Chen MJ, Vlcek L, Cummings PT (2012) Modeling the interaction between integrin-binding peptide (RGD) and rutile surface: the effect of cation mediation on Asp adsorption. Langmuir 28:2799–2811
Wu CY, Chen MJ, Skelton AA, Cummings PT, Zheng T (2013) Adsorption of arginine-glycine-aspartate tripeptide onto negatively charged rutile (110) mediated by cations: the effect of surface hydroxylation. ACS Appl Mater Interfaces 5:2567–2579
Wu H, Xu DG, Yang ML, Zhang XD (2016) Surface structure of hydroxyapatite from simulated annealing molecular dynamics simulations. Langmuir 32:4643–4652
Xu ZJ, Yang Y, Wang ZQ, Mkhonto D, Shang C, Liu ZP, Cui Q, Sahai N (2014) Small molecule-mediated control of hydroxyapatite growth: free energy calculations benchmarked to density functional theory. J Comput Chem 35:70–81
Yoshinari M, Kato T, Matsuzaka K, Hayakawa T, Shiba K (2010) Prevention of biofilm formation on titanium surfaces modified with conjugated molecules comprised of antimicrobial and titanium-binding peptides. Biofouling 26:103–110
Yucesoy DT, Hnilova M, Boone K, Arnold PM, Snead ML, Tamerler C (2015) Chimeric peptides as implant functionalization agents for titanium alloy implants with antimicrobial properties. JOM 67:754–766
Zahn D, Hochrein O (2003) Computational study of interfaces between hydroxyapatite and water. Phys Chem Chem Phys 5:4004–4007
Zhang Z, Fenter P, Cheng L, Sturchio NC, Bedzyk MJ, Predota M, Bandura A, Kubicki JD, Lvov SN, Cummings PT, Chialvo AA, Ridley MK, Benezeth P, Anovitz L, Palmer DA, Machesky ML, Wesolowski DJ (2004) Ion adsorption at the rutile-water interface: linking molecular and macroscopic properties. Langmuir 20:4954–4969
Zhao WL, Xu ZJ, Yang Y, Sahai N (2014) Surface energetics of the hydroxyapatite nanocrystal-water interface: a molecular dynamics study. Langmuir 30:13283–13292
Zhao WL, Xu ZJ, Cui Q, Sahai N (2016) Predicting the structure-activity relationship of hydroxyapatite-binding peptides by enhanced-sampling molecular simulation. Langmuir 32:7009–7022
Zheng T, Wu C, Chen M, Zhang Y, Cummings PT (2016) Molecular mechanics of the cooperative adsorption of a Pro-Hyp-Gly tripeptide on a hydroxylated rutile TiO_2(110) surface mediated by calcium ions. Phys Chem Chem Phys 18:19757–19764
Zhou H, Lee J (2011) Nanoscale hydroxyapatite particles for bone tissue engineering. Acta Biomater 7:2769–2781

Design Principles of Peptide Based Self-Assembled Nanomaterials

4

Rania S. Seoudi and Adam Mechler

Abstract

The ability to design functionalized peptide nanostructures for specific applications is tied to the ability of controlling the morphologies of the self-assembled superstructures. That, in turn, is based on a thorough understanding of the structural and environmental factors affecting self-assembly. The aim of designing self-assembling nanostructures of controlled geometries is achieved *via* a combination of directional and non-directional second order interactions. If the interactions are distributed in a geometrically defined way, a specific and selective supramolecular self-assembly motif is the result. In this chapter we detail the role of non-covalent interactions on the self-assembly of peptides; we will also discuss different types of peptide building blocks and design rules for engineering unnatural supramolecular structures.

Keywords

Programmed self-assembly • Hierarchical nanomaterials • Second order interactions • Unnatural peptides • Peptide building blocks • β-peptide folding

4.1 Introduction

The quest for new materials to meet the increasingly exacting requirements of rapidly changing industries from aerospace to regenerative medicine has led to intense activity in contriving non-conventional structures and properties (Feynman 1960; Adams and Barbante 2013; Garrett and Poese 2013; Quandt and Ozdogan 2010; Arico et al. 2005). One of the key demands towards novel materials is inherent functionality (Werner and Horne 2015; Laursen et al. 2015; Imperiali et al. 1999; Pagel and Koksch 2008; Wang et al. 2006a; Ryu et al. 2008; Care et al. 2015), primarily in the context of the so-called smart materials that are capable of responding to environmental

R.S. Seoudi • A. Mechler (✉)
Department of Chemistry and Physics, La Trobe Institute for Molecular Science, La Trobe University, Bundoora, VIC 3086, Australia
e-mail: A.Mechler@latrobe.edu.au

© Springer International Publishing AG 2017

A. Sunna et al. (eds.), *Peptides and Peptide-based Biomaterials and their Biomedical Applications*, Advances in Experimental Medicine and Biology 1030, https://doi.org/10.1007/978-3-319-66095-0_4

stimuli (Kopecek 2003; Lowik et al. 2010; Mano 2008; Seabra and Duran 2013; Sun et al. 2011), in applications as diverse as optical biosensors, molecular electronics and regenerative medicine (Scanlon and Aggeli 2008; Kim et al. 2015; Rad-Malekshahi et al. 2016; Wang et al. 2006b; Mart et al. 2006; Deming 2007; Stoop 2008; Tamerler and Sarikaya 2008; Toksoz et al. 2010; Nguyen et al. 2013). The ability of smart materials to respond to stimuli may be achieved through designing functional hierarchical systems: bulk materials composed of specific chemically or physically responsive nanostructures (Yoshida et al. 2006; Fairman and Akerfeldt 2005; Ju 2011).

Materials design often turns to nature for ideas. The strength of natural fibres such as silkworm (bombyx mori) silk turned attention to self-assembled systems: bioinspired materials that hold the promise to create hierarchical nanostructures (Fan et al. 2000; Quantock et al. 2015; Kumar et al. 2011; Bhattacharya et al. 2014) while offering the means of controlling the chemical functionality and morphology from nano- to macroscale structures (Bhushan 2009; Liu et al. 2013; Zhang 2002, 2003). Bioinspired materials are distinguished from synthetic polymeric materials by their assumed biocompatibility and biodegradability, however they are truly superior to polymeric materials in their programmable structural diversity and functionalizability (Joo et al. 2015; Reimers et al. 2015; Su and Wang 2015, Lee et al. 2001; Wang et al. 2004). Among the most versatile bioinspired materials are polypeptides (Zhao et al. 2014; MacEwan and Chilkoti 2014; Shen et al. 2015; Canalle et al. 2010; Numata 2015). Careful design of the peptide sequence can be used to control folding patterns, while chemical modification of the side chains can confer an active chemical or physical functionality on the polypeptide structures (Klinker and Barz 2015; Rasale and Das 2015; Engler et al. 2011; Zhang et al. 2014).

In traditional material science, fabrication of functional micro- and nanostructures is routinely achieved through 'top-down' methods, creating surface patterns with high-resolution optical lithography (Henzie et al. 2006; Leggett 2006; Colson et al. 2013). An alternative way of creating functional nanostructures is using self-assembly to build these structures 'bottom-up' from small molecule precursors (Cheng et al. 2006; Ariga et al. 2008; Zhang et al. 2009). Peptide-based bioinspired materials can be designed to form *via* self-assembly and thus hold the promise to implement bottom-up nanofabrication (Shimomura and Sawadaishi 2001; Shimizu 2003; Ariga et al. 2007; Koehler 2015). Self-assembly or bottom-up nanofabrication is based on thermodynamic principles: energy minimization drives small molecules to associate into large well-ordered structures that are held together only by non-covalent bonds (Zhang and Altman 1999; Mattia and Otto 2015; Levin et al. 2014; Kar et al. 2014). The non-covalent forces acting between the monomers are either of van der Waals (vdW) attraction, electrostatic forces, hydrogen bonding, or hydrophobic attraction (Zhao et al. 2008; Cao et al. 2013; Hobza et al. 2006). Because self-assembly occurs under equilibrium conditions, without any external stimuli, the system is sensitive to any environmental changes (Mann 2009; Grzelczak et al. 2010; Ma et al. 2010; Bowerman and Nilsson 2010). The environmental control parameters are temperature, pH, concentration and the physicochemical solvent characteristics (Koga et al. 2011; Rissanou et al. 2013; Chen 2005; Ryu and Park 2008; Lee et al. 2014; Tian et al. 2015). For example, changing the pH would change the protonation state of acidic ligands of the monomers, thus controlling the ability to form hydrogen bonds, leading to a change in the dimensionality of the oligomeric form (Victorov 2015). The ionic strength of solvent environment affects the strength and often the sign of the net force between charged molecules, for example, introducing salts to a solution of anionic peptides may lead to aggregation in which positive ions reduce the repulsion between the peptide molecules (Mendes et al. 2013). The dielectric constant of the solvent affects the strength of van der Waals interactions and thus affects the self-assembly as well (Bishop et al. 2009; Stendahl et al. 2006). Temperature can alter the hydrophobic/hydrophilic character of materials leading to folding or unfolding of oligopeptide structures (Leikin et al. 1995; Pochan et al. 2003), hence temperature variation can be used to design directed self-assembly through switching the assembly on/off. Electromagnetic effects can be used to the same end (Yu et al. 2014; Velichko et al. 2012; Wang et al. 2011; Park et al. 2013; Gilmartin et al. 2005). Due to these advantages, the design of bioinspired self-assembling nanostructures is at the forefront of contemporary materials science (Cao et al. 2013; Gazit 2007; Ariga et al. 2008) and has also been considered for various applications in medicinal and biochemistry (Kim et al. 2015; Seow and Hauser 2014; Bhattacharya et al. 2014).

A particularly active research area is the design of unnatural amino-acid based peptides aiming to achieve much greater structural polymorphism than that is possible in α-peptides (Werner and Horne 2015). β-peptides are the most commonly used unnatural peptides; these are similar to α-peptides in that β-peptides contain amide moieties that can form hydrogen bond network to stabilize secondary structures. They differ from α amino acids in the extra carbon atom in the backbone so that substitution is possible either on the C^2 or C^3 atom (Seebach et al. 2006; Wu and Wang 1999, Gellman 1998). Peptides based on beta amino acids (β-peptides) show an ability to fold into well-defined secondary structures (Horne 2015; Collie et al. 2015; Kortelainen et al. 2015). In principle, it is possible to design secondary structures using β-peptides that can be side chain functionalized without affecting the folding (Martinek and Fulop 2003; Gratais et al. 2015; Althuon et al. 2015). Thus, materials design based on unnatural β-peptides holds the promise to revolutionize bioinspired bottom-up nanofabrication. In this chapter, the means of controlling the folding and self-assembly of oligopeptides will be described, with examples of complex hierarchical structures.

4.2 Self-Assembly

Self-assembly is the process by which an initially disordered system becomes ordered through local non-covalent interactions (Whitesides et al. 1991; Lehn 2004; Whitesides and Grzybowski 2002). Self-assembly can be also templated in a structural sense, that may occur with imposing direction on the self-assembly process by an external force by magnetic (Tanase et al. 2002), electric (Smith et al. 2000) or flow fields (Huang et al. 2001). Because self-assembly happens *via* association of small molecules of nanoscale dimensions into more complex systems, the defects of such systems are minimized to atomistic or molecular scale (Mishra et al. 2011; Barth et al. 2005; Buehler and Ackbarow 2008). The incorporation of more than one association motif can encode into the molecular design the formation of secondary and/or hierarchical structures; this general form of supramolecular self-assembly (Pasini and Kraft 2004) offers great precision in structural materials design (Gazit 2007; Huang and Che 2015).

The formation of nanostructures with specific, predetermined shape, size and structure through self-assembly is one of the main objectives of materials science since this is the prerequisite of designing molecular devices. While some degree of success was achieved with the assembly of geometrically controlled nanoparticles (Abb et al. 2016; Bromley et al. 2008; Stupp et al. 2013), these systems lack the required versatility. Bioinspired self-assembly modelled on ligand-receptor interactions offers the necessary design rules (Kopecek and Yang 2009).

Specific self-assembly is prevalent in biological systems. While ligand-receptor binding yields only dimers, the mass of functional protein complexes such as a flagellum motor or a virus capsule reach megadalton to gigadalton sizes. Less specialized but much larger in assembly mass are linear oligomeric systems such as amyloid fibres causing Alzheimer's and Parkinson's diseases. Amyloid fibres have a complex β-sheet structure which is characterized by high stiffness and rigidity where hydrophilic, electrostatic, hydrogen bonding and π-π stacking interactions facilitate specific and selective self-assembly in a stable cross β-arrangement (Cinar et al. 2012). A common characteristic of these biological self-assembled systems is that multiple interactions are aligned along the structures, creating a geometrical pattern (Ulijn and Smith 2008; Zhang 2002). This inspired the invention of supramolecular chemistry that is also known as the chemistry of the non-covalent bond.

Supramolecular self-assembly arises from a synergy of multiple geometric, steric, ionic and directional bonds. It offers highly specific and selective recognition motifs that, when utilized in a bottom-up design, offers a great precision leading to a high-level manipulation of materials. Supramolecular assembly opened new avenues in the world of nanotechnology (Gazit 2007; Chen and Rosi 2010; Berl et al. 2000; Ramstrom et al. 2002), offering spontaneous formation of ordered structures at the nanoscale leading to macroscopic objects with nano-scale order (Gazit 2007). Supramolecular self-assembly can be classified into reversible and irreversible processes. The reversible supramolecular interactions involve a sequence of interactions that lead to the formation of supramolecular structures (Rajagopal and Schneider 2004). In irreversible assembly, the interactions occur at the same time leading to the formations of complex hierarchical structures (Rajagopal and Schneider 2004). Using these principles, protein and peptide based self-assembled materials were developed through several processes of molecular recognition, spontaneously arrange themselves to highly organized structures (Gazit 2007; Ulijn and Smith 2008; Lakshmanan et al. 2012).

4.2.1 Interactions That Drive Self-Assembly

Self-assembly can happen *via* a number of physical, also known as second order, interactions. Although those interactions form weak non-covalent bonds, the collective effect of them often leads to strong, stable structures. The competition between second order interactions as a function of environmental conditions gives rise to different structures (Mendes et al. 2013), hence it is important to understand those interactions and what are their fundamental characteristics.

The main interactions involved in self-assembly are hydrogen bonding, van der Waals interactions, solvophobic (e.g. hydrophobic) interactions and π-π stacking (Fig. 4.1).

In case of charged particles, electrostatic forces are also important, although in ionic solutions these are typically reduced in range and strength. A special category of interactions between colloidal systems are so-called entropic forces, including thermal fluctuation, steric, and hydration forces.

4.2.1.1 Hydrogen Bonding

While its exact nature is still debated, the hydrogen bond arises from the attraction between an electron-deficient hydrogen and a lone pair of electrons (e.g. of carbonyl oxygen). The hydrogen bond is directional, stronger, and shorter distance than a dipole-dipole interaction. It was hypothesized before that H-bond is a partial covalent bond, however no electron pair sharing was observed. Presently it is believed that H-bonding is a special case of polar interaction. Whereas the exact physical basis of H-bonding is not known, its strength does correlate to the dipole moment of the interacting moieties, and hence competition is possible for H-bonding sites; furthermore the attraction depends on the dielectric constant of the solvent.

The hydrogen bond is the one responsible for the formation of the three-dimensional macromolecular structures in what is called hydrogen bond polymerization (Israelachvili 2011). Hydrogen bonding is the strongest among the second-order interactions, however it is still a short range point-to-point interaction. Hydrogen bonding is a strong interaction, with bond energies ranging between 5-65 $kJmol^{-1}$ (Faul and Antonietti 2003) thus multiple hydrogen bonding moieties can drive self-assembly into robust supramolecular structures through specific and selective binding motifs due to the directionality and specificity of the bonding (Sijbesma and Meijer 1999). The cooperative effect of multiple hydrogen bonds, while also dependent on the solvent, is primarily defined by the nature of the donor and acceptor sites. Hydrogen bonding motif is used to direct the self-assembly mostly in apolar environment so as there is no competition for the binding sites from the solvent (Prins et al. 2001). The hydrogen bond is frequently used in the design of supramolecular structures, to build one-, two- or

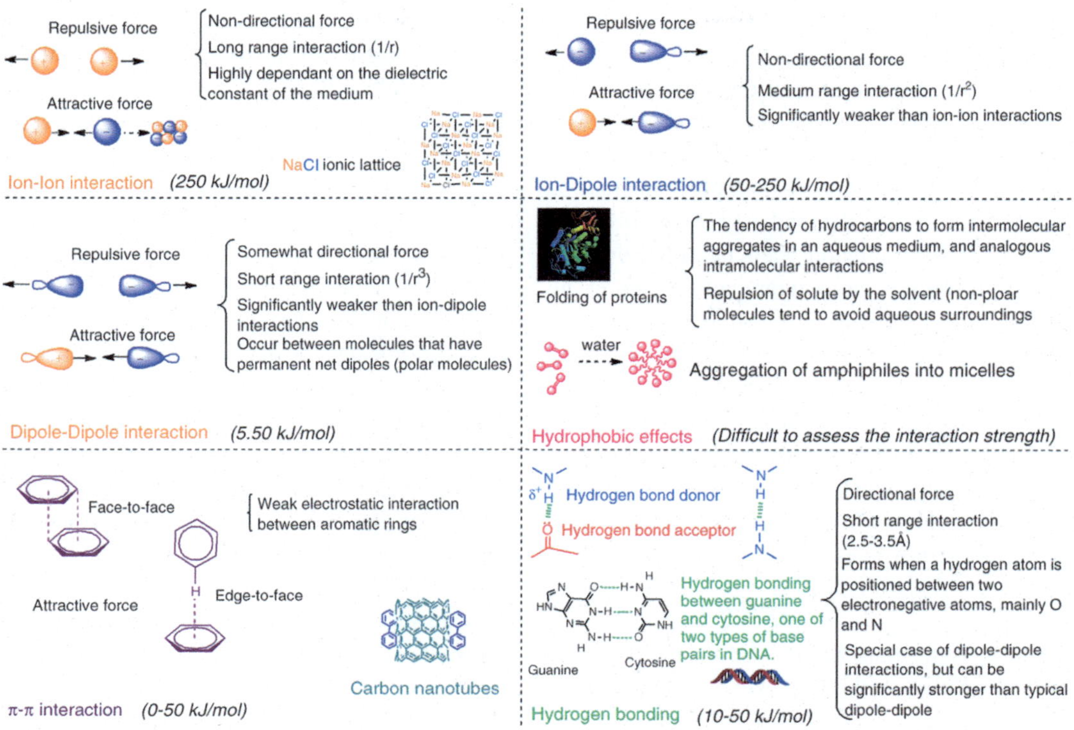

Fig. 4.1 Different types of non-covalent interactions involved in self-assembly (Mendes et al. 2013)

three-dimensional molecular crystals (Russell and Ward 1996).

4.2.1.2 Van der Waals Interactions

Also known as polar interactions, van der Waals (vdW) interactions exist between all molecules. Van der Waals interaction is not as strong as Columbic interaction or hydrogen bonding, however it may act on large surface areas thus giving van der Waals interaction the ability to influence structural assembly. There are three types of interactions that contribute to van der Waals forces: permanent dipole-dipole interactions (Keesom or orientation force), induced dipole-permanent dipole interaction which is also known as Debye induction force, and induced dipole-induced dipole interaction (London dispersion force, London 1937) that can occur between any species irrespective of polarity or charge. Although London dispersion force is inherently weak, it is nevertheless the reason behind the condensed (liquid) state of non-polar solvents.

Van der Waals interactions are not pairwise additive, but depend on the dimensionality of the system as well, becoming substantial if acting on a large surface area. The atomic arrangement and dimensionality were also shown to affect Van der Waals coefficient of carbon nanostructures (Gobre and Tkatchenko 2013).

Given the relative weakness of second-order interactions, it is important to consider the Boltzmann energy kT in the stability of self-assembled structures. Increasing the temperature weakens both hydrogen bonding and van der Waals interactions, which in case of polypeptides may lead to unfolding of secondary structures (Daura et al. 1998; Seebach et al. 2006; Gademann et al. 1999).

4.2.1.3 Solvophobic Interactions

The interaction of solutes with the solvent can take three different forms: if the solute is charged, strong orientation forces create a hydration shell; if the solute has hydrogen bond donors or acceptors, moieties of the molecule become embedded into the hydrogen bonding network of the solvent; and in case of apolar molecules, an exclusion shell is formed around the structure by the solvent. The latter gives rise to an apparent attraction between the solute molecules, known as solvophobic interaction. Solvophobic interaction can be expressed as the tendency of the solvent to minimize the exclusion shell by squeezing the apolar molecules into aggregates (Luo et al. 2015). The extent of solvophobic interactions depends on the strength of the hydrogen bond network of the solvent (Rodnikova 2007; Rodnikova and Barthel 2007).

Solvophobic interactions are long range, temperature dependent, entropically originated forces. The energy of the solvophobic interaction between two particles is:

$$E_{sp} = 2A_i\gamma \approx 4\pi r h_o \gamma \quad (4.1)$$

where h_o is the diameter, r is the distance between two particles, A_i is the Hamaker constant, and γ is the surface tension at the particle-solvent interface (Luo et al. 2015).

Self-assembly frequently involves solvophobic and solvophilic interactions. Hydrophobic moieties tend to minimize contacting surface areas with the solvent, thus particles would associate to minimize energy. If a hydrophilic part exists, it would tend to increase interfacial surface area with the solvent. The competition between the solvophobic/solvophilic interactions determines the overall structure that in the simplest amphiphilic systems, surfactants, induce the formation of micelle structures (Toksoz et al. 2010). Some designs of self-assembling peptides copy surfactant properties by introducing a hydrophobic "tail" attached to hydrophilic amino acids. Introducing such structures to water leads to the association of peptides into micellar nano-structures.

4.2.1.4 Electrostatic Interactions

Long-range nonselective strong Coulombic (electrostatic) interactions are either attractive or repulsive. Electrostatic interactions depend on the dielectric constant of the solvent. Coulombic forces act between charged groups. The importance of charge interactions in an electrolyte is much reduced for two reasons: the dielectric constant of the medium reduces the strength of the interaction (also leading to dissolution of salts in water), while solvated counterions shield the charges. For interactions between two surfaces:

$$w(D) = \frac{1}{4\pi\varepsilon\varepsilon_r}\frac{Q_1 Q_2}{D} \quad (4.2)$$

Where Q_1 and Q_2 are the interacting charges, ε is the vacuum dielectric permittivity, and ε_r is the dielectric constant of the medium in which the interaction takes place. Comparing the distance dependence of the electrostatic interaction (1/r), to Van der Waals interaction ($1/r^6$) it is clear that Van der Waals interactions act on much shorter range. Van der Waals and electrostatic interactions in solution are unified in the DLVO (Derjaguin- Landau- Verwey- Overbeek) theory, however in simple treatment it is usually suitable to consider each in separation.

4.2.1.5 Aromatic Interactions (π-π Stacking)

Aromatic-aromatic interactions occur between π-electron rich and π-electron deficient aromatic rings depending on the orientation of compounds such as edge-to-face, stacked and offset stacked rings. If aromatic rings within an oligomeric structure are oriented face-to-face, the attractive interaction imposes substantial stability on the self-assembled structures. Stacking interactions are hard to control in molecular design, hence it is usually combined with other non-covalent interactions to form specific structures (Klosterman et al. 2009). In nature β-amyloids assemble into fibres *via* the combination of hydrogen bonding and π-π stacking between phenyl rings (Yan et al. 2010; Gazit 2002).

Most of the non-covalent forces are hard to control due to non-directionality (except hydrogen bonding) and sensitivity to environmental conditions, which have to be considered in the molecular design. A way to overcome weakness or non-directionality of non-covalent interactions is through using a combination of non-covalent interactions in a geometrically defined pattern to drive interactions, which is essentially a realization of a supramolecular recognition motif.

4.3 Bioinspired Building Blocks of Self-Assembly

As it was outlined above, nature is the inspiration for building nanostructures using the self-assembly principle (Coppens 2005; Shu et al. 2011; Giuseppone 2012). Lipids, nucleic acids and amino acids are the main building blocks for building nanomaterials through the spontaneous organization into highly sophisticated systems using only noncovalent interactions (Ulijn and Smith 2008; Zhang et al. 2002). The former two have been extensively studied before; the focus of this chapter is on oligopeptide self-assembly.

Amino acids are the fundamental building blocks of nature's most versatile biopolymers: polypeptides. Polypeptides, or proteins, are molecular machines, having the most complex nanostructures in nature that allow them to build functional superstructures. These include transmembrane ion channels, essentially molecular valves; microtubules, the biochemical conveyor belts; and the flagellum rotor, nature's version of rotor-stator motion (Blair 2003; Mora et al. 2009; Moreau et al. 2014; Aryal et al. 2015). The origin of this structural and functional diversity is the ability of oligopeptides to fold into permanent nanostructures that are "coded" into the sequence, primarily helices and sheets, where the superstructures are based on these structural units.

Materials science has been slow in converting this biological versatility into technological solutions, largely since proteins are far too complex to design and manufacture. Small peptides, however, offer a simple means to build up a knowledge base that will one day allow the design of oligopeptide-based molecular machines (Gazit 2007; Ulijn and Smith 2008; Lakshmanan et al. 2012). The strategies of designing oligopeptide based self-assembling structures rely on the naturally occurring conformations of proteins such as β-sheet or α-helix as structural elements. The self-assembly motif might be based on amphiphilicity, salt bridges, hydrogen bonding, or even the use of aromatic π–stacking interactions (Zhang et al. 1993; Nikolic et al. 2011; Hauser et al. 2011).

Although peptides and proteins are made of only 20 "natural" amino acids, there are many more synthetic amino acids that can be prepared in laboratory. Such "unnatural" amino acids can be designed by substituting side chains and/or by insertion of extra carbon atoms in the backbone of the amino acids (at β and γ positions) (Seebach et al. 2006; Cheng et al. 2001). Thus, coded into the amino acid sequence, peptides can fold and self-assemble into nanotubes, helical ribbons, and fibrous scaffolds (Pizzey et al. 2008; Hamley 2014; Eom et al. 2015; Jadhav et al. 2014). The following part provides an overview of the common folding geometries that can serve as a basis for designing complex self-assembling nanostructures.

4.3.1 Secondary Structures of α-Peptides

α-peptides are best used for introducing the key concepts of peptide folding. α-peptide consists of alpha amino acids named so as the alpha carbon is substituted. Most natural amino acids are alpha type. The "design rules" of α-peptide folding are hence well known. The dominant folded structures are the α-helices and β-sheets (Woolfson and Ryadnov 2006). While the folding geometry largely depends on the nature of the side chains and the peptide sequence, there is enough flexibility in the design rules to provide the peptide structures with the needed functionality. Importantly, peptides have natural biochemical functions that can be utilized under unnatural conditions. There are several classes of membrane modulating/disrupting, signaling, and otherwise active peptides (Guillotte et al. 2016; Shuvaev et al. 2015; Fodale et al. 2014; Heyduk et al. 2010). Hence, structural and functional designs can go hand-in hand if the target application is biomedical; nano-technological applications, however, require novel design strategies.

4.3.1.1 β-Sheet

The dense hydrogen bonding network between the amide moieties in oligopeptide β-strands leads to the formation of the planar, "hairpin-like" folded structure known as β-sheet (Mendes et al. 2013). The "hairpin" orientation is preserved through in-plane intramolecular hydrogen bonds between the backbone amide moieties, with an interstrand distance of 5Å between strands (Nesloney and Kelly 1996), whereas the side chains are positioned approximately perpendicular to the plane of the β-sheet. The strands are rarely in a completely extended conformation hence beta sheets are "twisted". Importantly β-sheet can be formed between either parallel or antiparallel strands that satisfy the folding motif (Fig. 4.2).

While a very common folding in nature, β-sheets do not have highly exacting requirements in terms of amino acid residues, although different amino acids have different "propensities" to fold into beta sheets. It is best understood in terms of stabilizing factors, usually interactions between the side chains. β-sheets are stabilized by branching and cyclic side chains as well as charged residues. The latter is explained with pairwise interactions and salt bridge formation, however higher bond energy between adjacent residues does not necessarily lead to higher β-sheet stability.

Nevertheless, β-sheet stability depends on the side-chain pairwise interactions. As the β-sheet typically has a hydrophobic and a hydrophilic face, it is possible for the hydrophobic interaction to act as a stabilizing factor. The more hydrophobic cyclic and branched side chains have a stronger stabilizing effect, while β-sheets containing glycine and proline side chains have the lowest stability (Smith and Regan 1997; Minor and Kim 1994). Importantly, β-sheets can stack into sandwich-like superstructures (Nesloney and Kelly 1996) and hence β-sheet peptides have the ability to self-assemble into fibres, tapes/ribbons and fibrils.

4.3.1.2 α-Helix

Helical folding forms *via* the intramolecular hydrogen bond that occurs between the carbonyl oxygen C=O and the proton donor N-H of the amide between amino acid residues a given distance apart. In α-helix hydrogen bonding occurs between C=O of residue i and N-H of residue i+4 (Crisma et al. 2006) with a bond distance of 2.72Å and rise per turn 5.4 Å (Fig. 4.3) (Crisma et al. 2006; Doig 2008).

This α-turn structural motif is characterized by 3.6 residues per turn, or 13 carbon atoms in the backbone from H-bond donor to H-bond acceptor (Gunasekar et al. 2008). α-helix is a right- handed helix if formed from L-amino

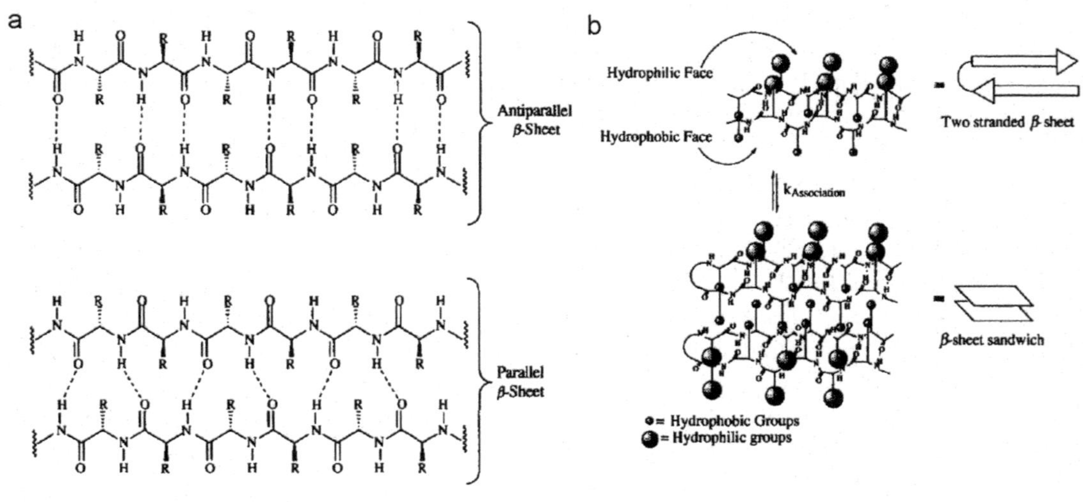

Fig. 4.2 (**a**) Structural representation for antiparallel and parallel β-sheets (**b**) Schematic representation of β-strands and β-sheets (Nesloney and Kelly 1996)

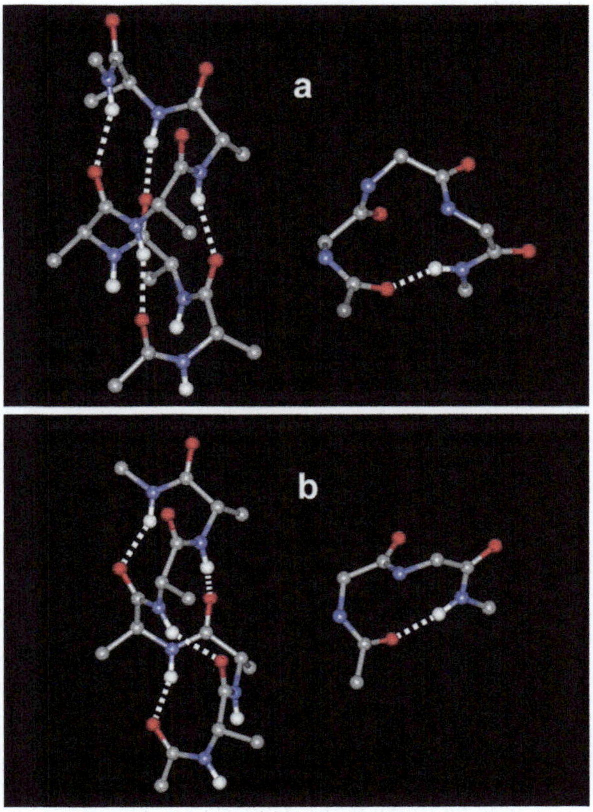

Fig. 4.3 Three dimensional illustration of the hydrogen bonding patterns of (**a**) α-helix and (**b**) 3_{10}-helix (Crisma et al. 2006)

acids. Elongated tightly stretched α-turns are considered to be an essential intermediate for the formation of a well-defined α-helix (Bolin and Millhauser 1999). Whereas five amino acids complete a helical turn, five residue long peptides do not form stable α-helices. For peptides containing only Aib (α-aminoisobutyric acid or $C^{\alpha,\alpha}$-dimethylglycine) seven residues are sufficient to form an α-helix, whereas for peptides made of combinations of natural amino acids, it is generally accepted that at least 13 residues are required to form a stable α-helix in solid state (Crisma et al. 2006).

α-helices may exhibit independent interactions between their side chains, such as ion-pair and charge-dipole interactions along the length of the helix due to the relative rigidity of the structure (Bromley and Channon 2011; Scholtz and Baldwin 1992). Both termini are stabilized by charged residues, negative at N-terminus and positive at C-terminus. The side chains of α-helix are spaced at 100° intervals along the axis of the helix, thus they do not align; at least 7 residues are required to line up two side chains over each other on the same helical face (Fig. 4.4 left) (Crisma et al. 2006).

Although it is generally believed that α-helices are stabilized by hydrophobic side chains, introducing polar side chains such as lysine or glutamine can promote helix solubility without major interference with the helix forming propensity. The polar side chains have to be introduced in such a way that no interactions occur between side chains. For example, if a polar side chain is introduced at position I, the next possible position of a polar side chain substitution is i+5 (Crisma et al. 2006; Monera et al. 1995).

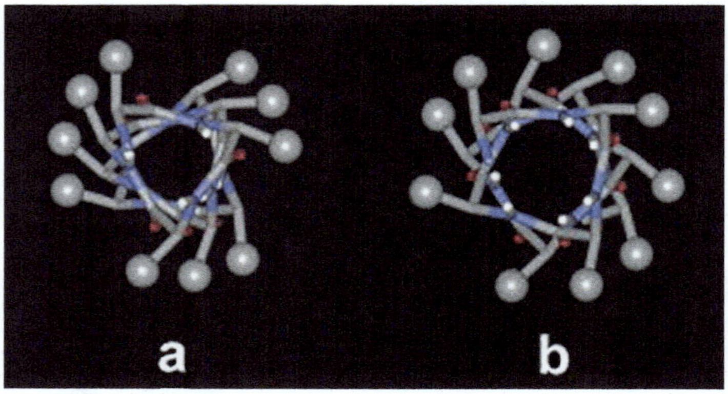

Fig. 4.4 Side chains distribution along (*left*) 3_{10}-helix, and (*right*) α-helix (Crisma et al. 2006)

The advantage of a helical peptide as a building block over the β-sheet peptide is the ability of side chain substitution of the α-helix. However, the scope of side chain modifications is constrained by side chain interactions that can affect α-helix formation (Padmanabhan et al. 1990; Banwell et al. 2009).

4.3.1.3 3_{10} Helix

The 3_{10} helix is distinguished by its intramolecular hydrogen bond between residues i and i+3 (Fig. 4.4 right). While 3_{10} helix has similar torsional angles to α-helix, yet 3_{10} helices are less stable compared to α-helix. 3_{10} helix has less favorable van der Waals energy and the hydrogen bond is not optimal due to the higher rise of the helix (~0.2 nm) and smaller helix diameter that introduces a higher conformational strain on the amino acid backbone. The integer number of residue per turn allows for the side chains of successive turns to align giving 3_{10} helix high symmetry along the main axis (Fig. 4.4 right). Interestingly, this theoretical symmetry is not seen in experimental data, instead side chains are slightly staggered (Crisma et al. 2006). Energetic analysis suggests the 3_{10} helix act as an intermediate during the mechanism of α-helix folding. While the α-helix is more common in nature, globular proteins often have short irregular 3_{10} helices near their termini (Crisma et al. 2006).

4.3.2 Secondary Structures of β-Peptides

α-peptides are well known since most – but not all – of the amino acids in natural proteins are alpha amino acids. Nature builds complex devices out of α-oligopeptides using a combination of secondary and tertiary folding, chaperones, and supramolecular binding; it has proven a hitherto insurmountable challenge to design even much simpler structures from α-amino acids. Long α-peptides suffer from reduced stability due to the exposed domains, and at least seven residues are required for two side chains to align on top of each other (Scholtz and Baldwin 1992). The low thermodynamic and biological stability of α-peptides have been major limitations in the use of α-peptides in biomedical applications (Mohle et al. 1999; Seebach and Gardiner 2008; Godballe et al. 2011). β-peptides offer solutions to all these problems. β-peptides are analogous to α-peptides, built from unnatural amino acids where one carbon is inserted into the backbone of the amino acid. In β-amino acids side chain substitution may happen on either C^2 (β^2-amino acids), or C^3 (β^3-amino acids) (Fig. 4.5). Unnatural β^3-amino acids are also known as β^3 homo amino acids in some works (Vaz et al. 2008; Pizzey et al. 2008; Gademann et al. 1999); in this chapter β^3 homo amino acids will be called β^3-amino acids, consistent with works by M-I Aguilar (Gopalan et al.

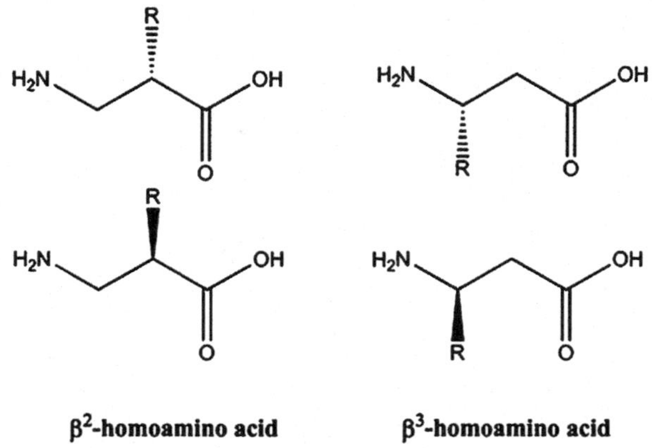

Fig. 4.5 Structures of unnatural β-amino acids (Gopalan et al. 2015)

2015; McDonald et al. 2014; Luder et al. 2016; Steer et al. 2002).

Beta peptides offer good alternative to α-peptides for structural design as they exhibit much higher metabolic and chemical stability while forming well-defined secondary structures (Chakraborty and Diederichsen 2005; Bruckner et al. 2003; Tomasini et al. 2013). Importantly, β-peptides fold into a larger variety of secondary structures compared to α-peptides (McGovern et al. 2012). Furthermore, folded β-peptides have been shown to assemble into well-defined three-dimensional superstructures that can be controlled to larger extent than α-peptides (Chakraborty and Diederichsen 2005).

Despite the fact that effective number of hydrogen bond donor-acceptor pairs decrease in β-peptides to one for each 4 chain atoms versus one for each 3 chain atoms in α-peptides, the folding stability increases moving from α, β, to γ peptides (Seebach et al. 2006) for unsubstituted backbones; yet substituted β-peptides showed even higher stability over γ-peptides (Gellman 1998). β-peptides are chemically similar enough to α-peptides that the existing understanding for controlling the folded structures and stability are easily compared between the two systems (Cheng et al. 2001). β-peptides showed similar ability to α-peptides to fold into helices and sheet-like conformations (DeGrado et al. 1999).

Similar to α-peptides, β-peptide folding is stabilized through intra-molecular hydrogen bonding between C=O and N-H moieties of the backbone amides (Gellman 1998). The carbonyl oxygens are either directed towards N or C terminus; hence the orientation of the amide dipole is approximately parallel to the main axis of the helix that results in exposed amides near the helix ends (Goodman et al. 2007). The helices are affected by side chain substitution that can happen on either β^2 or β^3 positions (Martinek and Fulop 2012). Depending on the position of substitution either left- or right-handed helices are formed (Fig. 4.6) (Seebach et al. 2006).

β-peptides can readily form stable helices with only four to six residues long sequences in contrast to the 13 amino acid long sequences required in the case of natural α-peptides for a permanent helix in protic solvents such as water and methanol (Wu and Wang 1998; Seebach and Gardiner 2008; Crisma et al. 2006). The presence of extra carbon in the backbone increases the distance and hence reduces the interactions between the side chains, allowing for easy side chain substitution without affecting the folding; conversely in α-peptides side chain substitution destabilizes the helix, at least five amino acid sequences are required between polar residues to avoid side chain interactions (Padmanabhan et al. 1990).

β-peptides have the ability to adopt various secondary structures such as 8, 10, 12, 14, 16, 18 and 20 helices (Fig. 4.7) (Cubberley and Iverson 2001), depending on the number of atoms involved in each turn (Cheng et al. 2001; Dado and Gellman 1994). Due to this rich variety of secondary structures, β-peptides are also known as foldamers. The presence of an additional carbon atom in the backbone provides β-peptides

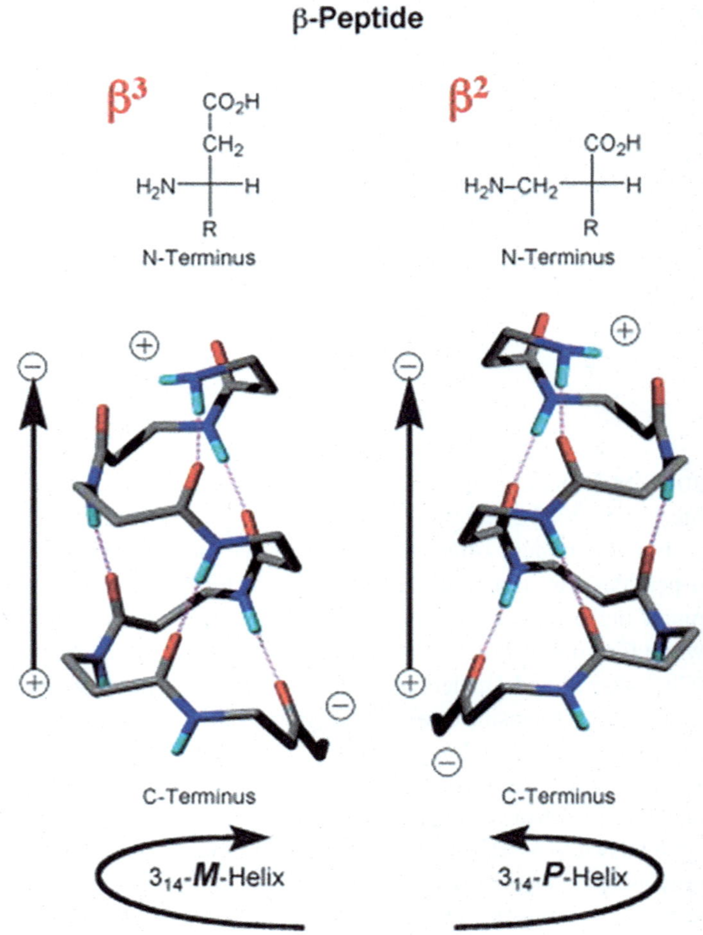

Fig. 4.6 *Right*- and *left*-handed β-helix (Seebach et al. 2006)

with larger conformational freedom than exists in case of α-peptides. Gauche conformation around torsion angle θ between C^2–C^3 (Fig. 4.8) is favored for folded β helix, where θ values are found to be near ±60° (Cheng et al. 2001). The unsubstituted backbone has high flexibility, where substitutions increase the conformational constraints (Cheng et al. 2001; Banerjee and Balaram 1997).

β-peptides exhibit large conformational freedom in the backbone that should make "adopting ordered conformation" entropically disfavored, especially in solutions (Applequist and Glickson 1971: Narita et al. 1986); yet it was seen that β-peptides frequently form ordered structures in spite of entropic and solvation effects (Hill et al. 2001; Mohle et al. 1999; Wu and Wang 1998).

The stability of β-helix depends on the steric factors and the interactions between the side chains that are the reasons for different helix formation between 2, 3, and 2/3 substitutions (Seebach and Matthews 1997).

4.3.2.1 14-Helix

Substitution on the carbon backbone at either of position 3 or position 2, and in some deprotected peptides in both 2,3 positions within the same residues usually result in the formation of 14-helix (Seebach et al. 2006; Cheng et al. 2001; Lelais and Seebach 2004; Wu et al. 2008; Seebach et al. 1997). Although both β^2 and β^3 -peptides with the same side chains fold into 14-helix, substitution on C^3 position is more stable compared to C^2 due to steric hindrance in the latter case (Seebach et al. 1998). Stabilized by the hydrogen

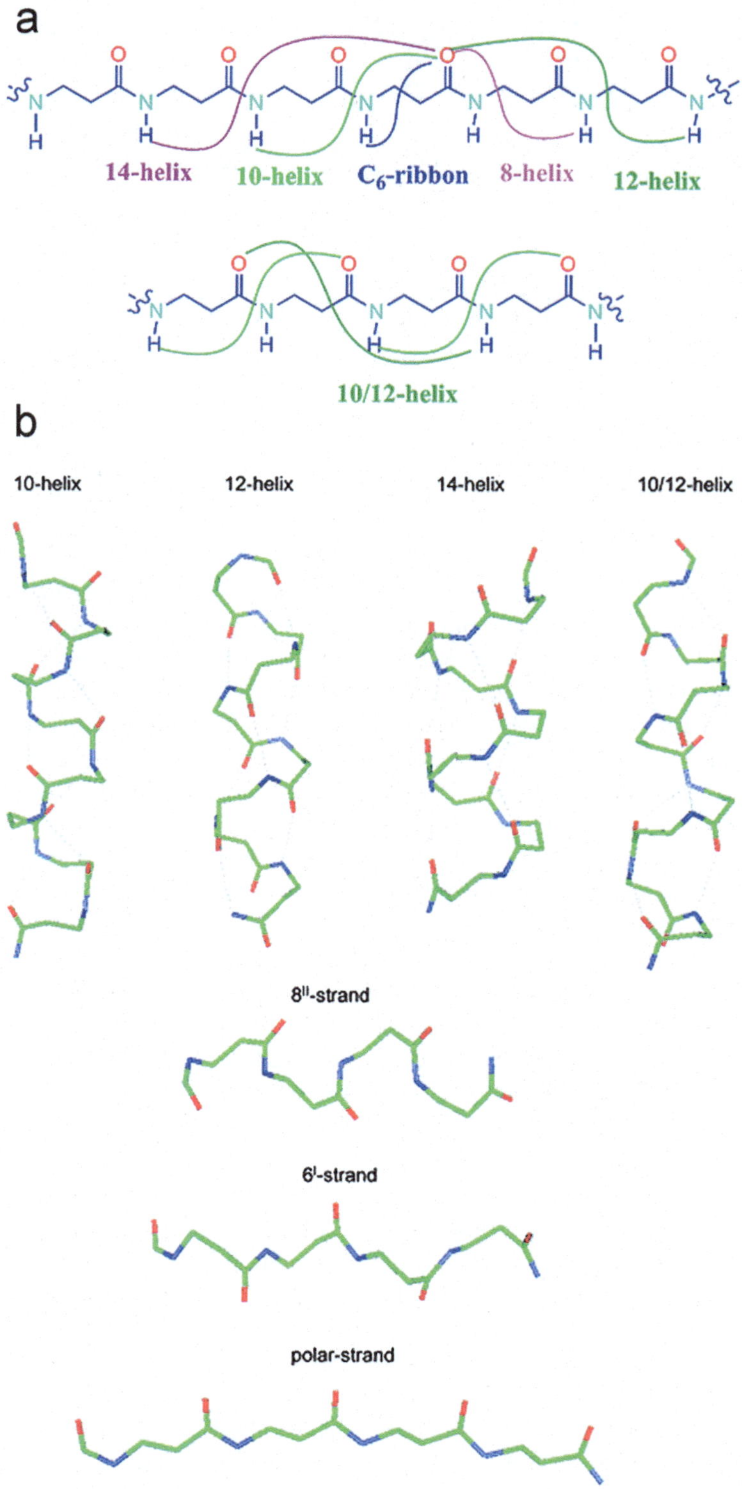

Fig. 4.7 (**a**) Backbone substitution patterns of β-peptides, (**b**) three-dimensional organization of β-peptides ((**a**) Reproduced by permission from Wu et al. (2008). Copyright (2008) American Chemical Society.) (**b**) (Martinek and Fulop 2003)

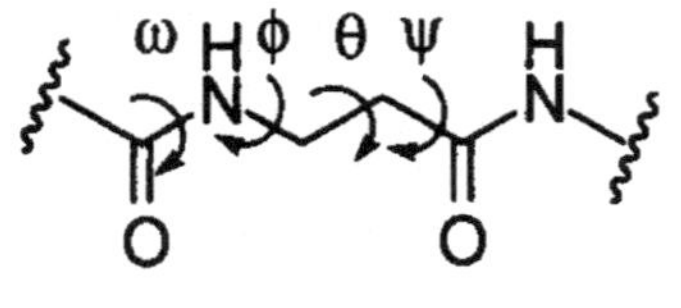

Fig. 4.8 Torsional angles of β-peptide conformations (Reproduced by permission from Cheng et al. 2001. Copyright (2001) American Chemical Society)

Table 4.1 Helical parameter for β-peptides

Helix nomenclature	Rise Å	Radius in Å	Number of residues per turn
14-helix	1.5	2.7	3
12-helix	2.2	2.3	2.5
10-helix	2.3	6	2.6
10/12	2.1	5.7	2.7
8	3	6.7	2
C_6-ribbon	4	6.8	1.7

bond between amide proton donor at position i and carbonyl oxygen at position (i+2) (Fig. 4.7) 14-helix is typically left-handed, although right-handed folding is also possible depending on the stereochemistry of the constituent amino acids (Cheng et al. 2001; Vaz et al. 2008). 14-helix differs from α-helix in having a larger radius (2.7 *vs*. 2.2 Å) (Table 4.1), and the carbonyl groups are directed towards the N-terminus (Goodman et al. 2007). The rise of 14-helix is slightly less than α-helix (a rise per residue of 1.56 Å for β-helix *vs* 1.5 Å for α-helix). While the pitch of the α-helix is 3.6 residues, the pitch of the 14-helix is nearly exactly 3 residues, thus allowing side chains to align at 120°, forming three functionally defined faces of the peptide (Seebach et al. 1996).

14-helices exhibit higher stability than α-analogues due to the propensity of β-substituted side chains that align laterally, i.e. normal to the main helix axis (Seebach et al. 1998; Cheng et al. 2001). The stability of the helices shows a chain length dependence, yet in methanol as few as 6 residues already form stable helices (Cheng et al. 2001; Niggli et al. 2012). For example, hexa $\beta^3(VAL)_2$-OH folds into a stable helix in pyridine and methanol (Seebach et al. 1996); where increasing the temperature (298-393K) has no effect on helix folding. Introducing rigidity in the backbone through cyclization enables even four residue peptides to form stable 14-helices in methanol or pyridine (Barchi et al. 2000): it was demonstrated with NMR that *trans*-2-aminocyclohexanecarboxylic acid (ACHC) tetramer folds into 14-helices in methanol (Barchi et al. 2000).

Helicity is often a surprisingly sensitive function of the chemical environment properties. Alanine, serine, and lysine rich β^3-hepta- and β^3-nona-peptides fold into 14-helix in methanol, yet low or no helicity is observed in aqueous medium (Abele et al. 1998). In polar solvents, the stability of the helix is a function of solvent molecules accessibility to the backbone amide moieties. Water as a highly polar protic solvent competes for hydrogen bonding with amide groups and thus it has a tendency to destabilize helices, unless the internal H-bonding is shielded from direct water access (Appella et al. 1999a). Hence to promote helix formation in water, different strategies have to be followed such as incorporating hydrophobic bulky branched side chains or introducing alternating polar and cyclic apolar residues along the helix in a way that one-third of the structure is hydrophilic and the rest consist of cyclic constrained backbone (Appella et al. 1999a). For example, the peptide that contain repeats of the [*trans*-2-ACHC/ *trans*-2-ACHC/ β^3-lysine] units showed the ability to form stable helix in aqueous solution compared to its isomeric analogues [*trans*-2-ACHC/ β^3-lysine/ *trans*-2-ACHC] that did not have the same helix forming propensity (Appella et al. 1999a; Raguse et al. 2001). At the same time, for this class of peptides there is overwhelming evidence for strong helicity in methanol (Raguse et al. 2002a; Epand et al. 2004).

Peptides containing cyclic constrained backbone have serious limitations in functionalizability (Raguse et al. 2003; Appella et al. 2000; Wipf and Wang 2000). Thus another strategy is used to stabilize helices by connecting the side chains of residues in adjacent turns of the helix either covalently or through salt bridge formation (Lee et al. 2007). The covalent crosslinking will make helices stable in most benign media, including aqueous environment; however this "stitching" is sensitive to the hydrolysis at extreme (both high and low) pHs (Vaz et al. 2008). Salt bridges form between anionic and cationic groups in adjacent turns of the helix. For example, the sequence of residues along the helix of H_2N-β^3(VEOrnVOrnEV)-COOH that contains Ornithine (Orn: a lysine analogue with a methylene removed from the backbone) forms helical turns where the glutamic acid is positioned next to ornithine along the helix permitting ionic interactions and hence salt bridge formation on both sides of the helix (Fig. 4.9) (Arvidsson et al. 2001). If arginine is used instead of ornithine in the sequence, e.g. in H_2N- β^3(VERVREV)-CO-NH_2, a lowered 14-helix propensity is observed. The difference in the length of the side chains may introduce a steric effect even if electrostatic interactions are theoretically possible between the side chains. As a simple way to overcome stability issues, a mixture of methanol and water was shown to enhance the helix stability (Rueping et al. 2004).

The presence of cyclic constrained amino acids at any position can stabilize the helix in the presence of any polar side chain combination. For example, the deca-peptide NH_3-(β^3 Tyr-β^3lys-ACHC-β^3Ser- β^3Ser-ACHC- β^3- β^3Ser-β^3Ser-ACHC- β^3lysine)-CO-NH_2 where the charged residues align at i, i+3 positions, and the non-aligned isomers all exhibit some degree of helicity in water in spite of the absence of cyclohexyl stacking (Lee et al. 2007).

Although many works emphasized the role of side chain branching at C^β position in helix stability, molecular dynamic simulations predicted that a hepta-β^3-peptide with alternating two residue motif alanine and leucine or three residue motif valine, alanine, and leucine would fold into 14-helix although with a lower helix propensity in the absence of valine residues, thus indicating that structural preferences are not affected by the branching at C^β position, although the helix forming propensity maybe lowered in absence of valine residues (Glattli et al. 2004). Hence, the structural role of side chain branching is still not entirely clear.

Although β^2-peptides also form 14-helices, the propensity to do so is comparatively weak when correlating to β^3-peptides; even in methanol the 14-helix is only detected at -20°C (Lelais

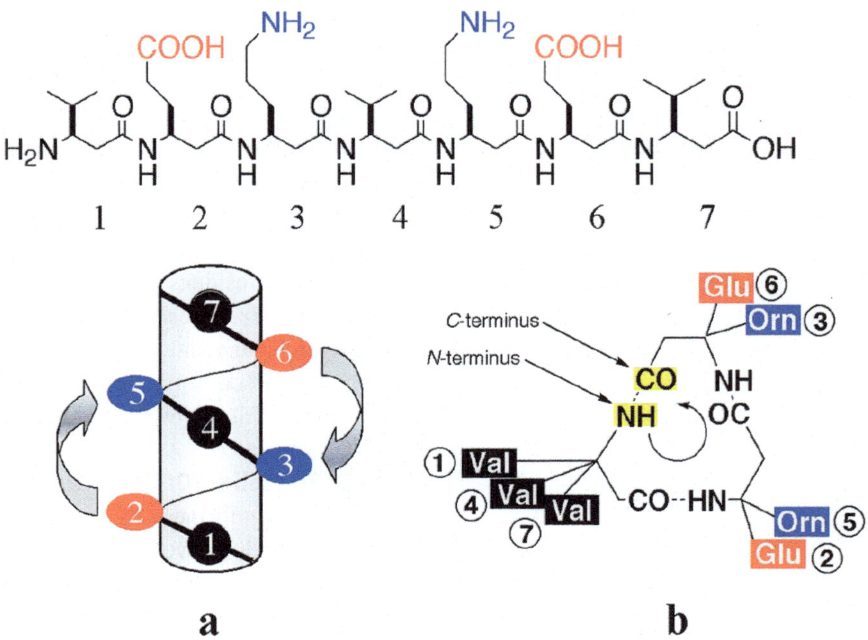

Fig. 4.9 Schematic representation of the pre-organization of side chains that allow salt bridge formation along the same face of the helix (Arvidsson et al. 2001)

and Seebach 2004). Mixed peptides made of β² and β³ amino acids have the ability to form 14-helices, for example the peptide β² (Val-Ala-Leu)-β³ (Val-Ala-Leu) forms 14-helix, yet the same but alternating β² and β³ residues with Boc-protected terminus show12/10 helix formation (Seebach et al. 1997).

4.3.2.2 12-Helix

With two additional methylene groups in the backbone, 12-helix has similar properties to the α-helix (Fig. 4.7). Unsubstituted β-peptides favor folding into 12-helix over any other type of helices, where helices are formed through the intramolecular hydrogen bond between C=O (i) and N-H (i+3) and 12-membered rings can form both in solid state and solution (Cheng et al. 2001). Importantly, even 4- residue peptides were able to show a reasonable population of helices when a water-soluble amino acid is included in the peptide sequence (Cheng 2004). Nevertheless, 12-helix stability is chain length dependent, the intensity of CD spectra of *trans*-ACPC (*trans*-2-aminocyclopenta carboxylic acid) based peptides increases from dipeptide up to hexa-peptide significantly (Appella et al. 1999b). While hexa-cyclization on the backbone (ACHC) of β-amino acids leads to 14-helix formation, penta-cyclization on the backbone (ACPC) yields 12-helix folding (Fig. 4.10) (Appella et al. 1997).

ACPC based peptides are hard to solubilize in aqueous media. A way to overcome low solubility is by cyclization with pyrrolidine rings instead of ACPC. The nitrogens in the rings are protonated thus allowing solubilization of the peptides in aqueous medium. (*2R*, *3R*)-aminoproline-pyrrolidine-based β-amino acids, when incorporated into hexa- β-peptides it promote 12-helix conformations in both methanol and water, with a higher propensity to form 12-helix in methanol (Porter et al. 2002). For example, hepta peptides that contain only amino acids (1*R*,2*R*)-*trans*-2-aminocyclopentane carboxylic acid (tans-2-ACPC), (3*S*,4*R*)-*trans*-3-aminopyrrolidine-4-carboxylic acid (*trans*-3,4-APC) and (*3R*)- β³-lysine (β³-Lys) in different ratios, the right balance of the ratio between the cyclic and acyclic β³-lysine residue can lead to 12-helix formation that is soluble either in methanol only or both methanol and water. The three peptides [tetra (*trans*-2-ACPC)-tri (*trans*-3,4-APC)], [tetra (*trans*-2-ACPC)-di (*trans*-3,4-APC)- (β³-Lys)], and [tetra (*trans*-2-ACPC)- (*trans*-3,4-APC)-di (β³-Lys)] with N-terminal *p*-methoxyphenacyl group all showed the ability to fold into 12-helix in methanol and although they gave weaker helical signal in water, they did show an ability to adopt 12-helix conformation. On the other hand, peptide that contains [tetra (*trans*-2-ACPC)-tetra (β³-Lys)] and N-terminal *p*-methoxyphenacyl group only folds into 12-helix in methanol. Thus, at least five of seven residues have to be constrained for peptide to adopt a 12-helix conformation in water (LePlae et al. 2002).

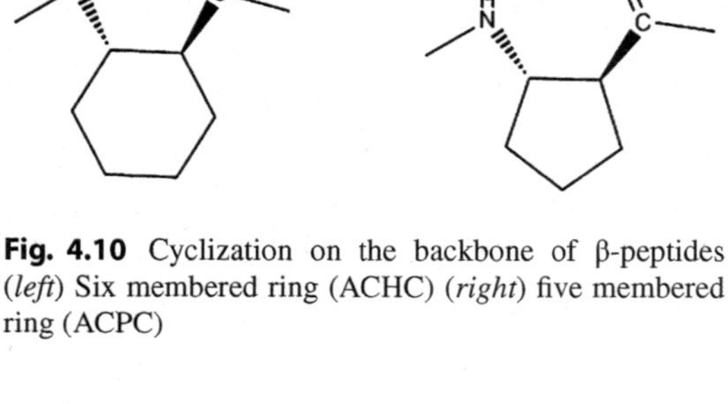

Fig. 4.10 Cyclization on the backbone of β-peptides (*left*) Six membered ring (ACHC) (*right*) five membered ring (ACPC)

Peptides containing cyclobutane constrained backbone also showed the ability to fold into 12-helix similar to amino-cyclopentane containing peptide depending on the number of cyclic residues: at least six residues are needed to enable the peptide to fold into 12-helix, whereas less than six cyclic residues will impart a tendency to fold into 8-helix. Yet, a tetra peptide with alternating cyclobutane and β-alanine residues would adopt 14-helix conformation (Fernandes et al. 2010; Torres et al. 2009; Izquierdo et al. 2004).

4.3.2.3 10-Helix, 10/12 or 12/10-Helix

10-helices are formed by an intramolecular hydrogen bond between i and i+2 residues (Fig. 4.7); the helix has μ-dihedral angle less than 60° (Fig. 4.8). It is a highly strained structure that is usually formed *via* oxetane cyclization of the backbone that stabilizes *cis* conformation (Wu et al. 2008; Claridge et al. 2001). However, 10-helix turns can be stable in combination with 12-turns to form a hybrid helix.

10/12 helix is the most intrinsically stable *unsubstituted* β-peptide helix as it is held by two types of hydrogen bond rings. The carbonyl oxygen C=O is alternatively pointing in opposite directions along the helix (Fig. 4.7) (Lelais and Seebach 2004). The 10-memberd rings are formed between (i, i+2), and the 12-membered rings are formed between (i+1 and i+3). 12/10-helices can also form from substituted peptides in three cases: β^2/β^3 on different residues, β^3/β^3 unlike substitution, and reciprocal substituted β^3/unsubstituted β peptides (Wu et al. 2008). The 10/12-helix has a low helix dipole moment as a result of alternating amide orientation (Cheng et al. 2001). Substitution at C^3 position in the first residue with C^2 or no substitution on the second residue leads to the formation of 10/12 helix, whereas C^2 substitution of the first residue and C^3 substitution of the second residue forms 12/10 helix (Cheng 2004; Rueping et al. 2002). Mixed β-peptides substituted on both 2,3 positions show strong tendency to form 10/12 helix; however 14-helix is formed if the peptides are N-terminally Boc-protected (Cheng 2004).

Peptides that fold into 14-helix can be tuned into 12/10 helix by altering substitution at the termini (Seebach et al. 1997). For example in case of the peptide described above, β^2(Val-Ala-Leu)-β^3(Val-Ala-Leu) with free termini adopts 14-helix conformation, whereas with Boc protected N-terminus and butanedioic acid at the carboxyl terminus the peptide folds into 12/10 helix (Seebach et al. 1997).

4.3.2.4 8-Helix

If the amide carbonyl group adopts gauche conformation, 8-helix can form with about 2 residues per turn with a hydrogen bond between amide i and carbonyl (i-2 (Fig. 4.7) (Abele et al. 1999; Wu et al. 2008). Introducing cyclopropane in the backbone of β-peptide promotes the formation of 8-helix folding. The hyper-conjugation between the cyclopropane and the carbonyl moieties leads to the preference for 8-helix, where the normal of the cyclopropane ring is aligned with the backbone and the C=O of the amide is in *cis* (bisecting) conformation, leading to the increase in the backbone rigidity (Wu et al. 2008; Abele et al. 1999; Seebach et al. 2006). *Trans, cis, and trans, trans* dipeptides based on cyclobutane (ACBC) show preference for 8-helix folding (Torres et al. 2009).

Other than the inclusion of cyclized amino acids, α-aminoxy peptides, in which C^{β} in the backbone is substituted with oxygen, also showed ability to fold into 8-helix (Li and Yang 2006). The preference for 8-helix is explained with the higher rigidity of these peptides in comparison to "regular" β-peptides due to the inclusion of oxygen in the backbone. It can be shown that for such a peptide, the 8 membered ring is the lowest energy structure due to the presence of stable N-O turns. Importantly, hydrophobic cyclic $\beta^{2,3}$-aminoxy residues with polar cyclic residues within the same peptide are not only water-soluble but also exhibit stable secondary structures in both methanol and aqueous medium in a wide pH range (Hao et al. 2015).

4.3.2.5 C_6-Ribbon

A theoretically predicted structure, C_6 ribbon is an intermediate structure between a helix and a sheet; it is depicted as a helix in (Fig. 4.7) (Wu et al. 2008). The 6-ribbon folding has about the same (calculated) energy in the gas phase as the 8-helix (Wu et al. 2008).

4.3.2.6 β Sheet-Like Structures (Parallel Sheets)

β-sheets result from different stereochemistries of the substitution on $\beta^{2,3}$ carbons, that is, (*2R*,3S) or (*3R*,2S) (Cheng et al. 2001; Wu and Wang 2000). The formation of β-sheets is associated with the existence of a substitution in forbidden axial position thus preventing β-peptides from folding into helices; instead β-like sheets are formed (Wu et al. 2008; Seebach et al. 2004). These sheets have anti C^2-C^3 torsion angle. Hence, unlike β-sheets of α-amino acids, the β-peptide carbonyl oxygens are oriented in one direction, thus providing the sheets with a net dipole moment (Fig. 4.11 top) (Krauthauser et al. 1997).

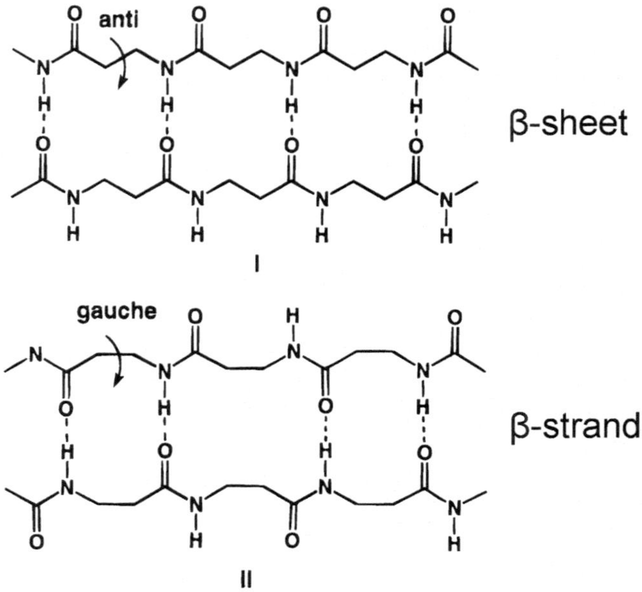

Fig. 4.11 *Top*, β-sheet with anti C^2-C^3 torsion angle, and *bottom*, β-strand with gauche C^2-C^3 torsion angle (Reproduced by permission from Krauthauser et al. 1997. Copyright (1997) American Chemical Society)

Similarly to α-peptides, the sheets may form by the parallel or antiparallel alignment of the strands. In planar β-strands the backbone has a "concave" geometry (Fig. 4.11 bottom), leading to an overall twisted geometry similar to α-peptide sheets. Nevertheless, β-strands have fewer twists than α-strands (Fig. 4.11 bottom) (Langenhan et al. 2003). While anti C^2-C^3 torsion has a net dipole moment due to carbonyl groups oriented in the same direction, gauche C^2-C^3 torsion lacks the net dipole as the carbonyl groups in that case orient in different direction along the strand.

4.3.3 Peptoids

Peptoids are a subclass of β-peptides that lack hydrogen atom donors in the backbone of the structure. The hydrogen of nitrogen amides are substituted with alkyl groups instead of substituting C^α or C^β side chains (Skovbakke et al. 2015; Reyes et al. 2015; Stephens et al. 2005), thus there is no possibility for an intramolecular hydrogen bond in the backbone. Peptoids fold into helix conformation as a result of side chain steric interactions. The tertiary amide in the backbone increases peptoid flexibility thus electronic n-π* interactions due to the presence of bulky cyclic side chains which offers a way to stabilize the secondary structures (Laursen et al. 2013, 2015; Olsen et al. 2008; Roy et al. 2008). The N-alkylation provides peptoids with resistance towards enzymatic degradation (Olsen 2010; Norgren et al. 2006; Roy et al. 2008). Peptoids up to 48 residues length can be synthesized using automated solid-phase protocol (Kirshenbaum et al. 1999; Olsen 2010). Peptoids form well-defined tertiary structures that enable them to mimic bioactive peptides (Laursen et al. 2015; Roy et al. 2008; Scott et al. 2008).

4.3.4 Hybrid α/β Peptides

The initial motivation for the design of hybrid peptides was to stabilize α-peptide turns through incorporation of unnatural *cis* or tarns-β-ACHC or *cis* β-ACC (Urman et al. 2007; Strijowski and Sewald 2004). Although incorporation of few β-amino acids did not alter the α-peptide conformation, increasing the number of β residues leads to the discovery of new folding conformations such as 13, 11 mixed 14/15, and 9/10 helices (Fig. 4.12) (De Pol et al. 2004; Baldauf et al. 2006).

Hybrid peptides can either adopt helical folding in which the hydrogen bond is pointing towards the N terminus such as 11 and 14/15 helices, or in which hydrogen bonds point

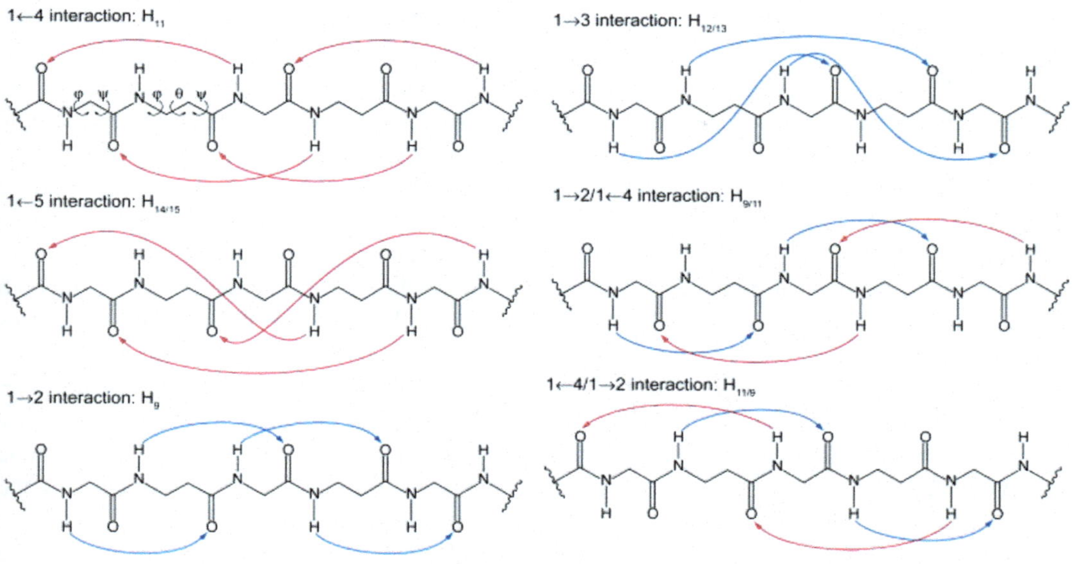

Fig. 4.12 Different types of helices formed by hybrid peptides, arrows indicate the directions of hydrogen bonds (Baldauf et al. 2006)

towards the C terminus such as 9 and 12/13 helices. In polar solvents hybrid peptides show stable 11/9 helix conformation and moderately stable 11 and 14/15 helix conformations. In all of the stable helices, the hydrogen bonds are either pointing towards the N-terminus or are alternating (Baldauf et al. 2006). Nevertheless, the major advantage of these new conformations is that it reduces the number of residues needed for the formation of stable helices in polar solvents to ~ 7 residues, compared to natural peptides that typically require at least 13 α-amino acid residues to form stable helices in solution (De Pol et al. 2004; Crisma et al. 2006).

If α-peptides form a hybrid with peptoids, a different family of hybrid peptides, "peptomers" is formed; these exhibit unusual extended conformations similar to protein catalytic centres and binding sites (Butterfoss et al. 2014).

4.4 Factors Affecting β-Peptide Folding

There is no clear consensus on the factors affecting β-peptide folding, however there are some known features characteristic of the helical conformation. In the following part some of the possible factors will be discussed in terms of introducing specific residues or a sequence motif.

14-helix is the most extensively studied among all β-peptides. Generally solvent environment has a complex effect: it weakens hydrogen bonding directly through its dipole moment and indirectly through its dielectric constant (Wolfenden 1978). If the polar solvent has a strong dipole moment, the solvent molecules may compete for the peptide hydrogen bond. Folding of peptide in water is a particularly challenging problem since the association of water to the carbonyl oxygen of the amide bond is stronger than its interaction with N-H, thus peptides should dissolve, and unfold, in water if water has direct access to the amide moieties (Ralston and De Coen 1974; Wolfenden 1978). Other than the solvent, the presence of polar residues also affects peptide folding as these may also hydrogen bond to the backbone (Ralston and De Coen 1974).

As detailed above, the folding of β-peptides depends on steric factors, and the strength of the intramolecular hydrogen bonds involved in the helix folding. Thus, helical β-peptide folding is promoted by the presence of branched bulky side chains; in contrast, in α-peptides non-branched side chains are preferred, and short side chains in particular. In both cases, helix folding is promoted by non-polar side chains (Padmanabhan and Baldwin 1991; Padmanabhan et al. 1990; Kritzer et al. 2005). The β-helix forming propensity in order of the substituted residues is β^3Ile > β^3Val ≈ β^3Ser ≥ β^3Tyr > β^3Ala ≥ β^3Leu (Kritzer et al. 2005; Abele et al. 1998).

Short polar side chains in the middle of the protein sequence interfere with helix formation by donating their hydrogen bond to the carbonyl group at the backbone of the structure, thus reducing the helix-forming propensity (Vijayakumar et al. 1999). A comparison between the effect of polar and charged side chains using serine and lysine residues in a hepta β^3-peptide (β^3Phy- β^3Ala- β^3X- β^3Phy- β^3Ala- β^3X- β^3Phy where X is Lys or Ser) showed that lysine disturbs the 14-helix formation much more than serine (Abele et al. 1998). The effect of serine residue depends on its position in the sequence. Serine can form intramolecular hydrogen bonds to available carbonyl oxygens of the backbone and is likely to stabilize the 14-helix content in water except when placed close to the N-terminus (Kritzer et al. 2005). Threonine has higher stabilizing effect, especially close to the termini; branching at C^β position and the resulting conformational flexibility may explain this difference.

Charged residues do not interfere directly with hydrogen bonding; electrostatic interactions can even promote helix folding due to the formation of salt-bridges between opposite charges, whereas inclusion of single charged residues may increase or decrease stability, depending on their charge and position in the sequence. For example, the presence of both glutamic acid and lysine decrease helicity in water (Rueping et al. 2004; Raguse et al. 2003) however lysine has the highest destabilizing effect at the C terminal region, whereas glutamic acid at the N-terminal region. The cause of the difference is unclear and it may be related to solvation effects of the opposite charges (Kritzer et al. 2005).

Thus, the known design rules can be summarized as follows. β-helices and α-helices are both stabilized by hydrophobic residues, α-helices prefer linear side chains, whereas β-helix stability is promoted by branching at the C^β position. Bulky side chains reduce the backbone flexibility and act as a shield protecting the amide backbone from the solvent interaction, thus promoting the helix stability. Introducing polar side chains have a weakening effect on the helix stability, yet the effect of polar residues differs according to their charge (- or +), the position from the peptide termini, the presence or absence of opposite charges, and the net charge. The presence of oppositely charged residues can stabilize the helix through the formation of salt bridges, positively charged residues are preferred to be near the N-terminus and negatively charged residues near C-terminus to promote helix stability.

4.5 Self-Assembled Oligopeptide Superstructures

The design of self-assembled peptide biomaterials starts with choosing a stable folding for the monomers, following well-established design rules as outlined above. These building blocks interact *via* second order interactions such as hydrogen bonding, hydrophobic, and van der Waals interactions to form superstructures that have the potential to create three dimensional networks (Boyle and Woolfson 2011; Tegoni 2014; Greenwald and Riek 2012). Peptide assemblies can be classified according to function, chemical structure, or usage (Castillo-León et al. 2011). The design of the peptide determines the way of assembly; the residues used in the peptide sequence provide each peptide with a unique character. Apolar amino acids such as alanine, isoleucine, leucine, methionine, or valine ensure that the self-assembly in polar solvents is driven mostly by the hydrophobic/solvophobic effect. Aromatic groups introduce π-π stacking interactions between the side chains. Hydrophilic amino acids, beyond increasing solubility, may also stabilize the structure through salt-bridge formation.

Self-assembled materials can be functionalized *via* incorporating specific residues in the monomer. For example, cysteine residues may form disulfide bonds that allow the chemical crosslinking of peptides, while also giving the structure a unique chemical reactivity to bind to noble metals (Durand et al. 2015; Chalker et al. 2009). Histidine residues coordinate to metal ions (Zastrow and Pecoraro 2013; Delort et al. 2006). Hydrogen bonding side chains: alcohols such as serine and threonine, or amides such as asparagine and glutamine can physically crosslink peptides. Charged residues often have the opposite effect reducing or prohibiting assembly through charge repulsion. These design rules, when implemented parallel to the folding criteria discussed above, give enormous flexibility in designing peptide based self-assembling materials (Gellman 1998; Appella et al. 1997, 1999a).

The self-assembly motifs can range from a simple amphiphilic character to specific and selective supramolecular recognition motifs. The design requirements towards the self-assembly motifs depend on the target geometry. Hence, in the following section some common superstructure morphologies are reviewed.

Self-assembly motifs for one dimensional structures must be geometrically defined: either hydrogen bonding patterns or specific hydrophobic non-selective interactions (Mandal et al. 2014). The fibres further interact through van der Waals and/or other secondary interactions forming three-dimensional structures, the simplest of which are bundles (Mandal et al. 2014). Fibres are ideal structural elements for building higher order superstructures: directional lateral interactions on opposite sides of fibres could yield two-dimensional ribbon-like structures or sheets, whereas geometrically controlled crosslinking has the potential to form controlled three dimensional networks.

Self-assembly of monomers in a longitudinal direction results in fibres. While typically only few nanometers in diameter, polypeptide fibres can grow to hundreds of micrometers in length. The best known natural fibres are worm and spider silks that are strong flexible fibres growing to meters length from beta sheet protein building blocks forming the cocoon, or web, of the silk-spinning species (Zhang 2002). Amyloidic (starch-like) fibres are pathologic structures, responsible for Alzheimer's and Parkinson's diseases (Majd et al. 2015; Breydo and Uversky 2015; Verma et al. 2015; Chiti and Dobson 2006) that also exhibit complex β-sheet structures of high stiffness and rigidity (Matsuurua 2014). Actin filaments, keratin, microtubules and similar fibrous structures are all self-assembled from specific precursor molecules in a selective and exclusive process (Guthold et al. 2007; Schaller et al. 2013; Courson and Rock 2010; Subramanian and Kapoor 2012). Typically hydrophobic, hydrogen bonding and π-π stacking interactions stabilize these fibres (Cinar et al. 2012; Bowerman and Nilsson 2012; Dehsorkhi et al. 2014; Fitzpatrick et al. 2013). Of these, naturally occurring amyloid fibers are among the best-defined biological self-assembled superstructures. The amyloid assembly is associated with the presence of large aromatic moieties suggesting that an important role is played by aromatic interactions in establishing the order and directionality of fibre formation (Fig. 4.13).

In amyloid sheets the strands are bound together by a hydrogen bonding network and it is generally believed that the sheets interact further *via* the aromatic π-π stacking of phenylalanine side chains aligned normal to the plane of the β-sheet (Gazit 2002, 2005; Bemporad et al. 2006; Profit et al. 2013; Cukalevski et al. 2012).

Yet, it was also argued that the amyloidic structure is an alternative conformation of any polypeptides and is not dependent on the peptide sequence (Fandrich and Dobson 2002). The α-helical "folded" protein structure and the amyloidic form are different energy minima; the thermodynamic stability of each is determined by the

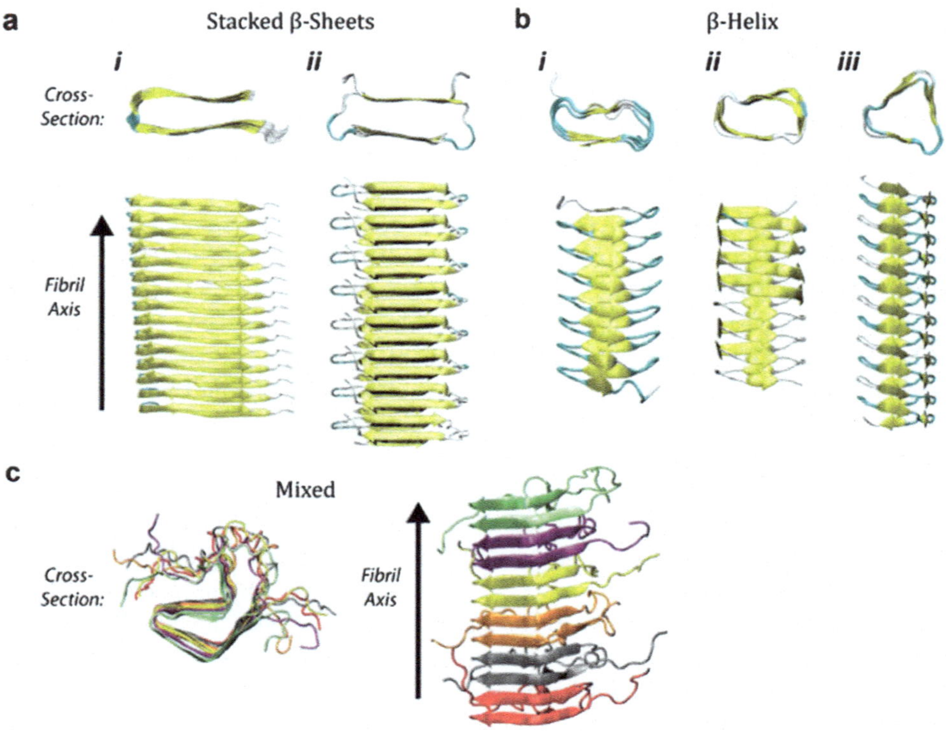

Fig. 4.13 Schematic of the amyloid beta stacking structure (Reproduced by permission from Solar and Buehler 2014 © IOP Publishing. Reproduced with permission. All rights reserved)

physico-chemical conditions such as pH and temperature (Fandrich and Dobson 2002; Chiti et al. 1999; Invernizzi et al. 2012).

Many aromatic-rich amyloidic proteins are implicated in pathogenic processes. The aromatic moieties promote the aggregation of proteins into the characteristic fibrous deposits. It is believed that increasing aromatic interactions increase the protein aggregation and cell toxicity. Alzheimer's, Parkinson's, and Huntington's diseases have been linked to an imbalance in the β-sheet production and clearance of the aggregates by the circulation, thus the accumulation of excess amyloidic proteins results in the pathologic effect (Zhao et al. 2012; Zhang et al. 2016; Sun et al. 2015).

While involved in the pathophysiology of the mentioned diseases, amyloid fibres are generally inert, metabolically and chemically resistant and hence frequently considered for applications in materials science. Yet, these peptides are large and the potential toxicity has not been dismissed completely. Hence, alternative ways are sought to design fibrous assemblies.

Tubular structures are a special form of one-dimensional material, exhibiting higher rigidity than fibres, with a potential use of the hollow center for the storage/delivery of a payload. Tubes are usually formed *via* hydrogen bonding or π-π stacking of aromatic short peptides (Ulijn and Smith 2008). Short structural elements often suffice to build nanotubes, for example, diphenylalanine peptides assemble into stable nanotubes that were used in nanowires applications (Hauser et al. 2014).

A special case of self-assembly is that of collagen fibres. Collagen attracted attention because in biology it is a major component of the extracellular matrix and connective tissues. It is formed from three left-handed identical protocollagen helices that are aligned along the same axis forming right-handed coiled coil super-helix; the coiled-coil assembly further bundles forming the collagen structure. Protocollagen helices usually contain glycine as each third residue that implements the collagen-folding motif; a collagen-like peptide has to contain the same motif (Woolfson

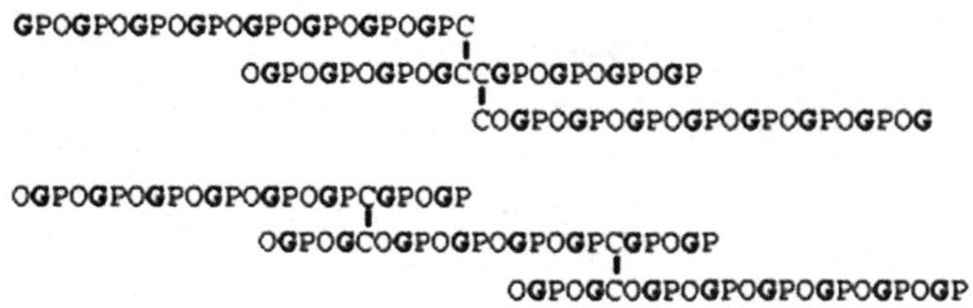

1

2

Fig. 4.14 The two crosslinked collagen structures using cysteine linkers (Koide et al. 2005)

2010). With a repeating residue sequence of X-Y-Gly, such as (Pro-Hyp-Gly)$_{10}$, the triple helix peptide assemble to branched filamentous structures through hydrophobic interactions driven by their alkyl chains (Fig. 4.14) (Koide et al. 2005).

Cysteine residues can be used as strengthening linkers due to cysteine ability to form disulfide-bridge thus linking the strands. The peptides normally entangle into a hydrogel but heat treatment promotes alignment into the stable triple helix upon cooling (Koide et al. 2005). A collagen-like peptide might be used as tissue scaffold in tissue engineering (Li and Yu 2013). The triple helix is able to encode insoluble cellular signals in synthetic gels (Stahl and Yu 2012). Encoding of G_3-$(POG)_8$ to PEG-based hydrogels through physical cross-linking can result in a scaffold with spatial patterns and gradients for cell-instructive cues.

4.6 Engineering Unnatural Supramolecular Structures

The previous examples have described artificial implementations of essentially biological structures, that is, bioinspired/biobased materials design. Deviating from these natural patterns is challenging as the design rules have to be invented, they cannot be copied from nature. Most of the work done on unnatural peptide self-assembly patterns has been done on amphiphiles; these are often based on ionic complementarity design principles thus this is discussed first.

One-dimensional self-assembly can be achieved in a fully artificial system by using ionic self-complementary peptides. These contain a repeated pattern of charged residues, where eliminating electrostatic repulsion happens through a distributed multipole interaction (Yang et al. 2009). This is often combined with overlapping hydrophobic pockets along the structures due to presence of hydrophobic residues placed between the charged residues (MacPhee and Woolfson 2004; Yang et al. 2009). The resulting ionic self-assembly leads to a ribbon-like stacked structure where the peptides align parallel to each other following a combination of hydrophobic and charge-complementary interactions: with one side of the assembly containing the charged residues, the other side is fully hydrophobic, thus the hydrophobic sides adhere forming a dimer. The ionic interactions between the charged residues on the other side of the structures result in further packing of the dimers, forming nanofiber scaffolds and gels (Fig. 4.15) (Zhang et al. 1995; Fung et al. 2003; Yang et al. 2007a, b, 2009).

4.6.1 Amphiphiles Based on α-Peptides

By far the largest number of works on oligopeptide design involve amphiphilic structures. Amphiphiles are molecules that contain both hydrophobic and hydrophilic zones. The molecules aggregate in a way that the hydrophobic moieties minimize contact with solvent molecules whereas the hydrophilic zones are solvated (Zhang and Wang 2011). The hydrophilic zones of amphiphilic peptides are implemented with charged and/or polar residues whereas the hydrophobic zone can be composed of hydrophobic amino acid residues or an aliphatic chain (Hamley 2011).

Assembly of amphiphiles is governed by two main driving forces: hydrophobic interactions between the apolar parts and electrostatic/hydration repulsion between the charged amino acid parts. The assembly can be tuned by changing the

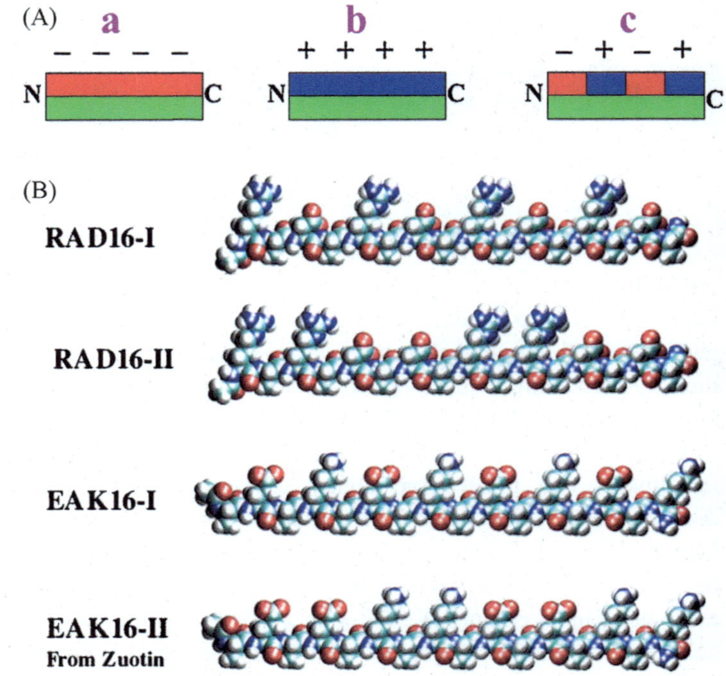

Fig. 4.15 Designing amphiphilic peptides that can further assemble into fibrous scaffolds (Yang et al. 2009)

spatial distribution and strength of hydrophobic and hydrophilic moieties. In pure hydrophobic interactions, only aggregates would form. Hydrophobic interactions can be weakened by introducing hydrophilic groups that are either charged or that form hydrogen bonds to water; this typically yields spherical micelles although the assembly is dependent on the molecular geometry. If the hydrophobic regions are bulky, regular micellar assembly is seldom possible. An increasing number of hydrophilic groups can also break the spherical geometry, forming long one-dimensional cylindrical micelles.

Given that protonation state of the "charged" residues has a substantial effect on amphiphilic character, the self-assembly of amphiphiles usually happens under a specific condition defined by pH, temperature, and the ionic environment. Unlike small molecule surfactants, in peptide amphiphiles the distribution of the hydrophobic/hydrophilic residues is varied, such as hydrophobic domains at both termini, alternating hydrophobic/hydrophilic residues or just one hydrophobic domain at one terminus, which provide a broader range of self-assembly motifs. Furthermore, peptide amphiphiles can have an additional zone of interaction: between the hydrophobic "tail" and hydrophilic "head" regions there is often a hydrogen bonding segment that alters the dominant form of self-assembly from micelles to sheet-like structures with the hydrogen bond linking the monomers. Conversely, the length of the hydrophobic tail mediates the curvature of the micellar assemblies (Aggeli et al. 2001). Since amphiphilic self-assembly has been extensively studied, a classification system exists.

Type I is the complementary ionic self-assembly motif: modules formed by alteration of anionic (–) and cationic (+) amino acid residues define the peptide properties. Positively charged lysine and the negatively charged glutamate are ideal building blocks. In this class the geometrical matrix can be tuned through the organization of the charged residues into "modules": module 1–+–+–+, module 2 ––++––++, and module 3 –––+++ (Zhang 2002; Zhang et al. 1995; Holmes et al. 2000). For example, VKVVKVKVKVK–VDPPT–KVKVKVKV-$CONH_2$.forms hydrogel at basic condition whereas Ac-QQRFEWEFEQQ-$CONH_2$ forms short fibrils with nematic properties (Koutsopoulos 2016).

Type II are switchable peptides that change conformation between β-sheets and α-helices as a function of environmental stimuli. These are also ionic complementary peptides containing the same alternating peptide residues but with a cluster of negatively charged residues at the N-terminus and a cluster of positively charged residues at the C- terminus. Peptides of this type are β-sheet formers at ambient temperature, whereas upon increasing the temperature the β-sheets break up and stable α- helical structures are formed instead (Zhang 2002; Zhang and Rich 1997; Altman et al. 2000).

Type III are beta sheet based amphiphilic peptides that can be designed to assemble in a surface-templated pattern. Such "molecular paints", although not amphiphilic peptides in the strictest sense, are nevertheless regularly included in this category (Zhang 2002). Molecular paints are based on a three zone peptide: (i) an anchor that is responsible for the formation of a covalent bond with the surface; (ii) a spacer that is responsible for the interactions between the molecules, controlling the flexibility and/or rigidity of the structure; (iii) a functional group recognized by cells or biological molecules. This type of self-assembled peptides enables engineering functional surfaces thus can be used for example in biomedical engineering (Zhang 2002)

Type IV are surfactant-like peptides. The self-assembly of surfactant-like molecules depends on the overall molecular geometry or shape factor (Fig. 4.16): if it is conical, the ideal assembled geometry is a spherical micelle with the hydrophilic residues facing outwards; if the molecular geometry is cylinder-like, a bilayer sheet is formed that may further roll into tubes or vesicles (Fig. 4.16).

Surfactant-like peptide designs rely on a sequence of 4 to 8 hydrophobic amino acid residues to the analogy of the "tail" groups of surfactants to drive self-assembly. The head group of these peptides is typically a short hydrophilic amino acid sequence of one or two residues. Surfactant-like peptides can assemble through tail-to-tail packing in parallel with intermolecular hydrogen bonding along the backbone, forming bilayers that further roll into either nanotubes or nanovesicles. Another possible way of surfactant-like peptides to self-assemble is a micelle-like geometry where the hydrophobic amino acid tails are directed towards the core and hydrophilic head groups forming the outer surface of the spherical or cylindrical structures. This latter type of surfactant-like peptides may have cationic or anionic head groups, such as peptide A_6K that can assemble into nanofibers, nanorods, and nanospheres depending on pH environment; mixtures with anionic surfactant-like peptide A_6D can form longer ordered nanofibres due to the formation of ion pairs (Tang et al. 2013; Qiu et al. 2009).

Micelle-forming molecules should have a purely conical geometry (Fig. 4.16); in practical terms, it is seldom achievable with peptide amphiphiles. Thus, the self-assembly of Type IV peptides more frequently leads to bilayer structure, that forms nanotubes or nanovesicles (Tang et al. 2013). The use of alanine residues as a tail leads to more stable structures due to the strong hydrophobic interactions compared to other hydrophobic residues such as glycine (Tang et al. 2013). The geometrical shape of surfactant-like peptides can be designed by choosing amino acids of different size to achieve geometrical compatibility, for example, while $AVK^{\pm}$ (Ac-AAAVVVK) had wedge-like shape, $AGK^{\pm}$(Ac-AAAGGGK) is an inverted wedge, and $A^6K^{\pm}$ (Ac-AAAAAAK) has a straight shape (Tang et al. 2013; Chen et al. 2009).

"Single tail" peptide-amphiphiles (Paramonov et al. 2006) belong to the family of surfactant-like peptides; in this case, the peptide includes an actual aliphatic tail, while hydrophobic residues act as a linker. The self-assembly in this type is not constrained by the bulk of the hydrophobic tail and thus micellar aggregates are more regularly achieved. If the hydrophilic areas of the peptide are accessible to water in the micellar assembly, hydrogen bonds can form to the solvent from peptide regions closer to the micelle core, and that may twist the hydrophilic groups towards the solvent, leading to loser, weaker assembly (Paramonov et al. 2006).

Reducing the bulk of the peptide sequence offers a way to design cylindrical molecular geometries to form stable flat lamellas bilayer assemblies (sheets). If a surfactant-like peptide is

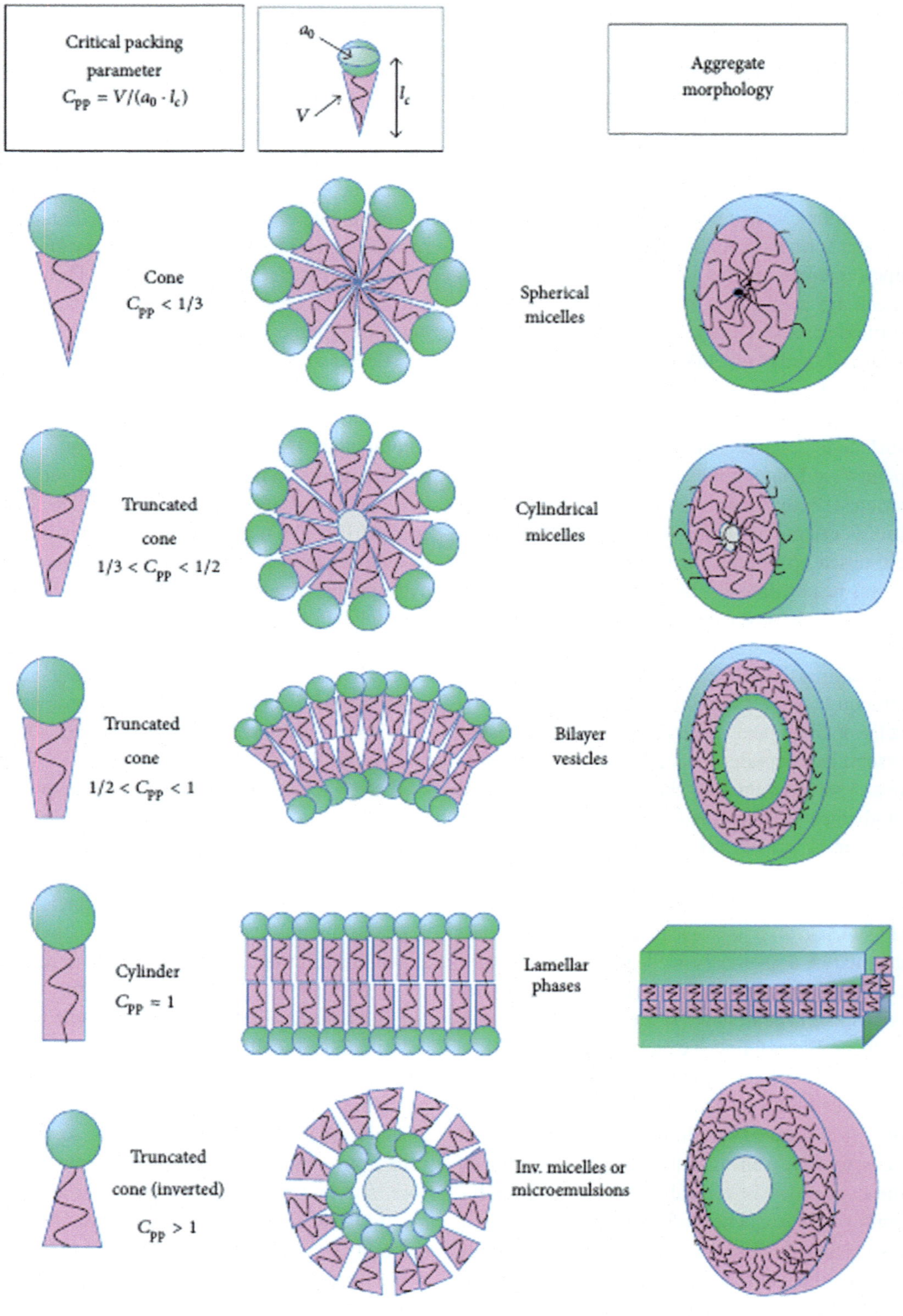

Fig. 4.16 Factors that affect geometrical shapes of amphiphile (Reproduced by permission from Lombardo et al. 2015. Copyright (2015) Hindawi)

designed from valine alternating with hydrophilic residues, for example C_{16}VEVE is introduced to an aqueous medium, valine surfaces associate with each other in order to minimize surface solution exposure. By that, a dimeric structural unit is formed with two hydrophobic tails. Thus, in an analogy to phospholipid membranes, flat bilayer forms instead of cylindrical nanofibres (Versluis et al. 2010; Cui et al. 2010). In contrast, cylindrical nanofibres are formed if the two valine residues are adjacent, e.g. C_{16}VVEE (Cui et al. 2009, 2010).

Amphiphiles form gels when the headgroup charges are screened, so by that amphiphiles would assemble into nanofibres or cylindrical structures (Hartgerink et al. 2002; He et al. 2014). In the absence of charge screening, the result of self-assembly in polar solvents yields micelles (Higashi and Koga 2008).

Short aromatic peptide sequences may also lead to the formation of nanotubes and vesicles (Reches and Gazit 2003, 2004; Zhang et al. 2003) through π-π stacking effects that imposes a supramolecular character on the amphiphile self-assembly (Zhang et al. 2003).

In case the hydrophilic head contains charged residues, the protonation state and hence the self-assembly is sensitive to pH: if the net charge is positive (basic residues), the self-assembly is preferred at acidic pH and vice versa for net negative charge. As most biological applications require that the peptides self-assemble at neutral pH, the challenge of delivering biological signal using such a system can be overcome by using mixed amphiphiles of opposite headgroup charges. In such a system, it was empirically shown that self-assembly takes place at neutral pH, thus it is possible to carry signal at neutral pH (Niece et al. 2003).

Peptide amphiphiles formed the basis of the first successful attempts to form biocompatible, highly ordered hierarchical structures and hence they have been used in biomedical applications such as bone and enamel regeneration (Mata et al. 2010; Huang et al. 2008, 2010; Sargeant et al. 2008), bio-regenerative medicine (Mart et al. 2006; Stupp et al. 2005; Stephanopoulos et al. 2013) and cancer therapy (Hartgerink et al. 2001; Mata et al. 2010; Standley et al. 2010). In an analogous application to regular surfactants, it is also possible to use peptide amphiphiles for dispersing hydrophobic solid phase, such as carbon nanotubes, in water; the peptide amphiphiles assemble on the surface of the hydrophobic particles (Arnold et al. 2005; Andersen et al. 2013).

4.6.2 Self-Assembly of α-Helical Building Blocks

Not much work has been done on the self-assembly of α-helical peptides due to a lack of suitable self-assembly motifs. α-helices are stabilized by local backbone hydrogen bonds with all side chains exposed along the helix; thus designing functional nanostructures from α-helix would be geometrically better defined than β-sheets where side chains mostly participate in/ hidden by the sheet stacking (Bromley and Channon 2011). Arranging α-residues to produce helices where all polar residues align on the same face of the helix yield amphiphilic helices, the most common motif for the self-assembly of helices into coiled-coil superstructures (Loo et al. 2012; Lupas 1996; MacPhee and Woolfson 2004). Either parallel or antiparallel, a number of helices twist into bundles with one of the helix side chains positioned in the core of the bundle leading to a tight packing. Consistently, the formation of coiled-coil superstructure requires optimal side chain interactions thus the side chains have to be at equivalent positions along the helix, that is typically achieved with a repeat pattern (Lupas 1996). Because α-helix contains 3.6 residues in each turn, hepta-repeat is required that ensures alignment of the residues after two turns. The rigidity of the coiled-coil structure allows for designing longer assemblies. Coiled-coil molecular recognition can occasionally form fibrillar filaments such as spectrin proteins that have the ability to assemble into two- and three- dimensional scaffolds (Herrmann et al. 2007). Helical peptides based on a charge complementary motif can also form a staggered coiled coil structure allowing an overlap between dimers, hence allowing for fibre formation (Pandya et al. 2000). If more than two helices form a parallel assembly through amphiphilic or charge complementary motif, nanotubes/barrels can be formed which might further assemble end-to-end into up to 100nm-long superstructures (Thomas et al. 2016; Thomson et al. 2014).

4.6.3 Amphiphiles Based on β-Peptides

As detailed before, β-peptides can adopt well-defined conformations and hence it is easy to implement similar amphiphilic motifs to the α-peptides. The assembly of amphiphilic β-peptides yields distinct structures that are different from the structures described in the previous section.

β-peptide amphiphilies are based on the 14-helical folding, where the chemistry of the three faces is controlled. The peptides can be globally amphiphilic when containing one hydrophilic face and two hydrophobic faces or non-globally amphiphilic if altering between hydrophobic and hydrophilic faces along the helix (Fig. 4.17) (Raguse et al. 2002b; Mondal et al. 2010a, b, c). The stability of the helix in the solution is maintained through introducing cyclo-

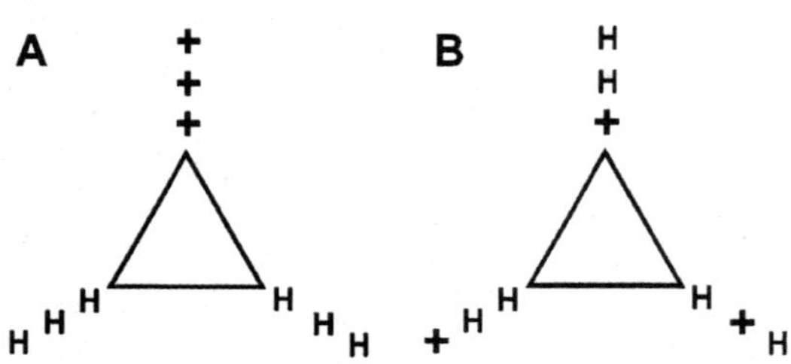

Fig. 4.17 Schematic representation of substitution patterns of amphiphilic 14-helix forming peptides suggested by Gellman et al (Reproduced by permission from Raguse et al. 2002b. Copyright (2002) American Chemical Society)

hexane constrained residues in the backbone (Raguse et al. 2001, 2002b; Mondal et al. 2010a).

Depending on the substitution pattern, β-peptides can assemble into either hollow cylinders or smaller globular structures (Mondal et al. 2010b). The nanopatterns of amphiphiles are easy to predict due to the aligning of side chains of β-peptides. Thus altering the position of hydrophilic groups along the parallel of the helix between two isomers would affect the lyotropic liquid-crystalline phase of both isomers (Pomerantz et al. 2008). The strength of the intermolecular interactions in this case is a good predictor of the three-dimensional self-assembly patterns even for isomeric peptides (Acevedo-Velez et al. 2011). A detailed study on the influence of aligning the functional groups along the helix confirmed that such a design allows for stacking the molecules. For example, when comparing a globally amphiphilic nonapeptide to its non-globally amphiphilic isomer (Fig. 4.18), in the first case the face of the like side chains interact to yield a free energy minimum. Hence, four molecules align parallel to each other forming rings that further stack forming hollow cylinders that grow into liquid crystals. In the isomer, while a tetrameric assembly is also possible, stacking would lead to one of the molecules turning perpendicular to the rest so the stack that is not favored energetically[206].

The motivation for much of the work described above is to design antimicrobial peptides, to the analogy of natural membrane disrupting peptides; the superstructures are not essential for such a role. In terms of antimicrobial action, the tubular structures were supposed to act as transmembrane channels; it was found that amphiphilies that do not possess net positive charge fail to act as antimicrobial peptides (Arvidsson et al. 2005). Although increasing chain length correlates to increasing antimicrobial activity (Hamuro et al. 1999), increasing 14-helix content does not influence the antimicrobial activity of these β-peptides (Raguse et al. 2002b).

The well-defined folding geometries and the structural, metabolic, and chemical stability of the secondary structure makes β-peptides well suited to replace natural proteins in many appli-

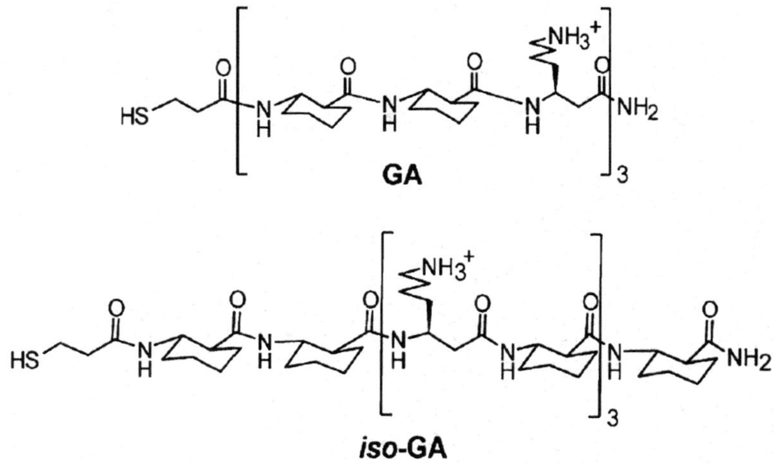

Fig. 4.18 Peptide amphiphiles proposed by Gellman et al. *top*, globally amphiphile peptides; *bottom*, non-golobally amphiphile peptides (Acevedo-Velez et al. 2011. Copy-right (2011) American Chemical Society)

cations such as antimicrobial polymers or as peptidomimetic oligomers (Tew et al. 2010; Patch and Barron 2002; Scott et al. 2008; Robinson 2011). Accordingly, an interesting application of the assemblies of amphiphilic β-peptides described above is as antimicrobial agents that act similar to short host defence peptides secreted by many organisms (Epand et al. 2008; Mondal et al. 2010c).

4.6.4 Head-to-Tail Supramolecular Self-Assembly of 14-Helical Peptides

The 14-helix geometry offers another way of implementing hierarchical superstructures. Given that in a 14-helix hydrogen bonding occurs between the carbonyl oxygen of the amide group of residue i and N-H of the amide group of residue i + 2, even though three residues complete a turn of the helix, these cannot have free termini at either end and hence, the least number of residues required to form a complete turn of the helix is five. Acylation of the N-terminus forms an amide and thus in that case four residues can already form an intramolecular hydrogen bond (Del Borgo et al. 2013). Importantly, this small modification also copies the intramolecular hydrogen bonding motif to the termini, offering three hydrogen bond donors at the C-terminus and three hydrogen bond acceptors at the N-terminus. This supramolecular self-assembly motif allows monomers to stack head-to-tail in a continuation of the helix (Fig. 4.19). This lead to the formation of fibres and fibre bundles of up to millimeter dimensions. The self-assembly was demonstrated for hexapeptides, tetrapeptides and tripeptides (Del Borgo et al. 2013; Seoudi et al. 2015a, b, 2016a, b).

4.7 Factors That Enable Designing Well-Defined Structures

Designing self-assembled structures from folded peptide monomers can be achieved through inducing non-covalent interactions between the individual molecules and/or molecular assemblies. While non-covalent interactions are inherently weak, the incorporation of multiple binding sites to form a supramolecular recognition motif can lead to robust superstructures (Rybtchinski 2011). Simultaneous control of multiple non-covalent interactions between the building blocks in a systematic way remains challenging but not impossible: the formation of nanostructures including nanorods, tubes (Qin et al. 2011; Shao et al. 2010), fibres (Krysmann et al. 2008; Dong et al. 2008), spheres, (Ma et al. 2010; Han et al. 2010) and donuts (Gao et al. 2009) have been successfully demonstrated.

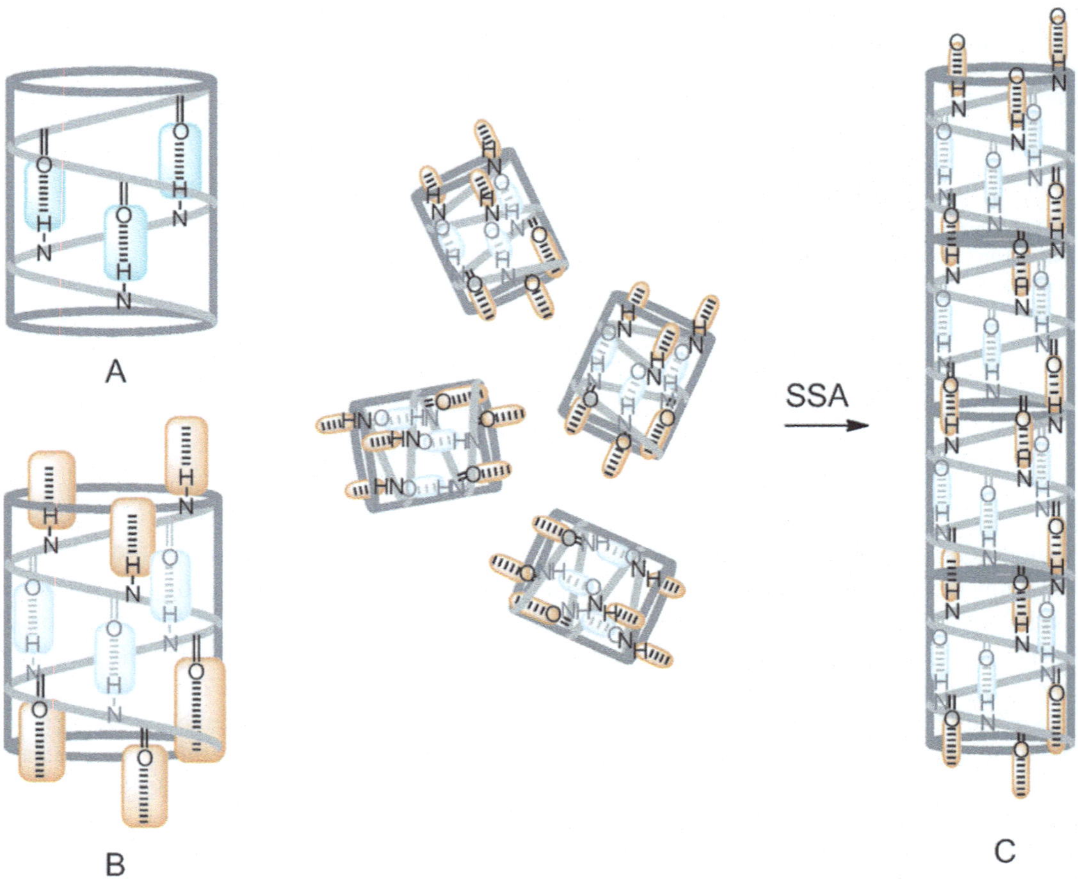

Fig. 4.19 (**a**) Intramolecular hydrogen bond network (**b**) Axial intermolecular hydrogen bond network, forming a supramolecular self-assembly motif (**c**) Continuous head-to-tail monomer stacking leads to long fibre formation (Del Borgo et al. 2013)

As the non-covalent interactions depend on the physical characteristics of the environment, changing the environmental conditions offers a way to control the self-assembly. Solvent characteristics affect non-covalent interactions through the dipole moment and protic/non-protic nature of the solvent molecules and the dielectric constant of the medium (Balamurugan and Muraleedharan 2012). These factors affect both hydrogen bonding and van der Waals interactions and hence the self-assembly process itself. Introducing residues that allow chemical cross-linking can further stabilize superstructures post-assembly. For example, cysteine residues form disulfide bonds under oxidizing conditions; it is possible to reverse the process by using a reducing agent (Hartgerink 2004; Hartgerink et al. 2002).

A typical problem of self-assembly design is controlled, preferably switchable gel formation. Gels are based on fibrous molecular assemblies that further interact to form a mesh structure to entrap solvent molecules. Hence, gelators require a strong primary and a weaker secondary binding motif. These interactions are dependent on the chemical environment, offering the means of switching the system between a crystalline-like and a gel-like state (Bartocci et al. 2015). An example of a peptide-based gelator is the dipeptide FF (phenylalanine-phenylalanine). It has got two ways of interacting: backbone to backbone hydrogen bonding and π-π stacking between the aromatic rings. Self-assembly takes place by a nucleation-growth mechanism triggered by evaporation of the solvent. Change from gel into crystalline state depends on the solvent polarity: apolar toluene allows for the self-assembly into organogel driven by π-π stacking between the aromatic rings, whereas polar ethanol solvent increases the parallel hydrogen-bonded β-sheet content due to the interference of ethanol and the gel changes into microcrystals (Zhu et al. 2010).

Given that non-covalent interactions have low binding energies, it is possible to control self-assembly using temperature. Increasing temperature causes disassembly of the aggregate, as the energy given to the system increases the kinetic energy of the monomers (Fenske et al. 2012). As in the previous example, phenylalanine dipeptide (FF) was switched between two forms using solvent polarity; assemblies of the same peptide also exhibit different morphologies as a function of temperature. While dissolution at room temperature in acetonitrile/water solution leads to the formation of a crystalline phase, heating the system and cooling again lead to a reorganization into nanowires and microtubes (Huang et al. 2014).

Temperature can also reduce hydrophobic interactions leading to increased hydration of oligopeptide structures. Regenerated silk fibroin (RSF) consists of alternating hydrophobic/hydrophilic blocks, the superstructure formed by which exhibited different morphologies depending on the incubating temperature. High temperature is associated with the aggregation of antiparallel β-sheets leading to the formation of protofilaments and globule-like features, whereas at lower temperatures coiled bead-like structures are dominant (Zhong et al. 2015).

The protonation state of acidic and basic amino acids such as lysine (K) or glutamic acid (E) are pH sensitive and thus including such residues in the sequence of a self-assembling system can result in a superstructure that is pH sensitive (Wang et al. 2009; Jiang et al. 2005). Most of the peptide amphiphiles contain such residues and so the overall charge of such peptides is also pH sensitive. Hence, if acidic residues are deprotonated, the electrostatic repulsion can suppress the self-assembly (Gossler-Schofberger et al. 2009). Lowering the pH leads to charge neutralization of the peptides and trigger the self-assembly, through reducing electrostatic repulsion (Jun et al. 2004). Most water soluble peptides are at least slightly amphiphilic. The hydrophilic residues often contain both positive and negative charges. As amphiphilic peptides assemble *via* a combination of electrostatic, hydrogen bonding, and hydrophobic interactions, using pH alter the charge thus enables controlling the self-assembly while also tuning hydrogen bond strength. For example, in the peptide ($C_{11}H_{19}O$-NH-VRGDV-COOH), in acidic medium the ionization state is lower hence the electrostatic repulsion between the head groups is much restrained and strong hydrogen bonds can be formed between the monomers; the result is a fibrous, entangled hydrogel structure. Increasing pH to neutral increases charge repulsion and disturbs hydrogen bonding, and hence fibre formation, yielding spherical aggregates (Jin et al. 2008). Another example is $(C_{17}H_{35}CO\text{-NH-GRGDG})_2KG$ that shows the formation of nanofibers at pH 3 that can be changed into aggregates of nanospheres upon increasing pH to 6 and further elevation of pH to 10 yielded lamellar layers (Qin et al. 2012).

It is not only the pH that can affect the electrostatic interactions in self-assembly, but the presence of salts as well. Counterions from the ionic environment may shield the charges of an oligopeptide which enhances the propensity of assembled oligomers. For example, the assembly of ionic–complementary peptide AEAEAKAKAEAEAKAK is dependent on the electrostatic interactions: while it always forms beta sheet fibres, it is possible to switch from individual fibres to fibrous networks using salts such as NaCl in an aqueous environment (Hong et al. 2005).

Structural design can also be used to control electrostatic interactions. Screening the electrostatic repulsion through introducing hydrophobic residues is frequently used to tune peptide assembly. For example, in β-amyloidic peptides increasing the hydrophobicity of peptide residues can alter peptide assembly from random coil into β-fibril in a physiological environment. Steric effects define the amount of charge shielding; the length and bulkiness of the hydrophobic side chains are key factors. Increasing the hydrophobicity of specific residues can clearly demonstrate this effect. For example, the difference between substituting the hydrophobic side chains valine, lysine, isoleucine, and cyclohexylalanine in X positions in the octapeptide (Ac-$(XKXK)_2$-NH_2) is seen in the self-assembly properties of the respective peptides. In case of (Ac-$(VKVK)_2$-NH_2), the small hydrophobic residues do not overcome the repulsion between the lysine residues thus no self-assembly happens even in high NaCl concentration (Bowerman et al. 2011). Replacing valine residues with more hydrophobic isoleucine, the peptides remain in a monomeric state, random coil conformation in 300mM NaCl that can, however, convert into β-sheets when increasing NaCl concentration to 700mM (Bowerman et al. 2011). On the contrary, substituting phenylalanine residues further extends the salt concentration range at which the peptide remains in an equilibrium between the monomeric and β-fibril form to 700 mM NaCl due to the interference of the π-π interactions with the formation of the β-sheets; β-fibrils are formed at 1000mM NaCl (Bowerman et al. 2011). Yet, further increasing hydrophobicity by using pentafluorophenylalanine (F5-Phe) yields a monomeric system in the absence of salts however already at 20 mM NaCl concentration β-fibrils are formed. Although cyclohexylalanine (Cha) is even more hydrophobic, substitution in the peptide yields β-fibrils only at 60mM NaCl concentration due to the steric effects of the bulky side chain. The results are a clear evidence of the role of the length and dimensions of the hydrophobic residues on shielding charge-charge repulsive interactions (Bowerman et al. 2011).

Moving beyond simple fibrous or spherical assembly geometries remains an enormous challenge. However, using beta peptides it is possible to create a longitudinal geometrical and chemical pattern coded into the peptide sequence, with the help of environmental control. Micrometer size superstructures can be obtained from hepta-(ACPC). This peptide folds into a 12-membered ring in THF; adding this solution to aqueous medium causes a sudden shift in the strength of the lateral hydrophobic interactions between the helical faces to allow association of helix faces leading to the formation of micrometer scale, homogeneous sized windmill-shaped structures (Fig. 4.20a) (Kwon et al. 2010).

Similarly, a tetrapeptide based on ACPC (CH_3COONH-$(ACPC)_4$-OBn) in mixture of methanol/water assemble into microtubes with rectangular cross-sections (Fig. 4.20c) through controlled solvent evaporation (Kim et al. 2012).

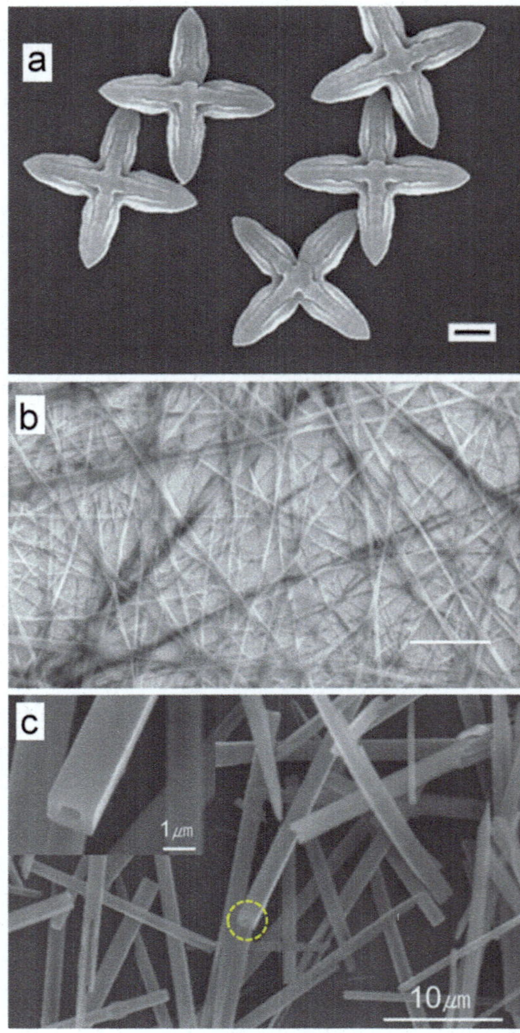

Fig. 4.20 (**a**) Windmill-shaped structures formed from self-assembly of hepta-ACPC in water (Kwon et al. 2010), (**b**) Straight fibres formed from self-assembly of tri-β amphiphilic peptide in a mixture of water and formic acid (Ishihara and Kimura 2010) (**c**) Microtubes with rectangular cross-sections formed by tetrapeptides in water/methanol mixture (Reproduced by permission from Kim et al. 2012. Copyright (2012) American Chemical Society http://pubs.acs.org/doi/full/10.1021/ja3088482)

A tri-β amphiphilic peptide that contains one ACHC and two β-glucosamino acids showed ability to assemble into rodlike crystals in a mixture of formic acid and diisopropyl ether. In this mixture the self-assembly showed a concentration dependence: at low concentrations rodlike structures were formed while increasing the concentration lead to the formation of straight fibres instead (Fig. 4.20b) (Ishihara and Kimura 2010).

The self-assembly of β-peptides can be organized by introducing nucleobases as recognition units. 14-helical peptides substituted at the N and C termini by guanine and cytosine, respectively, assemble through oligonucleotide-base pairing, forming band-like arrangements and higher order organized structures (Srivastava et al. 2009).

4.8 Conclusions

As it was demonstrated in this chapter, oligopeptides offer a rich library of building blocks and binding motifs for the design of bioinspired self-assembled materials. This variety is the main strength and weakness of the field at the same time, promising unmatched complexity in structure and function, yet limited by the understanding of the intra-and intermolecular interactions responsible for this complexity. Ultimately the ability to design functionalized peptide nanostructures for specific applications is tied to the ability of controlling the morphologies of the self-assembled superstructures, which is based on a thorough understanding of the structural and environmental factors affecting self-assembly. Much works was done on designing and stabilizing folding of oligopeptides, as well as adapting the basic principles of self-assembly to these geometrically defined building blocks. Thus, electrostatic and surfactant-like self-assembly has been successfully explored for oligopeptide superstructures. The next challenge is to improve the specificity and selectivity of the binding motifs: to implement truly supramolecular assembly, approximating the structural and functional complexity of proteins. This chapter has shown what is the state of the art; it is for the reader to take the next steps forward.

References

Abb S, Harnau L, Gutzler R, Rauschenbach S, Kern K (2016) Two-dimensional honeycomb network through sequence-controlled self-assembly of oligopeptides. Nat Commun 7:10335

Abele S, Guichard G, Seebach D (1998) (S)-β$^{(3)}$-homolysine- and (S)-β$^{(3)}$-homoserine-containing β-peptides: CD spectra in aqueous solution. Helv Chim Acta 81:2141–2156

Abele S, Seiler P, Seebach D (1999) Synthesis, crystal structures, and modelling of beta-oligopeptides consisting of 1-(aminomethyl) cyclopropanecarboxylic acid: Ribbon- type arrangement of eight- membered H-bonded rings. Helv Chim Acta 82:1559–1571

Acevedo-Velez C, Andre G, Dufrene YF, Gellman SH, Abbott NL (2011) Single-molecule force spectroscopy of beta-peptides that display well-defined three-dimensional chemical patterns. J Am Chem Soc 133:3981–3988

Adams FC, Barbante C (2013) Nanoscience, nanotechnology and spectrometry. Spectrochim Acta Part B 86:3–13

Aggeli A, Nyrkova IA, Bell M, Carrick L, McLeish TCB, Semenov AN, Boden N (2001) Exploiting peptide self-assembly to engineer novel biopolymers: Tapes, ribbons, fibrils and fibres. In: Aggeli A, Boden N, Zhang S (eds) Self-assembling peptide systems in biology, medicine and engineering. Springer, Dordrecht

Althuon D, Ronicke F, Furniss D, Quan J, Wellhofer I, Jung N, Schepers U, Brase S (2015) Functionalized triazolopeptoids – a novel class for mitochondrial targeted delivery. Org Biomol Chem 13:4226–4230

Altman M, Lee P, Rich A, Zhang SG (2000) Conformational behavior of ionic self-complementary peptides. Protein Sci 9:1095–1105

Andersen AJ, Robinson JT, Dai HJ, Hunter AC, Andresen TL, Moghimi SM (2013) Single-walled carbon nanotube surface control of complement recognition and activation. Acs Nano 7:1108–1119

Appella DH, Christianson LA, Klein DA, Powell DR, Huang XL, Barchi JJ, Gellman SH (1997) Residue-based control of helix shape in beta-peptide oligomers. Nature 387:381–384

Appella DH, Barchi JJ, Durell SR, Gellman SH (1999a) Formation of short, stable helices in aqueous solution by beta-amino acid hexamers. J Am Chem Soc 121:2309–2310

Appella DH, Christianson LA, Klein DA, Richards MR, Powell DR, Gellman SH (1999b) Synthesis and structural characterization of helix-forming beta-peptides: trans-2-aminocyclopentanecarboxylic acid oligomers. J Am Chem Soc 121:7574–7581

Appella DH, LePlae PR, Raguse TL, Gellman SH (2000) (R,R,R)-2,5-diaminocylohexanecarboxylic acid, a building block for water-soluble, helix-forming beta-peptides. J Org Chem 65:4766-4769

Applequist J, Glickson JD (1971) Conformation of poly-β-alanine in aqueous solution from proton magnetic

resonance and deuterium exchange studies. J Am Chem Soc 93:3276–3281
Arico AS, Bruce P, Scrosati B, Tarascon JM, Van Schalkwijk W (2005) Nanostructured materials for advanced energy conversion and storage devices. Nat Mater 4:366–377
Ariga K, Nakanishi T, Hill JP (2007) Self-assembled microstructures of functional molecules. Curr Opin Colloid Interface Sci 12:106–120
Ariga K, Hill JP, Lee MV, Vinu A, Charvet R, Acharya S (2008) Challenges and breakthroughs in recent research on self-assembly. Sci Tech Adv Mater 9:96
Arnold MS, Guler MO, Hersam MC, Stupp SI (2005) Encapsulation of carbon nanotubes by self-assembling peptide amphiphiles. Langmuir 21:4705–4709
Arvidsson PI, Rueping M, Seebach D (2001) Design, machine synthesis, and NMR-solution structure of a β-heptapeptide forming a salt-bridge stabilised 3_{14}-helix in methanol and in water. Chem Commun 649–650
Arvidsson PI, Ryder NS, Weiss HM, Hook DF, Escalante J, Seebach D (2005) Exploring the antibacterial and hemolytic activity of shorter- and longer-chain beta-alpha,beta-, and gamma-peptides, and of beta-peptides from $\beta^{(2)}$-3-aza- and $\beta^{(3)}$-2-methylidene-amino acids bearing proteinogenic side chains – A survey. Chem Biodivers 2:401–420
Aryal P, Sansom MSP, Tucker SJ (2015) Hydrophobic gating in ion channels. J Mol Biol 427:121–130
Balamurugan D, Muraleedharan KM (2012) Chemical environment as control element in the evolution of shapes – hexagons and rods' from an 11-helical $\alpha,\beta^{2,3}$-hybrid peptide. Soft Matter 8:11857–11862
Baldauf C, Gunther R, Hofmann HJ (2006) Theoretical prediction of the basic helix types in α,β-hybrid peptides. Biopolymers 84:408–413
Banerjee A, Balaram P (1997) Stereochemistry of peptides and polypeptides containing omega amino acids. Curr Sci 73:1067–1077
Banwell EF, Abelardo ES, Adams DJ, Birchall MA, Corrigan A, Donald AM, Kirkland M, Serpell LC, Butler MF, Woolfson DN (2009) Rational design and application of responsive α-helical peptide hydrogels. Nat Mater 8:596–600
Barchi JJ, Huang XL, Appella DH, Christianson LA, Durell SR, Gellman SH (2000) Solution conformations of helix-forming β-amino acid homooligomers. J Am Chem Soc 122:2711–2718
Barth JV, Costantini G, Kern K (2005) Engineering atomic and molecular nanostructures at surfaces. Nature 437:671–679
Bartocci S, Morbioli I, Maggini M, Mba M (2015) Solvent-tunable morphology and emission of pyrene-dipeptide organogels. J Pept Sci 21:871–878
Bemporad F, Taddei N, Stefani M, Chiti F (2006) Assessing the role of aromatic residues in the amyloid aggregation of human muscle acylphosphatase. Protein Sci 15:862–870
Berl V, Huc I, Khoury RG, Krische MJ, Lehn JM (2000) Interconversion of single and double helices formed from synthetic molecular strands. Nature 407:720–723
Bhattacharya P, Du D, Lin YH (2014) Bioinspired nanoscale materials for biomedical and energy applications. J R Soc Interface 11:14
Bhushan B (2009) Biomimetics: lessons from nature – an overview. Philos Trans A Math Phys Eng Sci 367:1445–1486
Bishop KJM, Wilmer CE, Soh S, Grzybowski BA (2009) Nanoscale forces and their uses in self-assembly. Small 5:1600–1630
Blair DF (2003) Flagellar movement driven by proton translocation. FEBS Lett 545:86–95
Bolin KA, Millhauser GL (1999) α and 3_{10}: the split personality of polypeptide helices. Acc Chem Res 32:1027–1033
Bowerman CJ, Nilsson BL (2010) A reductive trigger for peptide self-assembly and hydrogelation. J Am Chem Soc 132:9526–9527
Bowerman CJ, Nilsson BL (2012) Review self-assembly of amphipathic beta-sheet peptides: insights and applications. Biopolymers 98:169–184
Bowerman CJ, Liyanage W, Federation AJ, Nilsson BL (2011) Tuning β-sheet peptide self-assembly and hydrogelation behavior by modification of sequence hydrophobicity and aromaticity. Biomacromolecules 12:2735–2745
Boyle AL, Woolfson DN (2011) *De novo* designed peptides for biological applications. Chem Soc Rev 40:4295–4306
Breydo L, Uversky VN (2015) Structural, morphological, and functional diversity of amyloid oligomers. FEBS Lett 589:2640–2648
Bromley EHC, Channon KJ (2011) Alpha-helical peptide assemblies: giving new function to designed structures. Prog Mol Biol Transl Sci 103:231–275
Bromley EHC, Channon K, Moutevelis E, Woolfson DN (2008) Peptide and protein building blocks for synthetic biology: from programming biomolecules to self-organized biomolecular systems. ACS Chem Biol 3:38–50
Bruckner AM, Chakraborty P, Gellman SH, Diederichsen U (2003) Molecular architecture with functionalized β-peptide helices. Angew Chem Int Ed 42:4395–4399
Buehler MJ, Ackbarow T (2008) Nanomechanical strength mechanisms of hierarchical biological materials and tissues. Comput Methods Biomech Biomed Eng 11:595–607
Butterfoss GL, Drew K, Renfrew PD, Kirshenbaum K, Bonneau R (2014) Conformational preferences of peptide-peptoid hybrid oligomers. Biopolymers 102:369–378
Canalle LA, Lowik D, van Hest JCM (2010) Polypeptide-polymer bioconjugates. Chem Soc Rev 39:329–353
Cao MW, Cao CH, Zhang LJ, Xia DH, Xu H (2013) Tuning of peptide assembly through force balance adjustment. J Colloid Interface Sci 407:287–295

Care A, Bergquist PL, Sunna A (2015) Solid-binding peptides: smart tools for nanobiotechnology. Trends Biotechnol 33:259–268

Castillo-León J, Andersen KB, Svendsen WE (2011) Self–assembled peptide nanostructures for biomedical applications: advantages and challenges. In: Pignatello R (ed) Biomaterials science and engineering. InTechOpen

Chakraborty P, Diederichsen U (2005) Three-dimensional organization of helices: Design principles for nucleobase-functionalized β-peptides. Chem Eur J 11:3207–3216

Chalker JM, Bernardes GJL, Lin YA, Davis BG (2009) Chemical modification of proteins at cysteine: opportunities in chemistry and biology. Chem Asian J 4:630–640

Chen P (2005) Self-assembly of ionic-complementary peptides: a physicochemical viewpoint. Colloids Surf A Physicochem Eng Asp 261:3–24

Chen CL, Rosi NL (2010) Peptide-based methods for the preparation of nanostructured inorganic materials. Angew Chem Int Ed 49:1924–1942

Chen YZ, Qiu F, Lu YB, Shi YK, Zhao XJ (2009) Geometrical Shape of Hydrophobic section determines the self-assembling structure of peptide detergents and bolaamphiphilic peptides. Curr Nanosci 5:69–74

Cheng RP (2004) Beyond de novo protein design – de novo design of non-natural folded oligomers. Curr Opin Struct Biol 14:512–520

Cheng RP, Gellman SH, DeGrado WF (2001) β-peptides: from structure to function. Chem Rev 101:3219–3232

Cheng JY, Ross CA, Smith HI, Thomas EL (2006) Templated self-assembly of block copolymers: top-down helps bottom-up. Adv Mater 18:2505–2521

Chiti F, Dobson CM (2006) Protein misfolding, functional amyloid, and human disease. Annu Rev Biochem 75:333–366

Chiti F, Webster P, Taddei N, Clark A, Stefani M, Ramponi G, Dobson CM (1999) Designing conditions for in vitro formation of amyloid protofilaments and fibrils. Proc Natl Acad Sci USA 96:3590–3594

Cinar G, Ceylan H, Urel M, Erkal TS, Tekin ED, Tekinay AB, Dana A, Guler MO (2012) Amyloid inspired self-assembled peptide nanofibers. Biomacromolecules 13:3377–3387

Claridge TDW, Goodman JM, Moreno A, Angus D, Barker SF, Taillefumier C, Watterson MP, Fleet GWJ (2001) 10-Helical conformations in oxetane beta-amino acid hexamers. Tetrahedron Lett 42:4251–4255

Collie GW, Pulka-Ziach K, Lombardo CM, Fremaux J, Rosu F, Decossas M, Mauran L, Lambert O, Gabelica V, Mackereth CD, Guichard G (2015) Shaping quaternary assemblies of water-soluble non-peptide helical foldamers by sequence manipulation. Nat Chem 7:871–878

Colson P, Henrist C, Cloots R (2013) Nanosphere lithography: a powerful method for the controlled manufacturing of nanomaterials. J Nanomater 2013:948510

Coppens MO (2005) Scaling-up and -down in a nature-inspired way. Indust Eng Chem Res 44:5011–5019

Courson DS, Rock RS (2010) Actin cross-link assembly and disassembly mechanics for alpha-actinin and fascin. J Biol Chem 285:26350–26357

Crisma M, Formaggio F, Moretto A, Toniolo C (2006) Peptide helices based on alpha-amino acids. Biopolymers 84:3–12

Cubberley MS, Iverson BL (2001) Models of higher-order structure: foldamers and beyond. Curr Opin Chem Biol 5:650–653

Cui H, Muraoka T, Cheetham AG, Stupp SI (2009) Self-assembly of giant peptide nanobelts. Nano Lett 9:945–951

Cui HG, Webber MJ, Stupp SI (2010) Self-assembly of peptide amphiphiles: from molecules to nanostructures to biomaterials. *Biopolymers* 94:1–18

Cukalevski R, Boland B, Frohm B, Thulin E, Walsh D, Linse S (2012) Role of aromatic side chains in amyloid β-protein aggregation. ACS Chem Neurosci 3:1008–1016

Dado GP, Gellman SH (1994) Intermolecular hydrogen-bonding in derivatives of beta-alanine and gamma-amino buryric-acid model studies for the folding of unnatural polypeptide backbones. J Am Chem Soc 116:1054–1062

Daura X, Jaun B, Seebach D, van Gunsteren WF, Mark AE (1998) Reversible peptide folding in solution by molecular dynamics simulation. J Mol Biol 280:925–932

De Pol S, Zorn C, Klein CD, Zerbe O, Reiser O (2004) Surprisingly stable helical conformations in α/β--peptides by incorporation of *cis*-β-aminocyclopropane carboxylic acids. Angew Chem Int Ed 43:511–514

DeGrado WF, Schneider JP, Hamuro Y (1999) The twists and turns of β-peptides. J Pept Res 54:206–217

Dehsorkhi A, Castelletto V, Hamley IW (2014) Self-assembling amphiphilic peptides. *J. Pept. Sci.* 20:453–467

Del Borgo MP, Mechler AI, Traore D, Forsyth C, Wilce JA, Wilce MCJ, Aguilar MI, Perlmutter P (2013) Supramolecular self-assembly of *N*-acetyl-capped β-peptides leads to nano- to macroscale fiber formation. Angew Chem Int Ed 52:8266–8270

Delort E, Nguyen-Trung NQ, Darbre T, Reymond JL (2006) Synthesis and activity of histidine-containing catalytic peptide dendrimers. J Org Chem 71:4468–4480

Deming TJ (2007) Synthetic polypeptides for biomedical applications. Prog Polym Sci 32:858–875

Doig AJ (2008) Stability and design of α-helical peptides. In: Conn PM (ed) Progress in Molecular Biology and Translational Science, vol 83. Elsevier Academic Press, San Diego

Dong H, Paramonov SE, Hartgerink JD (2008) Self-assembly of α-helical coiled coil nanofibers. J Am Chem Soc 130:13691–13695

Durand KL, Ma XX, Xia Y (2015) Intra-molecular reactions between cysteine sulfinyl radical and a disul-

fide bond within peptide ions. Int J Mass Spectrom 378:246–254
Engler AC, Shukla A, Puranam S, Buss HG, Jreige N, Hammond PT (2011) Effects of side group functionality and molecular weight on the activity of synthetic antimicrobial polypeptides. Biomacromolecules 12:1666–1674
Eom JH, Gong J, Jeong R, Driver RW, Lee HS (2015) Parallelogram plate shaped foldecture from the controlled self-assembly of α/β-peptide foldamer. Solid State Sci 48:39–43
Epand RF, Raguse TL, Gellman SH, Epand RM (2004) Antimicrobial 14-helical β-peptides: potent bilayer disrupting agents. Biochemistry 43:9527–9535
Epand RF, Mowery BP, Lee SE, Stahl SS, Lehrer RI, Gellman SH, Epand RM (2008) Dual mechanism of bacterial lethality for a cationic sequence-random copolymer that mimics host-defense antimicrobial peptides. J Mol Biol 379:38–50
Fairman R, Akerfeldt KS (2005) Peptides as novel smart materials. Curr Opin Struct Biol 15:453–463
Fan HY, Lu YF, Stump A, Reed ST, Baer T, Schunk R, Perez-Luna V, Lopez GP, Brinker CJ (2000) Rapid prototyping of patterned functional nanostructures. Natur 405:56–60
Fandrich M, Dobson CM (2002) The behaviour of polyamino acids reveals an inverse side chain effect in amyloid structure formation. EMBO J 21:5682–5690
Faul CFJ, Antonietti M (2003) Ionic self-assembly: facile synthesis of supramolecular materials. Adv Mater 15:673–683
Fenske T, Korth HG, Mohr A, Schmuck C (2012) Advances in Switchable supramolecular nanoassemblies. Chem Eur J 18:738–755
Fernandes C, Faure S, Pereira E, Thery V, Declerck V, Guillot R, Aitken DJ (2010) 12-Helix folding of cyclobutane β-amino acid oligomers. Org Lett 12:3606–3609
Feynman RP (1960) There's plenty of room at the bottom. Eng Sci 23:22–36
Fitzpatrick AWP, Park ST, Zewail AH (2013) Exceptional rigidity and biomechanics of amyloid revealed by 4D electron microscopy. Proc Natl Acad Sci USA 110:10976–10981
Fodale V, Kegulian NC, Verani M, Cariulo C, Azzollini L, Petricca L, Daldin M, Boggio R, Padova A, Kuhn R, Pacifici R, Macdonald D, Schoenfeld RC, Park H, Isas JM, Langen R, Weiss A, Caricasole A (2014) Polyglutamine- and temperature-dependent conformational rigidity in mutant huntingtin revealed by immunoassays and circular dichroism spectroscopy. Plos One 9:31
Fung SY, Keyes C, Duhamel J, Chen P (2003) Concentration effect on the aggregation of a self-assembling oligopeptide. Biophys J 85:537–548
Gademann K, Jaun B, Seebach D, Perozzo R, Scapozza L, Folkers G (1999) Temperature-dependent NMR and CD spectra of β-peptides. On the thermal stability of β-peptide helices. Is the folding process of β-peptides non-cooperative? Helv Chim Acta 82:1–11
Gao J, Wang H, Wang L, Wang J, Kong D, Yang Z (2009) Enzyme promotes the hydrogelation from a hydrophobic small molecule. J Am Chem Soc 131:11286–11287
Garrett SL, Poese ME (2013) There's (still) plenty of room at the bottom. Appl Therm Eng 61:884–888
Gazit E (2002) A possible role for π-stacking in the self-assembly of amyloid fibrils. FASEB J 16:77–83
Gazit E (2005) Mechanisms of amyloid fibril self-assembly and inhibition. FEBS J 272:5971–5978
Gazit E (2007) Self-assembled peptide nanostructures: the design of molecular building blocks and their technological utilization. Chem Soc Rev 36:1263–1269
Gellman SH (1998) Foldamers: a manifesto. Acc Chem Res 31:173–180
Gilmartin BP, McLaughlin RL, Williams ME (2005) Artificial tripeptide scaffolds for self-assembly of heteromultimetallic structures with tunable electronic and magnetic properties. Chem Mater 17:5446–5454
Giuseppone N (2012) Toward self-constructing materials: a systems chemistry approach. Acc Chem Res 45:2178–2188
Glattli A, Seebach D, van Gunsteren WF (2004) Do valine side chains have an influence on the folding behavior of beta-substituted beta-peptides? Helv Chim Acta 87:2487–2506
Gobre VV, Tkatchenko A (2013) Scaling laws for van der Waals interactions in nanostructured materials. Nat Commun 4:6
Godballe T, Nilsson LL, Petersen PD, Jenssen H (2011) Antimicrobial β-peptides and α-peptoids. Chem Biol Drug Des 77:107–116
Goodman CM, Choi S, Shandler S, DeGrado WF (2007) Foldamers as versatile frameworks for the design and evolution of function. Nat Chem Biol 3:252–262
Gopalan RD, Del Borgo MP, Mechler AI, Perlmutter P, Aguilar MI (2015) Geometrically precise building blocks: the self-assembly of β-peptides. Chem Biol 22:1417–1423
Gossler-Schofberger R, Hesser G, Muik M, Wechselberger C, Jilek A (2009) An orphan dermaseptin from frog skin reversibly assembles to amyloid-like aggregates in a pH-dependent fashion. FEBS J 276:5849–5859
Gratais A, Pannecoucke X, Bouzbouz S (2015) 1,4 Addition of unprotected pyrrole onto chiral acrylamides: toward synthesis of new polypeptidic architectures. Org Biomol Chem 13:1082–1090
Greenwald J, Riek R (2012) On the possible amyloid origin of protein folds. J Mol Biol 421:417–426
Grzelczak M, Vermant J, Furst EM, Liz-Marzan LM (2010) Directed self-assembly of nanoparticles. ACS Nano 4:3591–3605
Guillotte M, Nato F, Juillerat A, Hessel A, Marchand F, Lewit-Bentley A, Bentley GA, Vigan-Womas I, Mercereau-Puijalon O (2016) Functional analysis of monoclonal antibodies against the *Plasmodium falciparum* PfEMP1-VarO adhesin. Malar J 15:16

Gunasekar SK, Haghpanah JS, Montclare JK (2008) Assembly of bioinspired helical protein fibers. Polym Adv Technol 19:454–468

Guthold M, Liu W, Sparks EA, Jawerth LM, Peng L, Falvo M, Superfine R, Hantgan RR, Lord ST (2007) A comparison of the mechanical and structural properties of fibrin fibers with other protein fibers. Cell Biochem Biophys. 49:165–181

Hamley IW (2011) Self-assembly of amphiphilic peptides. Soft Matter 7:4122–4138

Hamley IW (2014) Peptide nanotubes. Angew Chem Int Ed 53:6866–6881

Hamuro Y, Schneider JP, DeGrado WF (1999) *De novo* design of antibacterial β-peptides. J Am Chem Soc 121:12200–12201

Han TH, Ok T, Kim J, Shin DO, Ihee H, Lee HS, Kim SO (2010) Bionanosphere lithography via hierarchical peptide self-assembly of aromatic triphenylalanine. Small 6:945–951

Hao Y, An K, Zhang DW, Yang D (2015) A short helix formed by cyclic beta(2,3)-aminoxy peptides in protic solvents. Chem Asian J 10:2126–2129

Hartgerink JD (2004) Covalent capture: a natural complement to self-assembly. Curr Opin Chem Biol 8:604–609

Hartgerink JD, Beniash E, Stupp SI (2001) Self-assembly and mineralization of peptide-amphiphile nanofibers. Science 294:1684–1688

Hartgerink JD, Beniash E, Stupp SI (2002) Peptide-amphiphile nanofibers: a versatile scaffold for the preparation of self-assembling materials. Proc Natl Acad Sci USA 99:5133–5138

Hauser CAE, Deng RS, Mishra A, Loo YH, Khoe U, Zhuang FR, Cheong DW, Accardo A, Sullivan MB, Riekel C, Ying JY, Hauser UA (2011) Natural tri- to hexapeptides self-assemble in water to amyloid β-type fiber aggregates by unexpected α-helical intermediate structures. Proc Natl Acad Sci USA 108:1361–1366

Hauser CAE, Maurer-Stroh S, Martins IC (2014) Amyloid-based nanosensors and nanodevices. Chem Soc Rev 43:5326–5345

He B, Yuan X, Jiang DM (2014) Molecular self-assembly guides the fabrication of peptide nanofiber scaffolds for nerve repair. RSC Advances 4:23610–23621

Henzie J, Barton JE, Stender CL, Odom TW (2006) Large-area nanoscale patterning: chemistry meets fabrication. Acc Chem Res 39:249–257

Herrmann H, Bar H, Kreplak L, Strelkov SV, Aebi U (2007) Intermediate filaments: from cell architecture to nanomechanics. Nat Rev Mol Cell Biol 8:562–573

Heyduk E, Moxley MM, Salvatori A, Corbett JA, Heyduk T (2010) Homogeneous insulin and C-peptide sensors for rapid assessment of insulin and C-peptide secretion by the islets. Diabetes 59:2360–2365

Higashi N, Koga T (2008) Self-assembled peptide nanofibers. Adv Polym Sci 219:27–68

Hill DJ, Mio MJ, Prince RB, Hughes TS, Moore JS (2001) A field guide to foldamers. Chem Rev 101:3893–4011

Hobza P, Zahradnik R, Muller-Dethlefs K (2006) The world of non-covalent interactions: 2006. Collect Czech Chem Commun 71:443–531

Holmes TC, de Lacalle S, Su X, Liu GS, Rich A, Zhang SG (2000) Extensive neurite outgrowth and active synapse formation on self-assembling peptide scaffolds. Proc Natl Acad Sci USA 97:6728–6733

Hong Y, Pritzker MD, Legge RL, Chen P (2005) Effect of NaCl and peptide concentration on the self-assembly of an ionic-complementary peptide EAK16-II. Colloids Surf B 46:152–161

Horne WS (2015) Foldamers Biomimetic and built to order. Nat Chem 7:858–859

Huang ZH, Che SN (2015) Fabrication of Chiral materials via self-assembly and biomineralization of peptides. Chem Rec 15:665–674

Huang Y, Duan XF, Wei QQ, Lieber CM (2001) Directed assembly of one-dimensional nanostructures into functional networks. Science 291:630–633

Huang Z, Sargeant TD, Hulvat JF, Mata A, Bringas P, Koh CY, Stupp SI, Snead ML (2008) Bioactive nanofibers instruct cells to proliferate and differentiate during enamel regeneration. J Bone Miner Res:231995–232006

Huang Z, Newcomb CJ, Bringas P, Stupp SI, Snead ML (2010) Biological synthesis of tooth enamel instructed by an artificial matrix. Biomaterials 31:9202–9211

Huang RL, Wang YF, Qi W, Su RX, He ZM (2014) Temperature-induced reversible self-assembly of diphenylalanine peptide and the structural transition from organogel to crystalline nanowires. Nanoscale Res Lett 9:9

Imperiali B, McDonnell KA, Shogren-Knaak M (1999) Design and construction of novel peptides and proteins by tailored incorporation of coenzyme functionality. Top Curr Chem 202:1–38

Invernizzi G, Papaleo E, Sabate R, Ventura S (2012) Protein aggregation: mechanisms and functional consequences. Int J Biochem Cell Biol 44:1541–1554

Ishihara Y, Kimura S (2010) Nanofiber formation of amphiphilic cyclic tri-beta-peptide. J Pept Sci 16:110–114

Israelachvili JN (2011) Intermolecular and surface forces: Revised third edition. Academic, San Diego

Izquierdo S, Kogan MJ, Parella T, Moglioni AG, Branchadell V, Giralt E, Ortuno RM (2004) 14-helical folding in a cyclobutane-containing β-tetrapeptide. J Org Chem 69:5093–5099

Jadhav SV, Misra R, Gopi HN (2014) Foldamers to nanotubes: influence of amino acid side chains in the hierarchical assembly of α,γ^4-hybrid peptide helices. Chem Eur J 20:16523–16528

Jiang YG, Wang ZQ, Yu X, Shi F, Xu HP, Zhang X, Smet M, Dehaen W (2005) Self-assembled monolayers of dendron thiols for electrodeposition of gold nanostructures: toward fabrication of superhydrophobic/superhydrophilic surfaces and pH-responsive surfaces. Langmuir 21:1986–1990

Jin Y, Xu XD, Chen CS, Cheng SX, Zhang XZ, Zhuo RX (2008) Bioactive amphiphilic peptide derivatives with pH triggered morphology and structure. Macromol Rapid Commun 29:1726–1731

Joo JY, Amin ML, Rajangam T, An SSA (2015) Fibrinogen as a promising material for various biomedical applications. Mol Cell Toxicol 11:1–9

Ju HX (2011) Sensitive biosensing strategy based on functional nanomaterials. Sci China Chem 54:1202–1217

Jun S, Hong Y, Imamura H, Ha BY, Bechhoefer J, Chen P (2004) Self-assembly of the ionic peptide EAK16: the effect of charge distributions on self-assembly. Biophys J 87:1249–1259

Kar T, Mukherjee S, Das PK (2014) Organogelation through self-assembly of low-molecular-mass amphiphilic peptide. New J Chem 38:1158–1167

Kim J, Kwon S, Kim SH, Lee CK, Lee JH, Cho SJ, Lee HS, Ihee H (2012) Microtubes with rectangular cross-section by self-assembly of a short β-peptide foldamer. J Am Chem Soc 134:20573–20576

Kim S, Kim JH, Lee JS, Park CB (2015) β-sheet-forming, self-assembled peptide nanomaterials towards optical, energy, and healthcare applications. Small 11:3623–3640

Kirshenbaum K, Zuckermann RN, Dill KA (1999) Designing polymers that mimic biomolecules. Curr Opin Struct Biol 9:530–535

Klinker K, Barz M (2015) Polypept(o)ides: hybrid systems based on polypeptides and polypeptoids. Macromol Rapid Commun 36:1943–1957

Klosterman JK, Yamauchi Y, Fujita M (2009) Engineering discrete stacks of aromatic molecules. Chem Soc Rev 38:1714–1725

Koehler JM (2015) What are proteins teaching us on fundamental strategies for molecular nanotechnology? Nanotechnol Rev 4:145–160

Koga T, Matsui H, Matsumoto T, Higashi N (2011) Shape-specific nanofibers via self-assembly of three-branched peptide. J Colloid Interface Sci 358:81–85

Koide T, Homma DL, Asada S, Kitagawa K (2005) Self-complementary peptides for the formation of collagen-like triple helical supramolecules. Biorg Med Chem Lett 15:5230–5233

Kopecek J (2003) Smart and genetically engineered biomaterials and drug delivery systems. Eur J Pharm Sci 20:1–16

Kopecek J, Yang JY (2009) Peptide-directed self-assembly of hydrogels. Acta Biomater 5:805–816

Kortelainen M, Suhonen A, Hamza A, Papai I, Nauha E, Yliniemela-Sipari S, Nissinen M, Pihko PM (2015) Folding patterns in a family of oligoamide foldamers. Chem Eur J 21:9493–9504

Koutsopoulos S (2016) Self-assembling peptide nanofiber hydrogels in tissue engineering and regenerative medicine: progress, design guidelines, and applications. J Biomed Mater Res A 104:1002–1016

Krauthauser S, Christianson LA, Powell DR, Gellman SH (1997) Antiparallel sheet formation in β-peptide foldamers: Effects of β-amino acid substitution on conformational preference. J Am Chem Soc 119:11719–11720

Kritzer JA, Tirado-Rives J, Hart SA, Lear JD, Jorgensen WL, Schepartz A (2005) Relationship between side chain structure and 14-helix stability of $\beta^{(3)}$-peptides in water. J Am Chem Soc 127:167–178

Krysmann MJ, Castelletto V, McKendrick JE, Clifton LA, Hamley IW, Harris PJF, King SA (2008) Self-assembly of peptide nanotubes in an organic solvent. Langmuir 24:8158–8162

Kumar RJ, MacDonald JM, Singh TB, Waddington LJ, Holmes AB (2011) Hierarchical Self-assembly of semiconductor functionalized peptide alpha-helices and optoelectronic properties. J Am Chem Soc 133:8564–8573

Kwon S, Jeon A, Yoo SH, Chung IS, Lee HS (2010) Unprecedented molecular architectures by the controlled self-assembly of a β-peptide foldamer. Angew Chem Int Ed 49:8232–8236

Lakshmanan A, Zhang SG, Hauser CAE (2012) Short self-assembling peptides as building blocks for modern nanodevices. Trends Biotechnol 30:155–165

Langenhan JM, Guzei IA, Gellman SH (2003) Parallel sheet secondary structure in β-peptides. Angew Chem Int Ed 42:2402–2405

Laursen JS, Engel-Andreasen J, Fristrup P, Harris P, Olsen CA (2013) Cis-trans amide bond rotamers in β-peptoids and peptoids: evaluation of stereoelectronic effects in backbone and side chains. J Am Chem Soc 135:2835–2844

Laursen JS, Engel-Andreasen J, Olsen CA (2015) β-peptoid foldamers at last. Acc Chem Res 48:2696–2704

Lee CH, Singla A, Lee Y (2001) Biomedical applications of collagen. Int J Pharm 221:1–22

Lee MR, Raguse TL, Schinnerl M, Pomerantz WC, Wang XD, Wipf P, Gellman SH (2007) Origins of the high 14-helix propensity of cyclohexyl-rigidified residues in β-peptides. Org Lett 9:1801–1804

Lee CS, Lin CH, Hsieh WL, Chiao SM, Tung WC (2014) Self-assembling of peptides is tuned via an extended amphipathic helix. J Nanosci Nanotechnol 14:5568–5573

Leggett GJ (2006) Scanning near-field photolithography-surface photochemistry with nanoscale spatial resolution. Chem Soc Rev 35:1150–1161

Lehn JM (2004) Supramolecular chemistry: from molecular information towards self-organization and complex matter. Rep Prog Phys 67:249–265

Leikin S, Rau DC, Parsegian VA (1995) Temperature-favored assembly of collagen is driven by hydrophilic not hydrphobic interactions. Nat Struct Biol 2:205–210

Lelais G, Seebach D (2004) β^2-amino acids – syntheses, occurrence in natural products, and components of β-peptides. Biopolymers 76:206–243

LePlae PR, Fisk JD, Porter EA, Weisblum B, Gellman SH (2002) Tolerance of acyclic residues in the β-peptide

12-helix: access to diverse side-chain arrays for biological applications. J Am Chem Soc 124:6820–6821
Levin A, Mason TO, Adler-Abramovich L, Buell AK, Meisl G, Galvagnion C, Bram Y, Stratford SA, Dobson CM, Knowles TPJ, Gazit E (2014) Ostwald's rule of stages governs structural transitions and morphology of dipeptide supramolecular polymers. Nat Commun 5:8
Li X, Yang D (2006) Peptides of aminoxy acids as foldamers. Chem Commun:3367–3379
Li Y, Yu SM (2013) Targeting and mimicking collagens via triple helical peptide assembly. Curr Opin Chem Biol 17:968–975
Liu YD, Goebl J, Yin YD (2013) Templated synthesis of nanostructured materials. Chem Soc Rev 42:2610–2653
Lombardo D, Kiselev MA, Magazu S, Calandra P (2015) Amphiphiles self-assembly: Basic concepts and future perspectives of supramolecular approaches. Adv Cond Matter Phys 2015
London F (1937) The general theory of molecular forces. Trans Faraday Society 33:8–26
Loo Y, Zhang SG, Hauser CAE (2012) From short peptides to nanofibers to macromolecular assemblies in biomedicine. Biotechnol Adv 30:593–603
Lowik DWPM, Leunissen EHP, van den Heuvel M, Hansen MB, van Hest JCM (2010) Stimulus responsive peptide based materials. Chem Soc Rev 39:3394–3412
Luder K, Kulkarni K, Lee HW, Widdop RE, Del Borgo MP, Aguilar MI (2016) Decorated self-assembling β^3-tripeptide foldamers form cell adhesive scaffolds. Chem. Commun 52:4549–4552
Luo D, Yan C, Wang T (2015) Interparticle forces underlying nanoparticle self-assemblies. Small 11:5984–6008
Lupas A (1996) Coiled coils: new structures and new functions. Trends Biochem Sci 21:375–382
Ma ML, Kuang Y, Gao Y, Zhang Y, Gao P, Xu B (2010) Aromatic-aromatic interactions induce the self-assembly of pentapeptidic derivatives in water to form nanofibers and supramolecular hydrogels. J Am Chem Soc 132:2719–2728
MacEwan SR, Chilkoti A (2014) Applications of elastin-like polypeptides in drug delivery. J. Contrl Release 190:314–330
MacPhee CE, Woolfson DN (2004) Engineered and designed peptide-based fibrous biomaterials. Curr Opin Solid State Mater Sci 8:141–149
Majd S, Power JH, Grantham HJM (2015) Neuronal response in Alzheimer's and Parkinson's disease: The effect of toxic proteins on intracellular pathways. BMC Neurosci 16:13
Mandal D, Shirazi AN, Parang K (2014) Self-assembly of peptides to nanostructures. Org Biomol Chem 12:3544–3561
Mann S (2009) Self-assembly and transformation of hybrid nano-objects and nanostructures under equilibrium and non-equilibrium conditions. Nat Mater 8:781–792
Mano JF (2008) Stimuli-responsive polymeric systems for biomedical applications. Adv Eng Mater 10:515–527
Mart RJ, Osborne RD, Stevens MM, Ulijn RV (2006) Peptide-based stimuli-responsive biomaterials. Soft Matter 2:822–835
Martinek TA, Fulop F (2003) Side-chain control of β-peptide secondary structures – Design principles. Eur J Biochem 270:3657–3666
Martinek TA, Fulop F (2012) Peptidic foldamers: ramping up diversity. Chem Soc Rev 41:687–702
Mata A, Geng YB, Henrikson KJ, Aparicio C, Stock SR, Satcher RL, Stupp SI (2010) Bone regeneration mediated by biomimetic mineralization of a nanofiber matrix. Biomaterials 31:6004–6012
Matsuurua K (2014) Rational design of self-assembled proteins and peptides for nano- and micro-sized architectures. RSC Adv 4:2942–2953
Mattia E, Otto S (2015) Supramolecular systems chemistry. Nat Nanotechnol 10:111–119
McDonald CA, Payne NL, Sun GZ, Clayton DJ, Del Borgo MP, Aguilar MI, Perlmutter P, Bernard CCA (2014) Single β^3-amino acid substitutions to MOG peptides suppress the development of experimental autoimmune encephalomyelitis. J Neuroimmunol 277:67–76
McGovern M, Abbott N, de Pablo JJ (2012) Dimerization of helical β-peptides in solution. Biophys J 102:1435–1442
Mendes AC, Baran ET, Reis RL, Azevedo HS (2013) Self-assembly in nature: using the principles of nature to create complex nanobiomaterials. Wiley Interdiscip Rev Nanomed Nanobiotechnol 5:582–612
Minor DL, Kim PS (1994) Measurment of the β-sheet-forming propensities of amino-acids. Nature 367:660–663
Mishra A, Loo YH, Deng RH, Chuah YJ, Hee HT, Ying JY, Hauser CAE (2011) Ultrasmall natural peptides self-assemble to strong temperature-resistant helical fibers in scaffolds suitable for tissue engineering. Nano Today 6:232–239
Mohle K, Gunther R, Thormann M, Sewald N, Hofmann HJ (1999) Basic conformers in β-peptides. Biopolymers 50:167–184
Mondal J, Sung BJ, Yethiraj A (2010a) Sequence dependent self-assembly of β-peptides: insights from a coarse-grained model. J Chem Phys 132:7
Mondal J, Zhu X, Cui QA, Yethiraj A (2010b) Self-Assembly of β-peptides: insight from the pair and many-body free energy of association. J Phys Chem C 114:13551–13556
Mondal J, Zhu X, Cui QA, Yethiraj A (2010c) Sequence-dependent interaction of β-peptides with membranes. J Phys Chem B 114:13585–13592
Monera OD, Sereda TJ, Zhou NE, Kay CM, Hodges RS (1995) Relationship of sidechain hydrophobicity and α-helical propensity on the stability of the single-stranded amphipathic α-helix. J Pept Sci 1:319–329
Mora T, Yu H, Sowa Y, Wingreen NS (2009) Steps in the bacterial flagellar motor. PLoS Comp Biol 5:11

Moreau A, Gosselin-Badaroudine P, Chahine M (2014) Molecular biology and biophysical properties of ion channel gating pores. Q Rev Biophys 47:364–388

Narita M, Doi M, Kudo K, Terauchi Y (1986) Conformations in the solid-state and solubility properties of protected homooligopeptides of glycine and β-alanine. Bull Chem Soc Jpn 59:3553–3557

Nesloney CL, Kelly JW (1996) Progress towards understanding β-sheet structure. Biorg Med Chem 4:739–766

Nguyen LTH, Chen SL, Elumalai NK, Prabhakaran MP, Zong Y, Vijila C, Allakhverdiev SI, Ramakrishna S (2013) Biological, chemical, and electronic applications of nanofibers. Macromol Mater Eng 298:822–867

Niece KL, Hartgerink JD, Donners J, Stupp SI (2003) Self-assembly combining two bioactive peptide-amphiphile molecules into nanofibers by electrostatic attraction. J Am Chem Soc 125:7146–7147

Niggli DA, Ebert MO, Lin ZX, Seebach D, van Gunsteren WF (2012) Helical content of a β^3-octapeptide in methanol: molecular dynamics simulations explain a seeming discrepancy between conclusions derived from CD and NMR data. Chem Eur J 18:586–593

Nikolic A, Baud S, Rauscher S, Pomes R (2011) Molecular mechanism of β-sheet self-organization at water-hydrophobic interfaces. Proteins 79:1–22

Norgren AS, Zhang SD, Arvidsson PI (2006) Synthesis and circular dichroism spectroscopic investigations of oligomeric β-peptoids with α-chiral side chains. Org Lett 8:4533–4536

Numata K (2015) Poly(amino acid)s/polypeptides as potential functional and structural materials. Polym J 47:537–545

Olsen CA (2010) Peptoid-peptide hybrid backbone architectures. ChemBioChem 11:152–160

Olsen CA, Lambert M, Witt M, Franzyk H, Jaroszewski JW (2008) Solid-phase peptide synthesis and circular dichroism study of chiral β-peptoid homooligomers. Amino Acids 34:465–471

Padmanabhan S, Baldwin RL (1991) Straight-chain nonpolar amino-acids are good helix-formers in water. J Mol Biol 219:135–137

Padmanabhan S, Marqusee S, Ridgeway T, Laue TM, Baldwin RL (1990) Relative helix-forming tendencies of nonpolar amino-acids. Nature 344:268–270

Pagel K, Koksch B (2008) Following polypeptide folding and assembly with conformational switches. Curr Opin Chem Biol 12:730–739

Pandya MJ, Spooner GM, Sunde M, Thorpe JR, Rodger A, Woolfson DN (2000) Sticky-end assembly of a designed peptide fiber provides insight into protein fibrillogenesis. Biochemistry 39:8728–8734

Paramonov SE, Jun HW, Hartgerink JD (2006) Self-assembly of peptide-amphiphile nanofibers: the roles of hydrogen bonding and amphiphilic packing. J Am Chem Soc 128:7291–7298

Park S, Kwon TG, Lee SY (2013) Controlled self-assembly and alignment of organic-magnetic hybrid microrods. Powder Technol 250:46–51

Pasini D, Kraft A (2004) Supramolecular self-assembly of fibres. Curr Opin Solid State Mater Sci 8:157–163

Patch JA, Barron AE (2002) Mimicry of bioactive peptides via non-natural, sequence-specific peptidomimetic oligomers. Curr Opin Chem Biol 6:872–877

Pizzey CL, Pomerantz WC, Sung BJ, Yuwono VM, Gellman SH, Hartgerink JD, Yethiraj A, Abbott NL (2008) Characterization of nanofibers formed by self-assembly of β-peptide oligomers using small angle x-ray scattering. J Chem Phys 129:8

Pochan DJ, Schneider JP, Kretsinger J, Ozbas B, Rajagopal K, Haines L (2003) Thermally reversible hydrogels via intramolecular folding and consequent self-assembly of a *de Novo* designed peptide. J Am Chem Soc 125:11802–11803

Pomerantz WC, Yuwono VM, Pizzey CL, Hartgerink JD, Abbott NL, Gellman SH (2008) Nanofibers and lyotropic liquid crystals from a class of self-assembling β-peptides. Angew Chem Int Ed 47:1241–1244

Porter EA, Wang XF, Schmitt MA, Gellman SH (2002) Synthesis and 12-helical secondary structure of β-peptides containing (2R,3R)-aminoproline. Org Lett 4:3317–3319

Prins LJ, Reinhoudt DN, Timmerman P (2001) Noncovalent synthesis using hydrogen bonding. Angew Chem Int Ed 40:2382–2426

Profit AA, Felsen V, Chinwong J, Mojica ERE, Desamero RZB (2013) Evidence of π-stacking interactions in the self-assembly of $hIAPP_{22-29}$. Proteins 81:690–703

Qin SY, Xu XD, Chen CS, Chen JX, Li ZY, Zhuo RX, Zhang XZ (2011) Supramolecular architectures self-assembled from asymmetrical hetero cyclopeptides. Macromol Rapid Commun 32:758–764

Qin SY, Xu SS, Zhuo RX, Zhang XZ (2012) Morphology transformation via pH-triggered self-assembly of peptides. Langmuir 28:2083–2090

Qiu F, Chen YZ, Zhao XJ (2009) Comparative studies on the self-assembling behaviors of cationic and catanionic surfactant-like peptides. J Colloid Interface Sci 336:477–484

Quandt A, Ozdogan C (2010) Feynman, biominerals and graphene – Basic aspects of nanoscience. Commun Nonlinear Sci Numer Simulat 15:1575–1582

Quantock AJ, Winkler M, Parfitt GJ, Young RD, Brown DJ, Boote C, Jester JV (2015) From nano to macro: studying the hierarchical structure of the corneal extracellular matrix. Exp Eye Res 133:81–99

Rad-Malekshahi M, Lempsink L, Amidi M, Hennink WE, Mastrobattista E (2016) Biomedical applications of self-assembling peptides. Bioconj Chem 27:3–18

Raguse TL, Lai JR, LePlae PR, Gellman SH (2001) Toward β-peptide tertiary structure: self-association of an amphiphilic *14*-helix in aqueous solution. Org Lett 3:3963–3966

Raguse TL, Lai JR, Gellman SH (2002a) Evidence that the β-peptide *14*-helix is stabilized by β^3-residues with side-chain branching adjacent to the β-carbon atom. Helv Chim Acta 85:4154–4164

Raguse TL, Porter EA, Weisblum B, Gellman SH (2002b) Structure-activity studies of *14*-helical antimicrobial β-peptides: probing the relationship between conformational stability and antimicrobial potency. J Am Chem Soc 124:12774–12785

Raguse TL, Lai JR, Gellman SH (2003) Environment-independent *14*-helix formation in short β-peptides: striking a balance between shape control and functional diversity. J Am Chem Soc 125:5592–5593

Rajagopal K, Schneider JP (2004) Self-assembling peptides and proteins for nanotechnological applications. Curr Opin Struct Biol 14:480–486

Ralston E, De Coen JL (1974) Folding of polypeptide chains induced by the amino acid side-chains. J Mol Biol 83:393–420

Ramstrom O, Bunyapaiboonsri T, Lohmann S, Lehn JM (2002) Chemical biology of dynamic combinatorial libraries. Biochim Biophys Acta 1572:178–186

Rasale DB, Das AK (2015) Chemical reactions directed peptide self-assembly. Int J Mol Sci 16:10797–10820

Reches M, Gazit E (2003) Casting metal nanowires within discrete self-assembled peptide nanotubes. Science 300:625–627

Reches M, Gazit E (2004) Formation of closed-cage nanostructures by self-assembly of aromatic dipeptides. Nano Lett 4:581–585

Reimers K, Liebsch C, Radtke C, Kuhbier JW, Vogt PM (2015) Silks as scaffolds for skin reconstruction. Biotechnol Bioeng 112:2201–2205

Reyes FT, Kelland MA, Kumar N, Jia L (2015) First Investigation of the kinetic hydrate inhibition of a series of poly(β-peptoid)s on structure II gas hydrate, including the comparison of block and random copolymers. Energy Fuels 29:695–701

Rissanou AN, Georgilis E, Kasotaids E, Mitraki A, Harmandaris V (2013) Effect of solvent on the self-assembly of dialanine and diphenylalanine peptides. J Phys Chem B 117:3962–3975

Robinson JA (2011) Protein epitope mimetics as anti-infectives. Curr Opin Chem Biol 15:379–386

Rodnikova MN (2007) A new approach to the mechanism of solvophobic interactions. J Mol Liq 136:211–213

Rodnikova M, Barthel J (2007) Elasticity of the spatial network of hydrogen bonds in liquids and solutions. J Mol Liq 131:121–123

Roy O, Faure S, Thery V, Didierjean C, Taillefumier C (2008) Cyclic β-peptoids. Org Lett 10:921–924

Rueping M, Schreiber JV, Lelais G, Jaun B, Seebach D (2002) Mixed β^2/β^3-hexapeptides and β^2/β^3-nonapeptides folding to (*P*)-helices with alternating twelve- and ten-membered hydrogen-bonded rings. Helv Chim Acta 85:2577–2593

Rueping M, Mahajan YR, Jaun B, Seebach D (2004) Design, synthesis and structural investigations of a β-peptide forming a 3_{14}-helix stabilized by electrostatic interactions. Chem Eur J 10:1607–1615

Russell VA, Ward MD (1996) Molecular crystals with dimensionally controlled hydrogen-bonded nanostructures. Chem Mater 8:1654–1666

Rybtchinski B (2011) Adaptive supramolecular nanomaterials based on strong noncovalent interactions. ACS Nano 5:6791–6818

Ryu J, Park CB (2008) High-temperature self-assembly of peptides into vertically well-aligned nanowires by aniline vapor. Adv Mater 20:3754–3758

Ryu JH, Hong DJ, Lee M (2008) Aqueous self-assembly of aromatic rod building blocks. Chem Commun:1043–1054

Sargeant TD, Guler MO, Oppenheimer SM, Mata A, Satcher RL, Dunand DC, Stupp SI (2008) Hybrid bone implants: self-assembly of peptide amphiphile nanofibers within porous titanium. Biomaterials 29:161–171

Scanlon S, Aggeli A (2008) Self-assembling peptide nanotubes. Nano Today 3:22–30

Schaller V, Schmoller KM, Karakose E, Hammerich B, Maier M, Bausch AR (2013) Crosslinking proteins modulate the self-organization of driven systems. Soft Matter 9:7229–7233

Scholtz JM, Baldwin RL (1992) The mechanism of α-helix formation by peptides. Annu Rev Biophys Biomol Struct 21:95–118

Scott RW, DeGrado WF, Tew GN (2008) *De novo* designed synthetic mimics of antimicrobial peptides. Curr Opin Biotechnol 19:620–627

Seabra AB, Duran N (2013) Biological applications of peptides nanotubes: an overview. Peptides 39:47–54

Seebach D, Gardiner J (2008) β-peptidic peptidomimetics. Acc Chem Res 41:1366–1375

Seebach D, Matthews JL (1997) β-peptides: a surprise at every turn. Chem Commun 2015–2022

Seebach D, Ciceri PE, Overhand M, Jaun B, Rigo D, Oberer L, Hommel U, Amstutz R, Widmer H (1996) Probing the helical secondary structure of short-chain β-peptides. Helv Chim Acta 79:2043–2066

Seebach D, Gademann K, Schreiber JV, Matthews JL, Hintermann T, Jaun B, Oberer L, Hommel U, Widmer H (1997) 'Mixed' β-peptides: a unique helical secondary structure in solution. Helv Chim Acta 80:2033–2038

Seebach D, Abele S, Gademann K, Guichard G, Hintermann T, Jaun B, Matthews JL, Schreiber JV (1998) β^2- and β^3-peptides with proteinaceous side chains: synthesis and solution structures of constitutional isomers, a novel helical secondary structure and the influence of solvation and hydrophobic interactions on folding. Helv Chim Acta 81:932–982

Seebach D, Beck AK, Bierbaum DJ (2004) The world of β- and γ-peptides comprised of homologated proteinogenic amino acids and other components. Chem Biodivers 1:1111–1239

Seebach D, Hook DF, Glattli A (2006) Helices and other secondary structures of β- and γ-peptides. Biopolymers 84:23–37

Seoudi RS, Del Borgo MP, Kulkarni K, Perlmutter P, Aguilar MI, Mechler A (2015a) Supramolecular self-assembly of 14-helical nanorods with tunable linear and dendritic hierarchical morphologies. New J Chem 39:3280–3287

Seoudi RS, Dowd A, Del Borgo M, Kulkarni K, Perlmutter P, Aguilar MI, Mechler A (2015b) Amino acid sequence controls the self-assembled superstructure morphology of N-acetylated tri-β^3-peptides. Pure Appl Chem 87:1021–1028

Seoudi RS, Dowd A, Smith BJ, Mechler A (2016a) Structural analysis of bioinspired nano materials with synchrotron far IR spectroscopy. Phys Chem Chem Phys 18:11467–11473

Seoudi RS, Hinds MG, Wilson DJD, Adda CG, Del Borgo M, Aguilar MI, Perlmutter P, Mechler A (2016b) Self-assembled nanomaterials based on beta *(β^3)* tetrapeptides. Nanotechnology 27:9

Seow WY, Hauser CAE (2014) Short to ultrashort peptide hydrogels for biomedical uses. Mater Today 17:381–388

Shao H, Seifert J, Romano NC, Gao M, Helmus JJ, Jaroniec CP, Modarelli DA, Parquette JR (2010) Amphiphilic self-assembly of an n-type nanotube. Angew Chem Int Ed 49:7688–7691

Shen Y, Fu XH, Fu WX, Li ZB (2015) Biodegradable stimuli-responsive polypeptide materials prepared by ring opening polymerization. Chem Soc Rev 44:612–622

Shimizu T (2003) Bottom-up synthesis and morphological control of high-axial-ratio nanostructures through molecular self-assembly. Polym J 35:1–22

Shimomura M, Sawadaishi T (2001) Bottom-up strategy of materials fabrication: a new trend in nanotechnology of soft materials. Curr Opin Colloid Interface Sci 6:11–16

Shu LH, Ueda K, Chiu I, Cheong H (2011) Biologically inspired design. Cirp Ann Manuf Techn 60:673–693

Shuvaev VV, Brenner JS, Muzykantov VR (2015) Targeted endothelial nanomedicine for common acute pathological conditions. J Control Release 219:576–595

Sijbesma RP, Meijer EW (1999) Self-assembly of well-defined structures by hydrogen bonding. Curr Opin Colloid Interface Sci 4:24–32

Skovbakke SL, Larsen CJ, Heegaard PMH, Moesby L, Franzyk H (2015) Lipidated α-peptide/β-peptoid hybrids with potent anti-inflammatory activity. J Med Chem 58:801–813

Smith CK, Regan L (1997) Construction and design of β-sheets. Acc Chem Res 30:153–161

Smith PA, Nordquist CD, Jackson TN, Mayer TS, Martin BR, Mbindyo J, Mallouk TE (2000) Electric-field assisted assembly and alignment of metallic nanowires. Appl Phys Lett 77:1399–1401

Solar M, Buehler MJ (2014) Tensile deformation and failure of amyloid and amyloid-like protein fibrils. Nanotechnology 25

Srivastava R, Ray AK, Diederichsen U (2009) Higher aggregation of β-peptide networks controlled by nucleobase pairing. Eur J Org Chem 4793–4800

Stahl PJ, Yu SM (2012) Encoding cell-instructive cues to PEG-based hydrogels via triple helical peptide assembly. Soft Matter 8:10409–10418

Standley SM, Toft DJ, Cheng H, Soukasene S, Chen J, Raja SM, Band V, Band H, Cryns VL, Stupp SI (2010) Induction of cancer cell death by self-assembling nanostructures incorporating a cytotoxic peptide. Cancer Res 70:3020–3026

Steer DL, Lew RA, Perlmutter P, Smith AI, Aguilar MI (2002) β-amino acids: Versatile peptidomimetics. Curr Med Chem 9:811–822

Stendahl JC, Rao MS, Guler MO, Stupp SI (2006) Intermolecular forces in the self-assembly of peptide amphiphile nanofibers. Adv Funct Mater 16:499–508

Stephanopoulos N, Ortony JH, Stupp SI (2013) Self-assembly for the synthesis of functional biomaterials. Acta Mater 61:912–930

Stephens OM, Kim S, Welch BD, Hodsdon ME, Kay MS, Schepartz A (2005) Inhibiting HIV fusion with a β-peptide foldamer. J Am Chem Soc 127:13126–13127

Stoop R (2008) Smart biomaterials for tissue engineering of cartilage. Injury 39:S77–S87

Strijowski U, Sewald N (2004) Structural properties of cyclic peptides containing cis- or trans-2-aminocyclohexane carboxylic acid. Org Biomol Chem 2:1105–1109

Stupp SI, Donners J, Li LS, Mata A (2005) Expanding frontiers in biomaterials. MRS Bull 30:864–873

Stupp SI, Zha RH, Palmer LC, Cui HG, Bitton R (2013) Self-assembly of biomolecular soft matter. Faraday Discuss 166:9–30

Su K, Wang CM (2015) Recent advances in the use of gelatin in biomedical research. Biotechnol Lett 37:2139–2145

Subramanian R, Kapoor TM (2012) Building complexity: insights into self-organized assembly of microtubule-based architectures. Dev Cell 23:874–885

Sun TL, Qing GY, Su BL, Jiang L (2011) Functional biointerface materials inspired from nature. Chem Soc Rev 40:2909–2921

Sun XJ, Chen WD, Wang YD (2015) β-amyloid: the key peptide in the pathogenesis of Alzheimer's disease. Front Pharmacol 6:9

Tamerler C, Sarikaya M (2008) Molecular biomimetics: genetic synthesis, assembly, and formation of materials using peptides. MRS Bull 33:504–510

Tanase M, Silevitch DM, Hultgren A, Bauer LA, Searson PC, Meyer GJ, Reich DH (2002) Magnetic trapping and self-assembly of multicomponent nanowires. J Appl Phys 91:8549–8551

Tang CK, Qiu F, Zhao XJ (2013) Molecular design and applications of self-assembling surfactant-like peptides. J Nanomater 9

Tegoni M (2014) *De Novo* designed copper α-helical peptides: from design to function. Eur J Inorg Chem 2014:2177–2193

Tew GN, Scott RW, Klein ML, Degrado WF (2010) *De Novo* design of antimicrobial polymers, foldamers, and small molecules: from discovery to practical applications. Acc Chem Res 43:30–39

Thomas F, Burgess NC, Thomson AR, Woolfson DN (2016) Controlling the assembly of coiled–coil peptide nanotubes. Angew Chem Int Ed 128:999–1003
Thomson AR, Wood CW, Burton AJ, Bartlett GJ, Sessions RB, Brady RL, Woolfson DN (2014) Computational design of water-soluble α-helical barrels. Science 346:485–488
Tian XB, Sun FD, Zhou XR, Luo SZ, Chen L (2015) Role of peptide self-assembly in antimicrobial peptides. J Pept Sci 21:530–539
Toksoz S, Acar H, Guler MO (2010) Self-assembled one-dimensional soft nanostructures. Soft Matter 6:5839–5849
Tomasini C, Huc., Aitken DJ, Fulop F (2013) Foldamers. Eur J Org Chem 3408–3409
Torres E, Gorrea E, Da Silva E, Nolis P, Branchadell V, Ortuno RM (2009) Prevalence of eight-membered hydrogen-bonded rings in some bis(cyclobutane) β-dipeptides including residues with trans stereochemistry. Org Lett 11:2301–2304
Ulijn RV, Smith AM (2008) Designing peptide based nanomaterials. Chem Soc Rev 37:664–675
Urman S, Gaus K, Yang Y, Strijowski U, Sewald N, De Pol S, Reiser O (2007) The constrained amino acid β-Acc confers potency and selectivity to integrin ligands. Angew Chem Int Ed 46:3976–3978
Vaz E, Pomerantz WC, Geyer M, Gellman SH, Brunsveld L (2008) Comparison of design strategies for promotion of β-peptide 14-helix stability in water. ChemBioChem 9:2254–2259
Velichko YS, Mantei JR, Bitton R, Carvajal D, Shull KR, Stupp SI (2012) Electric field controlled self-assembly of hierarchically ordered membranes. Adv Funct Mater 22:369–377
Verma M, Vats A, Taneja V (2015) Toxic species in amyloid disorders: oligomers or mature fibrils. Ann Indian Acad Neurol 18:138–145
Versluis F, Marsden HR, Kros A (2010) Power struggles in peptide-amphiphile nanostructures. Chem Soc Rev 39:3434–3444
Victorov A (2015) Molecular thermodynamics of soft self-assembling structures for engineering applications. J Chem Technol Biotechnol 90:1357–1363
Vijayakumar M, Qian H, Zhou HX (1999) Hydrogen bonds between short polar side chains and peptide backbone: prevalence in proteins and effects on helix-forming propensities. Proteins 34:497–507
Wang YX, Robertson JL, Spillman WB, Claus RO (2004) Effects of the chemical structure and the surface properties of polymeric biomaterials on their biocompatibility. Pharm Res 21:1362–1373
Wang X, Liu F, Andavan GTS, Jing XY, Singh K, Yazdanpanah VR, Bruque N, Pandey RR, Lake R, Ozkan M, Wang KL, Ozkan CS (2006a) Carbon nanotube-DNA nanoarchitectures and electronic functionality. Small 2:1356–1365
Wang XY, Kim HJ, Wong C, Vepari C, Matsumoto A, Kaplan DL (2006b) Fibrous proteins and tissue engineering. Mater Today 9:44–53
Wang Y, Xu H, Zhang X (2009) Tuning the amphiphilicity of building blocks: controlled self-assembly and disassembly for functional supramolecular materials. Adv Mater 21:2849–2864
Wang MJ, Du LJ, Wu XL, Xiong SJ, Chu PK (2011) Charged diphenylalanine nanotubes and controlled hierarchical self-assembly. ACS Nano 5:4448–4454
Werner HM, Horne WS (2015) Folding and function in α/β-peptides: targets and therapeutic applications. Curr Opin Chem Biol 28:75–82
Whitesides GM, Grzybowski B (2002) Self-assembly at all scales. Science 295:2418–2421
Whitesides GM, Mathias JP, Seto CT (1991) Molecular self-assembly and nanochemistry: a chemical strategy for the synthesis of nanostructures. Science 254:1312–1319
Wipf P, Wang XD (2000) Diels-Alder approaches to ring-functionalized cyclic β-amino acids. Tetrahedron Lett 41:8747–8751
Wolfenden R (1978) Interaction of peptide-bond with solvent water-vapor-phase analysis. Biochemistry 17:201–204
Woolfson DN (2010) Building fibrous biomaterials from α-helical and collagen-like coiled-coil peptides. Biopolymers 94:118–127
Woolfson DN, Ryadnov MG (2006) Peptide-based fibrous biomaterials: some things old, new and borrowed. Curr Opin Chem Biol 10:559–567
Wu YD, Wang DP (1998) Theoretical studies of β-peptide models. J Am Chem Soc 120:13485–13493
Wu YD, Wang DP (1999) Theoretical study on side-chain control of the 14-helix and the 10/12-helix of β-peptides. J Am Chem Soc 121:9352–9362
Wu YD, Wang DP (2000) Theoretical study of $\beta^{2,3}$-peptide models. J Chin Chem Soc 47:129–134
Wu YD, Han W, Wang DP, Gao Y, Zhao YL (2008) Theoretical analysis of secondary structures of β-peptides. Acc Chem Res 41:1418–1427
Yan XH, Zhu PL, Li JB (2010) Self-assembly and application of diphenylalanine-based nanostructures. Chem Soc Rev 39:1877–1890
Yang H, Fung SY, Pritzker M, Chen P (2007a) Modification of hydrophilic and hydrophobic surfaces using an ionic-complementary peptide. Plos One 211
Yang H, Fung SY, Pritzker M, Chen P (2007b) Surface-assisted assembly of an ionic-complementary peptide: controllable growth of nanofibers. J Am Chem Soc 129:12200–12210
Yang YL, Khoe U, Wang XM, Horii A, Yokoi H, Zhang SG (2009) Designer self-assembling peptide nanomaterials. Nano Today 4:193–210
Yoshida M, Langer R, Lendlein A, Lahann J (2006) From advanced biomedical coatings to multi-functionalized biomaterials. Polym Rev 46:347–375
Yu J, Wang QR, Zhang X (2014) Effects of external force fields on peptide self-assembly and biomimetic silica synthesis. Appl Surf Sci 311:799–807

Zastrow ML, Pecoraro VL (2013) Designing functional metalloproteins: from structural to catalytic metal sites. Coord Chem Rev 257:2565–2588
Zhang SG (2002) Emerging biological materials through molecular self-assembly. Biotechnol Adv 20:321–339
Zhang SG (2003) Fabrication of novel biomaterials through molecular self-assembly. Nat Biotechnol 21:1171–1178
Zhang SG, Altman M (1999) Peptide self-assembly in functional polymer science and engineering. React Funct Polym 41:91–102
Zhang SG, Rich A (1997) Direct conversion of an oligopeptide from a β-sheet to an α-helix: a model for amyloid formation. Proc Natl Acad Sci USA 94:23–28
Zhang X, Wang C (2011) Supramolecular amphiphiles. Chem Soc Rev 40:94–101
Zhang SG, Holmes T, Lockshin C, Rich (1993) Spontaneous assembly of a self-complementary oligopeptide to form a stable macroscopic membrane. Proc Natl Acad Sci USA 90:3334–3338
Zhang SG, Holmes TC, Dipersio CM, Hynes RO, Su X, Rich A (1995) Self-complementary oligopeptide matrices support mammalian-cell attachment. Biomaterials 16:1385–1393
Zhang SG, Marini DM, Hwang W, Santoso S (2002) Design of nanostructured biological materials through self-assembly of peptides and proteins. Curr Opin Chem Biol 6:865–871
Zhang Y, Gu HW, Yang ZM, Xu B (2003) Supramolecular hydrogels respond to ligand-receptor interaction. J Am Chem Soc 125:13680–13681
Zhang Q, Liu SJ, Yu SH (2009) Recent advances in oriented attachment growth and synthesis of functional materials: concept, evidence, mechanism, and future. J Mater Chem 19:191–207
Zhang RJ, Zheng N, Song ZY, Yin LC, Cheng JJ (2014) The effect of side-chain functionality and hydrophobicity on the gene delivery capabilities of cationic helical polypeptides. Biomaterials 35:3443–3454
Zhang CW, Hang LT, Yao TP, Lim KL (2016) Parkin regulation and neurodegenerative disorders. Front Aging Neurosci 7:15
Zhao XB, Pan F, Lu JR (2008) Recent development of peptide self-assembly. Prog Nat Sci Mater Int 18:653–660
Zhao LN, Long HW, Mu YG, Chew LY (2012) The toxicity of amyloid β oligomers. Int J Mol Sci 13:7303–7327
Zhao LX, Li NN, Wang KM, Shi CH, Zhang LL, Luan YX (2014) A review of polypeptide-based polymersomes. Biomaterials 35:1284–1301
Zhong J, Liu XW, Wei DX, Yan J, Wang P, Sun G, He DN (2015) Effect of incubation temperature on the self-assembly of regenerated silk fibroin: a study using AFM. Int J Biol Macromol 76:195–202
Zhu PL, Yan XH, Su Y, Yang Y, Li JB (2010) Solvent-induced structural transition of self-assembled dipeptide: from organogels to microcrystals. Chem Eur J 16:3176–3183

Bioprinting and Biofabrication with Peptide and Protein Biomaterials

5

Mitchell Boyd-Moss, Kate Fox, Milan Brandt, David Nisbet, and Richard Williams

Abstract

The ability to fabricate artificial tissue constructs through the controlled organisation of cells, structures and signals within a biomimetic scaffold offers significant promise to the field of regenerative medicine, drug delivery and tissue engineering. Advances in additive manufacturing technologies have facilitated the printing of spatially defined cell-laden artificial tissue constructs capable of providing biomimetic spatiotemporal presentation of biological and physical cues to cells in a designed multicomponent structure. Despite significant progress in the field of bioprinting, a key challenge remains in developing and utilizing materials that can adequately recapitulate the complexities of the native extracellular matrix on a nanostructured, chemical level during the printing process. This gives rise to the need for suitable materials - particularly in establishing effective control over cell fate, tissue vascularization and innervation. Recently, significant interested has been invested into developing candidate materials using protein and peptide-derived biomaterials. The ability of these materials to form highly printable hydrogels which are reminiscent of the native ECM has seen significant use in a variety of regenative applications, including both organ bioprinting and non-organ bioprinting. Here, we discuss the emerging technologies for peptide-based bioprinting applications, highlighting bioink development and detailing bioprinter processors.

M. Boyd-Moss • R. Williams (✉)
School of Engineering, RMIT University, Melbourne, VIC, Australia

Biofab3D, St. Vincent's Hospital Melbourne, Fitzroy, VIC, Australia
e-mail: richard.williams@rmit.edu.au

K. Fox
School of Engineering, RMIT University, Melbourne, VIC, Australia

M. Brandt
Centre for Additive Manufacturing, School of Engineering, RMIT University, Melbourne, VIC, Australia

School of Engineering, RMIT University, Melbourne, VIC, Australia

D. Nisbet
Laboratory of Advanced Biomaterials, Research School of Engineering, The Australian National University, Canberra, ACT, Australia

© Springer International Publishing AG 2017

A. Sunna et al. (eds.), *Peptides and Peptide-based Biomaterials and their Biomedical Applications*, Advances in Experimental Medicine and Biology 1030, https://doi.org/10.1007/978-3-319-66095-0_5

Furthermore, this work presents application specific, peptide-based bioprinting approaches, and provides insight into current limitations and future perspectives of peptide-based bioprinting techniques.

Keywords
Bioprinting • Bioink • Biofabrication • Peptide • Protein biomaterials

Abbreviations

μCT	Micro computed tomography
10 T1/2	Mouse embryo fibroblasts
ADSC	Adipose derived stromal cells
AFA-LIFT	absorbing film-assisted laser-induced forward transfer
B16	Carcinoma cell line
BMSC	Bone marrow derived mesenchymal stem cells
C2C12	C2C12 myoblast cells
$CaCl_2$	Calcium chloride
CB[6]	Cucurbit[6]uril
DAH-HA	1,6-diaminohexane (DAH)-conjugated HA
dECM	Decellularized extracellular matrix
DPBS	Dulbecco's phosphate-buffered saline solution
ECM	Extracellular Matrix
GAG	Glycosaminoglycan
GelMA	Gelatin methacryloyl
GG	Gellan gum
H1ESCs	H1 embryonic stem cells
HA	Hyaluronic acid
HAC	Human articular chondrocytes
HAp	Hydroxyapatite
hASC	Human adipose-derived stem cells
HAVIC	Human aortic Valvular Interstitial cells
hCPC	Human cardiac progenitor cells
HEK293	Human embryonic kidney cell line
HepG2/C3A	Human liver cancer cells (hepatocellular carcinoma)
hFB	Human fibroblast cells
hKC	Human keratinocyte cells
hMSC	Human mesenchymal stem cells
HMVEC	Human microvascular endothelial cells
HUVEC	human umbilical vein endothelial cells
HNDF	Human neonatal dermal fibroblasts
HSC	Human bone marrow-derived mesenchymal stromal cells
hTMSC	Human inferior turbinate-tissue derived mesenchymal stromal cells
L6	Rat myoblast cell line
MAPLE-DW	Matrix-Assisted Pulsed Laser Evaporation Direct Writing
MC3T3 E1	Osteoblast precursor cells (Add to main article)
MCF-7	Breast adenocarcinoma cells
MG63	Osteoblast-like cells
MSC	Mesenchymal stem cells

mTgase	Microbial transglutaminise
NT2	Neural cells
PBS	Phosphate-buffered saline solution
PCL	Polycaprolactone
PEG	Poly(ethylene Glycol)
PEGDA	Poly(ethylene Glycol)-di-acrylate
PEGDMA	Poly(ethylene glycol) dimethacrylate
PEGMA	Poly(ethylene glycol) methacrylate
PEGTA	Poly(ethylene glycol)-tetra-acrylate
PEO	Poly(ethylene oxide)
PLA	Polylactic acid
PVA	Polyvinyl alcohol
Ru/SPS	Ruthenium/Sodium persulfate
SAP	Self-Assembled Peptide
SF-G	Silk-fibroin/Gelatin
SLA	Stereolithography
SMC	Smooth muscle cells
SPPS	Solid phase peptide synthesis
TGF-β	Transforming growth factor-β
VEGF	Vascular endothelial growth factor
VIC	Aortic valve leaflet interstitial cells

5.1 Introduction

Regenerative medicine is a rapidly advancing field which aims to replace functional tissue lost to disease, trauma or congenital anomalies (Sadtler et al. 2016). A key challenge for researchers is the requirement to develop an advanced, dynamic tissue engineered scaffold capable of adequately mimicking the complexities of the native extracellular matrix (ECM); providing the necessary behaviour cues to cells whilst promoting angiogenesis to enable the delivery of nutrients to developing tissue while facilitating waste removal (Chen and Liu 2016; Novosel et al. 2011; Fortunato et al. 2016). To achieve this, multidisciplinary techniques and insight from engineering, chemistry, material science, nanotechnology and biology are required to enable the tailored design of biomimetic scaffolds. These highly structured materials must reproduce aspects of the natural tissue environment; for instance, having the capacity to simultaneously providing physical support and biologically relevant molecules to cells. Effective materials should be capable of promoting and regulating regenerative cell behaviours, including survival, migration, proliferation, and differentiation down appropriate linages (Horgan et al. 2016).

The recent advancements of additive manufacturing technologies, material science, and cell biology have brought forth the ability to print complex artificial tissue-engineered scaffolds through bioprinting (Murphy and Atala 2014). This technology enables the precise placement of structural and biologically suitable materials containing living cells, biochemicals, and biologically relevant molecules within a well-defined geometry (Murphy and Atala 2014). The ability of these printed tissue scaffolds to have spatially defined regions of different material have the potential for the hitherto unavailable fabrication of tissue constructs capable of providing the necessary behavioural cues to cells, whilst also facilitating vascular network generation (Richards et al. 2017). Currently there are a number of bioprinting techniques being employed for the printing of tissue constructs based on commercially available or modified printers, including inkjet bioprinting, extrusion bioprinting, laser-assisted bioprinting, and stereolithography-based bioprinting. Figure 5.1 shows an idealised cartoon of the ability to present multiple cell types in a defined fashion with great fidelity.

These rapidly advancing techniques are being developed to print blends of biomaterials containing living cells in a liquid or hydrogel suspension, which are termed 'bioinks' (Hölzl et al. 2016). Ideally, bioinks should form tuneable, biocompatible scaffolds with adequate mechanical strength, rigidity, and shape fidelity; resemble the native ECM; and, facilitate large

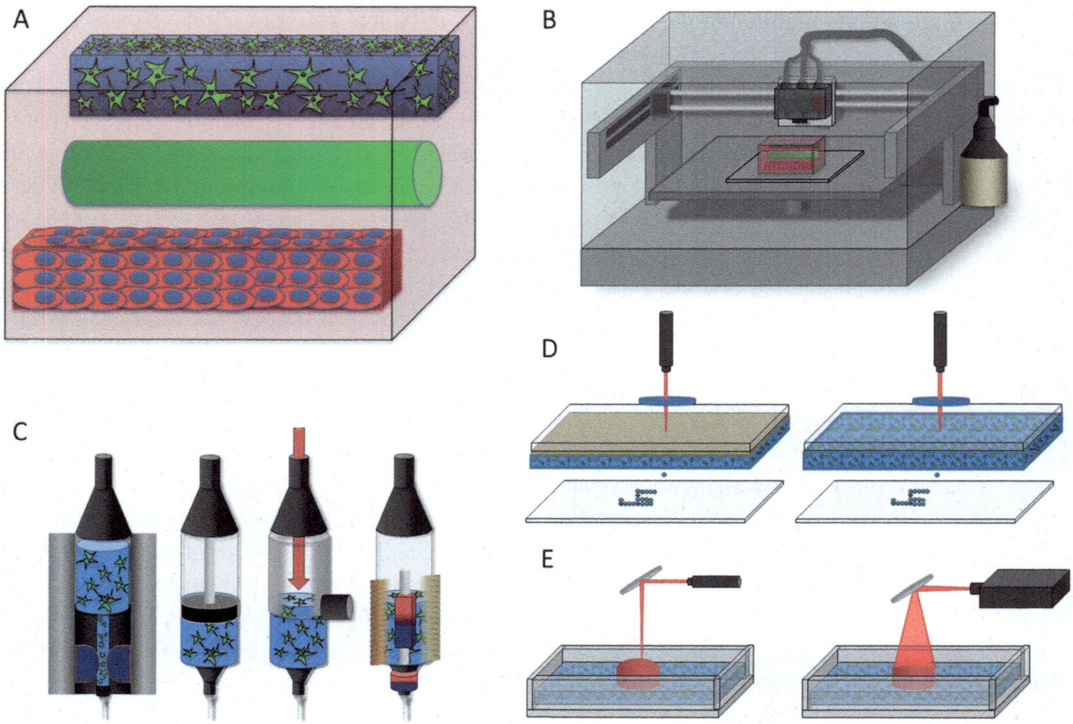

Fig. 5.1 Schematic showing working principles, bioprinting systems, and outcomes. (**a**) Cartoon of multi-component tissue construct showing different cellular and structural regions; top neural cells, middle rigid component, bottom blood vessel. (**b**) Basic bioprinter; (**c**) Inkjet printhead workings showing actuator in *dark blue*, Mechanical printhead showing plunger configuration, Pneumatic-driven printhead showing compressed air input with valve, and solenoid printhead showing solanoid valve (*left* to *right*); (**d**) AFA-LIFT laser-assisted bioprinter (*left*), MAPLE-DW laser-assisted printer (*right*); and (**e**) Beam-scanning SLA (*left*), Mask-image projection SLA

scale synthesis with minimal batch-to-batch variation (Loo et al. 2015). Furthermore, the bioink is required to protect cells from shear stresses whilst maintaining printability and nutrient transport post printing (Chimene et al. 2016). Understanding the biological, chemical, and mechanical requirements of cells is crucial for successful tissue growth, whilst understanding the material properties of bioinks is essential for ensuring printability, biocompatibility, and suitable biodegradability (Park et al. 2016).

The study of bioink material science is of particular interest for biological and engineering fields, including drug delivery, tissue engineering, and regenerative medicine (Gao and Cui 2016). Peptide and protein-based biomaterials in particular have been extensively researched for use in tissue engineering applications, due to their highly favourable inherent biological properties (Lee and Mooney 2001). This chapter will discuss the use of these as bioinks in a range of bioprinting applications, including adipose (Pati et al. 2014), bone (Zhou et al. 2016), cartilage (Rhee et al. 2016), cardiovascular (Pati et al. 2014; Duan 2017), and neural tissue regeneration (Zhu et al. 2016).

5.2 Peptide and Protein Based Bioinks

Peptide and protein hydrogels are formed when the molecules spontaneously self organise into (typically) nanofibrous structures that entangle to immobilise their surrounding solvent to form highly hydrated networks. These unique biomaterials have been predetermined as promising candidates for bioink formulation, due to their highly hydrated state (99% w/w water of their dry weight), ability to mimic the native tissue ECM, and propensity to undergo shear thinning before returning to their original shape (Buenger et al. 2012; Tibbitt and Anseth 2009; Jungst et al. 2016). This combined with their ability to shield cells from stress during printing makes these materials highly sought after for tissue bioprinting applications (Dubbin et al. 2016). Peptide and protein-based biomaterials may be natural (collagen, gelatin, fibrin)(Rice et al. 2013), synthetic (self-assembling peptides, SAPs) (Rodriguez et al. 2014), or may form part of a biosynthetic hydrogel (Collagen-PEG, PEG-Fibrinogen)(Sargeant et al. 2012; Dikovsky et al. 2006).

Natural hydrogels are used extensively in tissue engineering applications due to their inherent biological properties; however present challenges exist relating to batch-to-batch variation, risk of immune response and pathogen transfer (Nilasaroya et al. 2008; Nisbet and Williams 2012; Guo and Ma 2014). Synthetic hydrogels provide good mechanical properties, demonstrate high tuneability, and present no immunological concerns (Collins and Birkinshaw 2013; Guo and Ma 2014); however generally they have poor biodegradation and limited bioactivity unless modified (Lau and Kiick 2014; Guvendiren and Burdick 2013). Having said this, self-assembled peptides, (SAPs) are engineered from versatile molecular building blocks which under appropriate conditions, spontaneously organise to form an entangled nanoscale scaffold. These are easily synthesised through standard peptide synthesis approaches such as solid phase peptide synthesis (SPPS) (Li et al. 2015). The inherent biocompatibility and biodegradability of these molecules, combined with their batch-to-batch consistency and ease of synthesis have resulted in this material showing promise for biomedical applications (Li et al. 2015). Table 5.1 gives an overview of these classes of materials and their uses.

5.2.1 Natural Peptide/Protein-Based Bioinks

The native ECM contains a variety of fibrous proteins and polysaccharides which provide both structure and biochemical support to tissue (Bogdani et al. 2014). Natural peptide/protein-based hydrogels have been extensively used in tissue engineering and regenerative medicine due to their high bioactivity which provides suitable chemical, physical, and biological signals which are essential for supporting cell responses including survival, adhesion, differentiation and proliferation (Hunt et al. 2014). These materials also provide favourable degradation properties and are capable of producing hydrogels with high biocompatibility and biodegradability (Hunt et al. 2014); yet present a risk of pathogen transfer, immunogenic reaction and batch-to-batch variance (Guo and Ma 2014; Nilasaroya et al. 2008).

In tissue engineering and regenerative medicine, the central aim of scaffold development is to recapitulate the ECM. As such, many studies use of physical, chemical and enzymatic processes to remove cellular content from tissue and yield a decellularized extracellular matrix (dECM) (Pati et al. 2014; Jang et al. 2016). Other common protein-based biomaterials used for bioink formulation include proteins found in tissue, such as collagen, gelatin, fibrin, and fibroin/fibrinogen, which will also be discussed further below.

5.2.1.1 Decellularised Extracellular Matrix

The use of dECM materials in tissue engineering applications has gained significant momentum due to their inherent ability to mimic the complexities of the native ECM; providing both signals and structure to developing tissue. Currently, there is no gold standard decellularisation process, rather a variety of processes exist which may be categorised as chemical, physical, or biological (Gilbert et al. 2006). Chemical decellularisation processes make use of chemical agents, including acids and bases, hypotonic and hypertonic solutions, detergents, alcohols, and other solvents; physical processes take advantage of temperature, force and pressure, and non-thermal

Table 5.1 Peptide-based bioink classification, formulation, and application

	Base-material	Bioink	Application	References
Natural	Decellularised ECM	3% cartilage dECM, hASCs and hTMSCs	Cartilage regeneration	Pati et al. (2014)
		3% adipose dECM, hASCs and hTMSCs	Fat (adipose) tissue regeneration	Pati et al. (2014)
		3% heart dECM, rat myoblast cells (L6)	Cardiac muscle regeneration	Pati et al. (2014)
	Collagen/gelatin	10% wt GelMA, 0.6% wt collagen type I, Ru/SPS (0.2 mM/2 mM), MCF-7	Various tissues, (successful tests using HACs and HSCs	Lim et al. (2016)
		2 wt% collagen, 4 wt% alginate, MC3T3-E1	Osteogenic regeneration	Lee et al. (2015)
		2 wt% collagen, 4 wt% alginate, hASCs	Various tissues	Lee et al. (2015)
		5% collagen, crosslinked with 1 mM genpin (1 h), MG63s, hASCs	Hard-tissue regeneration	Kim et al. (2016)
		5% (w/v) gelatin, HEK293s, 3 U mL^{-1} mTgase	Angiogenesis/vasculagenesis	Irvine et al. (2015)
	Fibrin and fibrinogen	50unit mL^{-1} thrombin, 80 mM Ca^{2+}, HMVEC, (thrombin printed onto fibrinogen biopaper to form fibrin gel)	Angiogenesis/microvascular fabrication	Cui and Boland (2009)
	Fibrion	8 wt% silk fibroin, 15 wt% gelatin, hTMSCs, 500 units tyrosine.	Osteogenisis, chondrogenisis.	Das et al. (2015)
Semi-synthetic	SAPs	10 mg mL^{-1} ac-ILVAGK-NH_2, H1ESCs, hMSCs.	Gastrointestinal and skin regeneration	Loo et al. (2015)
	HA-gel-PEG	4 parts 2% Heprasil, 4 parts 2% Gelin-S, 1 part 8% PEGDA, 1 part 8% 8-arm PEG alkyne	Liver regeneration (toolkit shows various other tissues)	Skardal et al. (2015)
	Gel-PEO	3% gelatin, 2% PEO, HUVECS, HEK293s, 3 U mL^{-1} mTgase	Angiogenesis/vasculagenesis/various tissue	Irvine et al. (2015)

irreversible electroporation; whilst biological processes may employ enzymatic agents or non-enzymatic agents, including: chelating agents, toxins, serum and serine protease inhibitors (Crapo et al. 2011). The effectiveness of decellularisation is dependent on tissue source, density, organisation, composition, and finally, the desired final use of the tissue or organ (Crapo et al. 2011; Gilbert et al. 2006). In bioprinting applications, dECM constructs are often lyophilised before being ground into a powder and subsequently used to form a hydrogel (Pati et al. 2014; Jang et al. 2016).

Pati et al. have developed application specific dECM bioinks from decellularised adipose, cartilage and heart tissues (Pati et al. 2014). Decellularisation of the ECM material is completed through a combination of physical, chemical and enzymatic processes to yield decellularized tissue with ~98% cellular contents reduction and only 39 ± 15, 11 ± 1 and 6.7 ± 1.2 ng of DNA mg^{-1} of fat, cartilage and heart dECM, respectively. The required amount of ground dECM powder was weighed and solubilised in a solution of 0.5 M acetic acid with 10 mg of pepsin for 100 mg dECM – such that the final concentration of dECM was 3%. The pH of the solution was then raised to a physiological pH through dropwise addition of 10 M NaOH. Human adipose-derived stem cells (hASCs) and human inferior turbinate-tissue derived mesenchymal stromal cells (hTMSCs) were used to assess the effectiveness of adipose and cartilage dECM bioinks on adipogenic and chondrogenic differentiation respectively. Rat myoblast cells (L6) were used to verify functional enhancement of myoblasts in heart tissue-derived dECM (hdECM) bioinks. 10x concentrated culture media was added to the cell-free bioink at 1/10th volume before cultured cells are mixed into the bioinks at a concentration from 1 to 5 $x10^6$ cells mL^{-1}. Cell viability was found to be sufficiently maintained after printing, showing greater than 95% cell viability 24 h–incubation post-printing in media. Cell viability was found to be maintained above 90% when examined on days 7 and 14 with active cell proliferation evident. Gene expression analysis demonstrates increased expression in chondrogenic (SOX9 and COl2A1), cardiogenic (Myh6 and Actn1) and adipogenic (PPARγ and LPL) constructs when compared to that of the collagen bioink control. The group further expand on this work in a later study in which a photo-crosslinkable dECM gel was engineered through the addition of photoinitiator vitamin B2; resulting in a hydrogel which was approximately 33 times more stiff than the aforementioned dECM bioink (Jang et al. 2016).

5.2.1.2 Collagen and Gelatin

Both collagen and gelatin have been extensively investigated in tissue engineering applications due to their inherent biocompatibility, adhesive qualities, high porosity, high tensile strength and biodegradability; however their lack of rigidity, batch-to-batch variability and immunogenicity concerns limit their applicability in tissue engineering applications (Lynn et al. 2004; Rice et al. 2013). Collagen has a unique structure in which three individual polypeptide chains twist around each other to form a 3-stranded rope-like structure (Lee et al. 2001). Partial hydrolysis of collagen breaks down this structure into single-stranded, gelatin molecules (Hunt et al. 2014). Crosslinking of collagen and gelatin hydrogels is enabled through the use of gluteraldehyde, genipin, or water soluble carodiimides; or alternatively, non-covalently through the entanglement of fibres (Hunt et al. 2014). Due to their favourable bioactivity deriving from their natural source, recent approaches have focused on designing bioinks from these biomaterials.

A novel gelatin-based, photo-crosslinking bioink containing 10% wt Gelatin Methacryloyl (GelMA) and 0.6% wt collagen type I was described by Lim et al. (2016) In their study, two bioinks were engineered with different photoinitiators; the first contains the photoinitiator Irgacure 2959 (0.05% wt), whilst the second contains ruthenium/sodium persulfate (Ru/SPS, 0.2 mM/2 mM). The Irgacure 2959 containing bioink undergoes UV light (300–400 nm) enabled crosslinking while the second was crosslinked under visual light (400–450 nm). Using these bioinks, complex multilayered constructs (dome and human nose) were printed. Breast adenocarcinoma cells (MCF-7) encapsulated within the hydrogels show the visible light system to be superior for bioprinting applications, demonstrating >80% cell viability after day 1 in photoinitiator concentration tests (2 mM/20 mM), and ~90% for light intensity viability tests after 1 day (0.2 mM/2 mM concentration at intensity 100 mW cm^{-2} for 15 min). Contrastingly, the UV photo-crosslinking system shows ~50% (at 0.5% wt) and ~45% (at intensity 100 mW cm^{-2} and concentration 0.05% wt for 15 min) cell viability for the same tests – indicating that the crosslinking methodology is in some way damaging cells. Further cell types were tested including human articular chondrocytes (HACs) and human bone marrow-derived mesenchymal stromal cells (HSCs); in all cases the visual light system had superior cell viability compared to the UV light system.

Two collagen-based bioinks are developed by (Lee et al. 2015) The first bioink (CA-1) consists of collagen which has been cultured with preosteoblast cells for 1 day to allow for initial ECM deposition within the collagen bioink; whilst the second (CA-3) consists of collagen which has been cultured with preosteoblast cells for 3 days to allow for increased ECM deposition. Both inks were formulated as follows: 2% wt collagen was gelled in an incubator for 30 min before preosteoblast cells (1× 10^6 cells mL^{-1}) were cultured on the gelled collagen for various periods of time (1 day and 3 days). Mixtures were then combined with alginate (4 wt% in PBS) and cells in equal volumes at a density of 2.8 × 10^6 cells mL^{-1}. Other culture periods were also investigated however culture periods of 1 day and 3 days were found to present the best balance between printability and bioactivity. A pure alginate bioink was used as the control. An ionic crosslinking method was employed through aerosol dispersion of $CaCl_2$ onto printed constructs, resulting in alginate crosslinking. The compressive moduli of the printed inks were found to be 68 ± 15 kPa, 67 ± 13 kPa, and 69 ± 8 kPa for CA-1, CA-3, and alginate respectively. Unfortunately, these moduli's were considered too low and therefore had to be reinforced with Polycaprolactone (PCL) struts. Initial cell viability was sufficiently high with ~88% viability at 1 day. Cell proliferation rates were significantly higher on scaffolds fabricated from CA-3 compared to those of fabricated from either CA-1 or Alginate. From this study the collagen/ECM/alginate bioink was found to be effective in simultaneously presenting physical and biological cues required for tissue regeneration. Mechanically, the CA-3 bioink was able to protect cells from processing conditions, whilst biologically the ECM components of the ink provide a highly active cell-platform when compared to the alginate bioink.

Similarly, Kim et al. (2016) develop a collagen-based bioink which uses Genipin as a crosslinking agent. Type-I collagen was mixed at 3, 5, and 7 wt% with cells (1 × 10^6 cells mL^{-1} for MG63s and hASCs) to yield a bioink that was biocompatible and mechanically stable. Printed constructs were crosslinked through structure incubation in 0.1, 0.5, 1, 3 and 5 mM genipin solutions over different time periods (1, 6, 24, 48 h). Of these conditions, 5 wt % collagen crosslinked with genpin at concentration ~ 1 mM (over 1 h) was found to produce the best results. Using this ink, a cellblock of 21 × 21 × 12 mm was printed with a cell viability of >95%. *In vitro* testing of the printed construct shows increased viability, cell proliferation, and osteogenic activity compared to constructs printed using a control alginate-based bioink.

As the denatured form of collagen, a gelatin-based bioink developed by Irvine et al. (2015) takes advantage of enzymatic crosslinking through microbial transglutaminise (mTgase). Three gelatin concentrations were investigated (3%, 5%, 7%) along with three different mTgase concentrations (0.5, 1.5, 3 U mL^{-1}). Human umbilical vein endothelial cells (HUVECs) and HEK293s were encapsulated within the hydrogels. Lower concentrations of gelatin showed increased cellular spreading resulting in pseudopodia; whilst cells in the higher concentration (7%) gelatin bioinks maintained a round in shape. Constructs fabricated from bioinks with lower gelatin concentrations lacked sufficient mechanical strength, resulting in the inability to retain the printed structure. With this in mind, a 5% gelatin concentration was used for all subsequent printing applications. Together, these studies that gelatin and collagen hydrogels are shown to have significant promise for bioprinting applications owing to their favourable bioactivity and inherent ECM nanoarchitecture, however care needs to be taken to avoid cellular damage in the fabrication processes.

5.2.1.3 Fibrin and Fibrinogen

Fibrinogen is a soluble plasma protein which is converted to insoluble polymeric fibrin in response to vascular system damage, initiating clotting at the wound site (Doolittle 2001). The conversion of the precursor fibrinogen to fibrin is initiated by the serine protease thrombin (factor IIa) which plays a key role in haemostasis, inflammation and wound healing (Thiagarajan and Narayanan 2001). Fibrin is a major component of the provisional ECM formed during tissue repair following injury, and enables cell infiltration and anchoring at the wound site (Hsieh et al. 2017). Fibrin clots allow platelet, leukocyte and fibroblast infiltration and adherence in wound sites enabling natural wound repair; this, combined with the ease of fibrinogen purification from blood has resulted in fibrin development for a number of applications (Hsieh et al. 2017) - including tissue engineering (de la Puente and Ludeña 2014), drug delivery (Whelan et al. 2014) and clinically as a tissue sealant (Spotnitz 2010). The provisional fibrin matrix' also provides a physically supportive network to the wound site before eventually being replaced by natural ECM proteins, including collagen and fibronectin to form new tissue (Brown and Barker 2014). Recently, fibrin-forming hydrogels have been used as bioinks in bioprinting applications, enabling highly effective tissue regeneration owing to fibrins inherent wound healing properties.

Xu et al. (2006) investigate the effectiveness of fibrin in bioprinting applications. A thermal-based inject printer was modified to allow the layer-by-layer printing of neural (NT2) cells and fibrin gels, enabling development of 3D cellular structures for use in neural tissue engineering applications. The NT2 cell ink was formulated through the re-suspension of cultured NT2 neuronal precursor cell pellets in 0.5 mL of 3× Dulbecco's phosphate-buffered saline solution (DPBS) to obtain cell print suspensions of 2,000,000 cells mL^{-1}. The viability of this cell suspension post printing was approximately 74%. The fibrin gel was formulated by printing bovine thrombin (dissolved in 20 µm $CaCl_2$ to yield a final concentration of 20 IU mL^{-1}) onto a thin fibrinogen substrate. The gel was allowed 3–5 min to facilitate sufficient fibrin formation before a new cartridge containing NT2 neurons was loaded into the printer and neurons printed on top of the fibrin layer. Another thin layer of fibrinogen was then placed over the printed structure and the process was repeated. Results showed this printing method allowed the controlled deposition of patterns and structures of primary hippocampal and cortical neurons, whilst maintaining neuronal phenotypes and basic electrophysiological function.

The group further build on this work formulating a thrombin bioink containing human microvascular endothelial cells (HMVEC) which was printed onto a fibrinogen substrate to form cell laden fibrin channels suitable for microvascular formulation (Cui and Boland 2009). Various concentrations of fibrinogen, thrombin and Ca^{2+} were investigated and an optimal formulation was selected such that the final bioink contains: 1–8 million cells mL^{-1} HMVEC, 50 unit mL^{-1} thrombin, and 80 mM Ca^{2+} in 1× DPBS. This bioink was able to be effectively printed onto a 'biopaper' containing 60 mg mL^{-1} fibrinogen solution, enabling the simultaneous deposition of cells within a fibrin support gel. The cells were found to align within the channel during gel formation and effectively proliferate to form confluent lining along with the fibrin scaffold after 21 days. During proliferation, endothelial cells within the scaffold were shown to form tubular structures which show angiogenesis functionality. Furthermore, the fibrin-based bioinks demonstrate favourable biodegradability and mechanical properties highlighting their potential use in microvascular regeneration. Fibrin provides a novel natural biomaterial for use in bioprinting applications owing to its propensity to form a stable hydrogel upon thrombin and fibrinogen interaction. This enables the gentle encapsulation of cells within the printed scaffold and subsequently has attracted the interest of recent studies.

5.2.1.4 Fibroin

Fibroin is a natural fibrous protein existent in silk spun by silk-producing arthropods, including silkworms and spiders (Kundu et al. 2013). The protein constitutes one of two main proteins of silk. The other protein, sericin, is usually processed out of the natural material through alkali- or enzyme-based degumming processes. The composition of silk fibroin is dependent on its source. Silk fibroin sourced from *B. mori* is composed of a heavy (H) and a light (L) chain linked together by a disulphide bond, and another glycoprotein (P25) is also non-covalently linked to these chains (Tanaka et al. 1999); these fibroin proteins can form anti-parallel β-sheets (Kundu et al. 2013). Alternatively, silk produced by non-mulberry Saturniidae do not contain L chain or P25, instead they contain H-chain homodimers (Sehnal and Žurovec 2004); these proteins form β-sheets due to the higher Ala/Gly ratio and polyalanine blocks (Sehnal and Žurovecurovec 2004; Kundu et al. 2013). Silk fibroin offers numerous advantages for use as a biomaterial including its excellent inherent mechanical properties, good biocompatibility, biodegradability, ease of processing and ease of modification (Kundu et al. 2013). As such, these materials have been used for the regeneration of a variety of tissues, including bone, ligament, cartilage and tendon (Kasoju and Bora 2012).

Fibroin-based bioinks were developed by Das et al. (2015) in which hTMSCs were encapsulated in a silk-fibroin/gelatin (SF-G) biomaterial. The SF-G blend was prepared by adding 15 wt % of sterilised gelatin powder to 8% w/v SF solution, next hTMSCs were encapsulated within the SF-G solution to formulate the bioink. As two crosslinking mechanisms (enzymatic and physical) were investigated; the first incorporated tyrosinase (500 units) to enable enzymatic crosslinking; whilst the second was physically crosslinked through sonication at 50% amplitude for 10 s. When compared to an alginate control, both SF-G bioinks demonstrate superior mechanical properties in regards to stiffness and stability, with the enzymatic crosslinking mechanism resulting in the best stability over 21 days. However, cellular responses of hTMSCs show cell proliferation to be highest in the enzyme crosslinked bioink. Furthermore, the enzymatically crosslinked ink facilitates better cellular differentiation down chondrogenic and adipogenic lineages, whilst osteogenic differentiation favoured the physically crosslinked hydrogel. Cellular viability of the enzymatically crosslinked bioink was found to be maintained for over a month *in vitro* culture conditions. From this study, it was clear that the SF-G bioinks supported better cell proliferation, differentiation and viability than the control bioink, likely due to the increased bioactivity induced by the gelatin combined with the favourable mechanical properties of the silk fibroin.

5.2.2 Synthetic Peptide-Based Bioinks

Self-assembled peptide (SAP) hydrogels are a novel class of bioinspired materials capable of forming complex biomimetic structures through the spontaneous self-assembly of low-molecular weight peptide sequences (Horgan et al. 2016). The ability of SAPs to spontaneously form hydrogels containing intricate nanofibrous networks, combined with their ability to be synthesised through standard peptide synthesis approaches (including solid phase peptide synthesis) has resulted in success in tissue engineering applications (Nisbet and Williams 2012). As these peptides are synthesised from peptide derivatives or amino acids, they are inherently biocompatible and demonstrate good biodegradability (Li et al. 2015). Mechanically, SAPs offer high tunability; whilst biological functionalisation may be incorporated through the addition of well-known bioactive motifs (Nisbet and Williams 2012). One particular advantage of these materials is their ability to undergo reversible shear thinning, making them highly suited to minimally invasive tissue engineering and drug delivery, particularly when loaded with cells (Bruggeman et al. 2016). However, despite these materials being ideal candidates for bioprinting applications, very few studies have taken advantage of these synthetic peptide hydrogels for bioprinting applications.

Loo et al. detail a novel SAP bioink formulated from lysine-containing hexapeptides which self-assemble into stable nanofibrous hydrogels (Loo et al. 2015). Two hexapeptiddes (Ac-ILVAGK-NH2 and Ac-LIVAGK-NH2) were investigated and found to outperform various other SAPs in regard to gelation behaviour – requiring both lower peptide concentration and gelation time. Optimised formulations of the Ac-ILVAGK-NH2 and Ac-LIVAGK-NH2 hydrogels resulted in the development of bioink suitable formulation, capable of gelating within 5 and 30 seconds respectively. Furthermore, both gels were also shown to possess highly favourable mechanical properties with a stiffness of 40 kPA being achieved in Ac-ILVAGK-NH2 gels. Human mesenchymal stem cells (hMSCs) encapsulated within Ac-ILVAGK-NH2 gels were able to be successfully cultured and demonstrate uniform suspension within dispensed droplets. Using this bioink approach, multicellular constructs were able to be fabricated by sequential deposition of different cell types localised to specific domains as demonstrated through the deposition of a multi-domain construct which could serve as an *in vitro* model for skin. Furthermore, through the culturing of intestinal epithelial cells on the SAP hydrogel, the native phenotype of the cultured cells was maintained – even outperforming that of the gold standard Corning transwell membranes. Finally, the peptide scaffolds were also shown to be inherently biocompatible and stable, with subcutaneous implantations of hydrogel discs into healthy mice persisting for at least 2 months with minimal immune response and no capsule formation evident. This study highlights the potential for SAPs in bioprinting applications, however greater developments are still required to improve both the gelation speed and mechanical properties after printing of these materials.

5.2.3 Biosynthetic Bioinks

Although synthetic scaffolds offer the ability to precisely control the mechanical properties of a material, they often provide inadequate biological information for effective cell culture; alternatively, biologically derived scaffolds offer a variety of biological motifs to regulate cell activity which can make it difficult to precisely regulate specific cellular events (Almany and Seliktar 2005). Thus, combing the two material types can yield a biosynthetic scaffold which simultaneously provides the structural characteristics of synthetic materials whilst presenting the inherent biofunctionality of natural materials (Almany and Seliktar 2005). Despite this, few biosynthetic peptide-based bioinks have been developed to date.

Skardal et al. have developed a comprehensive hyaluronic acid-gelatin (HA-Gel) bioink toolkit comprised of a 2-crosslinker, 2-stage polymerisation technique (Skardal et al. 2015). The combination of Poly(ethylene glycol) (PEG)-based crosslinkers used facilitated the development of a highly modular toolkit suitable for a variety of tissue types. The bioinks were formulated through the addition of tissue-derived dECM-based solutions containing combinations of growth factors, collagens, glycosaminoglycans, and elastin, enabling the ability to target specific tissue regeneration. Using this model, primary liver spheroids were encapsulated and printed within a liver-specific bioink to create *in vitro* liver constructs which demonstrated high cell viability and tissue functionality through measurable albumin and urea output. The biosynthetic bioink toolbox places pays attention to both biochemical and physical properties required in tissue engineering applications and as such have a vast potential to make a large impact in a variety of applications.

A biosynthetic gelatin-Poly(ethylene oxide) (Gel-PEO) bioink was developed by Irvine et al. (2015) for '2D' printing applications. The bioink was formulated from 3% Type-A gelatin, 2% PEO, 3 U mL^{-1} mTgase, HUVECs, and human embryonic kidney cells (HEK293s). 2 U mL^{-1} mTgase was chosen to enable rapid gelation

(3 min for Gel-PEO bioink). The bioink showed consistent, even print lines with little ink bleeding, producing a well-defined grid pattern. Cellular florescence imaging showed increased cell proliferation and a high degree of spreading throughout the bioink. Biosynthetic peptide-based materials show potential in bioprinting applications due to their ability to demonstrate a versatile range of mechanical properties whilst maintaining bioactivity, however more research is required to increase the ease of synthesis and application specific modification of these materials (Fig. 5.2).

5.3 Bioprinting Techniques

Developments in nanotechnology and microfabrication has led to advances in additive manufacturing, enabling the ability to fabricate micrometer scale scaffolds and control the spatial distribution of cells more accurately than before (Wang et al. 2015). Among these developments lies the novel advanced biofabrication technique of bioprinting, which enables the printing of cell-laden inks, distinguishing this method from traditional additive manufacturing techniques in which the processing parameters were typically too harsh to support cell viability (Wang et al. 2015; Murphy and Atala 2014). The ability to print 3D, cells-embedded artificial tissue construct allows for highly intricate scaffold development, opening the possibilities for successful and complex tissue engineering approaches that were previously unavailable. Ultimately, bioprinting has the potential to (1) Create fully functional replacements for damaged or missing tissues in patients, and (2) Rapidly fabricate small-sized tissue models or organoids, for diagnostics, pathology modelling, and drug delivery (Skardal and Atala 2015).

Although not the focus of this chapter, some understanding of the design and operation of the printing processes is useful for context. Figure 5.1 shows schematic versions of each of these approaches. In bioprinting, the spatially con-

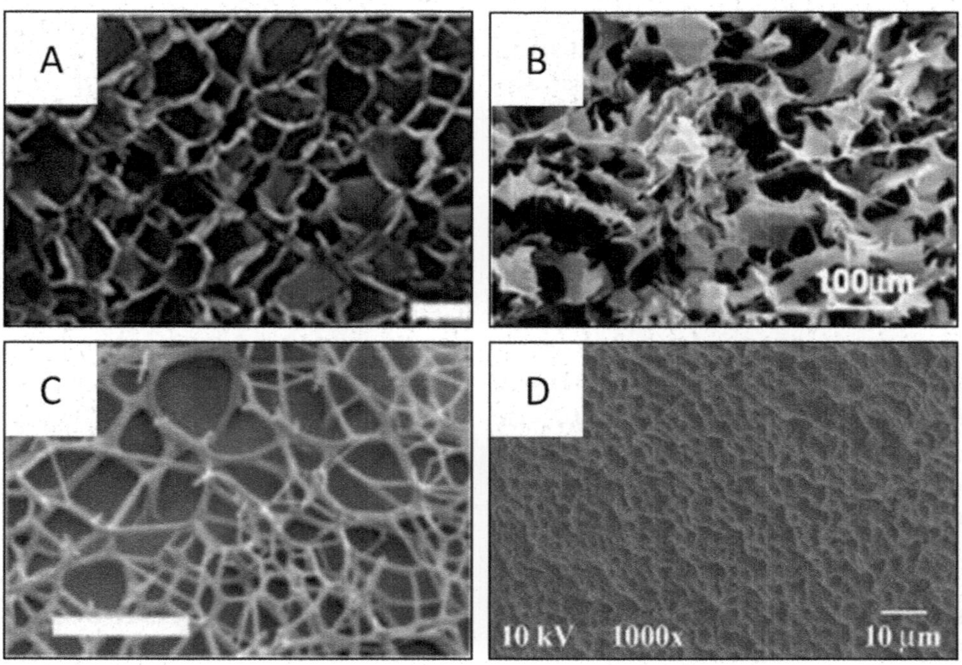

Fig. 5.2 Microstructure of common peptide-based materials. (**a**) SEM image GelMA (scale bar 100 µm); (**b**) SEM image silk Fibroin (scale bar 100 µm); (**c**) Cryo-TEM image Fmoc-FRGDF (SAP) hydrogel (scale bar 500 nm); and (**d**) GelMA/PEG hydrogel (scale bar 10 µm) (Reproduced with permission from Xiao et al. (2011), Tamada (2005); Smith et al. (2008) and Einerson et al. (2003) respectively)

trolled, layer-by-layer deposition of biological materials, biochemicals, and living cells yields 3D tissue constructs (Murphy and Atala 2014). There are four basic steps to this process: (1) a computerised 3D solid model is developed and converted to a suitable file format; (2) the file is then sent to the requisite additive manufacturing equipment (in this case, the bioprinter) where the file is then manipulated to control part orientation and scaling; (3) the structure is printed through layer-by-layer deposition, and finally; (4) maturation, where the materials undergo tissue culture conditions either *in vivo* or in a bioreactor (Huang et al. 2013; Murphy and Atala 2014). Different printing techniques depend on vastly differing material dispersion mechanisms (Tang et al. 2016). The four predominant bioprinting techniques currently being used for tissue engineering applications, these are: inkjet bioprinting, extrusion bioprinting, laser-assisted bioprinting, and stereolithography-based bioprinting (Skardal and Atala 2015). An overview of these techniques is given in Table 5.2, with references for further reading.

5.4 *In vitro* and *in vivo* Applications of Peptide-Based Inks

Bioprinting of artificial tissue constructs is set to revolutionise tissue engineering approaches allowing for rapid yet controlled deposition of cells, biomimetic structures and signals within a spatially-defined advanced tissue scaffold (Zhang and Zhang 2015). The potential capability of these approaches to provide immediate perfuseable vasculature, facilitating a greater depth of nutrient transport, allows for fabrication of tissue with greater thickness (Zhang et al. 2015). Furthermore, the use of cell-laden bioinks provides superior-distribution of biological components within a scaffold (Gao, Schilling, et al. 2015); whilst the spatial delivery of different bioinks using multi-head bioprinting technologies allows for the printing of gradient transition tissue constructs (Ren et al. 2016). Bioprinting using peptide-based bioinks has been extensively investigated and have found use in both organ-engineering and non-organ tissue engineering applications. Non-organ tissue engineering includes the printing of bone (Wüst et al. 2014; Levato et al. 2014; Duarte Campos et al. 2016), osteochondral (Gao et al. 2015a; Shim et al. 2016; Gao et al. 2015b), cartilage (Ren et al. 2016; Daly et al. 2016; Gurkan et al. 2014), vascular (Kolesky et al. 2014; Jia et al. 2016; Cui and Boland 2009; Guillotin et al. 2010), muscular (Choi et al. 2016), and neural tissue constructs (England et al. 2017) whilst organ regeneration includes the printing of cardiac (Pati et al. 2014; Jang et al. 2017; Zhang et al. 2016), cardiac valve (Duan et al. 2013; Duan et al. 2014), liver (Li et al. 2009; Bhise et al. 2016), and skin regeneration (Cubo et al. 2016) as well as organoid regeneration for use in drug screening applications (Li et al. 2009; Bhise et al. 2016). Table 5.3 highlights various applications where peptide-based bioinks have been used and details printing process and bioink selection.

5.4.1 Non-organ Tissue Regeneration

5.4.1.1 Bone-Tissue Regeneration

Bone is a bioceramic composite which is made up of organic collagen, inorganic hydroxyapatite (HAp), and water (Olszta et al. 2007). The hierarchical organisation of these materials results in the development of various bone types through the formation of mineralised collagen fibrils. The organisation of these fibrils is responsible for the determination of mechanical properties of bones, and therefore enables the formation of different bone types (compact and spongy) (Weiner and Wagner 1998). Bone tissue engineering is of particular interest when considering the regeneration of large orthopedic defects or the requirement to develop orthopedic implants that are mechanically more suitable to their biological environments (Burg et al. 2000).

Wüst and et al. (2014) developed a bioink designed specifically for bone-tissue regeneration applications. The bioink was composed of encapsulated hMSCs within an alginate and gela-

Table 5.2 Bioprinting processes showing working principle, materials, advantages and disadvantages

Bioprinting process	Technique	Working principle	Materials	Advantages	Disadvantages	References
Inkjet-based	Thermal-based	Thermally-vaporised bubble-induced pressure.	Low viscosity hydrogels and bioinks (relatively volatile)	High-speed printing,	Require relatively volatile materials	Vaezi et al. (2013) and Murphy and Atala (2014)
				Cost-effective,	Thermal and mechanical stresses,	
				Widely available	Low droplet directionality,	
					Nonuniform droplet size,	
					Prone to clogging,	
					Unreliable cell encapsulation	
	Acoustic-based	Acoustically induced force (piezoelectric through actuator volume change, ultrasound through ultrasound field)	Low viscosity hydrogels and bioinks	Controlled droplet size and direction,	High frequency induced cell damage and possible lysis	Murphy and Atala (2014)
				No thermal stress		
	Electrostatic-based	Actuator volume change through electrostatically induced pressure plate movement	Low viscosity hydrogels and bioinks	No thermal stress	Difficulty controlling droplet size	Kamisuki et al. (1998) and Nishiyama et al. (2008)
				Stable over a variety of materials		
Microextrusion-based	Pneumatic-driven	Air-pressure driven extrusion (valve-free and valve-based)	Low viscosity hydrogels and bioinks	Very high precision and accuracy (05 nL drops)	Possible delay due to compressed gas movement	Ozbolat and Hospodiuk (2016) and Murphy and Atala (2014)
				Simple mechanisms	Air required to be sterile	
				Force limited only by pressure capabilities of system	Increased system cost as accuracy increases	

	Mechanical-driven	Mechanical-induced pressure driven flow (screw and piston)	High viscosity hydrogels and bioinks	Direct control over bioink flow	Careful gear selection for screw-based approaches due to increased pressure and pressure drops along nozzle	Murphy and Atala (2014) and Ozbolat and Hospodiuk (2016)
				Able to facilitate high viscosity hydrogels	Lower force capabilities	
				Affordable, portable, easy to program, great special control		
	Solenoid-based	Pressure driven flow controlled by solenoid valve	Low viscosity hydrogels and bioinks	High accuracy (>1 μL drops)	Time delays between actuation and opening/sealing.	Ozbolat and Hospodiuk (2016) and Bammesberger et al. (2013)
				Advantageous for low viscosity inks particularly with ionic or UV-irradiation based crosslinking mechanisms	Vulnerable to factors affecting accuracy and reproducibility.	
					Particularly affected by change in material temperature yielding a change in viscosity.	
					Requires high actuation pressure to force material through open valve	

(continued)

Table 5.2 (continued)

Bioprinting process	Technique	Working principle	Materials	Advantages	Disadvantages	References
Laser-based	Laser-guided direct flow	Laser-controlled optical-trapping enabling biological molecule positioning	Limited materials with low-reflective index'	Simple mechanism	Low cell viability Low throughput Low-range of fabrication potential Movement efficiency greatly affected by material refractive index	Ozbolat et al. (2017), Odde and Renn (1999, 2000) and Ferris et al. (2013)
	Matrix-assisted pulsed laser evaporation direct-writing	Localised laser-induced pressure change resulting in hydrogel and cell deposition from 'ribbon' (hydrogel-coated) onto substrate	Non-transparent bioinks/ hydrogels with cell coating.	Low-powered lasers	Potential for UV induced cell damage. Ineffective with optically transparent inks	Murphy and Atala (2014), Ferris et al. (2013), Schiele et al. (2010) and Guillemot et al. (2011)
	Absorbing film-assisted laser-induced forward transfer	Localised laser-induced pressure change resulting in bioink deposition from 'ribbon' (film-coated) onto substrate	Moderate range of bioinks.	Offers UV protection to bioinks Can print transparent bioinks	Potential film-derived contamination.	Murphy and Atala (2014), Ferris et al. (2013), Schiele et al. (2010) and Guillemot et al. (2011)
Stereolithography-based	Beam-scanning	Photopolymerisation of bioinks through spatial control of scanning laser beam resulting in photosensitive ink irradiation and subsequent layer-by-layer deposition on build platform.	Photosensitive bioinks	Very high accuracy	Defocusing issues Risk of damage to cells (UV and chemical from photoinitiator) High costs	Vaezi et al. (2013), Cui et al. (2017) and Huang et al. (2013)
	Mask-image-projection	Photopolymerisation of bioinks through mask-controlled light irradiation, resulting in layer-by-layer deposition on build platform.	Photosensitive bioinks	High accuracy prints Rapid printing (whole layers at once)	Mask-associated challenges (accuracy, costs) Risk of damage to cells (UV and chemical from photoinitiator) High costs	Vaezi et al. (2013), Cui et al. (2017) and Huang et al. (2013)

Table 5.3 Application specific peptide-based bioprinting examples classified into non-organ and organ tissue regeneration

Application	Bioprinter classification	Technique/configuration	Ink formulation	References
Tissue regeneration				
Bone regeneration	Extrusion	Mechanical-based	hMSC laden hydroxyapatite/alginate/gelatin bioinks	Wüst et al. (2014)
	Extrusion	Mechanical-based	MSC laden PLA/collagen microcarriers within GelMA-Gellan Gum hydrogel. (microcarrier containing bioink)	Levato et al. (2014)
	Inkjet	–	hMSC laden agarose/collagen bioinks	Duarte Campos et al. (2016)
Osteochondral regeneration	Extrusion	Pneumatic and mechanical	hTMSC laden Atelocollagen bioink, hTMSC laden CB[6]-HA bioink, DAH-HA, PCL support ink	Shim et al. (2016)
	Inkjet	Thermal	hMSC laden PEG-GelMA bioink	Gao et al. (2015a)
	Inkjet	Thermal	hMSC laden Acrylated peptide (GRDS and GCRDGPQGIWGQDRCG)/PEG hydrogel	Gao et al. (2015b)
Cartilage regeneration	Extrusion	Mechanical-based	Chondrocytes laden collagen type II	Ren et al. (2016)
	Extrusion	Mechanical-based	BMSC laden alginate, agarose, PEGMA(BioINK™) and GelMA bioinks, PCL support ink	Daly et al. (2016)
	Extrusion	Pneumatic	hMSC laden GelMA bioink	Gurkan et al. (2014)
Vascular regeneration	Extrusion	Pneumatic-based	PDMS support-ink, Pluronic F127 fugative ink, GelMa (cell free) ink, 10 T1/2 laden GelMA bioink, HNDF laden GelMA bioink	Kolesky et al. (2014)
	Extrusion	Mechanical-based	HUVEC and MSC laden GelMA/alginate/PEGTA	Jia et al. (2016)
	Inkjet	Thermal	HMVEC laden thrombin bioink (forms fibrin bioink)	Cui and Boland (2009)
	Laser-assisted	AFA-lift	Eahy926 endothelial cell or B16 carcinoma cell laden thrombin/$CaCl_2$ bioink printed onto fibrinogen substrate.	Guillotin et al. (2010)
Muscle regeneration	Extrusion	Pneumatic	C2C12 myoblast laden mdECM bioink	Choi et al. (2016)

(continued)

Table 5.3 (continued)

Application	Bioprinter classification	Technique/configuration	Ink formulation	References
Neural regeneration	Extrusion	Pneumatic	Schwann cell laden fibrinogen/factor XIII/HA/PVA bioink	England et al. (2017)
Organ regeneration				
Cardiac regeneration	Extrusion	Pneumatic- and mechanical-driven	L6 laden hdECM bioinks	Pati et al. (2014)
	Extrusion	Pneumatic- and mechanical-driven	CPC and/or MSC laden hdECM,	Jang et al. (2017)
	Extrusion	Mechanical-driven	HUVEC laden alginate/GelMA	Zhang et al. (2016)
Cardiac valve regeneration	Extrusion	Mechanical-driven	SMC laden gelatin/alginate, VIC laden gelatin/alginate,	Duan et al. (2013)
	Extrusion	Mechanical-driven	HAVIC laden Me-HA/GelMA	Duan et al. (2014)
Liver regeneration	Extrusion	Mechanical-driven	ADSC laden gelatin/alginate/fibrinogen and hepatocyte laden gelatin/alginate/chitosan	Ii et al. (2009)
	Extrusion	Mechanical-driven	HepG2/C3A spheroid laden GelMA	Bhise et al. (2016)
Skin regeneration	Extrusion	Mechanical-driven	hFB laden fibrin gel	Cubo et al. (2016)

tin hydrogels prepared with differing amounts of HAp, and demonstrates adequate printability with an extrusion-based bioprinter. A two-step gelation method is employed which implements both reversible thermal- and irreversible chemical-crosslinking, enabling rapid gelation during printing and long-term stability post printing. To facilitate thermal gelation, the bioink solution was liquefied through heating to 40 °C prior to printing. After extrusion the bioink was allowed to cool through contact with a cooled substrate (~10 °C) leading to a temporary increase in ink rigidity. Once scaffold printing was completed it was immersed in a cold $CaCl_2$ solution to enable chemical crosslinking of alginate within the ink; resulting in irreversible crosslinking yielding a scaffold which demonstrates long-term stability. Three different bioink formulations were investigated: 8% HAp/2%Alginate/10%Gelatin, 4% HAp/2%Alginate/10%Gelatin, and 2%Alginate/10%Gelatin. The bioinks all demonstrate good cell viability (84–85%) 3 days post printing; however inclusion of HAp was found to increase bioink viscosity that leads to more difficult handling during printing. Interestingly, the addition of HAp was not found to significantly alter the stiffness of bioink constructs in this study, and hydrogels were shown to be ineffective in resembling the mechanical properties of bone tissue; despite this, two-phase printed structures were able to be printed and sufficiently imaged under μCT.

An alternative approach for the development of bioprinted bone-tissue engineering constructs was investigated by Levato and et al. in which MSC-laden polylactic acid (PLA) microcarriers were distributed within a GelMA-Gellan Gum (GelMA-GG) hydrogel (Levato et al. 2014). Using a 'green solvent-based' method, PLA microcarriers of average diameter 120 μm were fabricated and later functionalised with human recombinant collagen type-I to improve cell responsiveness. The hydrogel component of the bioink was formulated through GelMA incorporated with 0.1% w/v Iracure 2959 to facilitate photo-crosslinking. Furthermore, in an effort to improve bioink printability, 1% w/v gellan gum was added to the hydrogel solution. Microcarriers were either preseeded with rat derived MSCs and then supplemented at 30 mg mL^{-1} into hydrogel solutions (GelMA-GG MC-MSCs), or were supplemented alongside MSCs into the hydrogels (GelMA-GG MC's and MSCs); creating two separate bioinks. Cell viability post extrusion was found to be 60% after 1 day in GelMA-GG MC-MSCs, and 90% after 3 days. The addition of PLA microcarriers was also shown to increase the compressive modulus of GelMA-GG. Alizarin red staining intensely marked diffused areas around MC-MSCs complexes, indicating strongly mineralised areas; furthermore on a molecular level MC-MSCs gels showed the most OCN secretion and demonstrated matrix calcification suggesting the ability to form mature bone-like tissues. Despite the bioink not reaching sufficient compressive strengths for initial load-bearing applications, clinically relevant sized constructs were able to be printed with large cell numbers and high shape fidelity; furthermore, the bioinks demonstrated good printability and showed homogeneous cell distribution throughout the printed scaffold.

Duarte Campos et al. (2016) utilise inkjet bioprinting for the fabrication of bone-tissue engineering scaffolds. In combining agarose hydrogels with collagen type-I, a thermoresponsive bioink can be developed which demonstrates superior printability than pure collagen hydrogels. The agarose/collagen bioinks (containing MSCs) demonstrated high cell viability (>98%) 21 days post printing. Softer hydrogels with lower agarose concentrations and higher collagen concentrations were shown to demonstrate the highest degree of spreading with the longest elongation, as seen in ITGB3 staining; furthermore, through qPCR it was evident that genes COL1 and RUNX2 were expressed significantly more in low stiffness hydrogels (AG0.5-COL0.21) compared to other hydrogels. Finally, the increased cell spreading in softer hydrogels resulted in increased differentiation to osteoblast cells; leading the group to conclude osteogenic differentiation was improved in softer –agarose-collagen hydrogels due to less restricted cell spreading and branching. The study shows the challenges between selecting a correct balance between increased printability (high agarose bioinks) and increased oestrogenic differentiation (high collagen gels). Ultimately, these gels were found to be suitable for bone-tissue engineering bioprinting applications.

5.4.1.2 Osteochondral Regeneration

Osteochondral defects affect both articular cartilage and underlying bone tissues, therefore regenerative engineering approaches need to adequately address this interface transition, whilst supporting both bone and cartilage regeneration within their respective regions. As such, several strategies have been employed for osteochondral repair, including single –phase, layered and graded approaches (Nukavarapu and Dorcemus 2013). In such approaches, both structure and signalling are able to be modified to enable relevant cell differentiation and subsequent tissue regeneration.

A comprehensive 3D bioprinting process was presented by Shim et al. (2016) in which a multilayered 3D construct containing human mesenchymal stromal cells was developed for osteochondral-tissue regeneration. They made use of a number of hydrogels to provide both support and conducive cellular microenvironments for artificial tissue construct. The bioinks at the core of this study was based on atelocollagen and supramolecular HA, both of which were types of native ECM proteins. Atelocollagen containing hTMSCs and BMP-2 was used as the bone-forming bioink, whilst Cucurbit[6]uril-Hyaluronic acid (CB[6]-HA) containing hTMSCs and transforming growth factor-β (TGF-β) formed guest host interactions with another hydrogel, 1,6-diaminohexane (DAH)-conjugated HA (DAH-HA), to form the chondral layer. Pores are made available in the bone-forming portion of the 3D printed scaffold to allow for infiltration of vascularisation, whilst in the chondral section the PCL framework was fully interconnect to inhibit vascular invasion. High cell viability is evident in both printed bioinks; demonstrating 93% and 86% for atelocollagen and CB[6]/DAH-HA respectively. *In vitro* tests showed hTMSCs in the atelocollagen hydrogel exhibited a stretched morphology typical of the osteogenic phenotype at day 10, whilst hTMSCs within the CB[6]/DAH-HA bioink demonstrate spherical morphologies, characteristic of the chondrogenic phenotype. Additionally, atelocollagen bioink layers demonstrated an increase in ALP, COL-I, and OSX genes compared to those in the CB[6]/DAH-HA bioink layer, whilst the CB[6]/DAH-HA layers showed increases in ACAN, COL-II, and SOX9 compared to the atelocollagen layer at day 14. Furthermore, cells within atelocollagen constructs showed high levels of RUNX2 (a master transcription factor in osteoblasts), whereas cells cultured within CB[6]/DAH-HA hydrogels demonstrated high levels of COL-II (the major collagen in cartilage); RUNX2 was not detected in CB[6]/DAH-HA gels, nor was COL-II detected in atelocollagen hydrogels. These results demonstrated that the materials used were cytocompatible and capable of directing desired cell differentiation. *In vivo* tests on rabbit knee osteochondral defects reinforced these observations, with the printed scaffold yielding outstanding regenerative ability in both gross morphological and histological evaluations after 8 weeks implantation.

An alternative osteochondral bioprinting method was presented by Gao et al. (2015a, b) in which thermal-based inkjet printing was employed to print acrylated-peptide/PEG bioinks. The bioinks were prepared at formulation 10% w/v poly(ethylene glycol) dimethacrylate (PEGDMA), 1 mM acrylated GRGDS, 1 mM acrylated MMP-sensitive (GCRDGPQGIWGQDRCG) peptides, 0.05%w/v Irgacure, and 6×10^6 hMSCs mL^{-1}. Using this bioink, cylindrical artificial bone and cartilage tissue constructs were printed with 4 mm internal diameter. Printed constructs were then placed in a 24 well plate containing either 1 mL osteogenesis medium or 1 mL chondrogenesis medium with 10 ng mL^{-1} TGF- β and cultured for 21 days. Bioinks were compared to regular, PEG, PEG-GelMA and GelMA inks in regards to printability, with results showing the peptide-PEG bioinks demonstrate excellent printability with viscosities highly similar to that of regular ink. Alternatively, PEG inks with the addition of GelMA demonstrated poor printability due to the significant increase in viscosity resulting in frequent printhead clogging. The pure GelMA bioink was found to be unprintable. In regards to cell viability, PEG/peptide bioinks were shown to demonstrate improved cell viability after 24 h when compared to pure PEG hydrogels, with a ~90% cell viability in PEG/peptide hydrogels compared to ~83% in PEG hydrogels. The PEG-peptide scaffold demonstrated a lower compressive modulus than the PEG scaffold right after printing, however after osteogenic and chondrogenic differentiation, PEG/peptide scaffolds demonstrated significant increase in compressive moduli with an increase of 100% and 82% respectively from day 7 to 21. Alternatively, PEG hydrogels only showed a 44% and 38% increase from day 7 to 21 for osteogenisis and chondrogenisis samples respectively. A complex analysis of gene expression found the printed scaffolds placed in the osteogenic media expressed very high levels of COL10A1 and MMP13 genes, suggesting that the PEG-peptide scaffold strongly stimulates the potential endochondral bone formation; alternatively, scaffolds placed within the chondrogenic media demonstrates high chondrogenic differentiation factors, highlighting the ability for use in cartilage regeneration applications. Finally, the PEG/peptide bioink combined with the thermal inkjet bioprinting technique was found to promote homogeneous bone and cartilage regeneration, and thus, shows potential use in osteochondral regeneration applications.

A subsequent study by the same group investigated the use of an inkjet printable PEG/GelMA bioink for bone and cartilage tissue applications (Gao et al. 2015) Briefly, 10% w/v PEGDMA and 1.5% GelMA were mixed together with Irgacure 2959 (0.05% w/v) and subsequently seeded with hMSCs at 6×10^6 hMSCs mL^{-1}. The concentration of GelMA in the bioink was determined as the highest printable concentration. Once again, through comprehensive cell viability testing, mechanical testing, and gene expression, biochemical and histological analysis the PEG/GelMA bioink developed was found to encourage stem cell differentiation down desired linages. This study highlights the significant potential for PEG/GelMA based bioinks in osteochondral applications.

5.4.1.3 Cartilage Regeneration

Cartilage is a dense connective tissue with a highly ordered matrix consisting predominantly of collagens and proteoglycans (Daly et al. 2016). Depending on the composition of the matrix, cartilage is classified into three categories: elastic, hyaline, and fibrocartilage (Karlsen et al. 2016). Elastic cartilage is surrounded by perichondrium and characterized by the presence of chondrocytes immersed in an ECM (García-Martínez et al. 2017); in humans this tissue is found in the external ear, auditory tube, and epiglottis (Karlsen et al. 2016). Hyaline cartilage also contains primarily collagen type II; but rather is found in many load-bearing – particularly covering synovial joints as articular cartilage (Daly et al. 2016). Fibrocartilage contains both type I and II collagens, and was found extensively in tendons, vertebral disks and menisci (Daly et al. 2016; Karlsen et al. 2016). Due to the limited regenerative capabilities of cartilage and potential for chondral defects to induce early arthritis, cartilage regeneration was of significant interest to current tissue engineering approaches (Bhosale and Richardson 2008).

Ren et al. (2016) developed a gradient bioprinting method for the printing of zonal tissue constructs for cartilage engineering. Three collagen type-II based bioinks groups were formulated, each with differing cell densities (2×10^7, 1×10^7, and 0.5×10^7 cells mL^{-1}); for each bioink, two different types of constructs were prepared, one with a biomimetic gradient chondrocyte density and the other with a single cell density. For gradient scaffolds, the gradient concentrations were designed such that the total cell concentration through the whole construct equalled that of the homogenous scaffold. The biomimetic approach aims to mimic the native cell densities in zonal articular cartilage tissues; therefore, a construct was fabricated with areas of: 10% superficial layer (high cell density), 10% middle layer (medium cell density), and 80% deep layer (low cell density). Constructs were cultured for 0, 1, 2 and 3 weeks to investigate cell viability, reverse transcription PCR (RT-qPCR), biochemical assays, and histological analysis. Cell viability was found to be extremely high after 1 day, with an average live cell percentage of 93 ± 3% evident in all groups. A slight loss in cell numbers was evident over the 3 week period; however this was not statistically significant. Analysis of gene expression showed low expression of COL1A1 over the 3 weeks, whilst expression levels of COL2A1 and ACAN showed significant increase over the same time period. Average single-cell glycosaminoglycan (GAG) production was found to be highest in group B (1×10^7 cells mL^{-1} total concentration) after 3 weeks; whilst Alcian blue staining and immunohistochemical analysis of chondrocytes in different gradient constructs showed high positive staining for collagen type-II. This study demonstrated the potential for gradient-based bioprinting in zonal articular cartilage regenerations, with results highlighting the desired ECM deposition.

The use of bioprinting for fibrocartilage and hyline cartilage regeneration applications was investigated by Daly et al. (2016) in which 4 commonly used hydrogel bioinks (agarose, alginate, GelMA and PEGMA) were compared in respect to printing properties and potential to support either hyaline cartilage or fibrocartilage development. Each hydrogel was seeded with porcine-derived bone-marrow derived mesenchymal stem cells (BMSC) at density 2×10^7 cells mL^{-1}, before being used to form cylindrical artificial tissue constructs. Histological, immunohistological and biochemical analysis of constructs showed that over the 4 week culture period, alginate supported the highest GAG synthesis, whilst GelMA demonstrated the lowest GAG synthesis. Furthermore, Alginate and agarose hydrogels showed strong staining for collagen type-II, whilst GelMA and PEGMA stained weakly for collagen type-II, rather, stronger staining was evident for collagen type-I. These results suggest that the alginate and agarose bioinks support greater development of hyaline cartilage-like phenotypes, whilst fibrocartilage-like phenotypes were supported to a greater extent by GelMA and PEGMA-based bioinks. In printability investigations, GelMA (10% w/v combined with 0.05% w/v Irgacure 2959) demonstrated the greatest printability, whilst the PEGMA-based bioinks demonstrated poor printability. Alginate and agarose hydrogels presented moderate printability with higher filament spreading and variability compared to GelMA. Post printing cell-viability was demonstrated to be sufficiently high (~>80%) for all bioinks. The addition of a PCL support was found to reduce cell viability slightly, however viability of all groups remained sufficiently high (above 70%), demonstrating the capability to mechanically reinforce bioinks. The addition of PCL microfiber reinforcement in printed constructs was shown to significantly increase compressive stiffness of the scaffolds, yielding a printed scaffold which was within range of native hyaline tissue. Although this was higher than required for fibrocartilage tissue, modifications to fiber diameter and fiber spacing could result in a mechanically suitable construct. Finally, alginate- and agarose-based bioinks were shown to support the development of hyaline-like cartilage tissues whilst demonstrating sufficient printability; alternatively GelMA and PEGMA-based bioinks were shown to support fibrocartilage development, with GelMA exhibiting good printability properties. Through these studies, the use of peptide-based bioinks paired with suitable printing techniques are shown to be effective for cartilage engineering applications, enabling printing of anatomically mimetic tissue constructs which encourage healthy tissue regeneration.

5.4.1.4 Vasculature Repair

Due to the relation between narrowing blood vessels and heart disease, there is a significant need to develop vascular-tissue engineering approaches (Nerem and Seliktar 2001). Vascular tissue is comprised of three histologically distinct layers. The innermost layer (tunica interna) consists of a single layer of endothelial cells mounted on a basil lamina; below this lies the sub-endothelial fibro-elastic connective tissue layer and an organised internal elastic lamina (Pugsley and Tabrizchi 2000). The middle layer (tunica media) consists of predominantly smooth muscle cells and elastin fibers, and a layer of external elastic lamina; whilst the outermost layer (tunica adventitia) is composed of fibro-elastic connective tissuem(Pugsley and Tabrizchi 2000). Considering the multicomponent nature of these constructs, there is substantial difficulty in adequately recapitulating these vessels.

A novel 3D bioprinting method for the fabrication of tissue engineered structures was described by Kolesky et al. (2014) wherein cells, vasculature, and ECM mimic were all able to be printed to form an artificial tissue construct. The study took advantage of multi-head, extrusion-based bioprinting technique to print 4 differing inks: the first ink a 'fugitive' pluronic ink was designed to be gently removed (at <4 °C) post printing. This fugitive-ink acts to provide a temporary structure, which upon removal creates an open channel which was later seeded with human umbilical vein endothelial cells to form the preliminary vascular network. The Pluronic ink was made at a 40% wt Pluronic F127 concentration dissolved within deionised, ultrafiltrated water. A cell-free GelMA ink was prepared by dissolving 15% wt/v% GelMA powder in media. A photoinitiator (Irgacure 2959) was added to the solution at 0.3% wt; the addition of the photoinitiator enables photo-crosslinking under UV light. The second ink formulated was a GelMA based bioink; this ink was prepared with 15wt/v% GelMa dissolved in media with 2×10^6 human neonatal dermal fibroblasts (hNDF) mL^{-1}. A third bioink was prepared with the same formulation as the GelMA-HNDFs bioink, except HDMFs were replaced with mouse embryo fibroblast (10 T1/2) cells. Using these inks, a diverse range of vasculature structures were able to be printed within a tissue construct. A PDMS support-ink was also used to set a well-defined border for the whole tissue construct. Initially 1D, 2D and 3D structures were printed using the fugitive-ink surrounded and encapsulated by the GelMA cell-free bioink. Removal of the Fugative ink was found to be effective and may be done quickly at high fidelity even in the 3D vascular constructs; after removal, the open channels of the '2D' printed system were seeded with HUVECs at 1×10^7 cells mL^{-1} through injection before being cultured in media 48 h. After incubation, cells were found to demonstrate good cell viability (>95%) in live/dead staining confocal imaging. To further investigate the effectiveness of this printing technique and the bioinks formulated, a 3D construct containing both vasculature (fugitive-ink) and cell-laden constructs (HNDF containing GelMA bioink) was printed before being surrounded by the cell-free GelMA bioink. The fugitive-ink was once again removed and endothelial cells were seeded as before; the scaffold shows good cell viability for both inks in confocal live staining images, with each cell-type being clearly distinguishable within its designated printed zone. Finally, multiple cell-types were printed into a single tissue construct, and endothelial cells were seeded into the fugitive–ink removed channels. Confocal observations once again show good cell viability (81% for HNDFs and 82% for 10 T1/2 cells) after 7 days. Once again, there were clearly distinguishable regions correlating to the regions of different printed bioinks.

An extrusion-based bioprinting method was employed by Jia et al. (2016) for the fabrication of perusable vascular constructs in tissue engineering applications. Briefly, a bioink was developed from Alginate (2 or 3% w/v), GelMA (5 or 7% w/v) and poly(ethylene glycol)-tetra-acrylate (PEGTA, 2 or 3% w/v, which was seeded with HUVECs and hMSCs); additionally, a photoinitiator was included at 0.25% (w/v) to enable GelMA photo-crosslinking. Using a coaxial nozzle, the bioink was printed from the outer sheath; the inner sheath pumps a $CaCl_2$ solution to enable alginate crosslinking (ionic) inside the hollow vessel, whilst a spray of $CaCl_2$ solution was used to crosslink the external surface of the vessels. The alginate crosslinking provides immediate temporary structural stability to the hollow tubes. Next, covalent crosslinking occurs through GelMA photo-crosslinking, facilitating the development of a mechanically stable vessel. The vessel was then washed in mixture medium 3 times to remove excess $CaCl_2$ and photoinitiator, before being immersed in a CA^{2+}- chelating agent (EDTA) for 5 min to dissolve alginate. The PEGTA acts to combat the reduced mechanical strength post alginate removal, stabilising the crosslinked matrix. Printability investigations of the bioinks found the optimal ink formulation as follows: 7% (w/v) GelMA, 3% (w/v) Alginate,

2% (w/v) PEGTA; this ink was used for subsequent experimental observations. Cell viability was shown to be sufficiently high when UV exposure does not exceed 30s, demonstrating viability percentages above 80% after 1, 3, and 7 days. Cellular metabolic activity was also found to increase as a function of culture time, confirming the cytocompatibility of the approach. The vascular constructs also demonstrated high expression levels of α-SMA and CD31, both of which were crucial for vessel formation and stabilisation. The perfusability of vessels was proven through both confocal imaging and fluorescent microscopy. This work was extended through use of a tri-layered nozzle which enables variance in vessel diameter through the controlled alternation of bioink delivery from middle sheath to external sheath, providing a comprehensive approach to tissue engineered vascularisation.

Using a laser-assisted bioprinting approach Guillotin et al. (2010) investigate the biofabrication of microvascular tissue constructs through various bioink formulations. In this study, bioinks were formulated from alginate (containing glycerol), Matrigel, or a solution of thrombin; however due to the scope of this study, only the thrombin solution will be focused on. The thrombin solution was formulated containing cell-laden (Eahy926 or B16) thrombin (250 UI) and $CaCl_2$ (40 mM) which was printed onto a fibrinogen substrate. Using this approach, a stable construct was printed with microscale resolution compatible with microvascular dimensions. Cell-viability testing of Eahy926 cells demonstrates the presence of viable cells and good cell distribution. These studies highlight the potential for peptide-based bioprinting applications in vasculature regeneration, providing a key advance in addressing the urgent requirement to develop healthy vasculature within constructs.

5.4.1.5 Muscle Regeneration

Skeletal muscle is the most abundant muscle type and is responsible for various mechanisms, including voluntary and controlled body movement, facial expression and viscera protection (Lewis et al. 2009). Skeletal muscle is a hierarchical tissue composed of tightly bound muscular fascicles within an epimysium sheath. Each fascicle contains numerous tightly bound muscle fibres; each of which is a single, multinucleated cell (Lewis et al. 2009). Within these muscle cells lie myofibrils, formed from actin and myosin filaments (Schiaffino and Reggiani 2011). Skeletal-muscle regeneration offers promise to clinical treatments for tissue lost through trauma, or altered through disease.

Choi et al. (2016) employ pneumatic-driven extrusion for the bioprinting of functional skeletal muscle constructs. The bioink was formulated from decellularised skeletal muscle ECM seeded with C2C12 myoblasts, and compared to a collagen bioink in regard to mechanical properties, cell viability, myotube formation, and myogenic differentiation. Characterisation of muscle tissue-derived dECM (mdECM) verified that the cellular contents were removed whilst ECM components, including GAGs, collagen and laminin, were all retained. The bioink was found to possess excellent printability due to its thermally controlled gelation properties (owing to collagen content) and the ability of the material to undergo shear thinning; from this ink, a number of 3D cell-printed muscle constructs with varying shape, pores and architectures were able to be printed with uniform cell distribution and good cell viability. High cell viability (>90%) was evident after 24 h of printing and maintained over 14 days; furthermore, cells were found to begin longitudinally aligning over the culture period. Cell alignment was shown to be a function of line width and culture time; thinner printed lines (500 μm) demonstrated the greatest degree of cell alignment after induced differentiation, with 76 ± 2% alignment evident at day 7 compared to an initial value of 64 ± 9% at day 1. Mechanical testing of the mdECM bioink show improved stiffness, ultimate tensile strength, and elastic moduli compared to the control collagen hydrogel both at day 7 and day 14 of post-print culturing. Furthermore, the mechanical properties of both collagen and mdECM were found to improve greatly between days 7 and 14. Cell proliferation assays of printed constructs showed significantly increased levels of cell proliferation in mdECM-bioink printed structures compared to collagen bioink printed structures; furthermore, both collagen- and mdECM-bioinks demonstrate improved cell-proliferation as culture days progress from 1–7. Investigation of gene expression showed significantly greater myogenic gene expression of Myf5, MyoG, MyoD, and MHC in the mdECM bioink compared to the collagen control; indicating a higher level of cell stimulation toward myogenic maturation. A high level of myotube formation was evident in the mdECM printed constructs compared to the collagen control; with myotubes of greater length, width, area, and fusion index being evident in the mdECM bioink printed constructs. Finally, agrin preservation in mdECM was evident, providing potential for increased acetylcholine receptor cluster formation in 3D printed constructs. The bioink and printing process combination investigated in this study highlights the potential for mdECM bioink use in functional skeletal muscle development in tissue engineering applications, enabling the customised formation of muscle substrates for the treatment of muscular injuries.

5.4.1.6 Neural Regeneration

A nerve consists of a cell body (soma) and its extensions (axons and dendrites), the soma receives signals through its dendrites, before sending the signal down onward through the axon (Winter and Schmidt 2002). Nerves are groups of axons, and may form part of the central nervous system or peripheral nervous system (Winter and Schmidt 2002). In the peripheral nervous system, small injuries are able to be regenerated naturally; whilst larger injuries require surgical treatments (Schmidt and Leach 2003). The central nervous system is more complicated however in that the body inhibits repair, and as such there has been little success in spinal cord repair (Schmidt and Leach 2003).

Peptide based bioprinting was investigated by England et al. (2017) in which a pneumatic-driven extrusion printer was employed to print a fibrin-forming bioink formulated from fibrinogen, factor XII and HA. To enable nerve regeneration, Schwann cells were encapsulated within the hydrogel. The bioink was printed into a NaCL, $CaCl_2$, Polyvinyl Alcohol (PVA) and thrombin 'crosslinking' solution enabling fibrin formation whilst supporting the fabrication of multilayered fibrin-HA scaffolds. PVA concentration was found to be a crucial factor which affects material spreading and regulates strand uniformity, furthermore, PVA inclusion was found to compensate strand buoyancy; allowing for the formation of complex shapes; however, too much PVA results in poor adjacent layer attachment, subsequently resulting in lateral displacement of layers relative to each other. Fibrin fibres were found to orientate parallel to strand direction, with 97% of fibres orientated within ±10°, and 45% of fibres orientated within ±1° relative to the strand. Cell viability of Schwann cells within bioinks assessed immediately post printing showed high cell viability (~98%), which was maintained over 7 days; a high cell proliferation was also evident during the same period. Cell alignment was evident within printed constructs, with strands adopting elongated, bipolar morphologies; alternatively, no significant cell alignment was evident on control (plated) cultures of Schwann cells. Culturing of dorsal root ganglia on printed strands showed directional neurite growth compared to the control laminin-coated surface; furthermore, the dorsal root ganglia cultured on the fibrin-based strands demonstrated a greater average length of the longest dorsal root ganglia neurite. The findings of this study demonstrate the ability of naturally derived peptide bioink use in neural regeneration applications, enabling for printing of highly ordered autologous scaffolds (Fig. 5.3).

5.4.2 Organ Regeneration

5.4.2.1 Cardiac Regeneration

As heart-failure is the primary cause of death in industrialised countries; therefore there is significant need to develop successful and practical cardiac regeneration approaches. However cardiac tissue engineering approaches are complex as they are required to simultaneously facilitate heart function whilst facilitating cardiac tissue regeneration; thus, constructs should be (1) con-

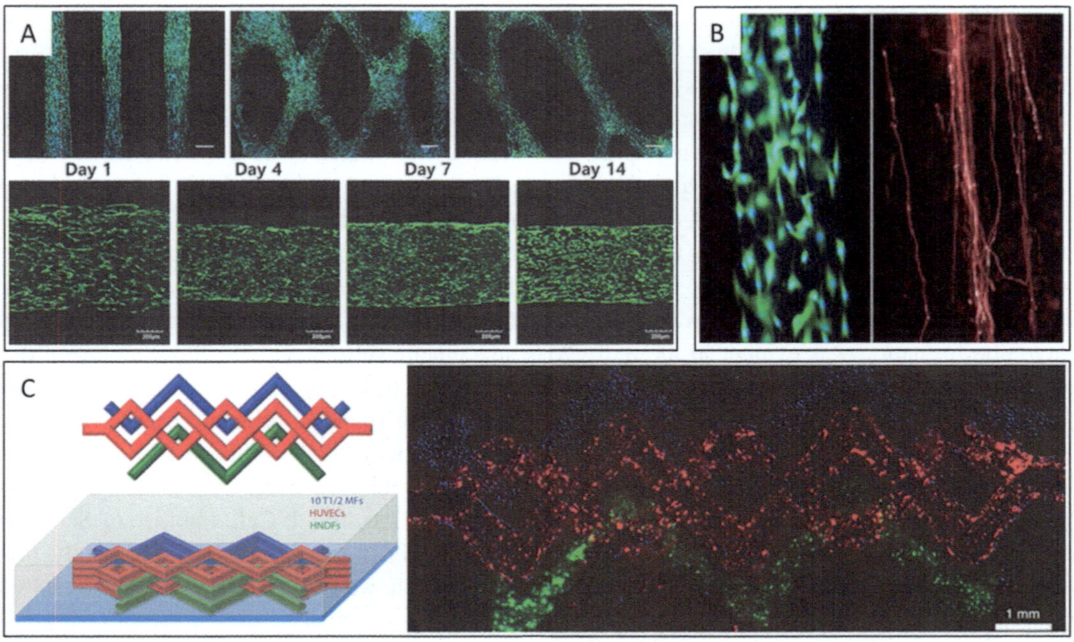

Fig. 5.3 Bioprinted non-organ tissue constructs: (**a**) Skeletal muscle constructs containing C2C12 myoblasts, [*top*] Fluorescence imaging showing high control over architecture, and [*bottom*] live dead staining of constructs at days 1, 4, 7, and 14; (**b**) Bioprinted fibrin nerve construct: showing [*left*] encapsulated Schwann cell alignment, [*right*] neurite elongation and alignment; (**c**) Bioprinting approach for multicomponent tissue fabrication: [*left*] Schematic view of printed 3D construct, and [*right*] composite fluorescent image showing 10 T1/2 fibroblasts (*blue*), HNDFs (*green*) and HUVECs (*red*) (Reproduced with permission from Choi et al. (2016), England et al. (2017), and Kolesky et al. (2014) respectively)

tractile, (2) electrophysiologic stability, (3) mechanically robust yet flexible, (4) vascularised, and (5) autologous (Zimmermann et al. 2004). Unfortunately however, such biomaterials currently do not exist.

Using a multi-head, extrusion-based deposition system, Pati et al. (2014) additively manufacture heart, cartilage and adipose tissue constructs *in vitro*; to do so they formulate a dECM bioink previously discussed (see Sect. 5.2.1.1). In addition, a synthetic PCL hydrogel bioink is used to provide structure and support to cartilage and adipose dECM constructs. A PCL framework was created and the dECM bioink was printed in every alternating gap between the PCL lines – making a porous structure. Printed heart tissue constructs contained no PCL supports. All dECM bioinks were then used to encapsulate cells; hASCs and hTMSCs were used to assess the effectiveness of adipose and cartilage dECM bioinks on adipogenic and chondrogenic differentiation respectively. L6 cells were used to confirm functionality of myoblasts in hdECM bioinks. Cell viability was found to be sufficiently maintained after printing, showing greater than 95% cell viability 24 h–incubation post-printing in media. Furthermore, cell viability was found to be maintained above 90% when examined on day 7 and 14 with active cell proliferation evident. Gene expression analysis demonstrates increased expression in chondrogenic (SOX9 and COl2A1), cardiogenic (Myh6 and Actn1) and adipogenic (PPARγ and LPL) constructs when compared to that of the collagen bioink control.

Extending on this work, the group develop a cardiac-specific approach wherein bioinks were formulated from hdECM bioinks (Jang et al. 2017). Three bioinks were formulated as follows: (I) 20 mg mL^{-1} hdECM and 0.02% (w/v) Vitamin B2 laden with cardiac progenitor cells (hCPCs); (II) 20 mg mL^{-1} hdECM, 0.02% (w/v) Vitamin

B2, and 10 μg mL^{-1} vascular endothelial growth factor (VEGF) laden with hMSCs; and (III) 20 mg mL^{-1} hdECM, 0.02% (w/v) Vitamin B2, and 10 μg mL^{-1} VEGF laden with hCPCs and hMSCs. Bioinks I and II were then used to make prevascularised stem-cell patches; initially, 2 layers of PCL was extruded as a support, next Bioink I and II were deposited alternatively, allowing for both muscular regeneration and vascular regeneration respectively. After printing, bioinks were exposed to UV for 30–60 s, allowing for vitamin B2 initiated crosslinking. Cell viability 10 min and 24 h post printing demonstrated high cell survival (~90–95%). The bioink design (both I and II) was found to recapitulate the cardiac tissue microenvironment, enhancing the structural maturation of cells. Furthermore, MSCs supplemented with VEGF were shown to promote vascular formation after 5 days in culture. *In vivo* tests demonstrated an improvement of cardiac function and a decrease of negative LV modelling when compared control, bioink I alone, and Bioink III developed patches; highlighting the potential of this approach for cardiac regeneration applications.

An alternative method for cardiac regeneration was investigated by Zhang et al. (2016) in which vasculature was first printed before being seeded with cardiomyoctyes, resulting in a highly vascularised tissue. Briefly, vasculature was printed through a HUVEC seeded alginate/GelMA bioink containing Irgacure 2959. A two-step crosslinking method was employed wherein ionic crosslinking of alginate was first induced through $CaCl_2$ interaction, before photoinduced crosslinking of GelMA. The constructs were then washed in phosphate-buffered saline solution (PBS) to remove excess $CaCl_2$. Next constructs were cultured for up to 33 days. After sufficient endothelium formation on the constructs (generally 15 days), Neonatal rat chondrocytes were seeded onto the scaffolds with spontaneous beating of cardiac tissues beginning after 48 h of culture. The method described was found to produce cardiac organoids with good cardiomyocyte maturation, alignment, and contraction, and provides potential for use in both cardiac-tissue engineering applications and drug screening applications.

5.4.2.2 Cardiac-Valve Regeneration

Heart valve disease is a substantial problem which can often result in valve dysfunction leading to further, potentially fatal cardiac or systemic complications (Zimmermann et al. 2004). These structures ensure unidirectional flow enabled through a unique arrangement of leaflets within a circular root; however, as these valves do not self-repair, there is a significant need to develop tissue engineering approaches which adequately regenerate healthy living tissues for surgical valve replacement (Duan et al. 2013; Zimmermann et al. 2004).

An alginate/gelatin bioink was developed by Duan et al. (2013) for use in cardiac-valve regeneration applications. The ink was initially investigated using a grid-like structure to assess its printability - with results showing prints of high accuracy (84.3 ± 10.9%), high viability (VICs, 84.6 ± 3.1% after 7 days) and good fidelity. The grid-like scaffolds were found to maintain geometry and mechanical integrity post printing, with an even distribution of cells evident in the structure. Mechanical properties of cell-free printed constructs were found to be comparable to that of non-calcified human aortic heart valve cusps (Elastic Modulus: 1.44 ± 0.30 and 1.98 ± 0.15 MPa respectively); however stiffness demonstrated a decrease with increasing culture time. Alternatively, aortic valve leaflet interstitial cell (VIC)-laden constructs initially demonstrated no significant difference in modulus (~1.1 MPa at day 0) which stayed consistent over culture time becoming statistically equivalent to cell-free samples after 7 day incubation. To print the aortic-valve constructs two bioinks were developed: the first contains smooth muscle cells (SMCs) and was used to print the root region; whilst the second contains VIC, and was used to fabricate the leaflet region of the valve. Cells were able to be printed at high density, and were shown to be heterogeneously distributed within the printed artificial tissue constructs; furthermore, the printed scaffold exhibits geometry comparable to the scanned original valve, with key anatomical features visible in the print. Cell viability was found to be sufficiently high, with an 81.4 ± 3.4% cell viability evident within the printed root (SMCs) component and a viability of 83.2 ± 4.0% in the leaflets (VICs). Finally, the encapsulated VIC and SMCs were shown to demonstrate good spreading and phenotype expression in culture.

The group further extend this work using an alternate hydrogel formulation, laden with human aortic vascular interstitial cells (HAVICs) (Duan et al. 2014). A variety of hydrogels were investigated, all based on formulations of ME-HA and –GelMA with differing concentrations (2%, 4%, or 6% and 6%, 10%, or 12% respectively). Irgacure 2959 (0.05% w/v) was also added to enable photo-crosslinking of GelMA. Mechanical testing of acellular hydrogels found the compressive moduli decreased with increasing GelMA concentration; whilst an increased ME-HA concentration resulted in a stiffer hydrogel. The viscosity of hydrogels was found to be too low for extrusion printing applications in 2% Me-HA and 6–10% GelMA hydrogels; whilst viscosity was too high in gels with 6% Me-HA and 10–12% GelMA, resulting in poor printability. For this reason, only gels containing 4% Me-HA and 6%, 10% and 12% GelMA were used for subsequent cell encapsulation investigations. Cell viability testing demonstrated >90% viability for all tested hydrogels (4% ME-HA/6% GelMA, 4% ME-HA/10% GelMA, 4% ME-HA/12% GelMA) on day 3 and day 7. Cells encapsulated within the softer hydrogel (12% ME-Gel) demonstrated the greatest degree of cell spreading compared to the other two gels (which demonstrated a more spherical shape) at day 7. Furthermore, all cells demonstrated significant increase in cell metabolic activity over the 7 day periods, corresponding with an increase in cell proliferation; however no significant increase was seen between hydrogel formulations. GAG content within the scaffolds was also found to increase between days 3 and 7 for all hydrogel formulations; once again there was no significant difference between each hydrogel formulation tested. Gene expression tests showed significant αSMA upregulation in 4% ME-HA/6% GelMA hydrogels; whilst all hydrogels demonstrated a significant increase in periostin and collagen I expression between day 3 and 7. Printability tests found the 4%Me-HA/10%GelMA hydrogel to demonstrate the best accuracy at 111.3 ± 7.2% (the printed construct slightly increased in area), whilst maintaining sufficiently high cell viability (>90%). For this reason, this ink was selected to print the aortic heart valves. A printed construct was successfully fabricated showing high similarity to the original CAD design. The heart-valve was constitute of two inks, the first was an acellular ink used to print the root, whilst the leaflets were printed using a HAVIC-laden bioink; both inks were formulated from a 4%Me-HA/10%GelMA hydrogel base. The printed heart valve was shown to demonstrate good viability, GAG increase, collagen deposition, and expression of both αSMA and VIM after 7-day culture; indicating the potential for peptide-based bioinks in heart-valve regeneration applications.

5.4.2.3 Liver Regeneration

The liver is a complex organ which plays an important role in metabolic activities. Essentially, the liver is composed of two branched vascular networks wand numerous hepatic lobules (Li et al. 2009). Blood supply plays a key role in transporting nutrients into the liver whilst removing waste (Li et al. 2009); therefore, there is a critical need for liver constructs to present a high degree of vascularisation.

Bioprinting in liver regeneration was investigated by Li et al. (2009) wherein an artificial liver construct was printed using two different inks. The first ink was designed for printing of vascular structures within the construct and therefore contains adipose-derived stromal cells (ADSCs) encapsulated within a gelatin/alginate/fibrinogen hydrogel (30%/5%/10% w/w); the second ink comprises the hepatic tissue and was formulated from hepatocytes encapsulated within a gelatin/alginate/chitosan (30%/5%/2% w/w) hydrogel. After assembly, the scaffold was crosslinked in a thrombin/$CaCl_2$/$Na_5P_3O_{10}$ solution. Using these bioinks, a functional liver construct was printed. Scanning electron microscopy images of the printed hydrogels (acellular) show the developed hydrogels to have physical architectures unlike that of either alginate or gelatin. After 2 weeks of culturing hepatocytes demonstrate an elliptical or round structure with ~23% containing double nuclei; suggesting hepatocytes were alive and proliferating. Immunofluorescence staining was positive for the both nuclei and the mature hepatocyte biomarker ALB; confirming cell-life and secretion ability. ADSCs were stained positive for the mature endothelial cell biomarker CD31; indicating endothelial cell differentiation in vascular printed constructs. However, few ADSCs demonstrated spindle morphologies - a key characteristic of endothelial cells; this may be due to insufficient culture time. To investigate functionality of the construct, the rate of increase in ALB, UREA, and ALT secreta in media was analysed. ALB content was found to increase steadily over 14 culture days, whilst ALT secretion was shown to increase during early stages before subsequently decreasing (after day 6). The steady increase in ALB confirms metabolic function. Alternatively, ALT secretion indicates acute liver damage; therefore it was suggested that the initial increase was due to hepatocyte isolation yielding an early inflammatory reaction, whilst the decrease may be a result the inflammatory reaction subsiding upon hepatocyte adaption to the new environment. The UREA content was found to initially increase before a slight decrease at day 14. Using this method, a liver-like construct is developed which upon culturing demonstrates preliminary functional tissue. The approach has potential application in tissue regeneration, physiological simulation, and drug screening systems.

A liver-on-a-chip platform containing bioprinted hepatic spheroids was presented by Bhise et al. (2016) wherein a GelMA bioink containing HepG2/C3A spheroids was bioprinted within a bioreactor platform. To ensure functional sustainability over long culture periods, the printed spheroids were cultured under continuous perfusion (200 μL h^{-1}) for 30 days, with findings demonstrating sufficient biomarker (albumin, transferrin, A1AT, and ceruloplasmin) secretion over cultured days; indicating sufficient metabolic activity and function. Drug toxicity tests were facilitated through spheroid exposure to acetaminophen within the bioreactor. Analysis of cellular activity shows that the acute hepatotoxic dose results in apoptosis of HEPG2 cells within spheroids, resulting in decreased biomarker secretion levels over time. The bioreactor platform paired with the bioprinted spheroids therefore was shown to respond to acute toxic drug doses; and therefore demonstrates the potential for bioprinting in organ engineering and drug-screening applications.

5.4.2.4 Skin Regeneration

Skin is the largest organ in the human body and serves as a protective barrier between the internal structures and the surrounding environment. This unique organ has three layers, the epidermis, dermis, and hypodermis. The epidermis is a thin layer attached to the basal membrane and primarily consists of keratinocytes and melanocytes (Böttcher-Haberzeth et al. 2010). The dermis is a vascularised layer which is rich in collagen and other ECM proteins excreted by fibroblasts; within this layer there are various other functional structures, including hair shafts and sweat glands (Böttcher-Haberzeth et al. 2010; MacNeil 2007). Alternatively, the hypodermis consists of mainly fat tissue which acts to insulate the body and provides an energy source (Böttcher-Haberzeth et al. 2010).

Bioprinting enabled skin regeneration was investigated by Cubo et al. (2016) wherein human bilayered skin constructs were printed using a fibrin-based bioink, containing fibroblasts and later seeded with keratinocytes. The bottom layer (dermal) of the construct was formed using a human fibroblast (hFB)-laden fibrin bioink, whilst the upper layer (epidermal) was formed through seeding of human keratinocyte cells (hKCs) on top of the initial dermal equivalent. *In vitro* tests showed well-spread hFBs within the dermal layer after 17 days; with immunofluorescence analysis confirming proper growth and spreading of hFBs through human VIM expression. Keratin K10 expression was detected in suprabasal cells; indicating normal skin synthesis. *In vivo* tests completed through grafting skin substrates on immuno deficient athymic mice resulting in development of grafted skin which was visually very similar to native human skin. Histological analysis found the regenerated skin presented a structure very similar to that of normal human skin, with all strata characteristic of normal skin easily identified. Immunofluorescence analysis of skin markers showed basal proliferation through K5 staining; whilst staining against human collagen VII shows high fluorescence at the dermo-epidermal junction demonstrating the presence of anchoring fibrils that bind together the epidermis and dermis. Analysis of K10 shows high expression which was characteristic of normal skin; whilst staining of filaggrin (a late differentiation marker) shows its presence in the granular layer. The presence of rete ridges suggests the developed skin was mature human skin rather than mouse skin, as mouse skin does not demonstrate the phenomenon. Immunostaining against human vimentin shows hFBs exclusively in the dermal region. Finally, analysis of hKC survival before and after printing shows acceptable viability, suggesting cells was not significantly affected during the detrimental conditions found with the printing process. This study highlights the potential of peptide-based bioprinting approaches in multicomponent tissue engineering applications (Fig. 5.4).

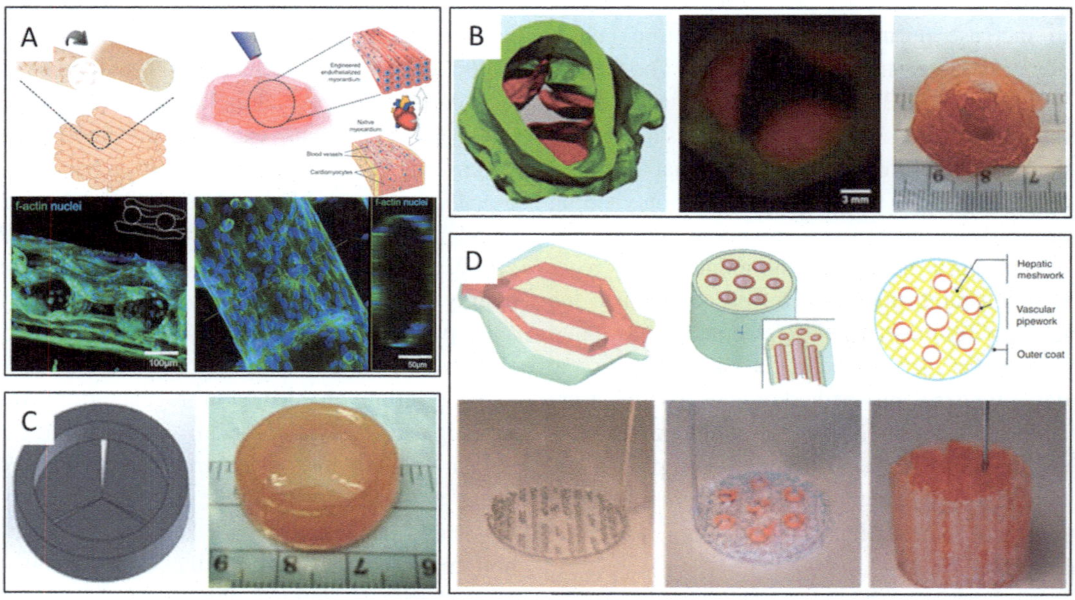

Fig. 5.4 Bioprinted organ tissue constructs. (**a**) Endothelialised myocardium: [*top*] Schematic process showing formation of endothelialised vascular bed (*left*) and cardiomyoctye seeding resulting in endothelialised myocardium formation (*right*), [*bottom*] Confocal fluorescence image showing cross-sectional area of three-layer scaffold at day 14 (*left*), high-resolution fluorescence imaging showing HUVEC distribution within a single microfiber at day 14 and 3D rendering of tubular structure at microfiber cross-section through dotted line (*right*); (**b**) Bioprinting of aortic valve conduit: [*left*] Aortic valve model reconstructed from micro-CT images (*Green* indicates valve root and red indicates valve leaflets), [*middle*] Fluorescent image of initial two bioprinted layers showing SMCs (labelled *green*) and VICs (labelled *red*), [*right*] printed aortic valve construct; (**c**) Bioprinting of heart valve conduit with encapsulation of HAVICs within leaflets: [*left*] Solidworks® designed heart valve model, [*right*] printed valve construct; (**d**) [*top*] Schematic of hybrid liver construct showing a cutaway view of the full model with branched vascular network with one-way inlet and outlet (*left*), middle section of hybrid liver construct model (*middle*), and single layer of model showing controlled architecture (*right*). [*bottom*] layer-by-layer fabrication process of liver construct (Reproduced with permission from Zhang et al. (2016), Duan et al. (2013), Duan et al. (2014) and Li et al. (2009), respectively)

5.5 Conclusion and Future Perspectives

The development of advanced artificial tissue-engineered constructs capable of effectively mimicking the intricacies of the spatially defined extracellular matrix within complex tissues would be of significant advance to the fields of tissue engineering, drug delivery and regenerative medicine. Currently, the requirement for vascularised, heterogenous constructs capable of simultaneously presenting biological and physical signals to cells is yet to be fully attained. Significant challenges facing the 3D bioprinting field are involved in bringing together the material, delivery and cellular aspects of the process. Recent developments in biofabrication made possible by additive manufacturing technology is a promising route to solve these issues due to their ability to spatially distribute cells, structures, and signals within an artificial biomimetic construct, enabling development of a variety of tissues types.

Despite the significant development of the equipment and 3D modelling software used in these systems, biomaterial selection remains a significant challenge; few bioinks adequately recapitulate the complexities of the native ECM. Novel application-specific materials are required to truly mimic the properties of the native ECM. The materials must provide suitable biodegradability, biocompatibility and mechanical properties to the scaffold, whilst enabling cell encapsulation and protection during printing processes. Furthermore, post-print crosslinking of these materials needs to be designed in such a way to minimise cell damage.

The spontaneous self-assembly and sol-gel transition of Peptide-based biomaterials have been extensively demonstrated to possess the easily incorporated chemical, physical and morphological charateristics that impart desirable biological properties ensuring high biocompatibility and good biodegradability. Natural-peptide based biomaterials have been extensively researched for bioprinting applications owing to their high bioactivity and resemblance to the native ECM; yet synthetic peptides have seen very little use in bioprinting applications. This is surprising, as the clinically desirable properties of being truely synthetic, with high purity, high shelf-life and batch-to-batch consistency has resulted in significant interest for use in moire traditional tissue engineering applications. Importantly, the ability of these materials to form hydrogels capable of encapsulating cells and biological factors whilst shielding cells from stresses encountered during printing makes them highly suitable for in bioink development. Biosynthetic peptide-based materials contain both natural and synthetic materials and are designed with the intent of combining both the favourable properties of natural materials (bioactivity, biocompatibility) with the favourable properties of synthetic materials (Good mechanical properties, ease of modification).

The development of effective bioinks from functionalised self-assembling peptides would be a practical way to increase bioink bioactivity whilst maintaining favorable properties associated with synthetic materials. The propensity of these peptides to spontaneously self-assemble into nanoarchitectures which mimic that of the ECM would not only provide a gentle cell-encapsulation mechanism; but would also provide printed tissue constructs with favorable nanoarchitectures for tissue growth. To achieve these bioinks, research needs to be placed into increasing the mechanical stiffness of hydrogels whilst reducing gelation time post printing.

Furthermore, the development of vasculature promoting bioinks which demonstrate immediate perfusability post printing would significantly improve current printing models. The ability of such systems to be paired with microfluidic devices capable of delivering fresh culture media and growth factors throughout the scaffold would facilitate oxygen and nutrient delivery to the whole tissue construct, ultimately resulting in healthy tissue development over greater scaffold area. This would enable the printing of large artificial tissue constructs facilitating large scale tissue regeneration which is currently unattainable. Finally, the improvement of current gradient transition approaches would enable improved printing of hard-to-soft tissue constructs (ie. Tendon-bone interface) resulting in effective, clinically relevant treatments.

Improvements in material selection, bioink design, and printing technologies are essential to enable successful and effective organ tissue regeneration. The complex nature of organs necessitates greater control over the spatial distribution of cells, structures and biological factors, and as such these need to be tightly controlled if functional tissue regeneration is to occur. Successful organ regeneration would greatly alleviate the current demand on donor organs and subsequently enable faster and more effective treatment methods. The development of peptide and protein bioinks along with current advancements in printing technology will help to unlock the potential of 3D bioprinting to transform the field of regenerative medicine.

References

Almany L, Seliktar D (2005) Biosynthetic hydrogel scaffolds made from fibrinogen and polyethylene glycol for 3D cell cultures. Biomaterials 26:2467–2477

Bammesberger SB, Kartmann S, Tanguy L, Liang D, Mutschler K, Ernst A, Zengerle R, Koltay P (2013) A low-cost, normally closed, solenoid valve for non-contact dispensing in the sub-μL range. Micromachines 4:9–21

Bhise NS, Manoharan V, Massa S, Tamayol A, Ghaderi M, Miscuglio M, Lang Q, Zhang YS, Shin SR, Calzone G (2016) A liver-on-a-chip platform with bioprinted hepatic spheroids. Biofabrication 8:014101

Bhosale AM, Richardson JB (2008) Articular cartilage: structure, injuries and review of management. Br Med Bull 87:77–95

Bogdani M, Korpos E, Simeonovic CJ, Parish CR, Sorokin L, Wight TN (2014) Extracellular matrix components in the pathogenesis of type 1 diabetes. Curr Diab Rep 14:1–11

Böttcher-Haberzeth S, Biedermann T, Reichmann E (2010) Tissue engineering of skin. Burns 36:450–460

Brown AC, Barker TH (2014) Fibrin-based biomaterials: modulation of macroscopic properties through rational design at the molecular level. Acta Biomater 10:1502–1514

Bruggeman KF, Rodriguez AL, Parish CL, Williams RJ, Nisbet DR (2016) Temporally controlled release of multiple growth factors from a self-assembling peptide hydrogel. Nanotechnology 27:385102

Buenger D, Topuz F, Groll J (2012) Hydrogels in sensing applications. Prog Polym Sci 37:1678–1719

Burg KJ, Porter S, Kellam JF (2000) Biomaterial developments for bone tissue engineering. Biomaterials 21:2347–2359

Chen F-M, Liu X (2016) Advancing biomaterials of human origin for tissue engineering. Prog Polym Sci 53:86–168

Chimene D, Lennox KK, Kaunas RR, Gaharwar AK (2016) Advanced bioinks for 3D printing: a materials science perspective. Ann Biomed Eng 44:2090–2102

Choi YJ, Kim TG, Jeong J, Yi HG, Park JW, Hwang W, Cho DW (2016) 3D cell printing of functional skeletal muscle constructs using skeletal muscle–derived bioink. Adv Healthc Mater 5:2636–2645

Collins MN, Birkinshaw C (2013) Hyaluronic acid based scaffolds for tissue engineering—a review. Carbohydr Polym 92:1262–1279

Crapo PM, Gilbert TW, Badylak SF (2011) An overview of tissue and whole organ decellularization processes. Biomaterials 32:3233–3243

Cubo N, Garcia M, del Cañizo JF, Velasco D, Jorcano JL (2016) 3D bioprinting of functional human skin: production and in vivo analysis. Biofabrication 9:015006

Cui X, Boland T (2009) Human microvasculature fabrication using thermal inkjet printing technology. Biomaterials 30:6221–6227

Cui H, Nowicki M, Fisher JP, Zhang LG (2017) 3D bioprinting for organ regeneration. Adv Healthc Mater 6:1601118

Daly AC, Critchley SE, Rencsok EM, Kelly DJ (2016) A comparison of different bioinks for 3D bioprinting of fibrocartilage and hyaline cartilage. Biofabrication 8:045002

Das S, Pati F, Choi Y-J, Rijal G, Shim J-H, Kim SW, Ray AR, Cho D-W, Ghosh S (2015) Bioprintable, cell-laden silk fibroin–gelatin hydrogel supporting multilineage differentiation of stem cells for fabrication of three-dimensional tissue constructs. Acta Biomater 11:233–246

de la Puente P, Ludeña D (2014) Cell culture in autologous fibrin scaffolds for applications in tissue engineering. Exp Cell Res 322:1–11

Dikovsky D, Bianco-Peled H, Seliktar D (2006) The effect of structural alterations of PEG-fibrinogen hydrogel scaffolds on 3-D cellular morphology and cellular migration. Biomaterials 27:1496–1506

Doolittle RF (2001) Fibrinogen and fibrin. In: eLS. Wiley, Chichester

Duan B (2017) State-of-the-art review of 3D bioprinting for cardiovascular tissue engineering. Ann Biomed Eng 45:195–209

Duan B, Hockaday LA, Kang KH, Butcher JT (2013) 3D bioprinting of heterogeneous aortic valve conduits with alginate/gelatin hydrogels. J Biomed Mater Res A 101:1255–1264

Duan B, Kapetanovic E, Hockaday LA, Butcher JT (2014) Three-dimensional printed trileaflet valve conduits using biological hydrogels and human valve interstitial cells. Acta Biomater 10:1836–1846

Duarte Campos DF, Blaeser A, Buellesbach K, Sen KS, Xun W, Tillmann W, Fischer H (2016) Bioprinting organotypic hydrogels with improved mesenchymal stem cell remodeling and mineralization properties for bone tissue engineering. Adv Healthc Mater 5:1336–1345

Dubbin K, Hori Y, Lewis KK, Heilshorn SC (2016) Dual-stage crosslinking of a gel-phase bioink improves cell viability and homogeneity for 3D bioprinting. Adv Healthc Mater 5:2488–2492

Einerson NJ, Stevens KR, Kao WJ (2003) Synthesis and physicochemical analysis of gelatin-based hydrogels for drug carrier matrices. Biomaterials 24:509–523

England S, Rajaram A, Schreyer DJ, Chen X (2017) Bioprinted fibrin-factor XIII-hyaluronate hydrogel scaffolds with encapsulated Schwann cells and their in vitro characterization for use in nerve regeneration. Bioprinting 5:1–9

Ferris CJ, Gilmore KG, Wallace GG (2013) Biofabrication: an overview of the approaches used for printing of living cells. App Microbiol Biotechnol 97:4243–4258

Fortunato TM, Beltrami C, Emanueli C, De Bank PA, Pula G (2016) Platelet lysate gel and endothelial progenitors stimulate microvascular network formation in vitro: tissue engineering implications. Sci Rep 6:25326

Gao G, Cui X (2016) Three-dimensional bioprinting in tissue engineering and regenerative medicine. Biotechnol Lett 38:203–211

Gao G, Schilling AF, Hubbell K, Yonezawa T, Truong D, Hong Y, Dai G, Cui X (2015a) Improved properties of bone and cartilage tissue from 3D inkjet-bioprinted human mesenchymal stem cells by simultaneous deposition and photocrosslinking in PEG-GelMA. Biotechnol Lett 37:2349–2355

Gao G, Yonezawa T, Hubbell K, Dai G, Cui X (2015b) Inkjet-bioprinted acrylated peptides and PEG hydrogel with human mesenchymal stem cells promote robust bone and cartilage formation with minimal printhead clogging. Biotechnol J 10:1568–1577

García-Martínez L, Campos F, Godoy-Guzmán C, del Carmen S-QM, Garzón I, Alaminos M, Campos A, Carriel V (2017) Encapsulation of human elastic cartilage-derived chondrocytes in nanostructured fibrin-agarose hydrogels. Histochem Cell Biol 147:83–95

Gilbert TW, Sellaro TL, Badylak SF (2006) Decellularization of tissues and organs. Biomaterials 27:3675–3683

Guillemot F, Guillotin B, Fontaine A, Ali M, Catros S, Kériquel V, Fricain J-C, Rémy M, Bareille R, Amédée-Vilamitjana J (2011) Laser-assisted bioprinting to deal with tissue complexity in regenerative medicine. MRS Bull 36:1015–1019

Guillotin B, Souquet A, Catros S, Duocastella M, Pippenger B, Bellance S, Bareille R, Rémy M, Bordenave L, Amédée J, Guillemot F (2010) Laser assisted bioprinting of engineered tissue with high cell density and microscale organization. Biomaterials 31:7250–7256

Guo B, Ma PX (2014) Synthetic biodegradable functional polymers for tissue engineering: a brief review. Sci China Chem 57:490–500

Gurkan UA, El Assal R, Yildiz SE, Sung Y, Trachtenberg AJ, Kuo WP, Demirci U (2014) Engineering anisotropic biomimetic fibrocartilage microenvironment by bioprinting mesenchymal stem cells in nanoliter gel droplets. Mol Pharm 11:2151–2159

Guvendiren M, Burdick JA (2013) Engineering synthetic hydrogel microenvironments to instruct stem cells. Curr Opin Biotechnol 24:841–846

Hölzl K, Lin S, Tytgat L, Van Vlierberghe S, Gu L, Ovsianikov A (2016) Bioink properties before, during and after 3D bioprinting. Biofabrication 8:032002

Horgan CC, Rodriguez AL, Li R, Bruggeman KF, Stupka N, Raynes JK, Day L, White JW, Williams RJ, Nisbet DR (2016) Characterisation of minimalist co-assembled fluorenylmethyloxycarbonyl self-assembling peptide systems for presentation of multiple bioactive peptides. Acta Biomater 38:11–22

Hsieh JY, Smith TD, Meli VS, Tran TN, Botvinick EL, Liu WF (2017) Differential regulation of macrophage inflammatory activation by fibrin and fibrinogen. Acta Biomater 47:14–24

Huang SH, Liu P, Mokasdar A, Hou L (2013) Additive manufacturing and its societal impact: a literature review. Int J Adv Manuf Technol 67:1191–1203

Hunt JA, Chen R, van Veen T, Bryan N (2014) Hydrogels for tissue engineering and regenerative medicine. J Mater Chem B 2:5319–5338

Irvine SA, Agrawal A, Lee BH, Chua HY, Low KY, Lau BC, Machluf M, Venkatraman S (2015) Printing cell-laden gelatin constructs by free-form fabrication and enzymatic protein crosslinking. Biomed Microdevices 17:1–8

Jang J, Kim TG, Kim BS, Kim S-W, Kwon S-M, Cho D-W (2016) Tailoring mechanical properties of decellularized extracellular matrix bioink by vitamin B2-induced photo-crosslinking. Acta Biomater 33:88–95

Jang J, Park H-J, Kim S-W, Kim H, Park JY, Na SJ, Kim HJ, Park MN, Choi SH, Park SH, Kim SW, Kwon S-M, Kim P-J, Cho D-W (2017) 3D printed complex tissue construct using stem cell-laden decellularized extracellular matrix bioinks for cardiac repair. Biomaterials 112:264–274

Jia W, Gungor-Ozkerim PS, Zhang YS, Yue K, Zhu K, Liu W, Pi Q, Byambaa B, Dokmeci MR, Shin SR, Khademhosseini A (2016) Direct 3D bioprinting of perfusable vascular constructs using a blend bioink. Biomaterials 106:58–68

Jungst T, Smolan W, Schacht K, Scheibel T, Jr G (2016) Strategies and molecular design criteria for 3D printable hydrogels. Chem Rev 116:1496–1539

Kamisuki S, Hagata T, Tezuka C, Nose Y, Fujii M, Atobe M (1998) A low power, small, electrostatically-driven commercial inkjet head. Proceed IEEE MEMS' 98:63–68

Karlsen TA, Jakobsen RB, Brinchmann JE (2016) Early molecular events during in vitro chondrogenesis. In:

Atkinson K (ed) The biology and therapeutic application of mesenchymal cells. Wiley, Hoboken
Kasoju N, Bora U (2012) Silk fibroin in tissue engineering. Adv Healthc Mater 1:393–412
Kim YB, Lee H, Kim GH (2016) Strategy to achieve highly porous/biocompatible macroscale cell blocks, using a collagen/genipin-bioink and an optimal 3D printing process. ACS Appl Mater Interfaces 8:32230–32240
Kolesky DB, Truby RL, Gladman A, Busbee TA, Homan KA, Lewis JA (2014) 3D bioprinting of vascularized, heterogeneous cell-laden tissue constructs. Adv Mater 26:3124–3130
Kundu B, Rajkhowa R, Kundu SC, Wang X (2013) Silk fibroin biomaterials for tissue regenerations. Adv Drug Deliv Rev 65:457–470
Lau HK, Kiick KL (2014) Opportunities for multicomponent hybrid hydrogels in biomedical applications. Biomacromolecules 16:28–42
Lee KY, Mooney DJ (2001) Hydrogels for tissue engineering. Chem Rev 101:1869–1880
Lee CH, Singla A, Lee Y (2001) Biomedical applications of collagen. Int J Pharm 221:1–22
Lee HJ, Kim YB, Ahn SH, Lee JS, Jang CH, Yoon H, Chun W, Kim GH (2015) A new approach for fabricating collagen/ecm-based bioinks using preosteoblasts and human adipose stem cells. Adv Healthc Mater 4:1359–1368
Levato R, Visser J, Planell JA, Engel E, Malda J, Mateos-Timoneda MA (2014) Biofabrication of tissue constructs by 3D bioprinting of cell-laden microcarriers. Biofabrication 6:035020
Lewis MP, Mudera V, Cheema U, Shah R (2009) Muscle tissue engineering. In: Meyer U, Meyer T, Handschel J, Wiesmann HP (eds) Fundamentals of tissue engineering and regenerative medicine. Springer, Berlin/Heidelberg
Li S, Xiong Z, Wang X, Yan Y, Liu H, Zhang R (2009) Direct fabrication of a hybrid cell/hydrogel construct by a double-nozzle assembling technology. J Bioact Compat Polym 24:249–265
Li R, Rodriguez A, Nisbet DR, Barrow CJ, Williams RJ (2015) Self-assembled peptide nanostructures for the fabrication of cell scaffolds. In: Castillo-Leon J, Svendsen WE (eds) Micro and nanofabrication using self-assembled biological nanostructures. Elsevier, Amsterdam
Lim KS, Schon BS, Mekhileri NV, Brown GC, Chia CM, Prabakar S, Hooper GJ, Woodfield TB (2016) New visible-light photoinitiating system for improved print fidelity in gelatin-based bioinks. ACS Biomater Sci Eng 2:1752–1762
Loo Y, Lakshmanan A, Ni M, Toh LL, Wang S, Hauser CA (2015) Peptide bioink: self-assembling nanofibrous scaffolds for three-dimensional organotypic cultures. Nano Lett 15:6919–6925
Lynn A, Yannas I, Bonfield W (2004) Antigenicity and immunogenicity of collagen. J Biomed Mater Res B Appl Biomater 71:343–354
MacNeil S (2007) Progress and opportunities for tissue-engineered skin. Nature 445:874–880
Murphy SV, Atala A (2014) 3D bioprinting of tissues and organs. Nat Biotechnol 32:773–785
Nerem RM, Seliktar D (2001) Vascular tissue engineering. Annu Rev Biomed Eng 3:225–243
Nilasaroya A, Poole-Warren LA, Whitelock JM, Martens PJ (2008) Structural and functional characterisation of poly (vinyl alcohol) and heparin hydrogels. Biomaterials 29:4658–4664
Nisbet DR, Williams RJ (2012) Self-assembled peptides: characterisation and in vivo response. Biointerphases 7:2
Nishiyama Y, Henmi C, Iwanaga S, Nakagawa H, Yamaguchi K, Akita K, Mochizuki S, Takiura K, Nakamura M (2008) Ink jet three-dimensional digital fabrication for biological tissue manufacturing: analysis of alginate microgel beads produced by ink jet droplets for three dimensional tissue fabrication. J Imaging Sci Technol 52:60201–60201–60201–60206
Novosel EC, Kleinhans C, Kluger PJ (2011) Vascularization is the key challenge in tissue engineering. Adv Drug Deliv Rev 63:300–311
Nukavarapu SP, Dorcemus DL (2013) Osteochondral tissue engineering: current strategies and challenges. Biotechnol Adv 31:706–721
Odde DJ, Renn MJ (1999) Laser-guided direct writing for applications in biotechnology. Trends Biotechnol 17:385–389
Odde DJ, Renn MJ (2000) Laser-guided direct writing of living cells. Biotechnol Bioeng 67:312–318
Olszta MJ, Cheng X, Jee SS, Kumar R, Kim Y-Y, Kaufman MJ, Douglas EP, Gower LB (2007) Bone structure and formation: a new perspective. Mater Sci Eng R Rep 58:77–116
Ozbolat IT, Hospodiuk M (2016) Current advances and future perspectives in extrusion-based bioprinting. Biomaterials 76:321–343
Ozbolat IT, Moncal KK, Gudapati H (2017) Evaluation of bioprinter technologies. Addit Manuf 13:179–200
Park JY, Gao G, Jang J, Cho D-W (2016) 3D printed structures for delivery of biomolecules and cells: tissue repair and regeneration. J Mater Chem B 4:7521–7539
Pati F, Jang J, Ha D-H, Kim SW, Rhie J-W, Shim J-H, Kim D-H, Cho D-W (2014) Printing three-dimensional tissue analogues with decellularized extracellular matrix bioink. Nat Commun 5:3935
Pugsley MK, Tabrizchi R (2000) The vascular system: an overview of structure and function. J Pharmacol Toxicol Methods 44:333–340
Ren X, Wang F, Chen C, Gong X, Yin L, Yang L (2016) Engineering zonal cartilage through bioprinting collagen type II hydrogel constructs with biomimetic chondrocyte density gradient. BMC Musculoskelet Disord 17:301
Rhee S, Puetzer JL, Mason BN, Reinhart-King CA, Bonassar LJ (2016) 3D bioprinting of spatially heterogeneous collagen constructs for cartilage tissue engineering. ACS Biomater Sci Eng 2:1800–1805
Rice JJ, Martino MM, De Laporte L, Tortelli F, Briquez PS, Hubbell JA (2013) Engineering the regenerative micro-

environment with biomaterials. Adv Healthc Mater 2:57–71
Richards D, Jia J, Yost M, Markwald R, Mei Y (2017) 3D bioprinting for vascularized tissue fabrication. Ann Biomed Eng 45:132–147
Rodriguez A, Wang T-Y, Bruggeman K, Horgan C, Li R, Williams RJ, Parish CL, Nisbet D (2014) In vivo assessment of grafted cortical neural progenitor cells and host response to functionalized self-assembling peptide hydrogels and the implications for tissue repair. J Mater Chem B 2:7771–7778
Sadtler K, Singh A, Wolf MT, Wang X, Pardoll DM, Elisseeff JH (2016) Design, clinical translation and immunological response of biomaterials in regenerative medicine. Nat Rev Mater 1:16040
Sargeant TD, Desai AP, Banerjee S, Agawu A, Stopek JB (2012) An in situ forming collagen–PEG hydrogel for tissue regeneration. Acta Biomater 8:124–132
Schiaffino S, Reggiani C (2011) Fiber types in mammalian skeletal muscles. Physiol Rev 91:1447–1531
Schiele NR, Corr DT, Huang Y, Raof NA, Xie Y, Chrisey DB (2010) Laser-based direct-write techniques for cell printing. Biofabrication 2:032001
Schmidt CE, Leach JB (2003) Neural tissue engineering: strategies for repair and regeneration. Annu Rev Biomed Eng 5:293–347
Sehnal F, Žurovec M (2004) Construction of silk fiber core in Lepidoptera. Biomacromolecules 5:666–674
Shim J-H, Jang K-M, Hahn SK, Park JY, Jung H, Oh K, Park KM, Yeom J, Park SH, Kim SW (2016) Three-dimensional bioprinting of multilayered constructs containing human mesenchymal stromal cells for osteochondral tissue regeneration in the rabbit knee joint. Biofabrication 8:014102
Skardal A, Atala A (2015) Biomaterials for integration with 3-D bioprinting. Ann Biomed Eng 43:730–746
Skardal A, Devarasetty M, Kang H-W, Mead I, Bishop C, Shupe T, Lee SJ, Jackson J, Yoo J, Soker S, Atala A (2015) A hydrogel bioink toolkit for mimicking native tissue biochemical and mechanical properties in bioprinted tissue constructs. Acta Biomater 25:24–34
Smith AM, Williams RJ, Tang C, Coppo P, Collins RF, Turner ML, Saiani A, Ulijn RV (2008) Fmoc-diphenylalanine self assembles to a hydrogel via a novel architecture based on π–π interlocked β-sheets. Adv Mater 20:37–41
Spotnitz WD (2010) Fibrin sealant: past, present, and future: a brief review. World J Surg 34:632–634
Tamada Y (2005) New process to form a silk fibroin porous 3-D structure. Biomacromolecules 6:3100–3106
Tanaka K, Inoue S, Mizuno S (1999) Hydrophobic interaction of P25, containing Asn-linked oligosaccharide chains, with the H-L complex of silk fibroin produced by *Bombyx mori*. Insect Biochem Mol Biol 29:269–276
Tang D, Tare RS, Yang L-Y, Williams DF, Ou K-L, Oreffo RO (2016) Biofabrication of bone tissue: approaches, challenges and translation for bone regeneration. Biomaterials 83:363–382
Thiagarajan P, Narayanan A (2001) Thrombin. In: eLS. John Wiley & Sons ltd, Chichester
Tibbitt MW, Anseth KS (2009) Hydrogels as extracellular matrix mimics for 3D cell culture. Biotechnol Bioeng 103:655–663
Vaezi M, Seitz H, Yang S (2013) A review on 3D micro-additive manufacturing technologies. Int J Adv Manuf Technol 67:1721–1754
Wang Z, Abdulla R, Parker B, Samanipour R, Ghosh S, Kim K (2015) A simple and high-resolution stereolithography-based 3D bioprinting system using visible light crosslinkable bioinks. Biofabrication 7:045009
Weiner S, Wagner HD (1998) The material bone: structure-mechanical function relations. Annu Rev Mater Sci 28:271–298
Whelan D, Caplice N, Clover A (2014) Fibrin as a delivery system in wound healing tissue engineering applications. J Control Release 196:1–8
Winter JO, Schmidt CE (2002) Biomimetic strategies and applications in the nervous system. In: Dillow A, Lowman A (eds) Biomimetic materials and design: biointerfacial strategies, tissue engineering, and targeted drug delivery. Marcel-Dekker, New York
Wüst S, Godla ME, Müller R, Hofmann S (2014) Tunable hydrogel composite with two-step processing in combination with innovative hardware upgrade for cell-based three-dimensional bioprinting. Acta Biomater 10:630–640
Xiao W, He J, Nichol JW, Wang L, Hutson CB, Wang B, Du Y, Fan H, Khademhosseini A (2011) Synthesis and characterization of photocrosslinkable gelatin and silk fibroin interpenetrating polymer network hydrogels. Acta Biomater 7:2384–2393
Xu T, Gregory CA, Molnar P, Cui X, Jalota S, Bhaduri SB, Boland T (2006) Viability and electrophysiology of neural cell structures generated by the inkjet printing method. Biomaterials 27:3580–3588
Zhang X, Zhang Y (2015) Tissue engineering applications of three-dimensional bioprinting. Cell Biochem Biophys 72:777–782
Zhang Y, Yu Y, Akkouch A, Dababneh A, Dolati F, Ozbolat IT (2015) In vitro study of directly bioprinted perfusable vasculature conduits. Biomater Sci 3:134–143
Zhang YS, Arneri A, Bersini S, Shin S-R, Zhu K, Goli-Malekabadi Z, Aleman J, Colosi C, Busignani F, Dell'Erba V, Bishop C, Shupe T, Demarchi D, Moretti M, Rasponi M, Dokmeci MR, Atala A, Khademhosseini A (2016) Bioprinting 3D microfibrous scaffolds for engineering endothelialized myocardium and heart-on-a-chip. Biomaterials 110:45–59
Zhou X, Zhu W, Nowicki M, Miao S, Cui H, Holmes B, Glazer RI, Zhang LG (2016) 3D bioprinting a cell-laden bone matrix for breast cancer metastasis study. ACS Appl Mater Interfaces 8:30017–30026
Zhu W, Harris BT, Zhang LG (2016) Gelatin methacrylamide hydrogel with graphene nanoplatelets for neural cell-laden 3D bioprinting. Conf Proc IEEE Eng Med Biol Soc 2016:4185–4188
Zimmermann W-H, Melnychenko I, Eschenhagen T (2004) Engineered heart tissue for regeneration of diseased hearts. Biomaterials 25:1639–1647

Peptides as Bio-inspired Molecular Electronic Materials

6

John Horsley, Jingxian Yu, Yuan Qi Yeoh, and Andrew Abell

Abstract

Understanding the electronic properties of single peptides is not only of fundamental importance to biology, but it is also pivotal to the realization of bio-inspired molecular electronic materials. Natural proteins have evolved to promote electron transfer in many crucial biological processes. However, their complex conformational nature inhibits a thorough investigation, so in order to study electron transfer in proteins, simple peptide models containing redox active moieties present as ideal candidates. Here we highlight the importance of secondary structure characteristic to proteins/peptides, and its relevance to electron transfer. The proposed mechanisms responsible for such transfer are discussed, as are details of the electrochemical techniques used to investigate their electronic properties. Several factors that have been shown to influence electron transfer in peptides are also considered. Finally, a comprehensive experimental and theoretical study demonstrates that the electron transfer kinetics of peptides can be successfully fine tuned through manipulation of chemical composition and backbone rigidity. The methods used to characterize the conformation of all peptides synthesized throughout the study are outlined, along with the various approaches used to further constrain the peptides into their geometric conformations. The aforementioned sheds light on the potential of peptides to one day play an important role in the fledgling field of molecular electronics.

Keywords

Peptides • Electron transfer • Bio-inspired • Molecular electronics • Electronic materials • Electrochemical methods

J. Horsley (✉) • J. Yu • Y.Q. Yeoh • A. Abell
ARC Centre of Excellence for Nanoscale BioPhotonics (CNBP), School of Chemistry and Physics, The University of Adelaide, Adelaide, SA 5005, Australia
e-mail: john.horsley@adelaide.edu.au

© Springer International Publishing AG 2017

A. Sunna et al. (eds.), *Peptides and Peptide-based Biomaterials and their Biomedical Applications*, Advances in Experimental Medicine and Biology 1030, https://doi.org/10.1007/978-3-319-66095-0_6

6.1 Introduction

6.1.1 The Role of Peptides in Molecular Electronics

Electron transfer pathways in proteins have evolved over millions of years to optimize processes related to energy conversion, so it follows that peptides are ideal candidates for the design and fabrication of molecular electronic components such as switches, diodes, and molecular wires (Sek 2013). Molecular electronics was first proposed in 1974, when Aviram and Ratner came up with the idea of a molecular rectifier (Aviram and Ratner 1974), however significant advances in the field have only been realized in recent years. In accordance with Moore's Law, the traditional "top-down" method of manufacture for semiconductor devices is facing many unavoidable challenges such as photolithography and etching, that can only be addressed by a "bottom up" approach (Maeda et al. 2013). Single molecules have attracted much interest as molecular components, where they are chemically bound between two metal electrodes to form a molecular junction. A bias voltage is then applied between the electrodes to probe the electronic properties of the single molecule in question (Sek 2013). The current/voltage properties of single-molecule junctions have been likened to those of conventional electronic devices (Benesch et al. 2009). The single-molecule approach to forming molecular junctions started with the discovery of scanning probe microscopy (SPM) in the early 1980s (McCreery and Bergren 2009), and has progressed rapidly with the onset of reliable methods such as conducting probe AFMs (CP-AFM) and STM-break-junctions (STM-BJ) (Marques-Gonzalez et al. 2013). Such nanoscale junctions have led to the use of molecules as prototypes for various electronic components, such as switches (Avellini et al. 2012; Darwish et al. 2012), rectifiers (Ding et al. 2014; Staykov et al. 2012) and transistors (Chen et al. 2012; Xu et al. 2013). Peptides have also recently been used in the successful design of single-molecule junctions (Uji et al. 2013). Intramolecular electron transfer in redox active peptides immobilized on self-assembled monolayers (SAMs) is analogous to a bias voltage applied through a single peptide sandwiched between two electrodes (Sek 2013). Much recent work has shown that reproducible intermolecular charge transfer utilizing π-π stacking is achievable in aromatic peptides bound between two metallic leads, providing careful design of the structure is realized (Smeu et al. 2009; Yew et al. 2011). Hence, systems utilizing both inter- and intra-molecular electron transfer will likely be useful features in future molecular electronic devices. Peptides are ideal for such a purpose, as they can be specifically designed to conform to well-defined secondary structures such as helices and β-strands, and have the capacity to self-assemble onto conductive surfaces. Peptides offer the opportunity to mimic nature for applications in molecular electronics. This "bottom up" approach for the fabrication of electronic components provides an opportunity to go beyond the inevitable physical limitations of conventional silicon-based electronics that are based on "top-down" approaches such as photolithography and etching, providing a major extension of the capabilities of conventional technology.

6.2 Electron Transfer in Peptides/Proteins

Electron transfer in proteins plays an important role in a wide range of processes at the cellular level, including photosynthesis (Bendall 1996; Smestad and Gratzel 1998). The associated movement of an electron from one species (donor) to another (acceptor) results in a change of oxidation states for both components. The precise mechanisms for this are at present in dispute, however two main modes are widely accepted, namely superexchange (also known as tunneling) and hopping (see Fig. 6.1). Superexchange is explained in simple terms if one imagines an electron situated on a donor in a narrow potential well that is separated by a distance from an electron acceptor. The area between the donor and the acceptor is an electrically insulating region and therefore acts as an obstacle for electron transfer. Classical physics

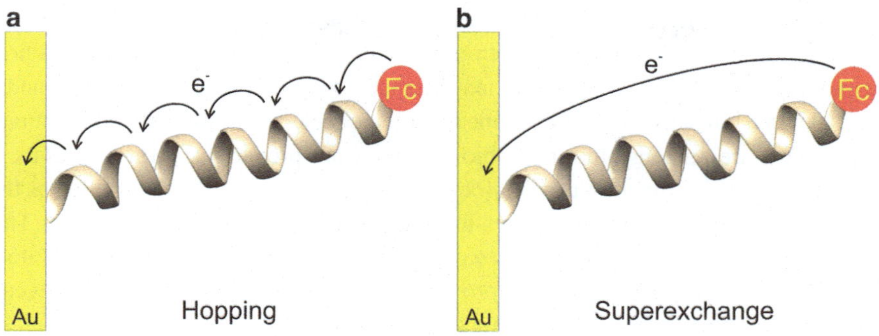

Fig. 6.1 Electron transfer across a helical peptide showing the two widely recognized mechanisms, **(a)** hopping and **(b)** superexchange (tunneling)

states that the electron cannot overcome this hurdle, but quantum mechanics indicates the possibility that the wave function of the electron infiltrates the insulating region to reach the potential well of an electron acceptor (Bendall 1996). The peptide bridge therefore takes on a virtual role, with a high probability that only an extremely small number of electrons occupy the bridge during the electron transport procedure (Petrov et al. 2003). With respect to the superexchange mechanism, the electron transport rate decreases exponentially with increased distance between the redox active moieties situated along the peptide chain (Morita and Kimura 2003). This means that tunneling is not an efficient method of electron transfer over long distances, for instance, greater than approximately 20 Å (Kai et al. 2008).

Electron transfer can also occur via a hopping mechanism, whereby an electron "hops" between the redox active moieties, using sites on the peptide chain that are coupled to each other electronically (Watanabe et al. 2005). For this to occur, the difference in the energy levels of the donor and the bridge separating donor and acceptor, must be negligible (Schlag et al. 2007). Recent theoretical studies have found that low-frequency rotation between neighboring amino acids in a peptide brings adjacent carbonyl groups into alignment, and hence into close proximity to one another. When the backbone dihedral angles reach this critical point, the energy barrier becomes almost negligible, and efficient charge transfer from one amino acid to the next can occur via a hopping mechanism (Schlag et al. 2000, 2007). A weak distance dependence is reported for the hopping mechanism, whereby the electron transfer rate decreases with a linear relationship when the distance between electron donor and acceptor is increased (Malak et al. 2004). In general, this mechanism is reported to occur over longer distances, for instance, greater than approximately 20 Å (Kai et al. 2008). As the rate of electron transfer decreases non-exponentially with distance, this long-range mechanism is thought to occur when the distance between the electron donor and acceptor is separated into shorter and therefore faster steps (Wang et al. 2009). It is widely believed that electron transfer utilizes both superexchange and hopping mechanisms, depending on factors such as the specific peptide architecture involved (Petrov et al. 2003).

Many proteins have evolved specifically for electron transfer from one redox active site to another (Long et al. 2005). In nature, a number of cofactors facilitate this electron transfer. However, some metal cofactors are separated by distances too great to promote efficient electron transfer via a superexchange mechanism (Seyedsayamdost et al. 2006). Situated between these electron donors and acceptors are well placed redox active amino acids, which are fundamental to charge separation and can act as "stepping stones" for electron transfer by exploiting a hopping mechanism. These electron transfer pathways in naturally occurring proteins require a sophisticated framework or architecture that is provided by well-defined secondary structures, such as helices and β-sheets, in order to locate the electron donors/acceptors in a precise and systematic manner (Morita et al. 2008). Just as these well-defined secondary structures are responsible for the biological activity in a protein, function and conformation are inextricably linked when it comes to electron transfer in proteins. Nature has optimized the design of electron transfer pathways in proteins, for instance enzymes such as cytochrome c oxidase, so that damage by free radicals and short-circuiting are prevented (Cordes and Giese 2009). The mechanism of electron transfer in aminoisobutyric acid (Aib) homo-oligomers has been shown to be defined by the extent of secondary structure (Yu et al. 2012). Hence, to further our fundamental understanding of electron transfer in proteins, it is important to consider systems that incorporate a well-defined secondary structure. The conformation of even the simplest protein can be labyrinthine, and as such, model synthetic peptides containing redox active moieties present as an ideal platform to investigate electron transfer in proteins.

6.2.1 Factors Affecting Electron Transfer in Peptides/Proteins

Many factors have been shown to influence electron transfer in peptides, including the distance separating the electron donor and acceptor (Malak et al. 2004; Mandal and Kraatz 2012), the dipole moment (Gatto et al. 2012; Lauz et al. 2012), secondary structure (Yu et al. 2012) and the nature of the constituent amino acid side chains (Gao et al. 2011; Wang et al. 2009). It has been suggested that as the method of charge transfer in proteins is dependent on many factors, it can vary greatly between electron hopping, hole hopping and tunneling (Han et al. 2012). Many researchers believe that electron transfer within a peptide is dependent on the distance between the electron donor and acceptor. Some are advocates for a tunneling mechanism (Antonello et al. 2003; Polo et al. 2005), whilst many others believe tunneling to occur over short distances, transferring to a hopping mechanism as the distance increases beyond a critical point (Arikuma et al. 2011; Morita and Kimura 2003). Sisido and co-workers investigated how the distance between electron donor and acceptor affected electron transfer. Poly-glutamate peptides were prepared with repeat units ($n = 0$–8), with a pyrenyl group incorporated as the electron donor in the centre of the α-helix. The electron transfer rates for each peptide exhibited an exponential dependence to distance between electron donor and acceptor. The distances, 3.9 Å ($n = 0$) and 15.9 Å ($n = 8$), are within the range that the superexchange mechanism is widely believed to operate in. An attenuation factor, which provides information on the conductibility of a species, of 0.66 A^{-1} was reported, which is in accordance with tunneling being the operative mechanism. A dependence between the electron transfer rate constant and the number of amino acid spacers in the peptide was found. It was also claimed that intramolecular hydrogen bonding is not only vital for the rigidity of the α-helical peptide, but enables electrons to take shortcuts between these bonds (Sisido et al. 2001). Another study using Aib homo-oligomer peptides found that the electron transfer rate increased, despite the distance between the electron donor and acceptor increasing by 3 Å. It was postulated that the intramolecular hydrogen bonds increase electronic coupling between donor and acceptor, and hence increase the rate of electron transfer (Antonello et al. 2003). Malak and co-workers investigated electron transfer using polyproline peptide bridges, ranging from $n = 0$–9. As proline does not contain an amide group, a polyproline peptide does not form intramolecular hydrogen bonds, and is unable to form a well-ordered helical structure. For the peptides with repeat units $n = 0$ to $n = 4$, a clear exponential decay of the electron transfer rate was found as the distance between donor and acceptor increased, indicating that the superexchange mechanism was operational. The distance between donor and acceptor for the peptide $n = 4$ is 18 Å, a figure between the 15–20 Å range that is believed to be transitional from a tunneling to a hopping mechanism. For the peptides $n = 5$ to $n = 9$, a weak distance dependence was reported, consistent with a hopping mechanism. The distance between donor and acceptor for these peptides is 21 Å ($n = 5$)

and 32 Å (n = 9) (Malak et al. 2004). Another study was conducted to investigate electron transfer in oligoglycine peptides. For the peptides containing two to five amino acid residues, tunneling was found to be the applicable mechanism, whilst for the hexapeptide, hopping may be the operative mode. The rate of electron transfer for the hexapeptide was determined to be faster than expected, and it was postulated that this may be due to a change in secondary structure from a polyglycine I conformation to a polyglycine II conformation (Sek et al. 2004). An electrochemical study was reported by our group on a series of Aib oligomers (n = 0–5) (Yu et al. 2012). The peptides (n = 0–2) gave rise to a random conformation, and the rate of electron transfer decreased exponentially. In comparison, the three structures (n = 3–5), possess well-defined intramolecular hydrogen bonding which defines their helical secondary structure. The rate of electron transfer for these structures (n = 3–5) was found to decrease non-exponentially with distance. This suggests that the transition from a superexchange to a hopping mechanism is the result of a change from a random structure to a well-defined helical conformation, and not merely a result of increased chain length (Yu et al. 2012) (see Fig. 6.2).

Arikuma and co-workers synthesized a helical peptide with 32 repeat units of (Ala-Aib), over 110 Å in length. Subsequent electrochemical results indicated a non-exponential distance dependence between donor and acceptor. This supported the involvement of the hopping mechanism for charge transfer along this long peptide, potentially using the amide groups co-ordinated in the intramolecular hydrogen bonding to facilitate electron transport. This is reported to occur when oxidation or reduction of the backbone amide groups is energetically favourable (Arikuma et al. 2011). Hopping is a thermally activated mechanism with an energy gap of approximately 0.2 eV, which is similar to that reported for DNA (Petrov et al. 2001).

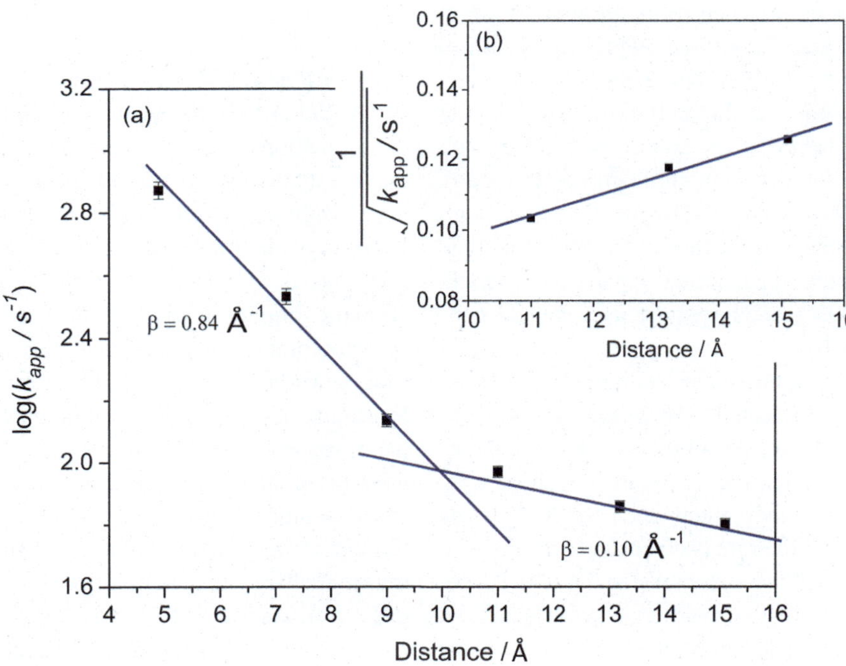

Fig. 6.2 Electrochemical data from a study reported on a series of oligomers of Aib (n = 0–5), showing (**a**) the transition between random structure (n = 0–2), the three data points on *left* of graph (5–9 Å), and helical structure (n = 3–5), the three data points on *right* of graph (11–15 Å), with the associated transition in electron transfer mechanism from superexchange to hopping. (**b**) Plot of the inverse of the square root of the electron transfer rate constant vs the iron to terminal nitrogen distance for the helical peptides (n = 3–5)

The influence of the dipole moment on electron transfer in peptides has also been considered. A helix has a dipole generated by an electrostatic field with a δ^+ charge at the *N*-terminus and a δ^- charge at the carboxyl end (about 3.5 D per amino acid residue) (Galoppini and Fox 1996). The rate and direction of electron transfer is affected by the electric field created by the dipole moment of a helical peptide (Yasutomi et al. 2005). Researchers have reported that electron transfer coupled with the dipole moment is faster than electron transfer against the dipole moment in helical peptides (Chaudhry et al. 2010; Galoppini and Fox 1996; Morita et al. 2000). The Kimura group synthesized two similar helical peptides containing (Leu-Aib) units, one with the redox active moiety (ferrocene) at the *N*-terminal and a disulfide group at the C-terminus, and the other at the opposite terminals. Both peptides formed well-ordered SAMs on a gold substrate, with the dipole moments oriented in opposite directions. The rate of electron transfer observed for the peptide containing the disulfide group attached to the gold substrate at the *N*-terminal, was consistently three times faster than that for the other peptide. It was postulated that this was due to the (δ^+) charge at the *N*-terminal lowering the energy barrier between the gold and the nearest amide group. The redox potential was not affected by the dipole moment, suggesting the role of the linker played a large part in accelerating electron transfer (Morita and Kimura 2003). Another study used helical peptides on gold surfaces with opposing dipoles, and found that the rates of electron transfer were not substantially affected. However, a six-fold increase in the rate was observed when the methylene chain linker was changed to a more conductive, π-bonded phenylene chain. It was concluded that the rate is determined by electron transfer through the linker (Watanabe et al. 2005). Nakayama and co-workers synthesized a 3_{10}-helical peptide with alternating (Ala-Aib) units with an attached oligo-phenylene ethynylene moiety, containing a terminal nitro group. Both integral parts are dipolar and hence both have different dipole magnitudes. It was proposed that stabilization of the dipole-dipole interaction between them may be reached through an antiparallel alignment, forming a planar composition. This would be conducive to self-assembly, possibly via π-π stacking (Nakayama and Kimura 2009). The nature of the constituent amino acid side chains also plays an important role in electron transfer in peptides. The peptide structure and specific amino acid spacers determine the electronic coupling that enables charge transfer to occur between electron donor and acceptor (Antonello et al. 2003). Redox active aromatic amino acids have also been proposed as through-space "stepping stones" for electron transfer (Cordes and Giese 2009; Giese et al. 2008). Studies on model peptides have shown that the rate of electron transfer increases significantly with the introduction of electron-rich side-chains (Cordes et al. 2008). Aromatic amino acids such as phenylalanine, tyrosine and tryptophan play vital roles in biology. Not only is tryptophan an essential amino acid used to manufacture serotonin and maintain normal brain and body function, it is believed to play a crucial role in electron transfer (Paredes et al. 2009). Both tryptophan and tyrosine have low oxidation potentials (Giese et al. 2009). Tyrosine acts as an electron donor in many biological signal transduction processes and in photosynthesis, where it readily undergoes deprotonation of its aromatic hydroxyl group, making it the most common radical found in proteins (Bollinger 2008; Cordes and Giese 2009). X-ray crystallography of a diverse range of proteins has disclosed the presence of oxidizable aromatic amino acids located between metal cofactors, thus advocating their role in the electron transfer process (Stubbe et al. 2003). Multiple sequence alignment of genomes from the respiratory oxidoreductase enzyme NDH1, have revealed the conservation of specific aromatic amino acids from simple prokaryotes through to man, that may serve as candidates for transient charge localization between metal clusters (Wittekindt et al. 2009). Positively charged amino acids are usually located in the neighboring environment, which may lower the energy levels of the aromatic amino acid's vacant orbitals, thus assisting electron transfer to proceed (Wittekindt et al. 2009). A study involving the biological enzyme, class I *Escherichia coli* ribonucleotide reductase, found that a relay shuttle

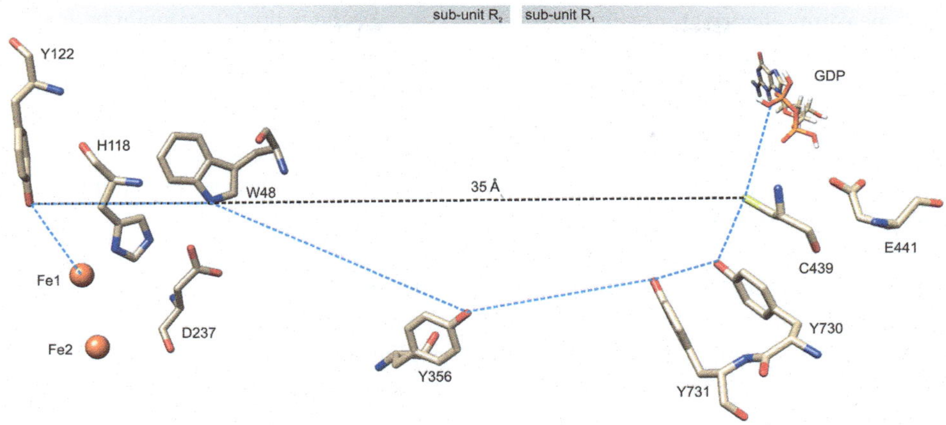

Fig. 6.3 The postulated electron relay chain in *Escherichia coli* ribonucleotide reductase enables long distance electron transfer over 35 Å from Cys[439] in the α_2 subunit to Tyr$^{122\bullet}$ in the β_2 subunit, utilizing aromatic amino acids in a hopping mechanism

comprising three tyrosine and one tryptophan residue was responsible for the rapid charge transfer spanning 35 Å (see Fig. 6.3). The rate of charge transfer, utilizing these redox active sites, was faster than a corresponding single-step tunneling reaction by at least four orders of magnitude (Stubbe et al. 2003). Efficient charge transfer through proteins over distances greater than approximately 20 Å is not achievable via a direct superexchange mechanism, as electron transfer rates decrease exponentially with increasing distance (Kai et al. 2008). Hence it is plausible that charge transfer occurs via a multi-step hopping mechanism, utilizing amino acids with felicitous redox potentials, to overcome this restriction (Cordes and Giese 2009).

A tryptophan residue has also been used as a relay station to facilitate electron transfer in a multi-step process in an azurin metallo-protein isolated from *Pseudomonas aeruginosa*. The copper redox centre was oxidized rapidly, with a charge transfer rate in excess of two orders of magnitude greater than that for single-step electron tunneling over the same distance (Shih et al. 2008). Kimura and co-workers determined that linearly spaced electron-rich naphthyl groups within 3_{10}-helical peptides increase the photocurrent by efficient electron hopping between the moieties, compared to reference peptides containing one or no naphthyl groups. The naphthyl group nearest the gold electrode was photo-excited, and a radical cation was subsequently formed when an electron was transferred from the naphthyl group to the gold. It was postulated that this radical cation then "hops" from the gold to the electron donor at the peptide terminus, using the aromatic naphthyl groups as "stepping stones" (Yanagisawa et al. 2004). A study by Cordes and co-workers investigated how aromatic side chains affect electron transfer by synthesizing peptides with an electron acceptor situated at the C-terminus, and a tyrosine residue functioning as an electron donor at the *N*-terminus. The peptide bridge consists of two triproline sequences, separated in the centre by one of four possible side chains [2x aromatic (di- and tri-methoxyphenylalanine) and 2x aliphatic]. The rate of electron transfer was determined to be in the order of 20–30 times faster when the central residue was aromatic, compared to the aliphatic amino acid side-chains, indicating intramolecular electron transfer over a distance of 20 Å by way of a two-step hopping mechanism, utilizing the aromatic side chains as "stepping stones" (Cordes et al. 2008). Recent work has shown that by incorporating tryptophan into lipid bilayers, the rate of electron transfer is increased by an order of magnitude, compared to that observed through a phospholipid membrane with no tryptophan interaction (Sarangi and Patnaik 2012). Amdursky used AFM as a probe to measure electrical conductivity in dipeptide networks, containing either two phenylalanine residues or one phenylalanine linked to a tryptophan. The dipeptide containing an FW sequence exhibited a five-fold increase in conductivity over that with the FF array (Amdursky 2013). Electron hopping has been shown to occur along a pathway exploiting three tryptophan residues in the helical enzyme DNA photolyase on a picosecond timescale (Lukacs et al. 2008). Computational studies by Chen and co-workers indicate that electron transfer between tyrosine and tryptophan not only requires specific conformations within the peptide, but the neighboring environment is critical to successful electron transfer. When the side chains of tyrosine and tryptophan are in close proximity, a hydrogen bond is formed between them and direct electron transfer can take place, involving proton-coupled electron transfer (PCET). A base may be used as a proton acceptor when the distance between them is too great, which forms a hydrogen bond with the tyrosine hydroxyl group. In many biological systems, basic groups such as histidine and lysine are located around these redox active sites and accessible for hydrogen bonding. It was postulated that charge transfer between the phenol group of the tyrosine residue and the indole of the tryptophan cation could occur via a hopping mechanism due to the presence of these basic groups (Chen et al. 2009). It is conceivable that any section of a structure with sufficient electron affinity could act as a through-space "stepping stone" for electron transfer (Horsley et al. 2014), and this will be discussed later in this chapter.

6.2.2 Electrochemical Methods

The investigation of electron transfer in peptides can be achieved by immobilization of peptides onto a metal electrode to form self-assembled monolayers (SAMs) (Arikuma et al. 2009; Morita and Kimura 2003; Yanagisawa et al. 2004), or in solution (Antonello et al. 2003; Polo et al. 2005). For electron transfer studies in solution it is necessary for the peptide to contain both an electron donor and an electron acceptor, with charge transfer between them triggered electrochemically or by UV radiation. An advantage of using SAMs for electrochemical studies is that only one redox active moiety is required, commonly ferrocene (Takeda et al. 2008). Marcus theory postulates that the electron transfer rate is reliant on the Gibbs free energy (ΔG), reorganization energy (λ), temperature (T) and the electronic coupling (H_{AB}) between an electron donor and acceptor (Eckermann et al. 2010). Hence SAMs have proven to be a great format for studying electron transfer, as each variable, ΔG, λ, T and H_{AB} can be controlled experimentally (Eckermann et al. 2010). Alkane-thiols can form poorly defined SAMs, however SAMs of helical peptides containing a terminal redox active moiety, covalently attached to a metal surface, have been shown to be highly ordered and an excellent model for investigating electron transfer (Yasutomi et al. 2004). Owing to the ramifications from the density of electronic states in Marcus Theory, accurate measurement of electron transfer rates through a peptide is subject to the correct choice of electrode material. Many metals and semi-conductors have been used as substrates for SAMs, such as silver, palladium, nickel and silicon. Gold has many advantages; it is inert and its strong affinity with sulphur enables the formation of stable thiol bonds between the modified peptide and substrate (Eckermann et al. 2010). Single walled carbon nanotubes (SWCNTs) have been shown to vertically align with the surface of a gold electrode through self-assembly, acting as molecular wires to enable direct electrical contact between the redox active moiety in a peptide and an electrode (Horsley et al. 2015) (see Fig. 6.4a, b). Functionalized SWCNTs have the added advantage of providing a greater surface density for the attachment of the peptide, together with a marked improvement in sensitivity and reproducibility of the electrochemical measurement over bare Au electrodes (Gooding et al. 2003a). The length of the nanotubes also allows the redox active moieties (eg. gold electrode and ferrocene) to be in close proximity, yet far enough apart to avoid direct colli-

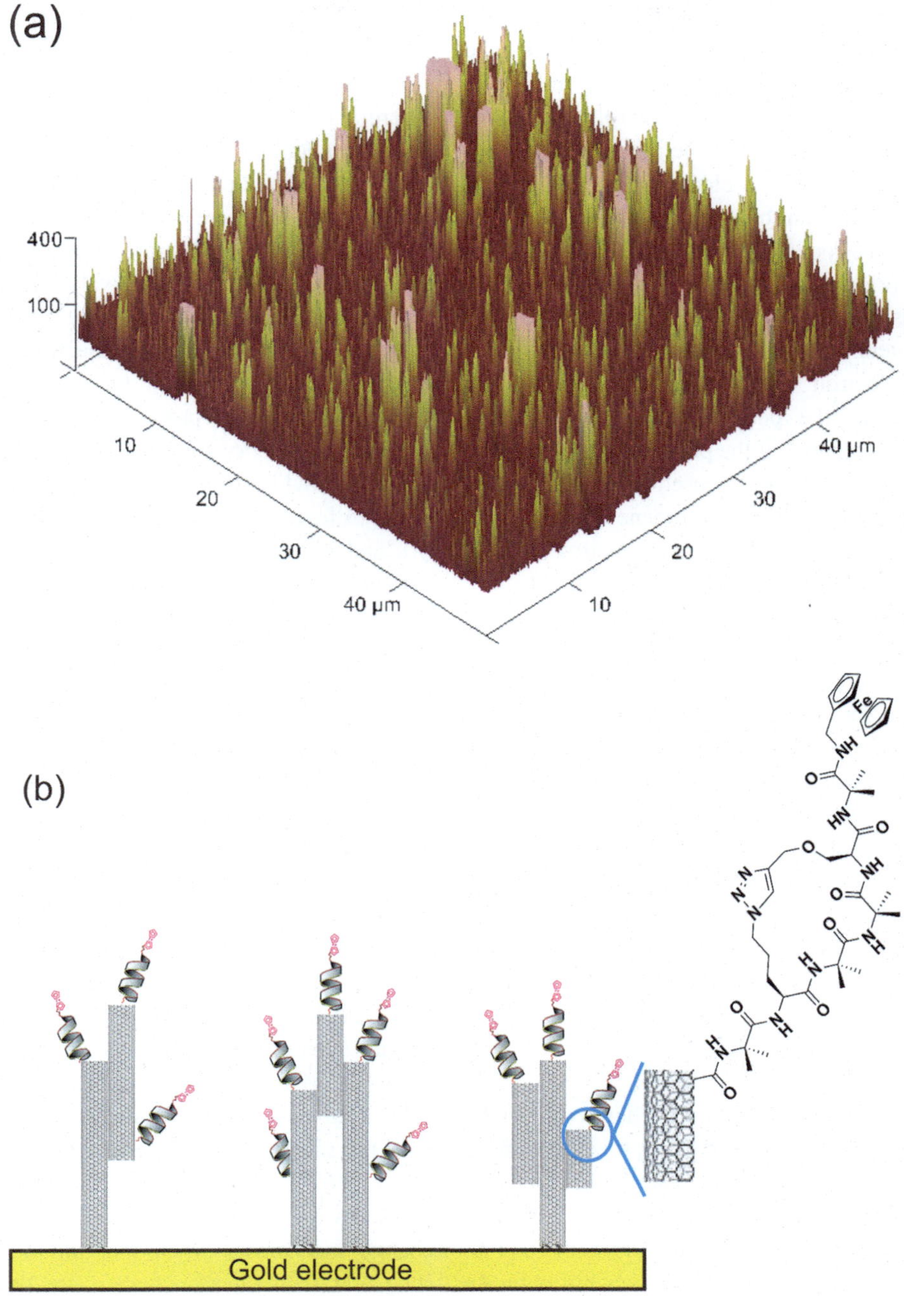

Fig. 6.4 (**a**) AFM image of a cross section of bundled, vertically aligned SWCNTs covalently bound to a gold electrode. (**b**) Schematic diagram depicting redox active helical peptides attached to SWCNTs, linked to a gold electrode

sion, making them ideal platforms for electrochemical research. This chemisorption method of forming SAMs also greatly reduces the risk of unspecific adsorption (Hermanson 2008).

Cyclic voltammetry is a common technique used to investigate electron transfer in peptides, and works by scanning the potential of an electrode and measuring the resulting current. A three-electrode cell, consisting of a working electrode (Au), counter electrode (Pt) and a reference electrode (Ag/AgCl), is typically employed. The immobilized peptide/electrode assembly is immersed into a conductive electrolyte solution. A potential is applied between the working electrode and the reference electrode, and the current measured between the working electrode and the counter electrode. The current (i) /potential (E) plot is referred to as a cyclic voltammogram, which is defined by two peak currents and two peak potentials, and provides information relating to electron transfer kinetics. The current increases as the peak reaches the oxidation/reduction potential of the peptide being analyzed. Ferrocene is a commonly used redox active moiety, with the oxidation of ferrocene to the ferrocenium ion (Fc/Fc^{+}) being a one-electron transfer process. As this is a reversible reaction, a similar peak is formed when the potential is reversed. Significant information, including the electron transfer rate constants (k_{ET}) for peptides, can be measured electrochemically using cyclic voltammetry. Following background subtraction, the surface concentration Γ (mol cm^{-2}) of the redox active peptides can be calculated from the area under the oxidation/reduction peaks of the cyclic voltammograms, as described by Laviron's formalism (Laviron 1979). The peak current (i_p) is related to the scan rate (v) as described by the equation:

$$i_p = \frac{n^2 F^2 A \Gamma v}{4RT} = \frac{nFQv}{4RT}$$

where Q is a charge (coulombs) derived from the peak area of the voltammogram, n is the number of electrons involved in the reaction, A is the surface area of the gold electrode (cm^2), R is the gas constant, and F is the Faraday constant (Yu et al. 2012). An observed linear relationship between the peak current and the scan rate demonstrates that the electron transfer reaction occurs via a surface bound species, that is, the observed electrochemical redox peaks arise exclusively from the peptides covalently anchored to the SWCNT/Au electrode (Bard and Faulkner 2000). The relationship between Ep (peak potential) and ln(v) is described by the following equation:

$$Ep = E^0 + \frac{RT}{\propto nF} \ln \frac{RTk_{ET}}{\propto nF} - \frac{RT}{\propto nF} lnv$$

where α is the transfer coefficient and E^0 is the formal potential. The electron transfer rate constant (k_{ET}) can then be extrapolated from the above equation. For electron transfer between a donor and acceptor, separated by a bridge (peptide chain), the system effectively has two electronic states. In the initial free energy state (reactant), for a one electron reaction, the electron is located on the donor, whereas the electron is situated on the acceptor in the final state (product). Marcus theory can be used to calculate k_{ET}, and also provide other important parameters such as Gibbs free energy (ΔG), reorganization energy (λ), and the electronic coupling constant (H_{AB}) between an electron donor and acceptor (Eckermann et al. 2010). The basic equation is as follows:

$$k_{ET} = \frac{2\pi}{h} \left| H_{AB} \right|^2 \frac{1}{\sqrt{4\pi\lambda k_b T}} \exp\left(-\frac{\left(\lambda + \lambda G^0\right)^2}{4\lambda k_b T} \right)$$

where k_b is the Boltzmann constant, h is Planck's constant, and T is absolute temperature. Reorganization energy is defined as the energy required for molecular rearrangement from the reactant state to the product state (Eckermann et al. 2010). The electronic coupling constant defines the strength of the force between the two moieties. Marcus theory assumes that the free energy curves for these two surfaces are the same, with the same curvature, except they are shifted (see Fig. 6.5).

Electron transfer may occur by the superexchange mechanism when the initial electronic state and final electronic state are degenerate, which occurs at the junction of these two energy surfaces (intersecting parabola) (Aubry 2007) (see Fig. 6.5). The probability of an electron to transfer from a donor to an acceptor decreases with increasing distance between the donor and acceptor. If the experimentally determined rate constant is much faster than the calculated superexchange rate, another mechanism may be responsible for electron transfer, for example, hopping.

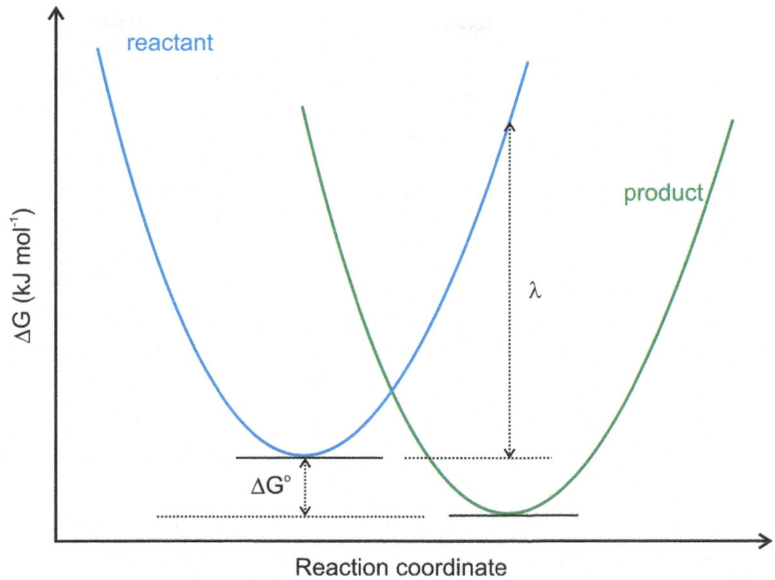

Fig. 6.5 Free energy curves for electron transfer, reactant (*left*) and product (*right*)

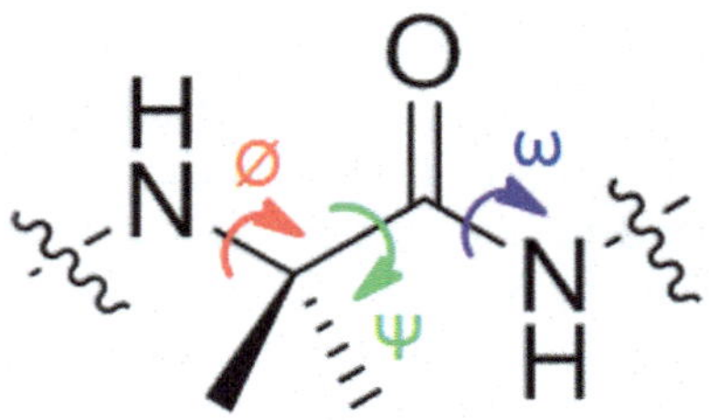

Fig. 6.6 Schematic diagram showing the position of dihedral angles in a peptide backbone. Phi (Ø) describes the rotation around the N-Cα bond, psi (Ψ) describes the rotation around the Cα-C(O) bond, and omega (ω) describes the rotation around the C(O)-N bond

6.3 Protein Secondary Structure

The primary structure of a protein is determined by its sequence of amino acids. This then folds into secondary, tertiary, and in some cases quaternary structure on the basis of a series of well-defined non-covalent interactions. The resulting three dimensional structures define the overall functionality of the protein (Bendall 1996). All proteins are polypeptides. There are two main secondary structures known as α-helices and β-pleated sheets, the structures of which are defined by a specific network of hydrogen bonds between the amide backbone carbonyl oxygen and amide hydrogens (Bendall 1996). The conformational freedom of a peptide is determined by the torsion angles of the backbone. Also known as the dihedral angles, these are defined by the atoms N-Cα-C(O) and N. Phi (Ø) describes the rotation around the N-Cα bond, psi (Ψ) describes the rotation around the Cα-C(O) bond, and omega (ω) describes the rotation around the C(O)-N bond (see Fig. 6.6). A Ramachandran plot is used to determine the allowed conformations in the Ø - Ψ plane for all secondary structures (Ramachandran and Sasisekharan 1968)

The alpha helix (α-helix) is the most prevalent secondary structure found in proteins. It is characterized by four amino acid residues per turn, with intramolecular hydrogen bonding occurring between the C=O of the residue in the (*i*) position and the amide hydrogen situated in the (*i* + 4) position. For an ideal right-handed α-helix, the dihedral angles should be approximately -60° (Ø) and -42° (Ψ). The 3_{10}-helix is a less common, but nonetheless important secondary structure in proteins (Toniolo and Benedetti 1991). It is characterized by three amino acid residues per turn, with intramolecular hydrogen bonding occurring between the C=O of the residue in the (*i*) position and the amide hydrogen situated in the (*i* + 3) position (Ousaka et al. 2007). Peptides comprising Aib residues, a naturally occurring hydrophobic amino acid found in antibiotics, display a large degree of 3_{10}-helical conformation. Previous studies have shown that stable 3_{10}-helical structures are formed when the peptide exceeds five residues in length (Maekawa et al. 2007; Morita et al. 2008). The backbone torsional angles of a 3_{10}-helix differ from those of an α-helix, with the 3_{10}-helix somewhat more elongated (Schievano et al. 2001). For an ideal right-handed 3_{10}-helix, the dihedral angles should be approximately -57° (Ø) and -30° (Ψ) (Jacobsen et al. 2011). The shorter, and hence stronger intramolecular hydrogen bonding in a 3_{10}-helix (i to $i + 3$) makes this structure more conductive than an α-helix (i to $i + 4$ hydrogen bonding) (Mandal and Kraatz 2012). Short 3_{10}-helices have been proposed as possible nucleation sites for the formation of helices during protein folding (Schievano et al. 2001). Studies have discovered an abundance of 3_{10}-conformations in trans-membrane helical proteins, including bacterial rhodopsins that are involved in signal transduction, ion channels and G-protein-coupled receptors (GPCRs) (Arrondo and Goñi 1999; Riek et al. 2001). The 3_{10}-helix is found in many enzymes responsible for electron transfer, including cytochrome c oxidase, the final molecule in the mitochondrial electron transfer chain (Beiszinger et al. 1998). While considerable research on electron transfer has been conducted on helical peptides, the possible role a β-strand plays in these processes has received limited attention (Horsley et al. 2015; Horsley et al. 2014; Langen et al. 1995; Sasaki et al. 2001; Shih et al. 2008; Yu et al. 2017). This structural motif has the carbonyl and amide groups positioned orthogonal to the backbone, thus minimizing steric hindrance between the side-chains (Abell et al. 2009a). An ideal, fully extended β-strand is defined by the bond angles −120° (Ø), 120° (Ψ) and 180° (ω) (Loughlin et al. 2004), however a broader scope extends to between $-160° < Ø < -100°$ and $90° < Ψ < 160°$ (Abell et al. 2009a). Two or more β-strands linked together via intermolecular hydrogen bonding form a β-sheet (either parallel, antiparallel, or mixed), which represents over 30% of all protein structure (Loughlin et al. 2004).

6.3.1 Characterization of Secondary Structure

Characterizing the conformation of peptide secondary structure is an important step, with a number of methods available, including Nuclear Magnetic Resonance (NMR), Fourier Transform Infrared Spectroscopy (FTIR) and Circular Dichroism (CD). ^{1}H NMR is a robust technique used to determine peptide secondary structure in solution. Each hydrogen atom in the peptide generates an individual signal which is distinctive of its chemical environment (Neuhaus 1993). The assignment of individual atoms in the peptide can then be realized. Used in conjunction with two-dimensional (2D) NMR (eg ROESY and COSY), the identity of the compound may be deduced, and most importantly, the geometric conformation revealed. Correlation Spectroscopy (COSY) is used to identify through-bond connectivities, that is, nuclei that are coupled to each other. Rotating-frame Overhauser Effect Spectroscopy (ROESY) is used to determine through-space correlations (to approximately 5 Å). A combination of 1D, 2D, and carbon (^{13}C NMR), enables a three-dimensional 'image' to be collected, and hence elucidates the particular structural geometry of the peptide. Specifically, a combination of consecutive NH (i) to NH ($i + 1$) correlations, CαH (i) to NH ($i + 1$) and medium range CαH (i) to NH ($i + 2$) ROESY connectivities can be used to determine 3_{10}-helical geometry (Neuhaus 1993), while CαH (i) to NH ($i + 4$) correlations are evident in an α-helical structure. The distance between the peaks of a multiplet signal is known as the coupling constant (J), which can also be used to differentiate secondary structures. The distance between the peaks of a doublet, representative of the correlation between an amide (i) and alpha (i) hydrogen (^{1}H NMR $J_{NHC\alpha H}$ coupling constant) in an α-helix is approximately 4.8 Hz, while that of a 3_{10}-helix is approximately 5.6 Hz (Smith et al. 1996). A combination of NH (i) to NH ($i + 1$), CαH (i) to NH ($i + 1$) and CβH (i) to NH ($i + 1$) ROESY correlations are indicative of a β-strand geometry, with ^{1}H NMR $J_{NHC\alpha H}$ coupling constants in the range of 8–10 Hz (Pehere and Abell 2012).

Molecules absorb specific energetic frequencies characteristic of their molecular structure, which can also be measured using FTIR spectroscopy. When a beam of infrared light is passed through an IR-active sample, that is, one with a change in the dipole moment, an infrared spectrum is produced. Absorbance occurs when the frequency of the IR radiation matches the transition energy of the vibrating bond or functional group. A transmittance spectrum is then generated, which pinpoints the position and intensity of peaks which are representative of specific chemical environments, thus disclosing details of the associated molecular structure. This is particularly relevant to peptides due to the sensitivity of amide bands to peptide structure (Haris and Chapman 1995). Hence Amide I and II bands are used extensively in determining peptide/protein secondary structure (Zhuang et al. 2009). The Amide I band for α- and 3_{10}-helices is in the order of 1650 cm^{-1} to 1655 cm^{-1} (Lakhani et al. 2011; Zhuang et al. 2009). For antiparallel β-sheets, the Amide I band ranges between 1612 cm^{-1} and 1640 cm^{-1}, with a small shoulder around 1685 cm^{-1}, while the Amide II band ranges between 1510 cm^{-1} and 1530 cm^{-1} (Zhuang et al. 2009). Furthermore, they should display N-H bond stretching in the region around 3281 cm^{-1} to 3302 cm^{-1} (Pehere et al. 2013b). CD spectroscopy measures the difference in absorbance of left and right circularly polarized light in an optically active peptide (Peeters et al. 1997). The ideal spectral range for determining secondary structure lies between 180 nm and 260 nm, with the spectrum of an α-helix displaying two minima at approximately 208 nm and 222 nm, with similar intensity. A 3_{10}-helix displays a strong negative minimum at approximately 205 nm (amide π to π* transition) and a far less intense minimum between 220 nm and 230 nm (amide n to π*) (Biron et al. 2002; Boal et al. 2007). Furthermore, a 3_{10}-helix should exhibit a slightly positive band at approximately 195 nm, of far less intensity than that of an α-helix (Toniolo et al. 1996). CD spectra can be ambiguous, and should only be used as a guide to structural conformation, in conjunction with other methods of characterization (Niggli et al. 2012).

6.3.2 Peptide Constraints

Significant research has been conducted on constraining linear peptides, via intramolecular cyclization, into macrocycles with well-defined secondary structure, for example helices and β-strands, predominantly for the enhancement of therapeutic agents (Boal et al. 2007; de Araujo et al. 2014; Jacobsen et al. 2011; Pedersen and Abell 2011). Peptides can be constrained via head to tail interactions, backbone to side-chain, or side-chain to side-chain. These structures show important biological properties, for example as potent and selective protease inhibitors (Abell et al. 2009b). A macrocyclic constraint introduced into a linear inhibitor of farnesyltransferase, resulted in a compound with a 20-fold improved potency (Driggers et al. 2008). Chen et al. introduced a constraint into a linear peptide inhibitor of the cancer target STAT 3, with the cyclic structure found to be three times more potent than the linear analogue (Chen et al. 2007).

There are several different synthetic approaches to constrain a peptide into its preferred conformation via cyclization of a linear precursor. Common methods include click chemistry (Huisgen cycloaddition), ring-closing metathesis (RCM) and lactamization. Irrespective of the method, cyclization is more readily achieved under conditions of high dilution, which serve to minimize side reactions such as dimer formation. The slow addition of reagents via a syringe pump can further enhance these reactions (Malesevic et al. 2004). In the early 1960s, the forerunner of the 'click reaction', the 1,3-dipolar Huisgen cycloaddition reaction was discovered, which couples organic azides with alkynes to form 1,2,3 triazoles (Huisgen 1963). This reaction gives mixtures of the 1,4- and 1,5-disubstituted adducts, and was not given much attention until nearly four decades later. In 2002, both the Sharpless and Meldal groups independently discovered that by using catalytic Cu(I) it was possible to enhance the rate by several orders of magnitude, whilst also achieving regioselectivity, resulting in essentially quantitative yields of 1,4-disubstituted 1,2,3-triazole product (Rostovtsev et al. 2002; Tornøe et al. 2002). Also referred to as "copper-catalyzed azide-alkyne cycloaddition" (CuAAC), it has become an important click reaction and a common method to form intramolecular macrocycles (Lopez et al. 2010). CuAAC can be performed either on solid phase (Ingale and Dawson 2011; Kapoerchan et al. 2008) or in solution (Chouhan and James 2011; Looper et al. 2006), with the product being thermodynamically and metabolically stable; resistant to both hydrolysis and oxidation. The 1,4-disubstituted 1,2,3-triazoles are similar to amide bonds insofar as both are planar, while also possessing a strong dipole moment (Holub and Kirshenbaum 2010). There is a notable difference in the distance between the R_1 and R_2 groups of both structures (Pedersen and Abell 2011), however the 1,4-disubstituted 1,2,3-triazole has previously been used as a surrogate for a *trans*-amide bond (Beierle et al. 2009). Linking residues (i to $i + 4$) in a linear α-helix, (i to $i + 3$) in a linear 3_{10}-helix, or residues (i to $i + 2$) in a linear β-strand using click chemistry, enhances secondary structure. Recent studies have indicated that while the 3_{10}-helix is a stable peptide structure, constraining the side-chains (i to $i + 3$) results in a more paradigmatic structure than the linear analogue (Jacobsen et al. 2011). For example, the X-ray crystal structure of a peptide cyclized side-chain to side-chain using click chemistry, showed that the dihedral angles deviated from an ideal 3_{10}-helix by no more than 2°, whereas the linear analogue was found to possess less helical content (Jacobsen et al. 2011). The triazole-containing constraint rigidifies the macrocycle, in turn reducing entropic loss due to structural pre-organization, as the reactive side-chains are held in close proximity. This also reduces the number of possible conformations available to the peptide (Holub et al. 2007). The reputed mechanism for the CuAAC reaction firstly involves the alkyne coordinating with the Cu(I) compound, with displacement of a ligand. The azide then coordinates to the Cu atom, displacing another ligand in the process. This results in the formation of a 6-membered intermediate, which contracts to form a triazole. Finally the Cu catalyst is cleaved, releasing the exclusive product, the 1,4-disubstituted 1,2,3-triazole (Angell and Burgess 2007).

RCM is another common and versatile method for intramolecular macrocyclization, and results in the formation of carbon-carbon bonds in the form of cyclic alkenes. Enhancement and rigidification of a 3_{10}-helical conformation in a peptide can be achieved through modification of the amino acid residues at the i and $i + 3$ positions, to carry terminal dienes. These can then be readily linked, as the alkenes are located in a proximal arrangement on the same side of the molecule (Boal et al. 2007). The Abell group have cyclized dienes located in the i and $i + 2$ positions to afford a peptide with a stable β-strand backbone (Abell et al. 2009a), a requirement for binding to many proteins. The conformation of this macrocycle, which has also been defined by its X-ray crystal structure, was shown to be a key intermediate of a potent inhibitor of the protease calpain 2, the over activation of which can lead to diseases such as Alzheimer's, stroke, and cataract formation (Abell et al. 2009a; Pehere et al. 2013a). The advent of ruthenium-based catalysts by Grubbs and co-workers has greatly improved RCM, with these reactions now benefiting from better stereoselectivity. Strategies such as microwave irradiation can also be used to promote ring closure and enhance yields (van Lierop et al. 2011). Metathesis was initially referred to as "stapling" due to the formation of the double bond in the macrocycle, however Hilinski and co-workers recently synthesized a structurally rigid helical peptide with multiple bridges which exhibited superior cell penetration, which they have termed as "stitched" (Hilinski et al. 2014).

Lactamization is another attractive method for cyclizing a linear precursor into a more stable lactam-bridged macrocycle (Lundquist and Pelletier 2002). The range of available amino acid residues with suitable side-chains, and the wide range of protecting groups available for amines and carboxylic acids, means that lactamization can be readily achieved in solution or on solid phase (Taylor 2002). Schievano et al. demonstrated, through the use of 2D NMR and molecular dynamics (MD) analysis, that by linking amino acid side-chain residues (i to $i + 3$) to form a lactam bridge, enhancement of the 3_{10}-helical conformation can be realized

(Schievano et al. 2001). Lactamization of an aspartic acid (*i*) and lysine (*i* + 4) residue in an extended linear peptide resulted in the stabilization of the α-helical conformation, with the ensuing compound showing improved potency and selectivity for the requisite k-opioid receptor (Lung et al. 1996). Kirby et al. used a lactam bridge to further constrain a peptide into its helical conformation, resulting in greatly improved specific binding affinity (K_i = 16 nM and EC_{50} = 12 nM) of neuropeptide receptors in human neuroblastoma cells (Kirby et al. 1997).

6.4 A Case Study: The Role of Peptides in Electronics

A more detailed understanding on exactly what controls the mechanisms and efficiency of charge transfer is required before the promise of peptides as molecular components can be fully realized. Despite the fact that secondary structure is known to play a key role in electron transfer, research relating to peptides containing an intramolecular macrocyclic constraint has been absent. Studies in this area are crucial to investigating the link between conformation and function, and we have recently presented electrochemical studies on a series of synthetic peptides to determine the effects that different macrocyclic side-bridge constraints, comprised of various ring sizes and chemical composition, have on electronic transport (Horsley et al. 2014, 2015; Yu et al. 2014, 2016, 2017).

Peptides **1–4** (see Fig. 6.7) contain Aib residues which are known to form predictable and stable 3_{10}-helical structures. Peptide **2** has a triazole-containing side-chain, while an alkene-containing side-chain is located at the same position in peptide **4**. The backbones of peptides **1** and **3** were further constrained into a 3_{10}-helix via Huisgen cycloaddition (click) (Jacobsen et al. 2011) and ring-closing metathesis (RCM) (Boal et al. 2007) respectively, resulting in additional conformational rigidity of their backbones.

Peptides **5–8**, as shown in Fig. 6.8, share a common β-strand conformation. Peptide **6** has a triazole-containing side-chain, while an alkene-containing side-chain is located in the same position in peptide **8**. The backbones of peptides **5** and **7** were further constrained into a β-strand conformation, via click and RCM respectively.

Peptides with either a β-strand or helical conformation were studied to cover a wide range of secondary structures, to fully investigate influences of backbone rigidity on electron transfer, with two different types of side-chain tethers investigated. Using a combination of NMR, IR and CD spectroscopy, the geometry of peptides **1–4** was confirmed as 3_{10}-helical, while that of peptides **5–8** was confirmed as β-strand. To further define their backbone geometries, the lowest energy conformers for the *N*-Boc protected analogues of **1–8** were determined by molecular modeling. The *N*-Boc protected peptides were used for this purpose as free amines are known to generate unrealistic electrostatic interactions, resulting in unstable lowest energy conformers (Burton et al. 1998). The models for the helical analogues of **1–4** indicated that the overall backbone lengths differed by no more than 11 picometers, while the mean intramolecular hydrogen bond lengths differed by no more than 6 picometers. The average dihedral angles for residues 1–6 in each of the analogues deviated from an ideal 3_{10}-helix by no more than 3.6° and 5.9° for Φ and ψ respectively. A similar result was found for the *N*-protected analogues of the β-strand peptides **5–8**, with their overall lengths differing by no more than 59 picometers. The molecular modelling studies, together with the various spectroscopic characterizations show that peptides **1–4** share remarkably similar 3_{10}-helical conformations, while peptides **5–8** exhibit a remarkably similar β-strand geometry.

The electrochemical properties of peptides **1–8** were investigated using cyclic voltammetry, with each peptide separately attached to vertically aligned single-walled carbon nanotube array/gold (SWCNTs/Au) electrodes. The cyclic voltammograms obtained for the helical peptides **1–4** reveal a pair of redox peaks characteristic of a one-electron oxidation/reduction reaction (Fc^+/Fc) (see Fig. 6.9). The formal potentials (E_o), electron transfer rate constants (k_{et}), and the sur-

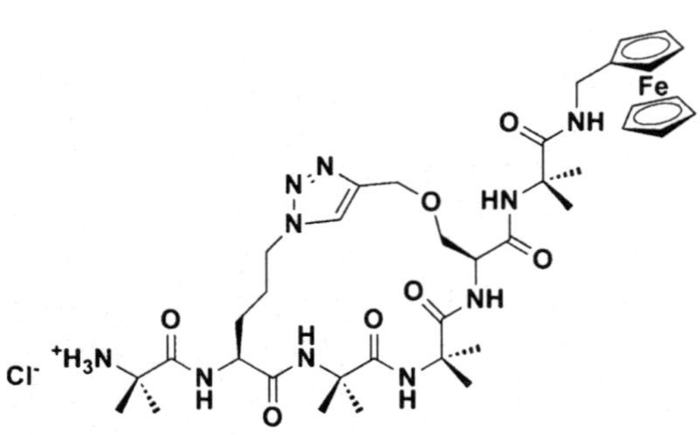

1

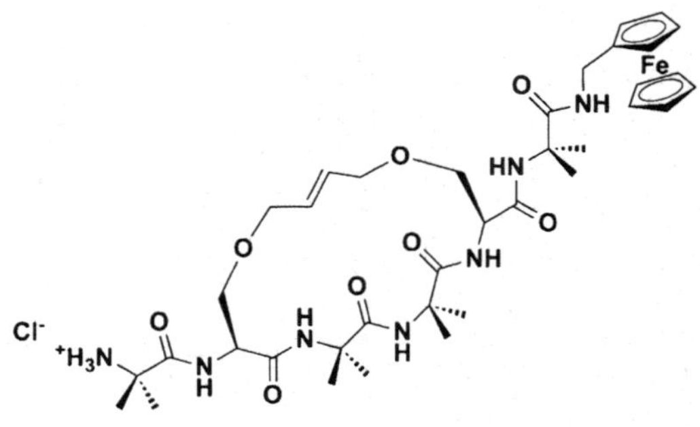

3

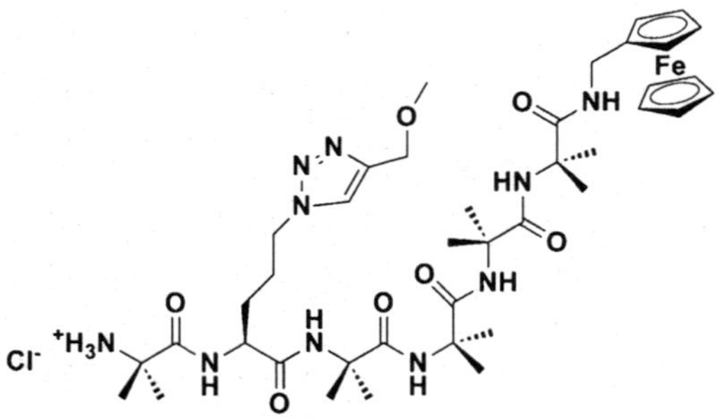

2

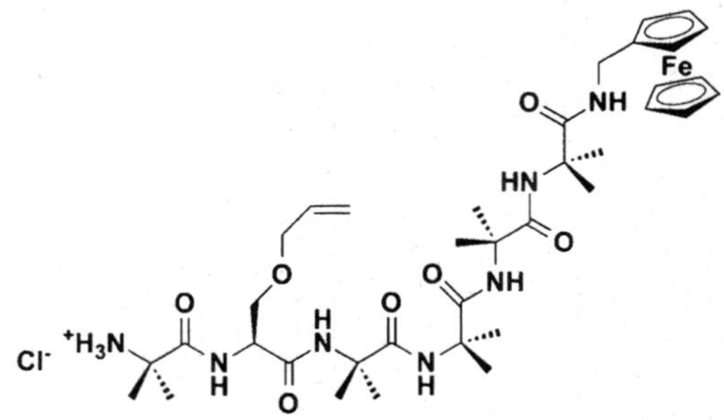

4

Fig. 6.7 Synthetic helical peptides **1–4**

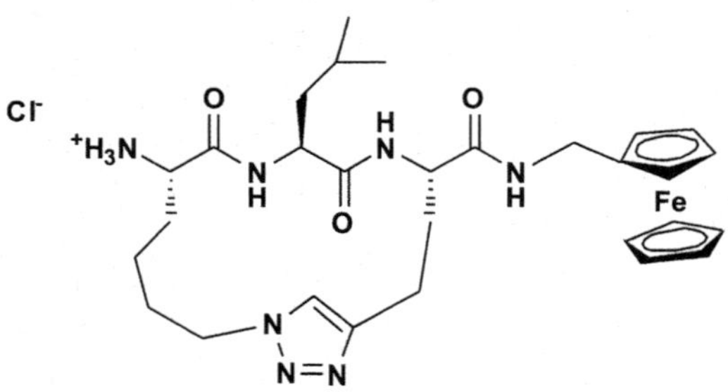

5

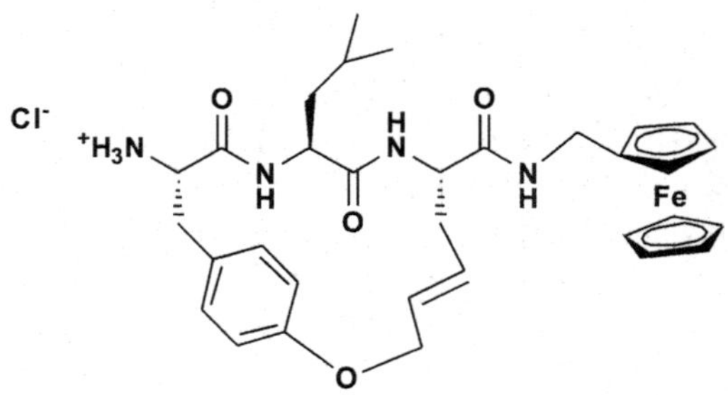

7

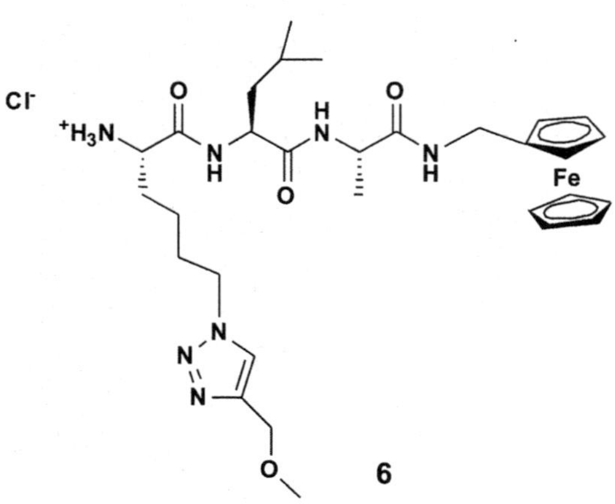

6

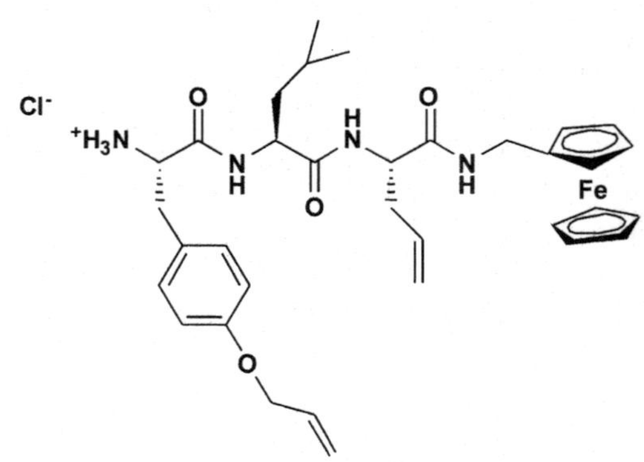

8

Fig. 6.8 Synthetic β-strand peptides **5–8**

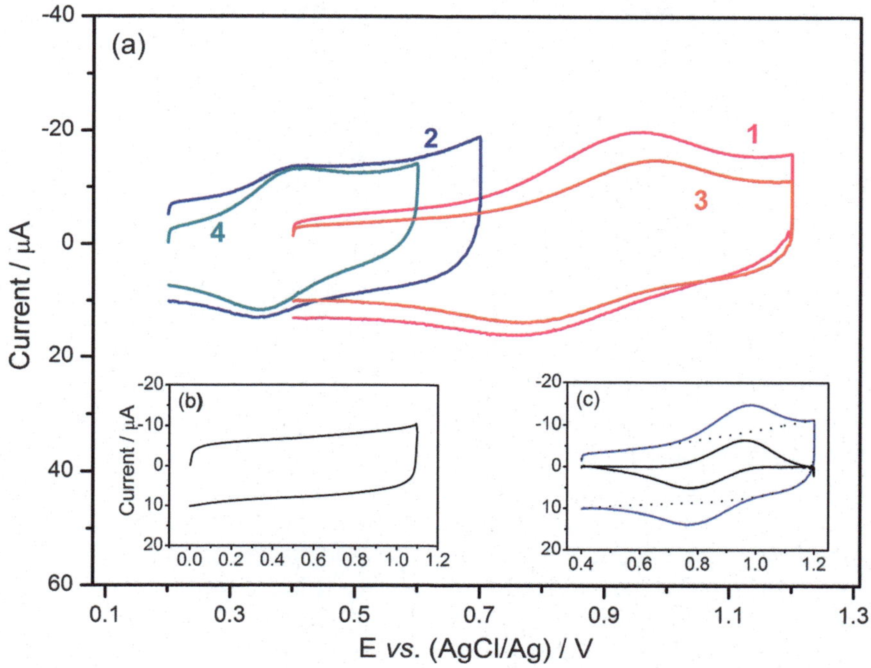

Fig. 6.9 (**a**) Cyclic voltammograms for helical peptides **1** (*pink*), **2** (*blue*), **3** (*red*) and **4** (*green*) immobilized on SWCNTs/Au electrodes taken at 5 V s^{-1} in 0.1 mol L^{-1} $TBAPF_6/CH_3CN$ solutions. (**b**) Typical cyclic voltammogram of SWCNTs/Au electrode taken at 5 V s^{-1} in 0.1 mol L^{-1} $TBAPF_6/CH_3CN$ solutions. (**c**) Background subtracted cyclic voltammogram (*black solid*) from the original curve (*blue solid*) with the background current (*dotted line*)

Table 6.1 Electron transfer rate constants (k_{et}), surface concentrations and formal potentials (E_o) for the helical peptides **1**–**4**

Peptide	Surface concentration (×10^{-10} mole.cm^{-2})	E_o (V *vs* AgCl/Ag)	k_{et}/ s^{-1}
1	3.76 ± 0.35	0.853	28.1 ± 3.6
2	2.52 ± 0.18	0.371	117.3 ± 9.9
3	4.37 ± 0.43	0.844	17.5 ± 1.5
4	4.02 ± 0.41	0.380	260.4 ± 25.3

face concentrations of the peptides are tabulated in Table 6.1.

The constrained peptides (**1** and **3**) and their linear analogues (**2** and **4**) exhibited considerably different formal potentials and electron transfer rate constants, despite sharing similar 3_{10}-helical geometry. The formal potentials (E_o) and electron transfer rate constants (k_{et}) fall into two distinct groups. The linear analogues displayed low formal potentials, with high electron transfer rate constants estimated to be 0.371 V and 117.3 s^{-1} for **2**, 0.380 V and 260.4 s^{-1} for **4** (see Table 6.1). The observed formal potentials for these linear peptides are similar to the formal potentials reported for other ferrocene-derivatized linear peptides attached to a gold surface, without carbon nanotubes (Arikuma et al. 2010; Mandal and Kraatz 2012). This further supports previous observations that carbon nanotubes have no significant effect on the electron transfer rate-limiting step (Gooding et al. 2003b; Yu et al. 2012). In contrast, the constrained peptides exhibited high formal potentials, with low electron transfer rate constants estimated to be

0.853 V and 28.1 s^{-1} for **1**, 0.844 V and 17.5 s^{-1} for **3**. Peptide **1**, with a triazole-containing side-bridge, shows a significant formal potential shift to the positive compared to the linear peptide **2**, of approximately 480 mV. A similar result was found in the formal potential of the cyclic peptide **3**, constrained by an alkene-containing tether, compared with that of the linear **4** (464 mV). These large formal potential shifts are significantly higher than other conformation-dependent structures, such as *cis-trans* cyclohexasilanes (110 mV) (Emanuelsson et al. 2014). These results show that oxidation/reduction of the redox-active ferrocene moieties in the constrained peptides is energetically much less favourable than those in the linear analogues. A significant decrease in the electron transfer rate constant was also noted upon the introduction of the constraints. The data from these compounds reveals an electron transfer rate constant for the constrained peptide **1** of 28.1 s^{-1}, 4-fold lower than that of the linear counterpart **2**. The observed electron transfer rate constant for the constrained peptide **3** was 17.5 s^{-1}, a remarkable 15-fold lower than that of the linear counterpart **4**. Previous studies have shown that electron transfer rate constants in peptides can vary greatly (Arikuma et al. 2010; Galka and Kraatz 2002), but not to such an extent as reported here.

The observed formal potentials (E_o) and electron transfer rate constants (k_{et}) for the β-strand peptides (**5–8**) also fall into two definitive groups. Once again the linear analogues displayed low formal potentials and high electron transfer rate constants, while the constrained peptides exhibited high formal potentials and low electron transfer rate constants (see Fig. 6.10, Table 6.2). The observed electron transfer rate constant for the constrained peptide **5** was found to be 10-fold lower than that of the linear peptide **6**, while that of the constrained peptide **7** is a remarkable 36-fold lower than that of the linear peptide **8** (see Table 6.2).

These results demonstrate an important link between backbone rigidity and the efficiency of electron transfer in peptides. They also further extend the generality of this effect, with similar results observed in both helical and β-strand peptides. The side-bridge constraint provides an additional reorganization energy barrier that impedes electron transfer within the peptide, in turn decreasing the charge transfer rate (Yu et al. 2014). This provides a unique approach to manipulate energy barriers and conductance in peptides. Despite the link between backbone rigidity and the efficiency of electron transfer in peptides, the electrochemical properties are also affected by the chemical composition and structural nature of the side-chains/bridges. A comparison of the data for the two linear hexapeptides (**2** and **4**) provides a measure of the influence of the electron rich alkene side-chains on the rate of electron transfer. These two peptides share a common 3_{10}-helical conformation and contain the same number of Aib residues. They are identical, only differing in the structure of the side-chain, as shown in Fig. 6.7. Peptide **2** has a triazole-containing side-chain, while an alkene-containing side-chain is located at the same position in peptide **4**. While the formal potentials for both peptides were similar, the observed electron transfer rate constant for **4** was more than double that of **2**. This observation is reinforced for the two β-strand linear peptides, with the electron transfer rate constant for **8** approximately double that of **6**. These results clearly demonstrate the ability

Table 6.2 Electron transfer rate constants (k_{et}), surface concentrations and formal potentials (E_o) for the β-strand peptides **5–8**

Peptide	Surface concentration (×10^{-10} mole.cm^{-2})	E_o (V *vs* AgCl/Ag)	k_{et} / s^{-1}
5	5.86 ± 0.54	0.825	22.5 ± 2.1
6	2.73 ± 0.26	0.349	223.2 ± 23.2
7	9.21 ± 0.89	0.676	11.7 ± 1.2
8	5.56 ± 0.31	0.408	421.4 ± 41.5

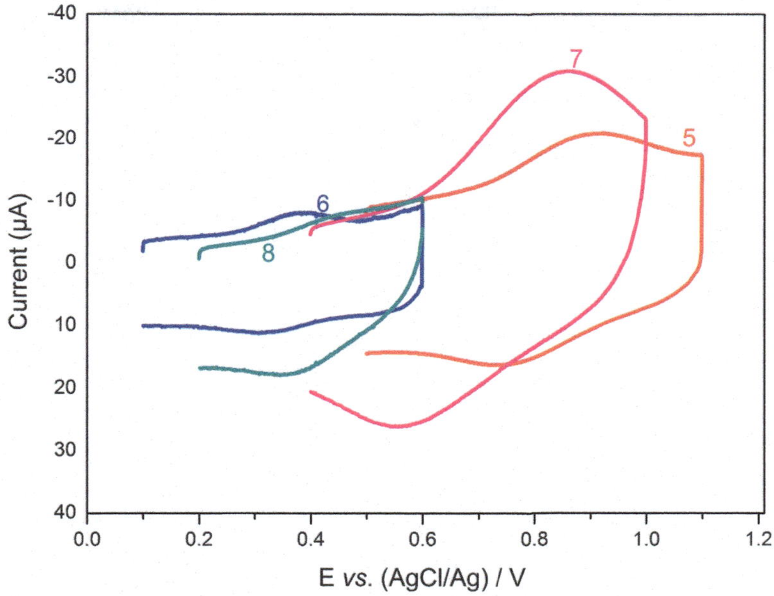

Fig. 6.10 Cyclic voltammograms for β-strand peptides **5** (*red*), **6** (*blue*), **7** (*pink*) and **8** (*green*) immobilized on SWCNTs/Au electrodes taken at 5 V s^{-1} in 0.1 mol L^{-1} $TBAPF_6$/CH_3CN solutions

of the electron-rich alkene and phenol groups to facilitate electron transfer through the peptide by acting as 'stepping stones'.

Each of the constrained peptides (**1**, **3**, **5** and **7**) recorded significant formal potential shifts to the positive (482, 464, 476 and 268 mV, respectively), and substantial decreases in the electron transfer rate constants (76%, 93%, 90% and 97%) compared to their unconstrained, linear counterparts **2**, **4**, **6** and **8**. The vast formal potential shifts and electron transfer rate constant drops in these peptides provide two distinct states (i.e. on/off). These states, brought about through side-bridge gating, afford a sizeable differential which is ideal for the design of molecular switches. Manipulating the chemical composition and backbone rigidity of peptides provides a new means to fine tune their electron transfer kinetics. This represents an important step for peptides in the design and fabrication of bio-inspired molecular-based electronic materials.

Acknowledgements This work is supported by the ARC Centre of Excellence for Nanoscale BioPhotonics (CNBP). The authors also gratefully acknowledge the assistance of the Australian National Fabrication Facility (ANFF).

References

Abell AD, Alexander NA, Aitken SG, Chen H, Coxon JM, Jones MA, McNabb SB, Muscroft-Taylor A (2009a) Synthesis of macrocyclic beta-strand templates by ring closing metathesis. J Org Chem 74(11):4354–4356

Abell AD, Jones MA, Coxon JM, Morton JD, Aitken SG, McNabb SB, Lee HYY, Mehrtens JM, Alexander NA, Stuart BG, Neffe AT, Bickerstaff R (2009b) Molecular modeling, synthesis, and biological evaluation of macrocyclic calpain inhibitors. Angew Chem Int Ed 48(8):1455–1458

Amdursky N (2013) Enhanced solid-state electron transport via tryptophan containing peptide networks. Phys Chem Chem Phys 15(32):13479–13482

Angell YL, Burgess K (2007) Peptidomimetics via copper-catalyzed azide-alkyne cycloadditions. Chem Soc Rev 36(10):1674–1689

Antonello S, Formaggio F, Moretto A, Toniolo C, Maran F (2003) Anomalous distance dependence of electron transfer across peptide bridges. J Am Chem Soc 125(10):2874

Arikuma Y, Takeda K, Morita T, Ohmae M, Kimura S (2009) Linker effects on monolayer formation and long-range electron transfer in helical peptide monolayers. J Phys Chem B 113(18):6256

Arikuma Y, Nakayama H, Morita T, Kimura S (2010) Electron hopping over 100 angstrom along an alpha helix. Angew Chem Int Ed 49(10):1800

Arikuma Y, Nakayama H, Morita T, Kimura S (2011) Ultra-long-range electron transfer through a self-assembled monolayer on gold composed of 120-angstrom-long alpha-helices. Langmuir 27(4):1530

Arrondo JLR, Goñi FM (1999) Structure and dynamics of membrane proteins as studied by infrared spectroscopy. Prog Biophys Mol Biol 72(4):367–405

Aubry S (2007) A nonadiabatic theory for electron transfer and application to ultrafast catalytic reactions. J Phys Condens Matter 19:255204

Avellini T, Li H, Coskun A, Barin G, Trabolsi A, Basuray AN, Dey SK, Credi A, Silvi S, Stoddart JF, Venturi M (2012) Photoinduced memory effect in a redox controllable bistable mechanical molecular switch. Angew Chem Int Ed 51(7):1611–1615

Aviram A, Ratner MA (1974) Molecular rectifiers. Chem Phys Lett 29(2):277–283

Bard AJ, Faulkner LR (2000) Electrochemical methods: fundamentals and applications, 2nd edn. Wiley, New York

Beierle JM, Horne WS, van Maarseveen JH, Waser B, Reubi JC, Ghadiri MR (2009) Conformationally homogeneous heterocyclic pseudotetrapeptides as three-dimensional scaffolds for rational drug design: receptor-selective somatostatin analogues. Angew Chem Int Ed 48(26):4725–4729

Beiszinger M, Sticht H, Sutter M, Ejchart A, Haehnel W, Rosch P (1998) Solution structure of cytochrome c6 from the thermophilic cyanobacterium *Synechococcus elongatus*. EMBO J 17(1):27–36

Bendall DS (1996) Protein electron transfer. BIOS Scientific Publishers, Oxford

Benesch C, Rode MF, Cizek M, Haertle R, Rubio-Pons O, Thoss M, Sobolewski AL (2009) Switching the conductance of a single molecule by photoinduced hydrogen transfer. J Phys Chem C 113(24):10315

Biron Z, Khare S, Samson AO, Hayek Y, Naider F, Anglister J (2002) A monomeric 3(10)-helix is formed in water by a 13-residue peptide representing the neutralizing determinant of HIV-1 on gp41. Biochemist 41(42):12687–12696

Boal AK, Guryanov I, Moretto A, Crisma M, Lanni EL, Toniolo C, Grubbs RH, O'Leary DJ (2007) Facile and E-selective intramolecular ring-closing metathesis reactions in 3(10)-helical peptides: a 3D structural study. J Am Chem Soc 129(22):6986–6987

Bollinger JM Jr (2008) Biochemistry – electron relay in proteins. Science 320(5884):1730–1731

Burton NA, Harrison MJ, Hart JC, Hillier IH, Sheppard DW (1998) Prediction of the mechanisms of enzyme-catalysed reactions using hybrid quantum mechanical molecular mechanical methods. Faraday Discuss 110:463–475

Chaudhry BR, Wilton-Ely J, Tabor AB, Caruana DJ (2010) Effect of peptide orientation on electron transfer. Phys Chem Chem Phys 12(34):9996

Chen J, Nikolovska-Coleska Z, Yang C-Y, Gomez C, Gao W, Krajewski K, Jiang S, Roller P, Wang S (2007) Design and synthesis of a new, conformationally constrained, macrocyclic small-molecule inhibitor of STAT3 via 'click chemistry'. Bioorg Med Chem Lett 17(14):3939–3942

Chen X, Zhang L, Zhang L, Wang J, Liu H, Bu Y (2009) Proton-regulated electron transfers from tyrosine to tryptophan in proteins: through-bond mechanism versus long-range hopping mechanism. J Phys Chem B 113(52):16681–16688

Chen Y-S, Hong M-Y, Huang GS (2012) A protein transistor made of an antibody molecule and two gold nanoparticles. Nat Nanotechnol 7(3):197–203

Chouhan G, James K (2011) CuAAC macrocyclization: high intramolecular selectivity through the use of copper–tris(triazole) ligand complexes. Org Lett 13(10):2754–2757

Cordes M, Giese B (2009) Electron transfer in peptides and proteins. Chem Soc Rev 38(4):892–901

Cordes M, Kottgen A, Jasper C, Jacques O, Boudebous H, Giese B (2008) Influence of amino acid side chains on long-distance electron transfer in peptides: electron hopping via "stepping stones". Angew Chem Int Ed 47(18):3461–3463

Darwish N, Diez-Perez I, Guo S, Tao N, Gooding JJ, Paddon-Row MN (2012) Single molecular switches: electrochemical gating of a single anthraquinone-based norbornylogous bridge molecule. J Phys Chem C 116(39):21093–21097

de Araujo AD, Hoang HN, Kok WM, Diness F, Gupta P, Hill TA, Driver RW, Price DA, Liras S, Fairlie DP (2014) Comparative alpha-helicity of cyclic pentapeptides in water. Angew Chem Int Ed 53(27):6965–6969

Ding W, Negre CFA, Palma JL, Durrell AC, Allen LJ, Young KJ, Milot RL, Schmuttenmaer CA, Brudvig GW, Crabtree RH, Batista VS (2014) Linker rectifiers for covalent attachment of transition metal catalysts to metal-oxide surfaces. ChemPhysChem 15(6):1138–1147

Driggers EM, Hale SP, Lee J, Terrett NK (2008) The exploration of macrocycles for drug discovery [mdash] an underexploited structural class. Nat Rev Drug Disc 7(7):608–624

Eckermann AL, Feld DJ, Shaw JA, Meade TJ (2010) Electrochemistry of redox-active self-assembled monolayers. Coord Chem Rev 254(15-16):1769–1802

Emanuelsson R, Löfås H, Wallner A, Nauroozi D, Baumgartner J, Marschner C, Ahuja R, Ott S, Grigoriev A, Ottosson H (2014) Configuration and conformation-dependent electronic structure variations in 1,4-disubstituted cyclohexanes enabled by a carbon-to-silicon exchange. Chem Eur J 20(30):9304–9311

Galka MM, Kraatz HB (2002) Electron transfer studies on self-assembled monolayers of helical ferrocenoyl-oligoproline-cystamine bound to gold. ChemPhysChem 3(4):356–359

Galoppini E, Fox MA (1996) Effect of the electric field generated by the helix dipole on photoinduced intramolecular electron transfer in dichromophoric alpha-helical peptides. J Am Chem Soc 118(9):2299

Gao J, Mueller P, Wang M, Eckhardt S, Lauz M, Fromm KM, Giese B (2011) Electron transfer in peptides: the

influence of charged amino acids. Angew Chem Int Ed 50(8):1926–1930
Gatto E, Porchetta A, Scarselli M, De Crescenzi M, Formaggio F, Toniolo C, Venanzi M (2012) Playing with peptides: how to build a supramolecular peptide nanostructure by exploiting helix center dot center dot center dot helix macrodipole interactions. Langmuir 28(5):2817–2826
Giese B, Graber M, Cordes M (2008) Electron transfer in peptides and proteins. Curr Opin Chem Biol 12(6):755–759
Giese B, Wang M, Gao J, Stoltz M, Müller P, Graber M (2009) Electron relay race in peptides. J Org Chem 74(10):3621–3625
Gooding JJ, Wibowo R, Liu YW, Losic D, Orbons S, Mearns FJ, Shapter JG, Hibbert DB (2003) Protein electrochemistry using aligned carbon nanotube arrays. J Am Chem Soc 125(30):9006–9007
Han B, Chen X, Zhao J, Bu Y (2012) A peptide loop and an alpha-helix N-terminal serving as alternative electron hopping relays in proteins. Phys Chem Chem Phys 14(45):15849–15859
Haris PI, Chapman D (1995) The conformational analysis of peptides using fourier transform IR spectroscopy. Biopolymers 37(4):251–263
Hermanson GT (2008) Bioconjugate techniques, 2nd edn. Academic, London
Hilinski GJ, Kim Y-W, Hong J, Kutchukian PS, Crenshaw CM, Berkovitch SS, Chang A, Ham S, Verdine GL (2014) Stitched α-helical peptides via bis ring-closing metathesis. J Am Chem Soc 136(35):12314–12322
Holub JM, Kirshenbaum K (2010) Tricks with clicks: modification of peptidomimetic oligomers via copper-catalyzed azide-alkyne [3 + 2] cycloaddition. Chem Soc Rev 39(4):1325–1337
Holub JM, Jang H, Kirshenbaum K (2007) Fit to be tied: conformation-directed macrocyclization of peptoid foldamers. Org Lett 9(17):3275–3278
Horsley JR, Yu J, Moore KE, Shapter JG, Abell AD (2014) Unraveling the interplay of backbone rigidity and electron rich side-chains on electron transfer in peptides: the realization of tunable molecular wires. J Am Chem Soc 136:12479–12488
Horsley JR, Yu J, Abell AD (2015) The correlation of electrochemical measurements and molecular junction conductance simulations in β-strand peptides. Chem Eur J 21:5926–5933
Huisgen R (1963) 1,3-dipolar cycloadditions. Past and future. Angew Chem Int Ed 2(10):565–598
Ingale S, Dawson PE (2011) On resin side-chain cyclization of complex peptides using CuAAC. Org Lett 13(11):2822–2825
Jacobsen O, Maekawa H, Ge NH, Gorbitz CH, Rongved P, Ottersen OP, Amiry-Moghaddam M, Klaveness J (2011) Stapling of a 3(10)-helix with click chemistry. J Org Chem 76(5):1228
Kai M, Takeda K, Morita T, Kimura S (2008) Distance dependence of long-range electron transfer through helical peptides. J Pept Sci 14(2):192–202
Kapoerchan VV, Wiesner M, Overhand M, van der Marel GA, Koning F, Overkleeft HS (2008) Design of azidoproline containing gluten peptides to suppress CD4+ T-cell responses associated with celiac disease. Bioorg Med Chem 16(4):2053–2062
Kirby DA, Britton KT, Aubert ML, Rivier JE (1997) Identification of high-potency neuropeptide Y analogues through systematic lactamization†. J Med Chem 40(2):210–215
Lakhani A, Roy A, De Poli M, Nakaema M, Formaggio F, Toniolo C, Keiderling TA (2011) Experimental and theoretical spectroscopic study of 310-helical peptides using isotopic labeling to evaluate vibrational coupling. J Phys Chem B 115(19):6252–6264
Langen R, Chang IJ, Germanas JP, Richards JH, Winkler JR, Gray HB (1995) Electron-tunneling in proteins - coupling through a bta-strand. Science 268(5218):1733–1735
Lauz M, Eckhardt S, Fromm KM, Giese B (2012) The influence of dipole moments on the mechanism of electron transfer through helical peptides. Phys Chem Chem Phys 14(40):13785–13788
Laviron E (1979) The use of linear potential sweep voltammetry and of a.c. voltammetry for the study of the surface electrochemical reaction of strongly adsorbed systems and of redox modified electrodes. J Electroanal Chem 100:263
Long YT, Abu-Rhayem E, Kraatz HB (2005) Peptide electron transfer: more questions than answers. Chem Eur J 11(18):5186
Looper RE, Pizzirani D, Schreiber SL (2006) Macrocycloadditions leading to conformationally restricted small molecules. Org Lett 8(10):2063–2066
Lopez MS, Supuran AJ, Poulsen CT, Poulsen SA (2010) Carbonic anhydrase inhibitors developed through 'click tailing'. Curr Pharm Des 16:3277–3287
Loughlin WA, Tyndall JDA, Glenn MP, Fairlie DP (2004) Beta-strand mimetics. Chem Rev 104(12):6085–6118
Lukacs A, Eker APM, Byrdin M, Brettel K, Vos MH (2008) Electron hopping through the 15 Å triple tryptophan molecular wire in DNA photolyase occurs within 30 ps. J Am Chem Soc 130(44):14394–14395
Lundquist JT, Pelletier JC (2002) A new tri-orthogonal strategy for peptide cyclization. Org Lett 4(19):3219–3221
Lung F-DT, Collins N, Stropova D, Davis P, Yamamura HI, Porreca F, Hruby VJ (1996) Design, synthesis, and biological activities of cyclic lactam peptide analogues of dynorphin A(1−11)-NH21. J Med Chem 39(5):1136–1141
Maeda H, Sakamoto R, Nishihara H (2013) Metal complex oligomer and polymer wires on electrodes: tactical constructions and versatile functionalities. Polymer 54(17):4383–4403
Maekawa H, Toniolo C, Broxterman QB, Ge N-H (2007) Two-dimensional infrared spectral signatures of 310- and α-helical peptides. J Phys Chem B 111(12):3222–3235

Malak RA, Gao ZN, Wishart JF, Isied SS (2004) Long-range electron transfer across peptide bridges: the transition from electron superexchange to hopping. J Am Chem Soc 126(43):13888

Malesevic M, Strijowski U, Bächle D, Sewald N (2004) An improved method for the solution cyclization of peptides under pseudo-high dilution conditions. J Biotechnol 112(1-2):73–77

Mandal HS, Kraatz H-B (2012) Electron transfer mechanism in helical peptides. J Phys Chem Lett 3(6):709–713

Marques-Gonzalez S, Yufit DS, Howard JAK, Martin S, Osorio HM, Garcia-Suarez VM, Nichols RJ, Higgins SJ, Cea P, Low PJ (2013) Simplifying the conductance profiles of molecular junctions: the use of the trimethylsilylethynyl moiety as a molecule-gold contact. Dalton T 42(2):338–341

McCreery RL, Bergren AJ (2009) Progress with molecular electronic junctions: meeting experimental challenges in design and fabrication. Adv Mater 21(43):4303–4322

Morita T, Kimura S (2003) Long-range electron transfer over 4 nm governed by an inelastic hopping mechanism in self-assembled monolayers of helical peptides. J Am Chem Soc 125(29):8732–8733

Morita T, Kimura S, Kobayashi S, Imanishi Y (2000) Photocurrent generation under a large dipole moment formed by self-assembled monolayers of helical peptides having an N-ethylcarbazolyl group. J Am Chem Soc 122(12):2850–2859

Morita T, Yanagisawa K, Kimura S (2008) Enhanced photocurrent generation by electron hopping through regularly-arranged chromophores in a helical peptide monolayer. Polym J 40(8):700–709

Nakayama H, Kimura S (2009) Chirally twisted oligo(phenyleneethynylene) by cyclization with alpha-helical peptide. J Org Chem 74(9):3462–3468

Neuhaus DE, P. A (1993) Methods in molecular biology, vol 17. Humana Press, Clifton

Niggli DA, Ebert M-O, Lin Z, Seebach D, van Gunsteren WF (2012) Helical content of a β3-octapeptide in methanol: molecular dynamics simulations explain a seeming discrepancy between conclusions derived from CD and NMR data. Chem Eur J 18(2):586–593

Ousaka N, Sato T, Kuroda R (2007) Intramolecular cross-linking of an optically inactive 310-helical peptide: stabilization of structure and helix sense. J Am Chem Soc 130(2):463–465

Paredes SB, Reiter C, Rodriguez R, A. (2009) Assessment of the potential role of tryptophan as the precursor of serotonin and melatonin for the aged sleep-wake cycle and immune function: *streptopelia risoria* as a model. Int J Tryptophan Res 2:23–36

Pedersen DS, Abell A (2011) 1,2,3-triazoles in peptidomimetic chemistry. Eur J Org Chem 2011(13):2399–2411

Peeters E, Christiaans MPT, Janssen RAJ, Schoo HFM, Dekkers HPJM, Meijer EW (1997) Circularly polarized electroluminescence from a polymer light-emitting diode. J Am Chem Soc 119(41):9909–9910

Pehere AD, Abell AD (2012) New beta-strand templates constrained by Huisgen cycloaddition. Org Lett 14(5):1330–1333

Pehere AD, Pietsch M, Gütschow M, Neilsen PM, Pedersen DS, Nguyen S, Zvarec O, Sykes MJ, Callen DF, Abell AD (2013a) Synthesis and extended activity of triazole-containing macrocyclic protease inhibitors. Chem Eur J 19(24):7975–7981

Pehere AD, Sumby CJ, Abell AD (2013b) New cylindrical peptide assemblies defined by extended parallel [small beta]-sheets. Org Biomol Chem 11(3):425–429

Petrov EG, Shevchenko YV, Teslenko VI, May V (2001) Nonadiabatic donor-acceptor electron transfer mediated by a molecular bridge: a unified theoretical description of the superexchange and hopping mechanism. J Chem Phys 115(15):7107–7122

Petrov EG, Shevchenko YV, May V (2003) On the length dependence of bridge-mediated electron transfer reactions. Elsevier 288:269–279

Polo F, Antonello S, Formaggio F, Toniolo C, Maran F (2005) Evidence against the hopping mechanism as an important electron transfer pathway for conformationally constrained oligopeptides. J Am Chem Soc 127(2):492

Ramachandran GN, Sasisekharan V (1968) Conformation of polypeptides and proteins. Adv Protein Chem 23:283

Riek RP, Rigoutsos I, Novotny J, Graham RM (2001) Non-α-helical elements modulate polytopic membrane protein architecture. J Mol Biol 306(2):349–362

Rostovtsev VV, Green LG, Fokin VV, Sharpless KB (2002) A stepwise Huisgen cycloaddition process: copper(I)-catalyzed regioselective "ligation" of azides and terminal alkynes. Angew Chem Int Ed 41(14):2596–2599

Sarangi NK, Patnaik A (2012) L-tryptophan-induced electron transport across supported lipid bilayers: an alkyl-chain tilt-angle, and bilayer-symmetry dependence. ChemPhysChem 13(18):4258–4270

Sasaki H, Makino M, Sisido M, Smith TA, Ghiggino KP (2001) Photoinduced electron transfer on beta-sheet cyclic peptides. J Phys Chem B 105(42):10416–10423

Schievano E, Bisello A, Chorev M, Bisol A, Mammi S, Peggion E (2001) Aib-rich peptides containing lactam-bridged side chains as models of the 3(10)-helix. J Am Chem Soc 123(12):2743–2751

Schlag EW, Sheu SY, Yang DY, Selzle HL, Lin SH (2000) Charge conductivity in peptides: dynamic simulations of a bifunctional model supporting experimental data. PNASUSA 97(3):1068–1072

Schlag EW, Sheu SY, Yang DY, Selzle HL, Lin SH (2007) Distal charge transport in peptides. Angew Chem Int Ed 46(18):3196–3210

Sek S (2013) Review peptides and proteins wired into the electrical circuits: an SPM-based approach. Biopolymers 100(1):71–81

Sek S, Sepiol A, Tolak A, Misicka A, Bilewicz R (2004) Distance dependence of the electron transfer rate through oligoglycine spacers introduced

into self-assembled monolayers. J Phys Chem B 108(24):8102–8105
Seyedsayamdost MR, Yee CS, Reece SY, Nocera DG, Stubbe J (2006) pH rate profiles of FnY356−R2s (n = 2, 3, 4) in *Escherichia coli* ribonucleotide reductase: evidence that Y356 is a redox-active amino acid along the radical propagation pathway. J Am Chem Soc 128(5):1562–1568
Shih C, Museth AK, Abrahamsson M, Blanco-Rodriguez AM, Di Bilio AJ, Sudhamsu J, Crane BR, Ronayne KL, Towrie M, Vlcek A, Richards JH, Winkler JR, Gray HB (2008) Tryptophan-accelerated electron flow through proteins. Science 320(5884):1760–1762
Sisido M, Hoshino S, Kusano H, Kuragaki M, Makino M, Sasaki H, Smith TA, Ghiggino KP (2001) Distance dependence of photoinduced electron transfer along alpha-helical polypeptides. J Phys Chem B 105(42):10407–10415
Smestad GP, Gratzel M (1998) Demonstrating electron transfer and nanotechnology: a natural dye-sensitised nanocrystalline energy converter. J Chem Educ 75(6):752–756
Smeu M, Wolkow RA, Guo H (2009) Conduction pathway of π-stacked ethylbenzene molecular wires on Si(100). J Am Chem Soc 131(31):11019–11026
Smith LJ, Bolin KA, Schwalbe H, MacArthur MW, Thornton JM, Dobson CM (1996) Analysis of main chain torsion angles in proteins: prediction of NMR coupling constants for native and random coil conformations. J Mol Biol 255(3):494–506
Staykov A, Li X, Tsuji Y, Yoshizawa K (2012) Current rectification in nitrogen- and boron-doped nanographenes and cyclophanes. J Phys Chem C 116(34):18451–18459
Stubbe J, Nocera DG, Yee CS, Chang MCY (2003) Radical initiation in the class I ribonucleotide reductase: long-range proton-coupled electron transfer? Chem Rev 103(6):2167–2202
Takeda K, Morita T, Kimura S (2008) Effects of monolayer structures on long-range electron transfer in helical peptide monolayer. J Phys Chem B 112(40):12840–12850
Taylor JW (2002) The synthesis and study of side-chain lactam-bridged peptides. Biopolymers 66(1):49–75
Toniolo C, Benedetti E (1991) The polypeptide 310-helix. Trends Biochem Sci 16(0):350–353
Toniolo C, Polese A, Formaggio F, Crisma M, Kamphuis J (1996) Circular dichroism spectrum of a peptide 3(10)-helix. J Am Chem Soc 118(11):2744–2745
Tornøe CW, Christensen C, Meldal M (2002) Peptidotriazoles on solid phase: [1,2,3]-triazoles by regiospecific copper(I)-catalyzed 1,3-dipolar cycloadditions of terminal alkynes to azides. J Org Chem 67(9):3057–3064
Uji H, Morita T, Kimura S (2013) Molecular direction dependence of single-molecule conductance of a helical peptide in molecular junction. Phys Chem Chem Phys 15(3):757–760
van Lierop BJ, Bornschein C, Jackson WR, Robinson AJ (2011) Ring-closing metathesis in peptides - the sting is in the tail! Aust J Chem 64(6):806–811
Wang M, Gao J, Muller P, Giese B (2009) Electron transfer in peptides with cysteine and methionine as relay amino acids. Angew Chem Int Ed 48(23):4232–4234
Watanabe J, Morita T, Kimura S (2005) Effects of dipole moment, linkers, and chromophores at side chains on long-range electron transfer through helical peptides. J Phys Chem B 109(30):14416
Wittekindt C, Schwarz M, Friedrich T, Koslowski T (2009) Aromatic amino acids as stepping stones in charge transfer in respiratory complex I: an unusual mechanism deduced from atomistic theory and bioinformatics. J Am Chem Soc 131(23):8134–8140
Xu Y, Cui B, Ji G, Li D, Liu D (2013) Effect of the orientation of nitro group on the electronic transport properties in single molecular field-effect transistors. Phys Chem Chem Phys 15(3):832–836
Yanagisawa K, Morita T, Kimura S (2004) Efficient photocurrent generation by self-assembled monolayers composed of 3(10)-helical peptides carrying linearly spaced naphthyl groups at the side chains. J Am Chem Soc 126(40):12780–12781
Yasutomi S, Morita T, Imanishi Y, Kimura S (2004) A molecular photodiode system that can switch photocurrent direction. Science 304(5679):1944–1947
Yasutomi S, Morita T, Kimura S (2005) pH-controlled switching of photocurrent direction by self-assembled monolayer of helical peptides. Journal of the Am Chem Soc 127(42):14564–14565
Yew SY, Shekhawat G, Wangoo N, Mhaisalkar S, Suri CR, Dravid VP, Lam YM (2011) Design of single peptides for self-assembled conduction channels. Nanotechnology 22:215606
Yu J, Zvarec O, Huang DM, Bissett MA, Scanlon DB, Shapter JG, Abell AD (2012) Electron transfer through α-peptides attached to vertically aligned carbon nanotube arrays: a mechanistic transition. Chem Comm 48(8):1132–1134
Yu J, Horsley JR, Moore KE, Shapter JG, Abell AD (2014) The effect of a macrocyclic constraint on electron transfer in helical peptides: a step towards tunable molecular wires. Chem Comm 50(14):1652
Yu J, Horsley JR, Abell AD (2016) Turning electron transfer 'on-off' in peptides through side-bridge gating. Electrochim Acta 209:65–74
Yu J, Horsley JR, Abell AD (2017) Exploiting the interplay of quantum interference and backbone rigidity on electronic transport in peptides: a step towards bio-inspired quantum interferometers. Mol Sys Des Eng 2:67–77
Zhuang W, Hayashi T, Mukamel S (2009) Coherent multidimensional vibrational spectroscopy of biomolecules: concepts, simulations, and challenges. Angew Chem Int Ed 48(21):3750–3781

Peptide-Based Materials for Cartilage Tissue Regeneration

7

Nurcan Hastar, Elif Arslan, Mustafa O. Guler, and Ayse B. Tekinay

Abstract

Cartilaginous tissue requires structural and metabolic support after traumatic or chronic injuries because of its limited capacity for regeneration. However, current techniques for cartilage regeneration are either invasive or ineffective for long-term repair. Developing alternative approaches to regenerate cartilage tissue is needed. Therefore, versatile scaffolds formed by biomaterials are promising tools for cartilage regeneration. Bioactive scaffolds further enhance the utility in a broad range of applications including the treatment of major cartilage defects. This chapter provides an overview of cartilage tissue, tissue defects, and the methods used for regeneration, with emphasis on peptide scaffold materials that can be used to supplement or replace current medical treatment options.

Keywords

Cartilage • Cartilage regeneration • Scaffold types • Peptide scaffolds • Self-assembly

N. Hastar • E. Arslan
Institute of Materials Science and Nanotechnology, National Nanotechnology Research Center (UNAM), Bilkent University, Ankara 06800, Turkey

M.O. Guler
Institute for Molecular Engineering, University of Chicago, Chicago, IL 60637, USA

Institute of Materials Science and Nanotechnology, National Nanotechnology Research Center (UNAM), Bilkent University, Ankara 06800, Turkey

A.B. Tekinay (✉)
Institute of Materials Science and Nanotechnology, National Nanotechnology Research Center (UNAM), Bilkent University, Ankara 06800, Turkey

Neuroscience Graduate Program, Bilkent University, Ankara 06800, Turkey
e-mail: atekinay@bilkent.edu.tr

© Springer International Publishing AG 2017

A. Sunna et al. (eds.), *Peptides and Peptide-based Biomaterials and their Biomedical Applications*, Advances in Experimental Medicine and Biology 1030, https://doi.org/10.1007/978-3-319-66095-0_7

7.1 Introduction

Cartilage is a connective tissue that structurally supports the body and assists in movement by creating low-friction platforms between the articular spaces of bone. Cartilaginous tissue does not contain any vascular and nervous elements, which severely limits its capacity for regeneration following injury (Ateshian 2007). Microfracture and autograft transplantation techniques are the current medical standards for the repair of cartilage damage. However, these methods are not fully capable of restoring the structural and functional integrity of cartilage tissue, necessitating the development of alternative procedures. The extracellular matrix (ECM) is responsible for the lubrication and articulation functions of cartilage. Therefore, ECM-mimetic intelligent and tunable scaffolds can serve as a temporary replacement for cartilage tissue while supplying the environmental conditions necessary for its regeneration. Modified polymers and peptides can also be utilized for the recruitment of mesenchymal stem cells to the damaged area, thereby promoting the formation of neo-cartilage tissue. However, these scaffolds must be biocompatible and biodegradable in order to minimize immunogenic responses, stimulate cell proliferation and integrate effectively with the surrounding tissue. In addition, they should have a porous structure to facilitate cellular migration and communication, and exhibit mechanical properties resembling native cartilage tissue to support newly formed tissues and provide the necessary signals for cellular recruitment and differentiation.

This chapter presents the features of cartilage tissue, its diseases, and the methods developed for promoting its regeneration. A broad overview of the advantages and disadvantages associated with both recent and established treatment options are discussed to provide further focus on the use of peptide bearing polymers and self-assembled peptide gels for cartilage engineering.

7.2 Cartilage Tissue

7.2.1 Structure of Cartilage Tissue

Cartilage tissue has a limited capacity for regeneration due to its avascular, aneural and alymphatic nature, which prevents the influx of oxygen, nutrients and biochemical signals that are required for effective wound repair. In addition, cartilage regeneration occurs through the activity of native chondrocytes and mesenchymal stem cells (MSCs), which have sparse populations and cannot facilitate the complete repair of large defects. Consequently, cartilage has a tendency to accumulate partially repaired defects with age, which progressively impairs tissue function and may result in degenerative joint diseases such as osteoarthritis, severely lowering the quality-of-life in the elderly population in particular.

Cartilage tissue is composed of cartilage-specific cells (chondrocytes) that constitute 3–5% of the adult articular cartilage by mass and are embedded in a highly dense ECM (Han et al. 2011). MSCs are also present in cartilage tissue, especially in the deep part of the perichondrium, where they encircle the cartilage and may differentiate into chondroblasts (Slomianka 2009). However, cartilage is mostly composed of ECM elements, and load-bearing, the main function of cartilage tissue, is provided mainly by the macromolecular components of dense ECM (Taipaleenmaki 2010). In addition to its mechanical integrity, cartilage tissue also provides joint lubrication and articulation, and the ECM microenvironment likewise plays crucial roles in these functions (Han et al. 2011). The cartilage ECM is composed mainly of a fibrillar collagen network that is supported by various proteoglycans (Han et al. 2011). Among the collagen fibrils, type II collagen is the primary structural component found in cartilage ECM. Other cartilage-associated collagens include types III, VI, IX, X, XI, XII and XIV, which all contribute to the mature extracellular microenvironment (Eyre 2002). Besides collagen fibrils, the cartilage ECM also contains proteoglycans such as aggrecan, chondroitin sulfate glycosaminoglycan, keratin sulfate glycosaminoglycan, hyaluronan, and glycoprotein lubricin (PRG4) (Lin et al. 2005). These cartilage-specific proteins are responsible for establishing the morphology and function of the tissue, and their malfunction contributes to the secondary expansion of cartilage defects. Consequently, the induction of proteoglycan synthesis is a promising approach for promoting the recovery of cartilage injuries.

7.2.2 Cartilage Diseases

Osteoarthritis is an inflammatory disease that is associated with the deterioration and erosion of articular cartilage, and occurs in response to a combination of genetic, metabolic and biochemical factors (Felson et al. 1997). The osteoarthritis (OA) is a pathological condition involving the interactions of cartilage, bone and synovium. Aging (and the associated mechanical wear in joints) is the main cause for osteoarthritis and other degenerative cartilage disorders, although joint problems can also manifest following traumatic accidents and in obese persons (Grotle et al. 2008). One major problem in the treatment of OA and similar diseases is the progressive loss of tissue integrity resulting from day-to-day activity and the bodily weight, as cartilage exhibits a minimal capacity of self-renewal that can easily be overwhelmed by such factors. Consequently, the pathophysiological formation of OA is predominantly due to the destabilization of the anabolic and catabolic dynamics of cartilage, which results in the progressive destruction of the tissue (Moreland 2003) This process is regulated by cytokines, growth hormones and enzymes (aggrecanases, matrix metalloproteinases, etc.) synthesized by chondrocytes and synovial cells. During OA formation, collagen degradation increases, proteoglycan content decreases, types of macromolecules change, and water content increases (Fig. 7.1, Heinegård and Saxne 2011; Belcher et al. 1997; Brandt et al. 2000; Moreland 2003). In addition, the amount and molecular size of hyaluronic acid in the synovial fluid are also reduced (Belcher et al. 1997). Such progressive degeneration is a very invasive process, and early intervention is key for its management: radiological osteoarthritic findings may develop in less than 10 years when the patients are not treated during the early period of injury (Prakash and Learmonth 2002). Even small (~10 mm) defects can result in the reduction of 50% or more of cartilage tissue in the joint, while larger (>10 mm) defects are known to provide an additional burden of 64% to the surrounding cartilage, as observed in a 14-year follow up study (Kock et al. 2008). Due to their severe and long-lasting effects, the treatment of cartilage diseases has been a subject of intensive study, although effective methods for promoting long-term tissue repair are yet to be discovered. Therefore, there is an urgent medical need for alternative treatment options and new multifunctional biomaterials that are able to respond to cartilage damage in a short period of time.

7.2.3 Clinical Treatment Strategies for Cartilage Regeneration

The microfracture technique is one of the most common treatment methods against cartilage damage. This method is based on the introduction of mesenchymal stem cells from lower bone into cartilage defects following the induction of local bleeding with the help of small needles. However, in the absence of a supportive artificial matrix, stem cells recruited in this manner generate fibrous tissue formations that are structurally and biomechanically dissimilar to the natural architecture of cartilage. In addition, successful treatment requires an accurate assessment of the extent of cartilage damage. In the clinic, arthroscopic debridement and lavage are performed for patients with less than 2 cm^2 of cartilage damage, as this method involves the cleaning of the site of injury and minimizes local pain resulting from frictional forces acting on damaged cartilage. This treatment method is generally more suitable for elderly patients, who are less mobile and do not run the risk of causing further damage to the site of injury through active movement (Gaissmaier et al. 2008). In contrast,

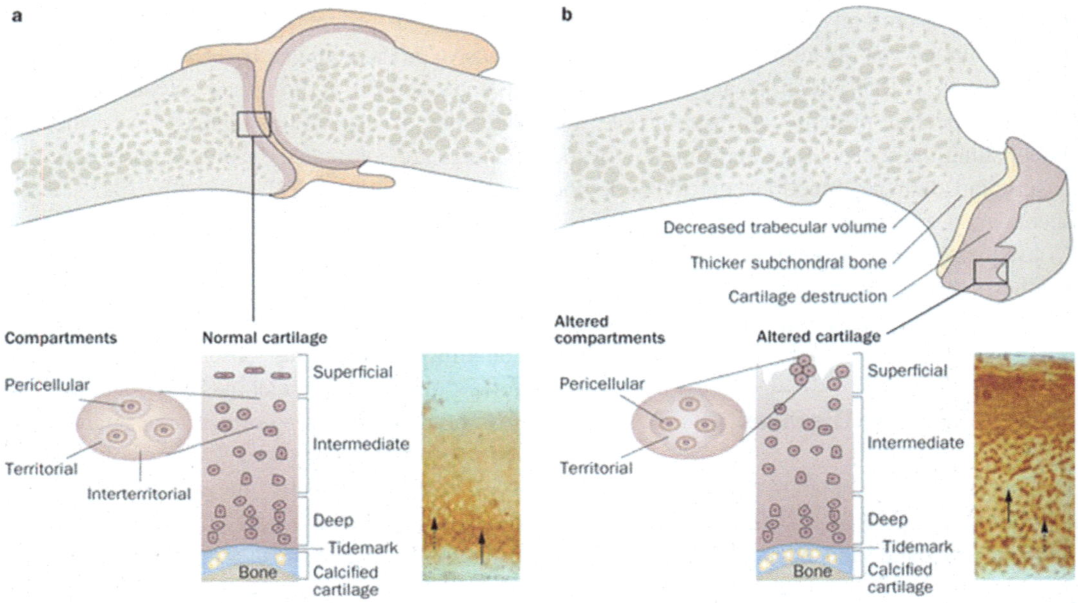

Fig. 7.1 (**a**) Representation of healthy joint of articular cartilage. Pericellular, territorial and interterritorial matrices form articular cartilage and each of them place at particular distance from the chondrocytes. Inset image indicates the immunohistochemistry of cartilage oligomeric matrix protein (COMP). While the territorial and pericellular matrices (*dashed arrow*) did not stained, the interterritorial matrix stains for COMP (*solid arrow*). (**b**) Representation of osteoarthritic joint. Although disease is at early stages, degeneration of cartilage partially, alteration in the underlying bone and cloning and reproduction of cells were observed in joints. No immunohistochemistry staining for COMP at interritorial matrix (*solid arrow*) and storage of newly synthesized COMP at pericellular matrix of the cartilage (*dashed arrow*) demonstrate the cloning of cells is not distinct yet (Reprinted from Heinegård and Saxne 2011)

the microfracture method is preferred for defect sizes in the range of 2–3 cm^2. This method is especially suitable for stimulating the migration of MSCs from bone marrow to cartilage, allowing new cartilage formation in full-thickness cartilage injuries (Kreuz et al. 2006). However, effective cartilage repair in this method is contingent upon the minimization of movement and mechanical loading at the affected site. In addition, mid- and long-term follow-up studies have shown that hyaline cartilage formation is limited and the newly formed tissue is dominated by fibrous cartilage following microfracture surgery: Although a hyaline-like cartilaginous matrix is initially formed at the site of injury, this material eventually acquires a fibrous character and deteriorates due to its inadequate mechanical and viscoelastic properties. Consequently, the joint gradually loses its functionality following surgery and must be replaced by prosthetics (Li et al. 2009). In this context, cell and tissue implantation studies are generally applied as a secondary treatment method in patients with cartilage damage greater than 2 cm^2. This treatment method is performed by transplanting chondral and osteochondral autografts in patients who do not respond well to lavage and microfracture treatments. In larger injuries, damaged cartilage tissue from non-load bearing joint regions may also be implanted. An important advantage of this method is the use of healthy, functional cartilage tissue, which can be obtained through arthroscopy. However, the implantation further damages the healthy tissue that surrounds the defect site, preventing the uniform integration of the autograft (Gillogly et al. 1998). Considering the disadvantages of these clinically applied treatment modalities, alternative treatment methods are needed for the effective recovery of cartilage injuries. Tissue engineering is a new and promising approach to enhance cartilage tissue regeneration, and a variety of biomaterials and cell culture methods have been developed to eliminate the deficiencies associated with conventional methods of cartilage repair.

7.3 Biomaterials for Cartilage Regeneration

Since cartilage has a limited capacity for regeneration, cells require additional biochemical and mechanical signals to effectively facilitate defect closure. Biomaterial scaffolds are next-generation tools for tissue engineering studies, and provide a combination of these signals in a 3D microenvironment to promote cellular attachment and proliferation. In addition, the highly porous structure of biomaterial scaffolds allows them to mediate the cell-cell interactions and signaling pathways that assists in cartilage repair (Ge et al. 2012). Biocompatibility and biodegradability are indispensable characteristics of all biomaterials designed for tissue applications, while flexibility, elasticity and customizable mechanical strength are required to emulate the natural environment of cartilage in particular (Pearle et al. 2005).

7.3.1 Natural and Synthetic Scaffolds

A broad range of both natural and synthetic biomaterials have been used to support cell adhesion, viability, proliferation, and differentiation in cartilage tissue engineering studies. Natural scaffolds can be extracted from live organisms and include materials such as collagens, GAGs and decellularized cartilage of animal origin. Foremost among these materials is hyaluronic acid, which is a carbohydrate-based, non-sulfated glycosaminoglycan that is found abundantly in cartilage tissue and promotes chondrocyte proliferation as well as chondrogenesis in MSCs. Many applications of this material exist in the literature: For example, a hyaluronic acid derivative biopolymer (Hyaff®-11) was seeded with autologous chondrocytes and shown to enhance the quality of newly formed cartilage tissue following its direct injection into full-thickness cartilage defects (Grigolo et al. 2001). Collagens (mostly type II) are also common in the natural cartilage ECM, and see frequent use in tissue engineering applications due to their lack of toxicity, ability to integrate rapidly into surrounding tissues, and role as a flexible platform for the attachment, proliferation and differentiation of cells. In addition, chondrocytes were shown to maintain their morphology and exhibit enhanced GAG production when seeded on porcine cartilage-derived type II collagen (Nehrer et al. 1997). Alginate and agarose are polysaccharide biomaterials that originate from seaweed and show good biocompatibility and cytocompatibility; however, their poor degradation kinetics and limited potential for functional modification limits their clinical use (Murphy and Sambanis 2001). Fibrin is another natural scaffold that is derived from blood and has been shown to modulate cartilage formation *in vivo* by facilitating the adhesion of chondrocytes (Fussenegger et al. 2003). Nevertheless, fibrin is mechanically inadequate for cartilage tissue engineering, and its degradation rate is unstable. Thus, it cannot provide a suitable environment as a scaffold for cartilage repair.

Batch-to-batch variations and donor requirements of natural scaffolds have led to the fabrication of synthetic scaffolds with controllable porosity, biodegradability and mechanical functions. Polymeric materials are typically preferred for the creation of synthetic materials due to their ease of fabrication and chemical modification. Poly-glycolic acid (PGA) is a popular material for cartilage autograft studies, and chondrocytes were shown to actively replace the polymer matrix with native ECM components when seeded onto PGA surfaces (Grande et al. 1997). In addition, PGA can form a copolymer with poly-lactic acid (PLA), and this material (poly-lactic-co-glycolic acid, PLGA) can be engineered to exhibit strong biocompatibility and precise control over degradation rates by adjusting the mixing ratios of PGA and PLA. PLGA offers a flexible environment for cartilage regeneration and was shown to mediate the differentiation of adipose-derived adult stem cells into heterogenic cartilage cell populations, which can be utilized for further cartilage engineering applications (Mehlhorn et al. 2009).

7.3.2 Composite Scaffolds

Hybrid scaffolds offer novel options for cartilage regeneration by combining the features of individual scaffold elements. For example, chondroitin sulfate (CS) is one of the essential components of cartilage ECM, and the addition of CS to a porous network will enhance the proliferation and differentiation of chondrocytes and provide a degree of bioadhesiveness. Likewise, methyacrylate and aldehyde groups can be integrated into scaffolds in order to establish additional bonds between biomaterials and tissue proteins. Poly (ε-caprolactone) (PCL) is a common material for such functionalization efforts, as it supports cell attachment, proliferation and ECM production, allowing it to mediate the repair of defect sites and formation of full-thickness cartilage, and the Food and Drug Administration (FDA) approves the clinical use of PCL (Garcia-Giralt et al. 2008). Due to the hydrophobic character of PCL, the cellular recognition and degradation of this polyester is relatively difficult compared to other biomaterials. The integration of polyethylene glycol (as methoxyl poly (ethylene glycol) methyl ether) to PCL enhances the bioactivity of the hydrogel by modulating its hydrophobicity and biodegradability. Graphene oxide (GO) can also be added to these scaffolds to create a more advanced biomaterial with high mechanical strength, conductivity and overall contact area for biological interactions (Lee et al. 2011). The resulting material is a highly porous, elastic, swellable scaffold that has the capacity to convey electricity and is not degraded easily in biological environments, allowing it to establish a suitable long-term environment for cartilage regeneration. (Liao et al. 2015)

Synthetic scaffolds can be further modified through the integration and controlled release of bioactive factors that play important roles in the regeneration process, such as growth factors and hormones. Platelet rich plasma (PRP) contains an elevated level of growth factors, yet its clinical use is limited due to its low mechanical strength and the burst release of its constituent growth factors (Fortier et al. 2011). However, PRP can be mixed with PCL and integrated into a gelatin membrane through the emulsion electrospinning method, producing a composite material that shows high bioactivity and can sustainably release growth factors to stimulate cellular attachment *in vitro* and support intrinsic cartilage regeneration *in vivo* (Liu et al. 2015).

In addition to soluble factors, cellular recruitment and integration are also crucial for the success of biomaterial scaffolds. Due to the limited availability of chondrocytes in cartilage tissue, many studies have focused on the use of other cell types to mediate cartilage maintenance and repair following injury. Neonatal chondrocytes (NChons) are the preferred allogeneic cell type for the regeneration of cartilage, but the limited amount of donors prevents their widespread use in clinical settings. Nevertheless, successful culturing environments of NChons were developed using biomaterial scaffolds, such as a 3D hydrogel composed of poly (ethylene glycol diacrylate) and chondroitin sulfate methacrylate. Despite this, however, tissue engineering applications typically rely on MSCs, which are available in abundance, proliferate readily under culture conditions, and can differentiate into chondrocytes. The maintenance of these cells in bioactive scaffolds is an effective means of inducing their differentiation into specific cell lines, and both natural and synthetic scaffolds have been used to support the adhesion, viability, proliferation and differentiation of MSCs. These scaffolds should have a specific set of features to succeed in this role: They must be biocompatible, exhibit minimal immunogenicity and toxicity, possess a microporous nature to ensure oxygen and nutrient transport, allow cells to remodel the scaffold matrix through biodegradation, and alter the stem cell differentiation process to result in a single, specific cell type (Tombuloglu et al. Tombuloglu et al. 2012). Consequently, mimicking the natural ECM of cartilage tissue is a complex process that requires the incorporation of various biomaterial components. In addition, MSCs can be derived from multiple sources and exhibit different phenotypes depending on their source. Bone-marrow stem cells (BMSCs) and adipose derived stem cells (ADSCs) are commonly used for cartilage repair, as they are abundant, easy to isolate and naturally inclined towards osteo-chondrogenic differentiation. Co-culture of multiple cell types is also possible, and large neocartilage nodules have been generated by increasing ADSC populations synergistically with NChons, suggesting that the spatial distribution of cells in 3D scaffolds and paracrine signaling between stem cells and chondrocytes are critical parameters for the management of cartilage tissue formation (Lai et al. 2013).

7.3.3 Peptide-Based Cartilage Engineering Scaffolds

7.3.3.1 Peptide-Edited Hydrogel Systems for Cartilage Regeneration

More effective scaffold materials can be developed through the incorporation of bioactive epitope-bearing peptide sequences into hydrogels, producing a multifunctional system that contains both the structural architecture and biological signals present in natural scaffolds. Collagen II, for example, is an abundantly found protein in the ECM of cartilage tissue, and the Streptococcal collagen-like II protein was modified with an HA-binding peptide (CGGGRYPISRPRKR) or CS binding peptide (CGGGYKTNFRRYYRF) (Chow et al. 2014) and an MMP7-sensitive crosslinker peptide (CGGGPLELRAGGGC) (Bahney et al. 2011) to create a versatile hydrogel providing a biodegradable and bioactive ECM-mimetic microenvironment. Chondrogenesis was significantly enhanced in human MSCs encapsulated in hydrogels containing the HA-binding peptide, while the CS-binding peptide-containing hydrogel increased MMP7 gene expression and ECM remodeling activity. As such, this multifunctional hydrogel was able to utilize the combined effect of various peptide moieties to stimulate chondrogenic differentiation, and could be used as a minimally invasive method for the repair of cartilage defects (Parmar et al. 2015).

Integrated systems can compensate the deficiencies of their individual components to form mechanically durable and bioactive epitope-bearing scaffolds. Demineralized bone matrix is flexible and strong, and presents a three-dimensional collagen scaffold. In addition, it is biocompatible, non-immunogenic and, due to its incorporation of various growth factors and cytokines, able to stimulate osteochondrogenic differentiation *in vitro* (Li et al. 2006) and *in vivo* (Gao et al. 2004). Together with chitosan, demineralized bone forms a mechanically stronger biphasic hydrogel. In addition, chitosan facilitates cell retention and promotes the integration of newly formed cartilage tissue to the defect site due to its sol-gel transition property. This hybrid system can be further modified by an MSC-affine peptide sequence (EPLQLKM), which increases the migration of MSCs to damaged cartilage tissue and allows the biphasic biomaterial to significantly enhance the repair of cartilage defects (Fig. 7.2) (Huang et al. 2014; Meng et al. 2015).

Articular cartilage tissue is composed of superficial, transitional and deep zones. Every zone contains different types of chondrocytes that secrete various amount of proteoglycans and type II collagen (Bhosale and Richardson 2008). Mimicking the layered native structure of articular cartilage may enhance the regeneration of this tissue, and both natural and synthetic biomaterial combinations can be used to form multi-layered functional scaffolds for this purpose. For example, it has been shown that a polyethylene glycol (PEG) based hydrogel, when combined with chondroitin sulfate (CS), matrix metalloproteinase sensitive peptide (MMP-pep) and hyaluronic acid scaffolds, can stimulate the formation of zone-specific chondrocytes from bone-marrow derived stem cells (BMSC) (Bhosale and Richardson 2008). The mechanical and biochemical stimuli provided by distinct scaffold regions led to the formation of specific zones exhibiting characteristic ECM compositions, and three zones of cartilage were created by the multi-layered hydrogel constructs. In particular, the

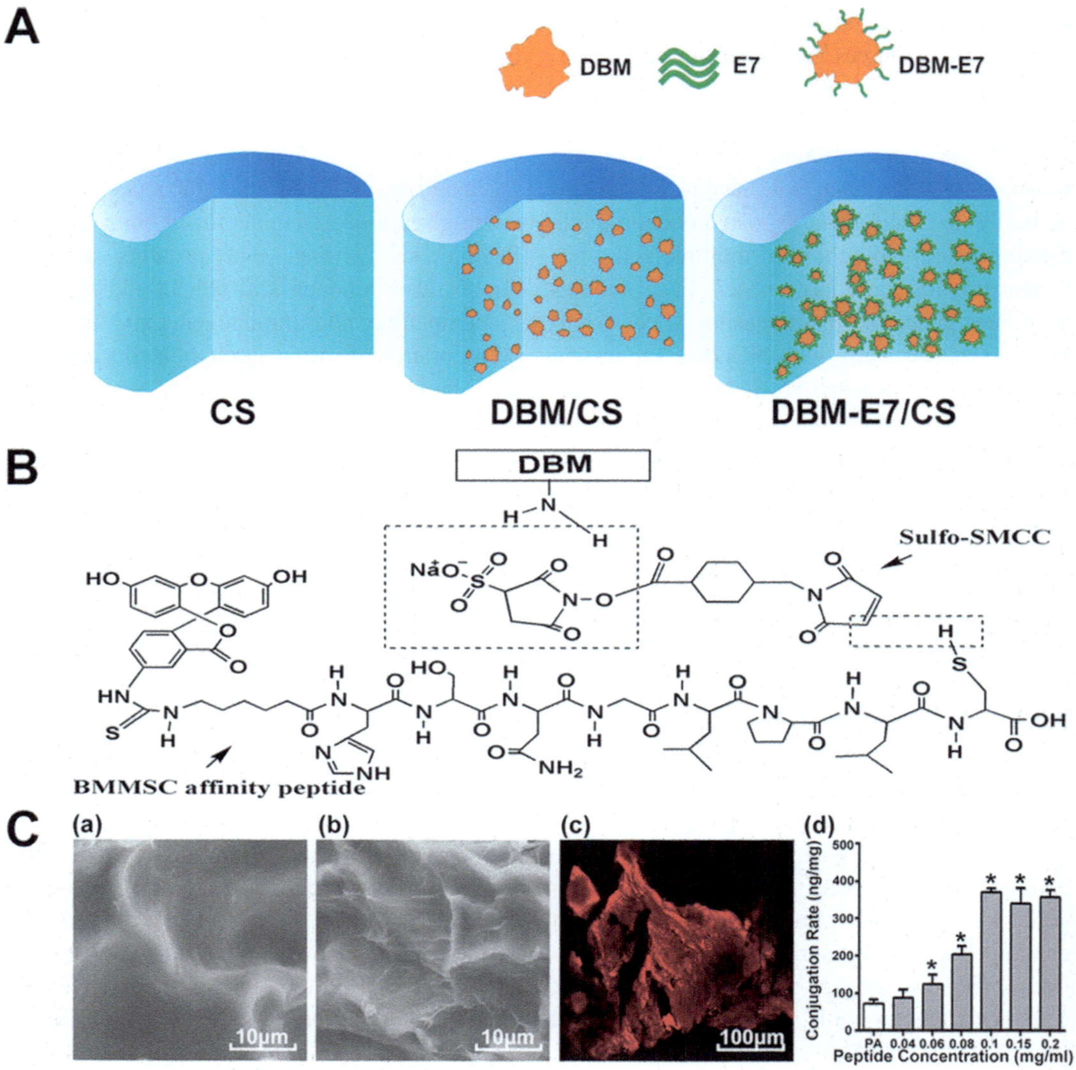

Fig. 7.2 (**a**) Representation of chitosan scaffold (CS), demineralized bone matrix (DBM) particles and chitosan hydrogel mixture (DBM/CS) scaffold, and combination of mesenchymal stem cells (MSCs) E7 affinity peptide and DBM particles within CS hydrogel scaffold. (**b**) Chemical structure of bone marrow-derived mesenchymal stem cells (BMMSCs) affinity peptide that is connected to DBM via sulfosuccinimidyl-4-(N-maleimidomethyl) cyclohexane-1-carboxylate (SMCC) cross-linker. (**c**) Visualization of (*a*) DBM and (*b*) DBM-E7 particles via SEM imaging and (*c*) rhodamine labeled DBM-E7 particles via confocal imaging. (*d*) Quantification of E7 peptide conjugation on DBM particle scaffold (*PA* physical adsorption; *$p < 0.05$ vs. PA) (Copyright © 2015, Rights Managed by Nature Publishing Group (Meng et al. 2015))

PEG:CS:MMP-pep combination differentiated stem cells into chondrocytes expressing elevated levels of collagen II and low levels of GAG, which is the feature of the superficial zone of cartilage, while the transitional zone was generated with the PEG:CS combination, and the PEG:HA scaffold promoted the formation of the deep zone. Collagen II expression decreased from the superficial to the deep zone, while the expressions of collagen X and proteoglycans were enhanced in this direction, which is similar to the situation observed under *in vivo* conditions and can be applied in clinical settings for the repair of full-thickness cartilage defects (Nguyen et al. 2011).

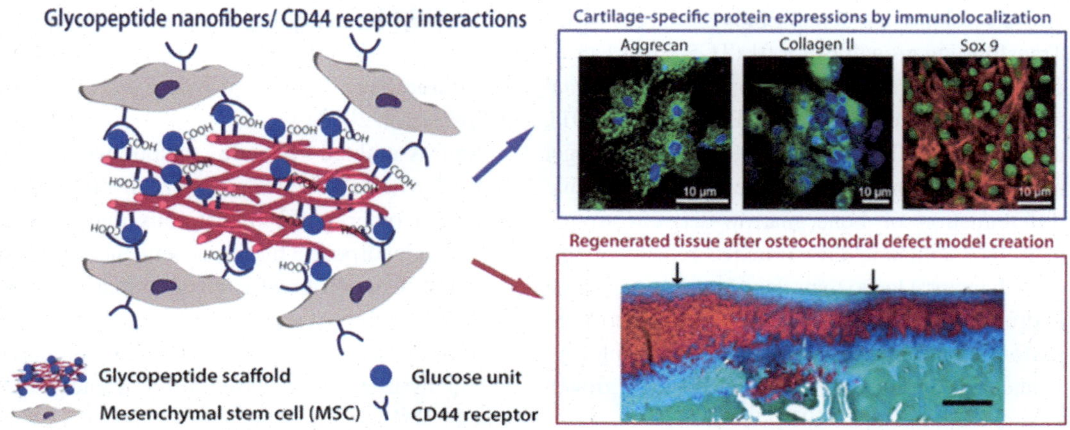

Fig. 7.3 Self- assembled glycopeptide scaffold mimicking hyaluronic acid. Commitment of mesenchymal stem cells to chondrogenesis was visualized by immunolocalization of cartilage specific proteins such as aggrecan, collagen II and Sox 9. Application of glycopeptide scaffold to osteochondral defected cartilage enhanced the regeneration of the tissue (Reprinted with permission from Ustun Yaylaci et al. 2016. Copyright 2016 American Chemical Society)

7.3.3.2 Self-Assembled Peptide Gels for Cartilage Regeneration

Tissue regeneration requires a complex set of biological signals that regulate the recruitment, adhesion, proliferation and differentiation of a broad range of cells. These signals are principally relayed through a fibrous extracellular matrix environment, and supramolecular materials are often designed to mimic this structure through noncovalent self-assembly mechanisms. Noncovalent interactions are preferred for scaffold formation due to their ability to be reversed under relatively mild conditions, which allows both external stimuli and cellular activity to remodel the material during growth and differentiation. In addition, noncovalent interactions are useful for the attachment of bioactive groups that replicate the functions of proteins, growth factors, cytokines and other biomolecules that play important roles during the tissue regeneration process. Peptide-based systems readily self-assemble under physiological conditions into biocompatible, biodegradable and bioactive epitope-bearing networks that mimic the ECM and support the proliferation, adhesion and migration of cells (Boekhoven and Stupp 2014). It was also shown in the study of Yaylaci et al. hyaluronic acid mimetic self-assembled peptide nanofiber gels enhance the chondrogenesis of MSCs and cartilage regeneration (Fig. 7.3) (Ustun Yaylaci et al. 2016). As such, self-assembled peptide gels have recently emerged as a class of next-generation biomaterials for tissue engineering applications.

Cell-seeded scaffolds for cartilage repair need to provide an appropriate environment for rapid cell division and cartilage-specific ECM deposition. Previously, chondrocytes encapsulated in alginate (Lemare et al. 1998) and agarose (Benya and Shaffer 1982) hydrogels were found to dedifferentiate or fail to preserve their morphologies across extended culturing and passaging periods. However, when encapsulated into self-assembled gels formed by KLD-12 (Ac-KLDLKLDLKLDL-Am) peptide molecules, chondrocytes were able to proliferate extensively while maintaining their phenotype and producing biomechanically functional ECM. The authors suggested that the utility of the system can be further enhanced by encapsulating growth factors within the gel matrix and investigating whether the newly formed cartilage is able to integrate into surrounding tissues (Kisiday et al. 2002). Cell fate can also be directed by controlling the release of soluble factors. Consequently, cellular behaviors can be regulated through the encapsulation of growth factors or cytokines into tunable biomaterial networks. Transforming growth factor β1 (TGF-β1) is a critical molecule for the differentiation of mesenchymal stem cells into chondrocytes (Barry et al. 2001), and the Ac-KLDLKLDLKLDL-Am peptide gel system was shown to stimulate the chondrogenic differentiation of bone marrow-derived stromal cells by facilitating the sustained release of this growth factor. Due to their flexible structures, these hydrogels could also be injected easily into non-uniform defect sites, making them promising candidates for the local delivery of cells and growth factors (Kopesky et al. 2014).

The sustained release of TGF-β1 was further regulated through the growth factor binding sequence HSNGLPL, which was discovered by phage display method and integrated into the structure of self-assembling TGFBPA hydrogels. These scaffolds were shown to exhibit a gradual release profile for TGF-β1, while rapid release was observed from control hydrogels. Human mesenchymal stem cells cultured on TGF-β1-encapsulated TGFBPA were viable and committed to the chondrogenic lineage. Further evaluation of the hydrogel was performed on an *in vivo* microfracture-treated chondral defect model, and TGF-β1 binding epitope density was found to strongly affect both the growth factor release profile and the chondrogenic regeneration capacity. In addition, TGFBPA was able to trigger cell differentiation and type II collagen expression even in the absence of encapsulated TGF-β1, possibly by retaining endogenously produced TGF-β1 (Shah et al. 2010).

Bioactivity of scaffold materials can also be enhanced by the combination of relatively large polymers with small self-assembled peptides. For example, the mixing of hydroxyapatite with the palmitoyl-$V_3A_3K_3$-Am peptide sequence was shown to result in the creation of a solid membrane at the interface between the two components, where the negatively-charged HA molecules were encapsulated by the positively-charged PA networks to form HA-filled sac structures. Viability analyses revealed that gel-filled sacs supported the survival of hMSCs and enhanced collagen type II expression in culture experiments with chondrogenic media supplemented with TGF-β1, suggesting that the sacs supply an environment conducive to the chondrogenic differentiation of mesenchymal stem cells (Capito et al. 2008).

Self-assembled peptide nanonetworks are generally more effective inducers of differentiation than other scaffold materials due to their ability to incorporate bioactive sequences in their structure. Chondrogenesis of bone marrow stem cells, for example, was found to be more extensive in self-assembling PA hydrogels compared to bulk hydrogels such as agarose. 3D encapsulation of BMSCs into RAD16-I (AcN- $(KLDL)_3$–CNH_2 and RAD16-I ($(RADA)_4$) promoted early chondrogenesis as a result of enhanced TGF-β1 signaling compared to agarose controls. Since the mechanical characteristics of both systems were similar, morphological differences during chondrogenesis can be attributed to sequence-specific signals. While cell-to-cell contact was present in the RAD16-I systems, multicellular clusters were observed only in the KLD12 hydrogel, and the integration of biochemical and biomechanical signals into the peptide scaffolds was found to allow precise control over the differentiation of MSCs (Kopesky et al. 2010).

7.4 Conclusion

Due to the low regenerative capacity of cartilage, even minor tissue defects can cause severe health problems without early intervention. While cartilage repair can be induced through the external administration of stem cells and growth factors, the lack of native blood vessels in cartilage requires the local delivery of the requisite cells and biomolecules, and scaffolds are ideal materials for this purpose. Invasive techniques such as autologous chondrocyte transplantation or microfractures can be used to stimulate cartilage repair, but are associated with severe disadvantages such as fibrosis and collateral tissue damage. Thus, recent advances in cartilage tissue engineering are encouraging from a clinical point of view. The facile synthesis and structural modification of injectable scaffolds is promising for the effective regeneration of large structural defects in cartilage. ECM-mimetic, porous and bioactive scaffolds are ideal materials for modulating the signaling networks responsible for cartilage formation and biointegration without eliciting undesirable immune complications. Further achievements in cartilage regeneration are expected to improve the quality of life of the elderly in particular, but it must nevertheless be noted that clinical studies involving peptide-based scaffolds are currently uncommon and that much work needs to be performed to suitably demonstrate the safety and efficiency of this new class of biomaterials for cartilage regeneration.

References

Ateshian GA (2007) Artificial cartilage: weaving in three dimensions. Nat Mater 6:89–90

Bahney C, Hsu C, Yoo J, West J, Johnstone B (2011) A bioresponsive hydrogel tuned to chondrogenesis of human mesenchymal stem cells. FASEB J 25:1486–1496

Barry F, Boynton R, Liu B, Murphy J (2001) Chondrogenic differentiation of mesenchymal stem cells from bone marrow: differentiation-dependent gene expression of matrix components. Exp Cell Res 268:189–200

Belcher C, Yaqub R, Fawthrop F, Bayliss M, Doherty M (1997) Synovial fluid chondroitin and keratan sulphate epitopes, glycosaminoglycans, and hyaluronan in arthritic and normal knees. Ann Rheum Dis 56:299–307

Benya P, Shaffer J (1982) Dedifferentiated chondrocytes reexpress the differentiated collagen phenotype when cultured in agarose gels. Cell 30:215–224

Bhosale A, Richardson J (2008) Articular cartilage: structure, injuries and review of management. Br Med Bull 87:77–95

Boekhoven J, Stupp S (2014) 25th anniversary article: supramolecular materials for regenerative medicine. Adv Mater 26:1642–1659

Brandt KD, Smith GN, Simon LS (2000) Intraarticular injection of hyaluronan as treatment for knee osteoarthritis: what is the evidence? Arthritis Rheum 43:1192–1203

Capito R, Azevedo H, Velichko Y, Mata A, Stupp S (2008) Self-assembly of large and small molecules into hierarchically ordered sacs and membranes. Science 319:1812–1816

Chow L, Armgarth A, St-Pierre J, Bertazzo S, Gentilini C, Aurisicchio C, McCullen S, Steele J, Stevens M (2014) Peptide-directed spatial organization of biomolecules in dynamic gradient scaffolds. Adv Healthc Mater 3:1381–1386

Eyre D (2002) Collagen of articular cartilage. Arthritis Res 4:30–35

Felson DT, Zhang Y, Hannan MT, Naimark A, Weissman B, Aliabadi P, Levy D (1997) Risk factors for incident radiographic knee osteoarthritis in the elderly: the Framingham study. Arthritis Rheum 40:728–733

Fortier L, Barker J, Strauss E, McCarrel T, Cole B (2011) The role of growth factors in cartilage repair. Clin Orthop Relat Res 469:2706–2715

Fussenegger M, Meinhart J, Hobling W, Kullich W, Funk S, Bernatzky G (2003) Stabilized autologous fibrin-chondrocyte constructs for cartilage repair in vivo. Ann Plast Surg 51:493–498

Gaissmaier C, Koh JL, Weise K (2008) Growth and differentiation factors for cartilage healing and repair. Injury 39:88–96

Gao J, Knaack D, Goldberg V, Caplan A (2004) Osteochondral defect repair by demineralized cortical bone matrix. Clin Orthop Relat Res 427:S62–S66

Garcia-Giralt N, Izquierdo R, Nogues X, Perez-Olmedilla M, Benito P, Gomez-Ribelles J, Checa M, Suay J, Caceres E, Monllau J (2008) A porous PCL scaffold promotes the human chondrocytes redifferentiation and hyaline-specific extracellular matrix protein synthesis. J Biomed Mater Res A 85:1082–1089

Ge Z, Li C, Heng B, Cao G, Yang Z (2012) Functional biomaterials for cartilage regeneration. J Biomed Mater Res A 100:2526–2536

Gillogly SD, Voight M, Blackburn T (1998) Treatment of articular cartilage defects of the knee with autologous chondrocyte implantation. J Orthop Sports Phys Ther 28:241–251

Grande D, Halberstadt C, Naughton G, Schwartz R, Manji R (1997) Evaluation of matrix scaffolds for tissue engineering of articular cartilage grafts. J Biomed Mater Res 34:211–220

Grigolo B, Roseti L, Fiorini M, Fini M, Giavaresi G, Aldini N, Giardino R, Facchini A (2001) Transplantation of chondrocytes seeded on a hyaluronan derivative (Hyaff®-11) into cartilage defects in rabbits. Biomaterials 22:2417–2424

Grotle M, Hagen KB, Natvig B, Dahl FA, Kvien TK (2008) Obesity and osteoarthritis in knee, hip and/or hand: an epidemiological study in the general population with 10 years follow-up. BMC Musculoskelet Disord 9:132

Han L, Grodzinsky A, Ortiz C (2011) Nanomechanics of the cartilage extracellular matrix. Ann Rev Mater Res 41:133–168

Heinegård D, Saxne T (2011) The role of the cartilage matrix in osteoarthritis. Nat Rev Rheumatol 7:50–56

Huang H, Zhang X, Hu X, Shao Z, Zhu J, Dai L, Man Z, Yuan L, Chen H, Zhou C, Ao Y (2014) A functional biphasic biomaterial homing mesenchymal stem cells for in vivo cartilage regeneration. Biomaterials 35:9608–9619

Kisiday J, Jin M, Kurz B, Hung H, Semino C, Zhang S, Grodzinsky A (2002) Self-assembling peptide hydrogel fosters chondrocyte extracellular matrix produc-

tion and cell division: implications for cartilage tissue repair. Proc Nat Acad Sci USA 99:9996–10001
Kock NB, Smolders JM, Van Susante JL, Buma P, Van Kampen A, Verdonschot N (2008) A cadaveric analysis of contact stress restoration after osteochondral transplantation of a cylindrical cartilage defect. Knee Surg Sports Traumatol Arthrosc 16:461–468
Kopesky P, Vanderploeg E, Sandy J, Kurz B, Grodzinsky A (2010) Self-assembling peptide hydrogels modulate *in vitro* chondrogenesis of bovine bone marrow stromal cells. Tissue Eng Part A 16:465–477
Kopesky PW, Byun S, Vanderploeg EJ, Kisiday JD, Frisbie DD, Grodzinsky AJ (2014) Sustained delivery of bioactive TGF-β1 from self-assembling peptide hydrogels induces chondrogenesis of encapsulated bone marrow stromal cells. J Biomed Mater Res A 102:1275–1285
Kreuz PC, Steinwachs MR, Erggelet C, Krause SJ, Konrad G, Uhl M, Südkamp N (2006) Results after microfracture of full-thickness chondral defects in different compartments in the knee. Osteoarthrithis Cartilage 14:1119–1125
Lai J, Kajiyama G, Smith R, Maloney W, Yang F (2013) Stem cells catalyze cartilage formation by neonatal articular chondrocytes in 3D biomimetic hydrogels. Sci Rep 3:3553
Lee W, Lim C, Shi H, Tang L, Wang Y, Lim C, Loh K (2011) Origin of enhanced stem cell growth and differentiation on graphene and graphene oxide. ACS Nano 5:7334–7341
Lemare F, Steimberg N, Le Griel C, Demignot S, Adolphe M (1998) Dedifferentiated chondrocytes cultured in alginate beads: restoration of the differentiated phenotype and of the metabolic responses to interleukin-1β. J Cell Physiol 176:303–313
Li X, Jin L, Balian G, Laurencin C, Anderson D (2006) Demineralized bone matrix gelatin as scaffold for osteochondral tissue engineering. Biomaterials 27:2426–2433
Li WJ, Chiang H, Kuo TF, Lee HS, Jiang CC, Tuan RS (2009) Evaluation of articular cartilage repair using biodegradable nanofibrous scaffolds in a swine model: a pilot study. J Tissue Eng Regen Med 3:1–10
Liao J, Qu Y, Chu B, Zhang X, Qian Z (2015) Biodegradable CSMA/PECA/graphene porous hybrid scaffold for cartilage tissue engineering. Sci Rep 5:9879
Lin Y, Luo E, Chen X, Liu L, Qiao J, Yan Z, Li Z, Tang W, Zheng X, Tian W (2005) Molecular and cellular characterization during chondrogenic differentiation of adipose tissue-derived stromal cells in vitro and cartilage formation in vivo. J Cell Mol Med 9:929–939
Liu J, Nie H, Xu Z, Guo F, Guo S, Yin J, Wang Y, Zhang C (2015) Construction of PRP-containing nanofibrous scaffolds for controlled release and their application to cartilage regeneration. J Mater Chem B 3:581–591
Mehlhorn A, Zwingmann J, Finkenzeller G, Niemeyer P, Dauner M, Stark B, Sudkamp N, Schmal H (2009) Chondrogenesis of adipose-derived adult stem cells in a poly-lactide-co-glycolide scaffold. Tissue Eng Part A 15:1159–1167
Meng Q, Man Z, Dai L, Huang H, Zhang X, Hu X, Shao Z, Zhu J, Zhang J, Fu X, Duan X, Ao Y (2015) A composite scaffold of MSC affinity peptide-modified demineralized bone matrix particles and chitosan hydrogel for cartilage regeneration. Sci Rep 5:17802
Moreland LW (2003) Intra-articular hyaluronan (hyaluronic acid) and hylans for the treatment of osteoarthritis: mechanisms of action. Arthritis Res Ther 5:54–67
Murphy C, Sambanis A (2001) Effect of oxygen tension and alginate encapsulation on restoration of the differentiated phenotype of passaged chondrocytes. Tissue Eng 7:791–803
Nehrer S, Breinan H, Ramappa A, Shortkroff S, Young G, Minas T, Sledge C, Yannas I, Spector M (1997) Canine chondrocytes seeded in type I and type II collagen implants investigated *in vitro*. J Biomed Mater Res 38:95–104
Nguyen L, Kudva A, Saxena N, Roy K (2011) Engineering articular cartilage with spatially-varying matrix composition and mechanical properties from a single stem cell population using a multi-layered hydrogel. Biomaterials 32:6946–6952
Parmar P, Chow L, St-Pierre J, Horejs C, Peng Y, Werkmeister J, Ramshaw J, Stevens M (2015) Collagen-mimetic peptide-modifiable hydrogels for articular cartilage regeneration. Biomaterials 54:213–225
Pearle AD, Warren RF, Rodeo SA (2005) Basic science of articular cartilage and osteoarthritis. Clin Sports Med 24:1–12
Prakash D, Learmonth D (2002) Natural progression of osteo-chondral defect in the femoral condyle. Knee 9:7–10
Shah RN, Shah NA, Del Rosario Lim MM, Hsieh C, Nuber G, Stupp SI (2010) Supramolecular design of self-assembling nanofibers for cartilage regeneration. Proc Natl Acad Sci U S A 107:3293–3298
Slomianka L (2009) Blue histology – skeletal tissues – cartilage. www.lab.anhb.uwa.edu.au/mb140/CorePages/Cartilage/Cartil.htm
Taipaleenmaki H (2010) Factors regulating chondrogenic differentiation. University of Turku, Finland
Tombuloglu A, Tekinay AB, Guler MO (2012) Materials for articular cartilage regeneration. Recent Pat Biomed Eng 5:187–199
Ustun Yaylaci S, Sardan Ekiz M, Arslan E, Can N, Kilic E, Ozkan H, Orujalipoor I, Ide S, Tekinay AB, Guler MO (2016) Supramolecular GAG-like self-assembled glycopeptide nanofibers induce chondrogenesis and cartilage regeneration. Biomacromolecules 17:679–689

Peptides and Drug Delivery

8

Kavisha R. Ulapane, Brian M. Kopec,
Mario E.G. Moral, and Teruna J. Siahaan

Abstract

Peptides have been used as drugs to treat various health conditions, and they are also being developed as diagnostic agents. Due to their receptor selectivity, peptides have recently been utilized for drug delivery to target drug molecules to specific types of cells (*i.e.* cancer cells, immune cells) to lower the side effects of the drugs. In this case, the drug is conjugated to the carrier peptide for directing the drug to the target cells (*e.g.* cancer cells) with higher expression of a specific receptor that recognizes the carrier peptide. As a result, the drug is directed to the target diseased cells without affecting the normal cells. Peptides are also being developed for improving drug delivery through the intestinal mucosa barrier (IMB) and the blood-brain barrier (BBB). These peptides were derived from intercellular junction proteins such as occludins, claudins, and cadherins and improve drug delivery through the IMB and BBB via the paracellular pathways. It is hypothesized that the peptides modulate protein-protein interactions in the intercellular junctions of the IMB and BBB to increase the porosity of paracellular pathways of the barriers. These modulator peptides have been shown to enhance brain delivery of small molecules and medium-sized peptides as well as a large protein such as 65 kDa albumin. In the future, this method has the potential to improve oral and brain delivery of therapeutic and diagnostic peptides and proteins.

Keywords

Targeted drug delivery • Peptide-drug conjugate • Peptide-particle conjugate • Blood-brain barrier • Intestinal mucosa barrier • Modulator of intercellular junctions • Brain delivery

K.R. Ulapane
Department of Chemistry, The University of Kansas,
2095 Constant Avenue, Lawrence, KS 66047, USA

B.M. Kopec • M.E.G. Moral
Department of Pharmaceutical Chemistry, The University of Kansas,
2095 Constant Avenue, Lawrence, KS 66047, USA

T.J. Siahaan (✉)
Department of Chemistry, The University of Kansas,
2095 Constant Avenue, Lawrence, KS 66047, USA

Department of Pharmaceutical Chemistry, The University of Kansas,
2095 Constant Avenue, Lawrence, KS 66047, USA
e-mail: siahaan@ku.edu

© Springer International Publishing AG 2017

A. Sunna et al. (eds.), *Peptides and Peptide-based Biomaterials and their Biomedical Applications*,
Advances in Experimental Medicine and Biology 1030, https://doi.org/10.1007/978-3-319-66095-0_8

Abbreviations

AJ adherens junction
ANG angiopep
APaseP aminopeptidase P
APP amyloid precursor protein
BBB blood brain barrier
BBMEC bovine brain microvessel endothelial cells
CAD *cis*–asconitic anhydride–DOX
C–CPE *clostridium perfringens enterotoxin*
CNS central nervous systems
CPP cell penetrating peptide
CPT camptothecin
DOX doxorubicin
DSDS drug self-delivery system
DSPE-PEG 1,2-distearoyl-sn-glycero-3-phosphoethanolamine-N-methoxy-(polyethylene glycol)
EC-1 extracellular domain-1
ECM extracellular matrix
EGF epidermal growth factor
EGRF epidermal growth factor receptor
EGTA ethyleneglycol-bis-(β-aminoethyl ether)-N, N′-tetraacetic acid
ERK extracellular signal-regulated kinases
FACS fluorescence-activated cell sorting
FDA Food and Drug Administration
FITC fluorescein isothiocyanate
GABA γ-aminobutyric acid
Gd-DTPA gadopentetic acid
GIP glucose-dependent insulinotropic peptide
GLP-1 glucagon-like peptide-1
GMB glioblastoma multiforme
GnRH gonadotrophin-releasing hormone
HEK human embryonic kidney cells
HIV-1 human immunodeficiency virus 1
HPMA (2-hydroxypropyl) methacrylamide
ICAM-1 intercellular adhesion molecule-1
IMB intestinal mucosa barrier
JAM junction adhesion molecules
LFA-1 lymphocyte function-associated antigen-1
LHRH luteinizing hormone-releasing hormones
LRP-1 lipoprotein-related protein 1
MDCK Madin-Darby Canine Kidney Epithelial Cells
MDR1 multi drug resistance-1
MEK mitogen-activated protein kinase
MMAF monomethyl auristatin F
MMPs matrix metalloproteinases
MRI magnetic resonance imaging
NB nanobubbles
NLS nuclear localization signal
NMR nuclear magnetic resonance
NRP-1 neuropilin-1
OCT octreotide
ODN octadecanol
OVCAR-3 ovarian cancer cell-3
PAMAM-PAsp-PEG poly(amidoamine)-b-poly(aspartic acid)-b-poly(ethylene glycol)
pAntp antennapedia homeodomain protein of Drosophila
PEG polyethylene glycol
PET positron emission tomography
pHLIP pH low insertion peptide
PI3K phosphoinositide 3-kinase
PTX paclitaxel
RES reticuloendothelial system
RGD Arg-Gly-Asp
SA stearylamine
SCCHN squamous cell carcinoma of head and neck
SI secretase inhibitor
siRNA small interfering RNA
STTR somatostatin receptors
TAMRA tetramethylrhodamine
TAT *trans*-activating transcriptional activator
TEER *trans*-epithelial electrical resistance
TfR transferrin receptor
TJ tight junction
TPP triphenylphosphonium
ZO *zonula occludin*
Zot *zonula occludens toxin*

8.1 Introduction

Peptides have been successfully developed as therapeutic and diagnostic agents because of their selectivity to bind the respective target receptors (Kaspar and Reichert 2013; Uhlig et al. 2014; Fosgerau and Hoffmann 2015). Currently, there are more than 60 peptide drugs approved by the US Food and Drug Administration (FDA). Thus, development of peptide drugs has increased significantly in the past decades and, as of today, approximately 140 peptides are in clinical trials as potential drugs. In addition, more than 500 therapeutic peptides are in preclinical development (Fosgerau and Hoffmann 2015). Available peptide drugs which include oxytocin, calcitonin, octreotide, and exenatide are being used to treat various conditions. Some bioactive peptides have been derived from endogenous substances; however, some peptides were derived from truncation of the active region(s) of the parent proteins. For example, opioid peptides such as enkephalins, endorphins, and dynorphins that are found in the brain have been used as drugs. Oxytocin, an endogenous hormone released by the posterior pituitary, is a cyclic peptide synthesized in the paraventricular nucleus of the hypothalamus (Bell et al. 2014). Oxytocin and its analogs work as neurotransmitters in the brain to facilitate breastfeeding, induce labor, and treat postpartum hemorrhage. Calcitonin peptide is a hormone produced by the thyroid gland to control calcium and potassium levels in the blood. A synthetic salmon-calcitonin peptide has been used to treat osteoporosis; this peptide was first developed as a nasal spray (Tella and Gallagher 2014). Tumors can be treated by octreotide, which inhibits the release of growth hormones (Broder et al. 2015). Type-2 diabetes is treated successfully with exenatide (Table 8.1), which is a derivative of a glucagon-like peptide-1 agonist (GLP-1; Table 8.1) (Knop et al. 2017). Exenatide was developed to increase the *in vivo* half-life because GLP-1 was ineffective in clinical trials for diabetes treatment due to its short half-life. Exenatide binds to GLP-1 receptor and regulates glucose metabolism and insulin secretion. Both GLP-1 and glucose-dependent insulinotropic peptide (GIP) hormones are produced upon ingestion of food to stimulate insulin secretion; however, only GLP-1 causes insulin secretion in a diabetic state.

Bioactive peptides can also be derived from the active region(s) of large functional proteins. One example is "Arg-Gly-Asp" (RGD), which is derived from the sequence of extracellular matrix (ECM) proteins such as fibronectin, fibrinogen, vitronectin, von Willebrand factor, laminin, and collagen (Ruoslahti and Pierschbacher 1987). RGD sequence on the ECM is recognized by various integrin receptors on the cell surface for cell adhesion to ECM. The ECM-integrin binding is essential in various disease processes such as thrombosis, angiogenesis and tumor metastasis (Ruoslahti 1994). In thrombosis, the process of vascular blood clotting prevents the normal blood flow from the heart, which involves platelet aggregation. Platelet aggregation results from interactions of fibrinogen and platelets, which are mediated by recognition of RGD sequences on the α- and γ-subunits of fibrinogen by gpIIb/IIIa integrin receptors on platelet surfaces (Dunehoo et al. 2006). Therefore, RGD peptides (*e.g.* integrilin or eptifibatide) and peptidomimetics (*e.g.* aggrastat or tirofiban) have been used as antithrombic agents in the clinic. Integrilin and aggrastat are selective and potent ligands for gpIIb/IIIa receptors, blocking platelet aggregation during thrombosis. Angiogenesis in solid tumors can be inhibited by RGD peptides (Table 8.1), which are designed to bind to cell-surface $\alpha_V\beta_3$ and $\alpha_V\beta_5$ integrins that are overexpressed during tumor angiogenesis (Mas-Moruno et al. 2010). Other cell adhesion peptides, such as those derived from intercellular adhesion molecule-1 (ICAM-1) and lymphocyte function-associated antigen-1 (LFA-1) receptors, have been shown to inhibit T-cell adhesion by adhering to epithelial and endothelial cells during inflammation (Yusuf-Makagiansar et al. 2002).

In addition to their use as drugs, bioactive peptides have been used as targeting moieties for the delivery of drug payloads (*e.g.* anticancer and anti-inflammatory) to specific types of cells in tissues, and ultimately, to reduce their adverse side effects. Some of these peptides are internalized by their respective receptors into cells via a

Table 8.1 Peptide names and sequences

Peptide	Sequence
Exenatide	HGEGTFTSDLSKQMEEEAVRLFIEWLKNGGPSSGAPPPS-NH_2
GLP-1	HDEFERHAEGTFTSDVSSYLEGQAAKEFIAWLVKGR - NH_2
ALOS4	Cyclo1,9(CSSAGSLFC)
RGD-1	Cyclo(RGDyK)
Transportan	GWTLNSAGYLLGKINLKALAALAKKIL
Penetratin	RQIKIWFQNRRMKWKK
PAF-26	RKKWFW
Octreotide	Cyclo2,7(fCFwKTCT)
GnRH-1	EHWSYkLRPG-NH_2
GnRH-2	EHWSHkWYPG-NH_2
GnRH-3	EHWSHDWKPG-NH_2
pHLIP	AAEQNPIYWARYADWLFTTPLLLLDLALLVDADEGTCG
MMP-hexapeptide	PVGLIG
ANG	TFFYGGSRGKRNNFKTEEY
ANG-SI	TFFYGGSRGKRNNFK–EVN-sta-VAEF
ANG-PEG-S	TFFYGGSRGKRNNFK–PEG–EVN-sta-VAEF
ANG-TAT	TFFYGGSRGKRNNFK–TEEYGRKKRRQRRRPPQQ
Biotin-ANG	TFFYGGSRGKRNNF([Biotin)TEEY
Biotin-ANG-TAT	TFFYGGSRGKRNNFK(Biotin)TEEYGRKKRRQRRRPPQQ
T10	HAIYPRH
ERK	MPKKKPTPIQLNP
T10-ERK	HAIYPRH–GGCG–MPKKKPTPIQLNP
PEGA	Cyclo1,10(CPGEPEGAGC)
A54	AGKGTPSLETTP
G3-C12	ANTPCGPYTHDCPVKR
KLA	D(KLAKLAK)$_2$
RGD-2	Cyclo(RGDfC)
Pep-1	Cyclo1,9(CGEMGWVRC)
EGRF	YHWYGYTPQNVI
CPP	CKRRMKWKK
YSA	YSAYPDSVPMMS
RGD-3	CRGDK
OCC2	GVNPQAQMSSGYYYSPLLAMC(Acm)SQAYGSTYLNQYIYHYC(Acm)TVDPQE; Acm = Acetamido methyl
OP_{90-135}	DRGYGTSLLGGSVGYPYGGSGFGSYGSGYGYGYGYGYGYGGYTDPR-NH_2
OP_{90-103}	DRGYGTSLLGGSVG
Lip-OP_{90-103}	Lipid-C12-DRGYGTSLLGGSVG
C1C2	SSVSQSTGQIQSKVFDSLLNLNSTLQATR-NH_2
HAV6	Ac-SHAVSS-NH_2
ADTC5	Cyclo1,7Ac-(CDTPPVC)NH_2
cIBR7	Cyclo1,8(CPRGGSVC)
cLABL	Cyclo1,12(PenITDGEATDSGC)
C-CPE	SSYSGNYPYSILFQKF
AT-1002	FCIGRL
PN-78	FDFWITP
PN-159	KLALKLALKALKLAALKLA-NH_2

receptor-mediated endocytosis process. Others have been conjugated to drug-loaded nanoparticles for specific delivery to corresponding cell targets of the peptide. Receptor selective peptides were also investigated as diagnostic agents by conjugating them to dyes, radioisotopes, or magnetic resonance imaging (MRI) contrast agents.

The ultimate goal of targeted drug delivery is to direct a drug to diseased cells or organs (*e.g.* cancer cells), while avoiding normal cells. This results in a drug construct with lower side effects than the free form of its parent drug. This chapter describes the roles of peptides in drug delivery, including the use of peptides as: (a) peptide-drug and peptide-particle conjugates for targeting molecules to a specific type of cells and (b) modulators of biological barriers for improving the oral and brain delivery of drugs and diagnostic agents.

8.2 Peptide-Drug Conjugates for Targeted Drug Delivery

Drug targeting methods are normally explored to reduce side effects by directing toxic drugs to cells involved in disease states, leaving normal cells minimally affected. Conjugation of the drug to its peptide carrier (targeting agent) can be done directly or through a chemical linker. Thus, as a peptide carrier selectively binds to a specific receptor on the surface of targeted cells (*i.e.* cancer cells), it carries along the drug or diagnostic molecule with it. As an example, cancer cells have a certain upregulated receptor(s) (*e.g.* HER-2, EGFR) compared to normal cells. These upregulated receptors become distinguishing and exploitable features to selectively direct populations of conjugated drug molecules to accumulate in cancer cells over normal cells. Binding of the conjugate to the target receptor is followed by cellular uptake of the ligand-receptor complex into the early endosomes via receptor-mediated endocytosis (Majumdar and Siahaan 2012). From the endosomes, the conjugate reaches the lysosomes where the drug is released as a result of lowered pH and/or enzyme degradation in the lysosomes. The rate of release of the drug can be controlled by designing the appropriate linker between the drug and the peptide (Hamann et al. 2002; Majumdar and Siahaan 2012; Buyuktimkin et al. 2016). This method has been successfully applied in antibody-drug conjugates such as Adcetris® and Kadcycla® to treat cancer patients (Leal et al. 2014; Buyuktimkin et al. 2016). Peptides are smaller than proteins, and can be rapidly synthesized using solid-phase methods. Unlike proteins, most peptides have only primary and secondary structures. Thus, most peptides do not suffer as much from physical instability as do proteins, which often leads to the formation of aggregates that generate immunogenicity. Due to their small size, peptide-drug conjugates are potentially less immunogenic than protein-drug conjugates. Although their formulation remains challenging, peptide conjugate formulation is usually less complicated than that of protein-drug conjugates.

In designing a peptide-drug conjugate, a functional group (*i.e.* amine, carboxylic acid, alcohol, or thiol) within the structure of the drug can be used to link the drug directly to a targeting peptide's N- or C-termini, or a side-chain functional group of the Lys, Asp, Glu, Ser, Thr, or Cys residue(s) on the peptide. This direct drug-peptide linkage may be in the form of an amide, ester, or thioether bond. In many cases, the drug may be conjugated to the peptide using a molecular linker (*e.g.* PEG, maleimido, etc.). Other than bridging drug and targeting components of the construct, the vital role of the linker is to provide a distance between the peptide and the drug, which is often crucial to the overall activity and potency of the construct. In one respect, ample distance between the drug and targeting-peptide components prevents interference (or steric hindrance) in binding to their respective receptors. This is because both drug and peptide have a molecular surface recognized by their respective receptors for biological activity. Normally, the conjugation is done at the functional group away from the bioactive region of the peptide or the drug. In addition, the linker can be designed to control the release of the drug from the conjugate upon reaching the cell targets within the tissue. Premature release of the drug in the systemic or lymphatic circulation before reaching the respective target cells can be harmful, and ultimately, defeats the purpose of drug-targeting.

8.2.1 Peptide-Drug Conjugates for Cancer Therapy and Diagnostics

Chemotherapy remains the treatment of choice for cancer patients (Song et al. 2015). Most chemotherapeutic agents are cytotoxic and kill not only cancer cells but also normal cells in the body. Some drugs have poor solubility, are highly toxic, or cannot cross the cancer cell membranes into the intracellular space. Most cancer cells eventually generate resistance to anticancer drugs after multiple treatments. One of the drug-resistance mechanisms is due to the overexpression of efflux pumps *(i.e.* Pgp, MRP, MDR1) that expel the anticancer drug from the cancer cell membranes. Therefore, there is a need to develop an alternative method to deliver drugs into cancer cells and overcome drug resistance by avoiding the efflux pumps. One way to increase drug penetration across cell membranes is by utilizing receptor-mediated endocytosis mechanisms. Thus, drugs have been conjugated to peptides, proteins, nanocarriers, carbon nanotubes, and dendrimers. As an example, paclitaxel (PTX), which is widely used to treat breast, ovarian, testicular, and cervical cancers, is known to have poor water solubility (Majumdar et al. 2016). Thus, conjugation to a targeting peptide via a polyethylene glycol (PEG) linker increases solubility and improves selectivity to target cancer cells.

8.2.1.1 Cell Adhesion Peptides

Cell adhesion peptides have been used to target drugs and radioisotopes for cancer treatments and diagnostics. RGD peptides are cell adhesion molecules that have been extensively explored as carriers in peptide-drug conjugates. Certain cyclic RGD peptides bind selectively to $\alpha_V\beta_3$ and $\alpha_V\beta_5$ integrin receptors, which are upregulated during angiogenesis in tumors. Thus, these selective cyclic RGD peptides have been used to deliver radioisotopes such as ^{18}F, ^{99m}Tc, ^{125}I, or ^{64}Cu as cancer diagnostic agents. For example, ^{18}F-containing galactose was incorporated to RGD-1 peptide (Table 8.1) via an amide bond to the D-Lys side chain (Haubner et al. 2005). The ^{18}F-labeled conjugate binds selectively to upregulated $\alpha_V\beta_3$ integrin receptors on tumor vasculature as observed by positron emission tomography (PET). Thus, the conjugate can be used as an imaging tool to detect angiogenesis and tumor metastasis *in vivo* (Haubner et al. 2005). The levels of $\alpha_V\beta_3$ receptors on human tumor cells have been detected with ^{18}F-labeled conjugates and observed using PET. It was found that the detected levels of $\alpha_V\beta_3$ receptors were similar to those detected using immunohistochemistry (Beer et al. 2006). Accumulation of ^{18}F-labeled RGD-1 peptide in human tumors showed intra- and inter-variability of conjugate accumulation due to different levels of $\alpha_V\beta_3$ across different individuals (Haubner et al. 2005). The various levels of $\alpha_V\beta_3$ found in humans can be used to predict populations of cancer patients most likely to respond to treatments with RGD-anticancer drug conjugates. RGD peptides have been used to selectively deliver the anticancer drugs doxorubicin (DOX) (Arap et al. 1998) and PTX (Chen et al. 2005) to cancer cells *in vitro* and *in vivo*. RGD-1-DOX conjugate suppressed the growth of breast cancer xenographs in mice better than DOX alone, suggesting that cyclic RGD peptide improves the targeting of DOX to breast cancer cells *in vivo* (Arap et al. 1998).

ALOS4 peptide (Table 8.1) is a non-RGD peptide that also binds to $\alpha_V\beta_3$ integrin. The peptide was linked to camptothecin (CPT) and fluorescein isothiocyanate (FITC) via a GABA linker to give ALOS4-CPT and ALOS4-FITC, respectively (Redko et al. 2016). FACS analysis showed a strong binding of ALSO4-FITC to WM-266-4, a malignant melanoma cell line. *In vivo* studies confirmed that tumor-bearing WM-266-4 cells in mice intravenously administered with ALOS4-FITC showed accumulation of ALOS4-FITC specifically in tumors rather than organs as observed after 24 hours. Next, ALOS4-CPT enhanced drug cytotoxicity to tumor cells better than CPT and other anticancer drugs. CPT activity has been known to be deactivated by the hydrolytic opening of a vital lactone ring in its structure. In contrast to its free form, the stability of this lactone ring is increased in the ALOS4-CPT conjugate (Redko et al. 2016). At 10 μM,

CPT alone kills high percentages of both malignant WM-266-4 and non-malignant HEK-293 tumor cells (human embryonic kidney cells) while ALOS4-CPT kills 70% of malignant WM-266-4 cells compared to 30% of the non-malignant HEK-293 cells. The activity of the conjugate is dose-dependent (Redko et al. 2016). These results affirm that ALOS4 selectively targets and delivers conjugates to malignant tumor cells.

8.2.1.2 Cell-Penetrating Peptides

Generally, most peptides cannot readily cross the cell membranes due to their physicochemical properties; however, cell-penetrating peptides (CPPs with 6–30 amino acids) are capable of crossing membranes and entering the cytoplasm of the cells. A detailed mechanism of the cellular uptake of CPPs is not well understood, but it may take place either by direct translocation or by endocytosis. The early CPPs identified were *trans*-activating transcriptional activator (TAT) peptides, derived from human immunodeficiency virus 1 (HIV-1), and antennapedia homeodomain protein of drosophila (pAntp). These long sequences have been reduced to 6–7 amino acid peptides, which maintain similar cell penetrating behavior. CPPs have been used for cellular delivery of small drug molecules (*i.e.* doxorubicin, methotrexate and taxol), proteins, nucleic acids, and contrasting agents. Apart from the natural CPPs, synthetic and semi-synthetic CPPs, including the chimeric 27-amino acid transportan, penetratin, and PAF-26, were designed to facilitate drug delivery (Table 8.1).

8.2.1.3 Peptide Hormone for Drug Delivery

Peptide hormones such as octreotide (OCT), gonadotrophin-releasing hormone (GnRH), and epidermal growth factor (EGF) peptides (Table 8.1) have been investigated for delivering drugs to cancer cells. GnRH receptors are overexpressed in malignant tumor of ovarian, breast, prostate cancers as part of the paracrine/autocrine regulatory system of malignant tumors (Bajusz et al. 1989; Limonta et al. 2003; Muranyi et al. 2016). OCT and other somatostatin peptides bind to somatostatin receptors (STTRs) especially STTR2, which are upregulated in breast, cervical, colon, lung, ovarian cancers cells (Hejna et al. 2002). OCT peptide has a long half-life in systemic circulation with good tissue penetration due to its uptake by the STTR2 receptor. OCT has also been used for targeting radiotherapies (Muranyi et al. 2016). PTX-OCT conjugate was designed to improve the biological properties of PTX and overcome the issue of cancer resistance. Ovarian cancer is treated with PTX, but normally through multiple sessions/doses, often causing the emergence of drug resistance (Chen et al. 2016). Localization of OCT peptide after delivery has been monitored using FITC-labeled OCT peptide (FITC-OCT), injected into nude mice bearing a xenografted tumor. Localization of FITC-OCT on the xenografted tumor confirmed abnormally high levels of STTR2 receptor expression in tumors. PTX-OCT also suppressed tumor growth in mice xenografts better than in those treated with free PTX, OCT, and mixtures of PTX + OCT (Chen et al. 2016). This result demonstrates the selectivity of PTX-OCT to tumor cells on the basis of high expression of STTR2. In addition, the conjugate downregulates the expression of multi drug resistance-1 (MDR1) protein.

GnRH or LHRH peptides effectively deliver anticancer drugs such as DOX and CPT to cancer cells (Nagy et al. 1996; Dharap 2003). FITC-labeled GnRH analogues have been used to compare targeting efficiencies of GnRH-1, GnRH-2, and GnRH-3 peptides (Table 8.1) in human breast, colon, pancreas, and prostate cancer cells to that in the non-tumor cell line such as Madin-Darby canine kidney epithelial (MDCK) cells. This study also revealed that human pharynx tumor cells similarly overexpress GnRH receptors on cell surfaces such as human breast, colon and prostate cancer cell lines. In contrast, pancreatic tumor cells (BxPC-3) do not present GnRH-1 receptors on their membranes (Muranyi et al. 2016). As expected, GnRH peptides are internalized by tumor cells via active transport mechanisms. Although different cancer cell lines vary in their uptake properties for the three different GnRH peptides, uptake by all tumor cells was significantly higher than in the control MDCK cell-line, thus indicating the role of GnRH receptor upregulation in tumor cells.

8.2.1.4 pH Low Insertion Peptide (pHLIP)

A pH low insertion peptide (pHLIP; Table 8.1) was developed as a pH-dependent cell-penetrating peptide for drug delivery (Burns et al. 2017). pHLIP is a water-soluble membrane peptide that interacts weakly with cell membranes at neutral pH; however, when the cell surface is slightly acidic, the pHLIP peptide is inserted into the cell membranes as a stable transmembrane α-helix. Its primary sequence is characterized by acidic residues (*i.e.* Asp or Glu) that can be protonated at the low extracellular pH observed in tumors. In testing the concept, six pHLIP derivatives were conjugated to monomethyl auristatin F (MMAF) to make pHLIP-MMAF conjugates and the MMAF is attached to the pHLIP C-terminus via a S-S bond that can be cleaved in the cytoplasm (Burns et al. 2017). The efficacy of six pHLIP-MMAF conjugates was evaluated *in vitro* against cultured cancer cells to find the lead conjugate. *In vivo*, the lead conjugate showed significant therapeutic efficacy in mouse models without overt toxicities. pHLIP-MMAF was localized in cancer cells and inhibited the proliferation of cancer cells in a pH-selective and concentration-dependent manner.

8.2.1.5 MMP Peptides

Tumor cells have a high expression of MMPs (MMP-2 and MMP-9) that are important in tumor proliferation and metastasis (Paez Pereda et al. 2000). One MMP-hexapeptide, PVGLIG, has a high binding affinity to matrix metalloprotease-2 (MMP-2) enzyme. Conjugated to PTX at the C-terminus of the MMP peptide via an ester bond, PTX-MMP was found to deliver PTX in a tumor-specific manner (Huang et al. 2016). Incubation of PTX-MMP with MMP2, as well as with cancer cells (*i.e.* HT-1080 and U87MG), releases PTX from the conjugate. PTX release was higher in HT-1080 and U87MG cells compared to negative control cells (*i.e.* Hep-2 and Hep G2), suggesting the involvement of MMPs in both cancer cells (Huang et al. 2016). PTX-MMP shows significantly higher cytotoxicity in HT-1080 and U87MG cells compared to PTX alone, with no difference in toxicity between PTX-MMP and PTX on Hep-2 and Hep G2 cells, which have low expression of MMPs. Mice implanted with HT-1080 or U87MG cells have a higher survival rate when treated with PTX-MMP compared to those treated with PBS, PTX, and the MMP hexapeptide (Huang et al. 2016). These results support a role for the peptide and MMP-2 in the activity of the PTX-MMP conjugate against tumor cells.

8.2.1.6 A Combination of Peptides

A combination of two peptides has been used to target drugs to certain cells. Angiopep (ANG) peptide has been used alone or in combination with other peptides (*e.g.,* TAT peptide) to deliver drugs to neuronal cells. ANG peptide (Table 8.1) was derived from the ligand of a low-density lipoprotein-related protein 1 (LRP1) receptor that is involved in the uptake and processing of amyloid precursor protein (APP) in the intracellular compartment inside endosomal vesicles (Kim et al. 2016). LRP-1 has been shown to mediate transport of various ligands across the BBB (Li et al. 2016). To prove the concept, ANG peptide alone was conjugated to β-secretase inhibitor (SI) (*i.e.* ANG-SI and ANG-PEG-SI; Table 8.1) for endosomal delivery of neuronal cells to inhibit the formation of amyloid-beta (Aβ) (Barve et al. 2016; Kim et al. 2016). Neuroblastoma cells internalize the ANG-SI conjugate better than SI peptide alone, suggesting that the uptake is through receptor-mediated endocytosis. Conjugation of ANG to an SI peptide alters the recognition of the ANG peptide by LRP1 receptors because the uptake of the ANG-SI conjugate is unaffected in the decrease of LRP1 receptors. This suggests the involvement of another receptor in the uptake of the conjugate (Kim et al. 2016).

A combination of ANG and TAT peptides was used to deliver PTX as a conjugate (ANG-TAT-PTX; Table 8.1) across the BBB, and this conjugate was developed to treat glioblastoma brain tumor (Li et al. 2016). ANG-TAT-PTX is expected to bind and be internalized by the LRP-1 receptor across the BBB. In previous studies, a conjugate of PTX with angiopep-2 and -3 (ANG1005) has been shown to cross the BBB and was investigated in clinical trials (Regina et al. 2008; Thomas et al. 2009). The cellular uptake of ANG-TAT by U87 glioblastoma cells was higher than that of ANG alone (Li et al. 2016). It is interesting to find that, although ANG-TAT and TAT-ANG were both internalized by U87 glioblastoma cells, only ANG-TAT crossed the BBB (Li et al. 2016). The brain delivery studies were done using Biotin-ANG-TAT (Table 8.1), which was detected in brain tumor tissue. Biotin-ANG-TAT has significantly higher deposition (1.8 times) than Biotin-ANG; in this case, the TAT peptide improved brain tumor uptake. ANG-TAT-PTX-treated mice with implanted U87 glioblastoma cells in the brain have better survival rate than diseased mice treated with ANG-PTX or PTX alone (Li et al. 2016). Therefore, TAT peptide is important in improving the conjugate brain delivery.

A combination of T10 and extracellular signal-regulated kinases (ERK) peptides (Table 8.1) was also used to deliver DOX molecule to breast cancer cells to overcome drug resistance (Sheng et al. 2016). T10 peptide binds to and can be internalized by transferrin receptor (Tfr), which is overexpressed in tumor cells. ERK peptide can prevent activation of ERK by inhibiting phosphorylation and its binding to mitogen-activated protein kinase (MEK). T10 peptide was conjugated to ERK peptide via a spacer (GGCG), and the thiol group on the Cys residue was linked to DOX to give T10-ERK-DOX conjugate. The DOX cellular uptake in MCF7/ADR cancer cells was increased when attached to the conjugate. The conjugate reversed the drug resistance by downregulating Pgp expression and inhibiting ERK phosphorylation. Although T10-DOX delivered DOX to MCF7/ADR cancer cells and suppressed MCF7/ADR xenograft in nude mice, T10-ERK-DOX had better efficacy than T10-DOX in suppressing growth of MCF7/ADR tumor xenografts in nude mice (Sheng et al. 2015). This indicates that a combination of two peptides with different mechanisms improves the outcome of tumor suppression activity.

The concept of dual peptide targeting was also applied to PEGA peptide (Table 8.1) that binds a membrane-bound proline-specific aminopeptidase P (APaseP). APaseP is expressed approximately 100-fold higher in vasculature and malignant lesions in breast cancer than in normal tissues (Cordova et al. 2016). Thus, PEGA-TAT-TAMRA conjugate was used to evaluate cellular delivery and localization of the peptide in breast cancer cells (Cordova et al. 2016). The conjugate was internalized by cancer cells *in vitro* and *in vivo* in tumor xenografts. Although conjugation of PEGA to TAT resulted in reduced selectivity for PEGA to APaseP, the overall results of uptake and localization of the dual peptide conjugate showed selective delivery to breast cancer tissue (Cordova et al. 2016). Thus, this dual peptide has the potential to delivery cytotoxic drugs to tumor cells *in vivo*.

8.3 Peptide-Particle Conjugates for Drug Delivery

Nanoparticles are another emerging technology to improve drug delivery to a specific type of cells. One potential advantage of nanoparticles is that they can be used to store the drug and deliver it in a controlled-release fashion. The drug release can be coupled to the different redox conditions between the extra- and intra-cellular environments of the cell because of elevated concentrations of reductive substances in tumor cells, which differentiate them from normal cells (McEligot et al. 2005; Jones 2010). Certain types of nanoparticles are generated due to self-assembly and micelle formation of the components in water because of their low critical micelle concentration. The micelles normally have high drug encapsulation efficiency. One example is PEGylated chitosan-based glycolipid, which can form a redox-responsive nanocarrier system called A54-PEG-CSO-ss-SA. The nanoparticles were studded with A54 peptide (Table 8.1) conjugated to a PEG moiety. The nanoparticles were loaded with DOX and directed to human hepatoma cells by A54 peptide (Liu et al. 2016). The PEG moiety also serves to increase the *in vivo* half-life of nanoparticles by avoiding uptake by the reticuloendothelial (RES) system. *In vitro* and *in vivo* studies of nanoparticles show that DOX can be released via reduction of the disulfide bond depending on the amount of reductive substances in the tumor cells (Liu et al. 2016).

Recently, some efforts have been shifted from targeting drugs to specific types of cells to targeting them to subcellular organelles (*e.g.* the nucleus or mitochondria). In this case, the drug delivery systems are decorated with ligands that are specific for subcellular compartments, including the nuclear localization signal (NLS) and the lipophilic triphenylphosphonium (TPP) cation. Both NLS and TPP can penetrate the nucleus because of their high affinity for nuclear pore complexes and can anchor to mitochondria via electrostatic interactions (Smith et al. 2003; Kang et al. 2010). Previously, drugs conjugated with NLS or TPP failed to reach the nucleus or mitochondria because the design of these conjugates was not favorable for entering cancer cells from the extracellular space (Jensen et al. 2003; Callahan and Kopecek 2006). To overcome this problem, a (N-(2-hydroxypropyl) methacrylamide (HPMA) polymeric delivery system was conjugated to G3-C12 peptide (Table 8.1), a galectin-3-targeting ligand. The ligand was used for cellular uptake by cancer cells as well as subcellular mitochondria inside the cells (Sun et al. 2017). An antibiotic KLA peptide (Table 8.1) was also conjugated to HMPA to give a G3-C12-HPMA-KLA delivery system. The *in vitro* studies showed increased receptor-mediated internalization into PC-3 cells with overexpressing galectin-3. Moreover, the specific binding between galectin-3 and the G3-C12 peptide directed HPMA-KLA conjugates to the mitochondria with enhanced cytotoxicity. An *in vivo* study revealed that the G3-C12 peptide significantly enhanced the tumor accumulation of the polymer conjugate, exhibiting the best therapeutic efficacy and an improved survival rate in animals (Sun et al. 2017).

Carboplatin has been used to treat ovarian cancer; however, the uptake of carboplatin by ovarian cancer cells becomes poor because of drug resistance upon multiple treatments of cancer cells. Thus, the poly(amidoamine)-*b*-poly(aspartic acid)-*b*-poly(ethylene glycol) (PAMAM-PAsp-PEG) system was designed to improve carboplatin delivery to ovarian cancer cells (OVCAR-3). The nanoparticles utilize RGD-2 peptide (Table 8.1) to direct them to OVCAR-3 cells that have overexpression of cell surface $\alpha_V\beta_3$ and $\alpha_V\beta_5$ integrin receptors. Carboplatin molecules were attached to the polymer via a coordination complex with two carboxylic acid on the poly-aspartic acid chains tethered to the polymer. The release of carboplatin was pH-dependent, and 88% of carboplatin was released from the polymer over 50 h at pH 5.5, while only 18% of carboplatin was released over 50 h at pH 7.4. To track the cellular uptake and localization of the polymer in OVCAR-3 cells, Cy5-dye was also connected to the particles via PEG linker (Wang et al. 2016). The results showed that the particles containing RGD-2 peptide were efficiently internalized by the cells compared to particles without RGD-2 peptide. The targeted particles have significantly higher toxicity to the cells than carboplatin alone. It was proposed that carboplatin was occurring in the lysosome due to pH change and protonation of the carboxylic acid of the Asp residues (Wang et al. 2016).

A new drug self-delivery system (DSDS) was designed as nanocarrier for delivering PTX; in this case, PTX was conjugated to octadecanol via a disulfide bond to produce PTX-ODN. The PTX-ODN can self-assemble to form nanoparticles. Then, Pep-1 (Table 8.1) recognized by overexpressed interleukin-13 receptor α2 (IL-13Rα2) was used to direct the particles to glioblastoma multiforme (GBM) and for crossing the BBB and blood-brain-tumor barrier (Jiang et al. 2017). In this case, the Pep-1-PEG-DSPE conjugate is used to incorporate Pep-1 on the DSDS. The PTX-loaded nanoparticles were engulfed by IL-13Rα2 receptor-mediated endocytosis into glioblastoma cells and disintegrated in the endosomes to release the PTX-ODN component (Jiang et al. 2017). The disulfide bond of PTX-ODN was reduced in the endosomes by glutathione to release PTX (Jiang et al. 2017). To follow the uptake and movement of the nanoparticles inside U87MG cells, the particles were labeled with coumarin-6 fluorophore. It was confirmed that the nanoparticles were internalized by U87MG cells in a receptor-mediated manner. *In vivo*, the nanoparticles can be detected in the U87MG glioma brain tumor grafted in nude mice (Jiang et al. 2017). Brain tumor mice treated with Pep-1-PTX-nanoparticles showed a higher survival population than those treated with vehicle, taxol, and PTX-octadecanol conjugate (Jiang et al. 2017).

Nanosize particles (PEG-EGFR-PTX) were constructed using branched PEG conjugated to PTX and an epidermal growth factor receptor (EGRF) peptide (Table 8.1). The nanoparticles were designed to improve PTX delivery to cancer cells overexpressing EGFR (Majumdar et al. 2016). The roles of PEG were to increase drug-water solubility and half-life of the particles. The abilities of PEG-EGFR-PTX, PTX-PEG, and PTX to inhibit cell growth were evaluated in squamous cell carcinoma of the head and neck (SCCHN) cell line. The results showed that the IC_{50}s of PEG-EGFR-PTX, PTX-PEG, and PTX were 21.74, 8.05, and 1.47 nM, respectively (Majumdar et al. 2016). The lower activity of PEG-EGFR-PTX compared to PTX alone may be due to the less efficient uptake of the PEG-EGFR-PTX particles rather than to passive diffusion of PTX. Unfortunately, the toxicities of PEG-EGFR-PTX particles and PTX were not compared between EGFR overexpressing cancer cells and normal cells to prove particle targeting by EGFR peptide. Therefore, it is difficult to evaluate the usefulness of the particles in treating tumors *in vivo*.

A conjugate of peptide in nanobubbles (NBs) was designed to deliver small interfering RNA (siRNA) molecules, which have high specificity for the oncogenic mRNA in cancer cells. siRNA molecules are known to have unfavorable physicochemical properties (*e.g.,* size and anionic charges) for partitioning and crossing the cellular membranes to enter the intracellular space and exert their activity. To overcome this problem, Myc siRNA was conjugated to CPP (Table 8.1) to give CPP-Myc siRNA, which is encapsulated in ultrasound sensitive NBs. Ephrin peptide (YSA peptide, Table 8.1) was attached to the surface of NBs using 1,2-distearoyl-sn-glycero-3-phosphoethanolamine-N-methoxy-(polyethylene glycol) (DSPE-PEG) to give CPP-siRNA/YSA-NB (Xie et al. 2015). YSA peptide selectively binds to overexpressed EphA 2 protein on the cell surface (Xie et al. 2015). After cellular uptake, the CPP-Myc siRNA was released from NBs upon exposure to ultrasound that induced apoptosis of MCF-7 breast cancer cells *in vitro*. CPP-Myc siRNA/YSA-NB administered in conjunction with ultrasound significantly suppressed tumor growth of MCF-7 xenografts in mice compared to without ultrasound (Xie et al. 2015). The CPP-Myc siRNA/YSA-NB + ultrasound was significantly more effective than controls CPP-NC-siRNA/YSA-NB + ultrasound or CPP-siRNA alone, suggesting that YSA and ultrasound improved the efficacy of the CPP-Myc siRNA.

To reduce premature drug release, prodrug nanomedicine was developed to increase stability and solubility and to reduce injection of inactive carriers (Song et al. 2015). Clinical trials of prodrug carriers such as albumin-PTX have shown promise. When formulating prodrug nanomedicine, PEGylation of the drug is often used as carrier because PEGylation is neither toxic nor immunogenic, and can improve the half-life of the drug in circulation (Song et al. 2015). A cleavable linker between the drug and the carrier is also an essential component. The *cis*-asconitic anhydride-DOX (CAD) was conjugated to a PEG group via an amide bond to make a PEG-CAD prodrug. Then, the PEG-CAD prodrug was conjugated to RGD-3 peptide (Table 8.1), which selectively binds to neuropilin-1 (NRP-1) receptors. NRP-1 receptors have been shown to be overexpressed in tumor vessels as well as in many human cancer cell lines. The conjugation utilized a thiol-ene reaction to give an acid-labile prodrug called PEG-CAD-RGD-3. Due to the amphiphilic nature of PEG-CAD-RGD-3, it assembled into nanoparticles in water. During *in vitro* and *in vivo* administration, the release of DOX was triggered in acidic pH, but was restricted in neutral pH environment. Compared to DOX alone, the presence of RGD-3 peptide improved nanomedicine endocytosis and cytotoxicity into tumor cells. In Balb/c mice, PEG-CAD-RGD-3 nanomedicine has shown prolonged accumulation of DOX in tumors (Song et al. 2015).

8.4 Peptide Modulation of Biological Barriers to Improve Drug Delivery

Biological barriers such as the intestinal mucosa barrier (IMB) and blood-brain barrier (BBB) are present to protect the body from infections entering into the systemic circulation and the central nervous systems (CNS), respectively (Deli 2009). IMB is composed of a single layer of epithelial cells at the luminal side of the gastrointestinal tract followed by the lamina propria and the *muscularis mucosae*. The BBB is comprised of the luminal and abluminal membranes of the brain capillary endothelium as the major route for molecules (drugs and diagnostic agents) to enter the brain (Pardridge 2012). The IMB and BBB function as selective filters to allow needed substances and nutrients to enter the systemic circulation or brain, respectively, while preventing unwanted substances such as toxins from crossing the barriers. The gastrointestinal tract, skin, kidney, and lung barriers are made up of epithelial cells while the BBB microvessels are composed of endothelial cells. The delivery of molecules through the IMB and BBB is normally via the transcellular and paracellular pathways. Passive diffusion of drugs through the transcellular pathway depends on physicochemical properties of the drugs, and the passive diffusion of these drugs is normally regulated by Lipinski's rules of five. In general, peptide and protein drugs cannot cross the transcellular pathways due to their size, hydrophilicity, and hydrogen bonding potential. However, some hydrophilic small and large molecules (*e.g.* peptides and proteins) can cross the biological barriers via the transcellular pathway using receptor-mediated transporters.

Alternatively, drug molecules could cross the IMB and BBB via the paracellular pathway, where molecules pass through the intercellular space between the cells (O'Donnell and Maddrell 1983; Laksitorini et al. 2014). The paracellular space or intercellular junctions are connected by cell-cell adhesion proteins, forming a contiguous membrane connection (Anderson and Van Itallie 2009; Van Itallie and Anderson 2014). Therefore, there is a size limit for molecules to cross the paracellular pathway; normally, only ions and molecules with hydrodynamic radius <11 Å can cross this pathway (Lutz and Siahaan, 1997b). This limitation is imposed by the tight junctions that are mediated by cell-cell adhesion proteins such as occudins, claudins, and junction adhesion molecules (JAMs), which act as a fence to prevent free diffusion of molecules. Below the tight junctions there are adherens junctions (AJ), which are mediated by nectin and calcium-binding cadherins (Zheng et al. 2006). Beneath the adherens junctions lie the desmosomes, which are composed of desmoglein and desmocollin proteins; these proteins are also part of the cadherin family of cell-cell adhesion molecules with calcium-dependent binding properties (Garrod and Chidgey 2008).

Modulation of the intercellular junctions of the IMB and BBB have shown promise in enhancing paracellular permeation of molecules. A hypertonic mannitol solution is used clinically to deliver anticancancer drugs to treat terminally ill brain tumor patients (Neuwelt et al. 1979, 1984, 1987). This method is called osmotic delivery because the hypertonic solution shrinks the BBB vascular endothelial cells and modulates the intercellular junctions to increase their porosity. Various chemicals as specific and nonspecific junction modulators (*i.e.* sodium caprate, sodium decanoate, oleic acid, ethyleneglycol-bis-(β-aminoethyl ether)-N, N′-tetraacetic acid (EGTA)) have successfully improved penetration of molecules through *in vitro* models of biological barrier (Lutz and Siahaan 1997a, b). Due to uncontrolled paracellular opening, many toxic and unwanted side effects were observed with some of these methods. Thus, much research is focused on designing synthetic peptides that selectively modulate the protein-protein interactions in the intercellular junctions to improve paracellular permeation of delivered molecules.

8.4.1 Peptide Modulation of Tight Junction Proteins

One way to improve delivery of drug molecules via paracellular pathways of IMB and BBB is by modulating the interactions of cell-cell adhesion proteins in the intercellular junctions to increase the porosity of the paracellular pathways (Laksitorini et al. 2014; Bocsik et al. 2016). Several peptides derived from occludins have been synthesized and evaluated for this purpose. Occludins are 60 kDa membrane proteins that are involved in maintaining tight junction integrity. They are composed of four transmembrane domains, three cytoplasmic domains, and two extracellular loops of approximately similar size. One of these extracellular loops contains more Tyr and Gly residues (Gonzalez-Mariscal et al. 2003). The OCC2 peptide (Table 8.1) derived from extracellular loop 2 has been shown to modulate the tight junctions of A6 cell monolayers; the peptide lowers the transepithelial electrical resistance (TEER) values of the monolayers (Wong and Gumbiner 1997). OCC2 also enhances the penetration of paracellular markers such as inulin, dextran 3000, and dextran 40,000 across the A6 cell monolayers, indicating that the peptide increases paracellular porosity. OP_{90-135} peptide (Table 8.1) derived from the first loop of occludin can lower the TEER values of Caco-2 cell monolayers, a model for IMB (Tavelin et al. 2003). The peptide also enhanced the transport of a paracellular marker, ^{14}C-mannitol, across the Caco-2 cell. A smaller OP_{90-103} peptide (Table 8.1) has better modulatory activity than the parent OP_{90-135} in Caco-2 cell monolayers. In addition, Lip-OP_{90-103} peptide (Table 8.1) that is an N-terminus lipid-alkylated peptide has about 11 times higher modulatory activity than the parent OP_{90-103}.

Besides the occludins, claudins (Cldn-1, -2, -3, and -4) are also responsible for forming tight junctions; claudins have a transmembrane structure similar to that of occludins (Gonzalez-Mariscal et al. 2003; Anderson and Van Itallie 2009; Van Itallie and Anderson 2014). They have a short cytoplasmic N-terminus, two extracellular loops, and a C-terminal cytoplasmic domain. Both occludins and claudins interact via their C-terminus to zonula occludin-1 (ZO-1), ZO-2, and ZO-3 to stabilize the cytoskeleton membranes of the tight junctions (Schneeberger and Lynch 2004). Knocking down the expression of Cldn-1, Cldn-4, occludin, and ZO-1 increases the paracellular permeation of molecules and ions across the cell monolayers (Van Itallie and Anderson 2014). This result confirms their importance in maintaining the tight junctions. A 29-amino acid C1C2 peptide derived from the extracellular loop-1 of claudin-1 can enhance the permeation of small and large paracellular markers (*i.e.* Lucifer Yellow and FITC-Dextran 10 KDa) across the cell monolayers (Zwanziger et al. 2012). *In vivo*, the peptide increases the brain delivery of tetrodoxin and enkephalin peptides, suggesting that the peptide modulates the tight junctions of the BBB. The proposed mechanism of action of C1C2 peptide is via binding to claudins followed by induction of claudin endocytosis into the cytoplasmic domain (Staat et al. 2015). Therefore, this internalization lowers the population of claudin in the tight junction to make the tight junctions looser.

8.4.2 Peptide Modulators of Adherens Junction Proteins

Peptides derived from the extracellular domain-1 (EC1) domain of E-cadherin (*i.e.* HAV and ADT peptides) have been shown to modulate the intercellular junctions of MDCK and Caco-2 cell monolayers (Makagiansar et al. 2001; Sinaga et al. 2002). HAV and ADT peptides have been shown to inhibit E-cadherin-mediated cell-cell adhesion of single cells of bovine brain microvessel endothelial cells (BBMEC) as well as the intercellular junctions of BBMEC monolayers (Lutz and Siahaan 1997a, b; Pal et al. 1997). HAV6 and/or ADTC5 peptides (Table 8.1) increase the *in vivo* brain delivery of molecules in mice and/or rats; the delivered molecules include paracellular markers (*i.e.* ^{14}C-mannitol, 25 kDa IRdye800cw-Polyethylene glycols or PEG), anticancer drugs (*i.e.* ^{3}H daunomycin, Glu-CPT), efflux pump substrates (*i.e.* rhodamine 800 (R800), ^{3}H daunomycin), magnetic resonance imaging (MRI) enhancing agents (*i.e.* gadopentetic acid or Gd-DTPA), peptides (*i.e.* IRdye800cw-cLABL and cIBR7; Table 8.1) and proteins (*i.e.* 65 kDa galbumin) (Kiptoo et al. 2011; On et al. 2014; Laksitorini et al. 2015; Alaofi et al. 2016; Ulapane et al. 2017). The brain depositions of radioactive molecules such as ^{14}C-mannitol and ^{3}H-daunomycin were detected and quantified in the brain homogenates using a radioactive counter, while the quantity of brain deposition of R800, IRdye800cw-PEG, and IRdye800cw-cLABL was determined in the intact isolated brain using near IR fluorescence imaging. MRI was used in living animals to detect the brain distribution of Gd-DTPA and galbumin. Finally, the amounts of brain-delivered non-labeled Glu-CPT and cIBR7 peptide in rats were detected using LC-MS/MS. The duration of modulation of the *in vivo* BBB for small molecules was less than 1 h for HAV6 peptide and between 2 and 5 h for ADTC5 peptide (On et al. 2014; Laksitorini et al. 2015). However, the duration of BBB modulation by HAV6 and ADTC5 was too short for delivering a large molecule such as 65 kDa galbumin, less than 10 min for HAV6 and from 10–40 min for ADTC5 peptide (Ulapane et al. 2017). The results suggest that the peptides create small pores in the intercellular junctions with a long-time duration compared to a short duration for large pores in the BBB intercellular junctions. The results also indicate that the BBB modulation is reversible.

The mechanism of action of HAV and ADT peptides is potentially due to their binding to E-cadherin to inhibit cadherin-cadherin interactions in the intercellular junctions of the BBB. Using nuclear magnetic resonance spectroscopy (NMR) and molecular docking studies, HAV and ADT peptides were shown to bind at different sites on the EC1 domain of E-cadherin (Alaofi et al. 2017). It is proposed that the HAV6 peptide binds to the EC1 domain and inhibits the binding of the EC1 domain from one E-cadherin to the EC2 domain of another cadherin on the same membranes, which is the *cis*-cadherin interaction. In contrast, ADTC5 peptide binds to the EC1 domain to prevent the *trans*-EC1 domain from swapping between two E-cadherins from opposite cell membranes or *trans*-cadherin interactions.

8.4.3 Other Peptide Modulators Tight Junctions

Bocsik et al. have shown that C-CPE, AT-1002, PN-78, and PN-159 peptides (Table 8.1) could modulate the intercellular junctions of the IMB and BBB in cell culture models (Bocsik et al. 2016). These peptides were not derived from the sequence of proteins from the intercellular junctions of the IMB and/or the BBB. C-CPE and AT-1002 peptides were respectively derived from *clostridium perfringens enterotoxin* (C-CPE) and zonula occludens *toxin* (Zot). Both peptides modulate the penetration of molecules through the paracellular pathways of IMB and BBB *in vitro* and/or *in vitro* (Sonoda et al. 1999; Bocsik et al. 2016). PN-78 and PN-159 peptides were discovered using phage display and they increased the paracellular permeation of molecules through the lung epithelial cell monolayer (Herman et al. 2007). C-CPE, AT-1002, PN-78, and PN-159 modulate the intercellular junctions of Caco-2 cell and brain endothelial monolayers as models of IMB and BBB, respectively. This junction modulation was reflected in the lowering of *trans*-epithelial/endothelial electrical resistance (TEER) upon peptide treatment. The paracellular transport of fluorescein across the Caco-2 cell monolayers was enhanced by C-CPE, AT-1002, and PN-159 but not PN-78 peptides. However, all four peptides enhanced the penetration of albumin across the Caco-2 cell monolayers. The paracellular permeation of both fluorescein and albumin across the BBB cell monolayers was increased by AT-1002, PN-78, and PN-159 but not C-CPE. Thus, these peptides can be used to deliver drug molecules across the IMB and the BBB in *in vivo* studies.

8.5 Conclusion

Peptides have been successfully developed as drugs. Now, peptides have been extensively investigated to deliver drugs to specific cells to lower their unwanted side effects. Similarly, conjugation of peptides to labeled molecules or atoms was shown to be useful for potential diagnostic agents to locate diseased cells within the body using various detection methods such as MRI and PET. Finally, delivery across the intestinal mucosa and the blood-brain barrier can also be enhanced by modulation of the protein-protein interactions in the intercellular junctions of these barriers using peptides. Modulation of the BBB using cadherin peptides can enhance the brain delivery of small-to-large molecules to the brains of living animals. Thus, modulation of the intercellular junctions can be exploited to clinically deliver drug and diagnostic molecules through the IMB and BBB in the future.

Acknowledgements We would like to thank the National Institute of Neurological Disorders and Stroke (NINDS), National Institutes of Health (NIH) for funding our research with R01-NS075374 grant. M.E.G.M thanks the NIH for NIH for an IRACDA postdoctoral fellowship (5K12GM063651). We would like to thank Nancy Harmony for proofreading this manuscript.

References

Alaofi A, On N, Kiptoo P, Williams TD, Miller DW, Siahaan TJ (2016) Comparison of linear and cyclic His-Ala-Val peptides in modulating the blood-brain barrier permeability: impact on delivery of molecules to the brain. J Pharm Sci 105:797–807

Alaofi A, Farokhi E, Prasasty VD, Anbanandam A, Kuczera K, Siahaan TJ (2017) Probing the interaction between cHAVc3 peptide and the EC1 domain of E-cadherin using NMR and molecular dynamics simulations. J Biomol Struct Dyn 35:92–104

Anderson JM, Van Itallie CM (2009) Physiology and function of the tight junction. Cold Spring Harb Perspect Biol 1:a002584

Arap W, Pasqualini R, Ruoslahti E (1998) Cancer treatment by targeted drug delivery to tumor vasculature in a mouse model. Science 279:377–380

Bajusz S, Janaky T, Csernus VJ, Bokser L, Fekete M, Srkalovic G, Redding TW, Schally AV (1989) Highly potent analogues of luteinizing hormone-releasing hormone containing D-phenylalanine nitrogen mustard in position 6. Proc Natl Acad Sci U S A 86:6318–6322

Barve A, Jain A, Liu H, Jin W, Cheng K (2016) An enzyme-responsive conjugate improves the delivery of a PI3K inhibitor to prostate cancer. Nanomedicine 12:2373–2381

Beer AJ, Haubner R, Sarbia M, Goebel M, Luderschmidt S, Grosu AL, Schnell O, Niemeyer M, Kessler H, Wester HJ, Weber WA, Schwaiger M (2006) Positron emission tomography using [^{18}F]Galacto-RGD identifies the level of integrin $\alpha_V\beta_3$ expression in man. Clin Cancer Res 12:3942–3949

Bell AF, Erickson EN, Carter CS (2014) Beyond labor: the role of natural and synthetic oxytocin in the transition to motherhood. J Midwifery Womens Health 59:35–42

Bocsik A, Walter FR, Gyebrovszki A, Fülöp L, Blasig I, Dabrowski S, Ötvös F, Tóth A, Rákhely G, Veszelka S, Vastag M, Szabó-Révész P, Deli MA (2016) Reversible opening of intercellular junctions of intestinal epithelial and brain endothelial cells with tight junction modulator peptides. J Pharm Sci 105:754–765

Broder MS, Beenhouwer D, Strosberg JR, Neary MP, Cherepanov D (2015) Gastrointestinal neuroendocrine tumors treated with high dose octreotide-LAR: a systematic literature review. World J Gastroenterol 21:1945–1955

Burns KE, Hensley H, Robinson MK, Thévenin D (2017) Therapeutic efficacy of a family of pHLIP-MMAF conjugates in cancer cells and mouse models. Mol Pharm 14:415–422

Büyüktimkin B, Stewart Jr J, Tabanor K, Kiptoo P, Siahaan TJ (2016). Protein and peptide conjugates for targeting therapeutics and diagnostics to specific cells. In: Wang B, Hu L, Siahaan TJ (eds) Drug delivery: principles and applications, 2nd. Wiley, Hoboken

Callahan J, Kopecek J (2006) Semitelechelic HPMA copolymers functionalized with triphenylphosphonium as drug carriers for membrane transduction and mitochondrial localization. Biomacromolecules 7:2347–2356

Chen X, Plasencia C, Hou Y, Neamati N (2005) Synthesis and biological evaluation of dimeric RGD peptide-paclitaxel conjugate as a model for integrin-targeted drug delivery. J Med Chem 48:1098–1106

Chen X, Zhang XY, Shen Y, Fan LL, Ren ML, Wu YP (2016) Synthetic paclitaxel-octreotide conjugate reversing the resistance of A2780/Taxol to paclitaxel in xenografted tumor in nude mice. Oncotarget 7:83451–83461

Cordova A, Woodrick J, Grindrod S, Zhang L, Saygideger-Kont Y, Wang K, DeVito S, Daniele SG, Paige M, Brown ML (2016) Aminopeptidase P mediated targeting for breast tissue specific conjugate delivery. Bioconjug Chem 27:1981–1990

Deli MA (2009) Potential use of tight junction modulators to reversibly open membranous barriers and improve drug delivery. BBA-Biomembranes 1788:892–910

Dharap SS, Qiu B, Williams GC, Sinko P, Stein S, Minko T (2003) Molecular targeting of drug delivery systems to ovarian cancer by BH3 and LHRH peptides. J Control Rel 91:61–73

Dunehoo AL, Anderson M, Majumdar S, Kobayashi N, Berkland C, Siahaan TJ (2006) Cell adhesion molecules for targeted drug delivery. J Pharm Sci 95:1856–1872

Fosgerau K, Hoffmann T (2015) Peptide therapeutics: current status and future directions. Drug Discov Today 20:122–128

Garrod D, Chidgey M (2008) Desmosome structure, composition and function. Biochim Biophys Acta 1778:572–587

Gonzalez-Mariscal L, Betanzos A, Nava P, Jaramillo BE (2003) Tight junction proteins. Prog Biophys Mol Biol 81:1–44

Hamann PR, Hinman LM, Beyer CF, Lindh D, Upeslacis J, Flowers DA, Bernstein I (2002) An anti-CD33 antibody-calicheamicin conjugate for treatment of acute myeloid leukemia. Choice of linker. Bioconjug Chem 13:40–46

Haubner R, Weber WA, Beer AJ, Vabuliene E, Reim D, Sarbia M, Becker KF, Goebel M, Hein R, Wester HJ, Kessler H, Schwaiger M (2005) Noninvasive visualization of the activated $\alpha_V\beta_3$ integrin in cancer patients by positron emission tomography and [^{18}F]Galacto-RGD. PLoS Med 2:e70

Hejna M, Schmidinger M, Raderer M (2002) The clinical role of somatostatin analogues as antineoplastic agents: much ado about nothing? Ann Oncol 13:653–668

Herman RE, Makienko EG, Prieve MG, Fuller M, Houston ME Jr, Johnson PH (2007) Phage display screening of epithelial cell monolayers treated with EGTA: identification of peptide FDFWITP that modulates tight junction activity. J Biomol Screen 12:1092–1101

Huang C, Yi X, Kong D, Chen L, Min G (2016) Controlled release strategy of paclitaxel by conjugating to matrix metalloproteinases-2 sensitive peptide. Oncotarget 7:52230–52238

Jensen KD, Nori A, Tijerina M, Kopecková P, Kopecek J (2003) Cytoplasmic delivery and nuclear targeting of synthetic macromolecules. J Control Release 87:89–105

Jiang Y, Wang X, Liu X, Lv W, Zhang H, Zhang M, Li X, Xin H, Xu Q (2017) Enhanced antiglioma efficacy of ultrahigh loading capacity paclitaxel prodrug conjugate self-assembled targeted nanoparticles. ACS Appl Mater Interfaces 9:211–217

Jones DP (2010) Redox sensing: orthogonal control in cell cycle and apoptosis signalling. J Internal Med 268:432–448

Kang B, Mackey MA, El-Sayed MA (2010) Nuclear targeting of gold nanoparticles in cancer cells induces DNA damage, causing cytokinesis arrest and apoptosis. J Am Chem Soc 132:1517–1519

Kaspar AA, Reichert JM (2013) Future directions for peptide therapeutics development. Drug Discov Today 18:807–817

Kim JA, Casalini T, Brambilla D, Leroux JC (2016) Presumed LRP1-targeting transport peptide delivers beta-secretase inhibitor to neurons *in vitro* with limited efficiency. Sci Rep 6:34297

Kiptoo P, Sinaga E, Calcagno AM, Zhao H, Kobayashi N, Tambunan US, Siahaan TJ (2011) Enhancement of drug absorption through the blood-brain barrier and inhibition of intercellular tight junction resealing by E-cadherin peptides. Mol Pharm 8:239–249

Knop FK, Bronden A, Vilsboll T (2017) Exenatide: pharmacokinetics, clinical use, and future directions. Expert Opin Pharmacother 18:555–571

Laksitorini M, Prasasty VD, Kiptoo PK, Siahaan TJ (2014) Pathways and progress in improving drug delivery through the intestinal mucosa and blood-brain barriers. Ther Deliv 5:1143–1163

Laksitorini MD, Kiptoo PK, On NH, Thliveris JA, Miller DW, Siahaan TJ (2015) Modulation of intercellular junctions by cyclic-ADT peptides as a method to reversibly increase blood-brain barrier permeability. J Pharm Sci 104:1065–1075

Leal M, Sapra P, Hurvitz SA, Senter P, Wahl A, Schutten M, Shah DK, Haddish-Berhane N, Kabbarah O (2014) Antibody-drug conjugates: an emerging modality for the treatment of cancer. Annals NY Acad Sci 1321:41–54

Li Y, Zheng X, Gong M, Zhang J (2016) Delivery of a peptide-drug conjugate targeting the blood brain barrier improved the efficacy of paclitaxel against glioma. Oncotarget 7:79401–79407

Limonta P, Montagnani MM, Marelli M, Motta M (2003) The biology of gonadotropin hormone-releasing hormone: role in the control of tumor growth and progression in humans. Front Neuroendocrinol 24:279–295

Liu N, Tan Y, Hu Y, Meng T, Wen L, Liu J, Cheng B, Yuan H, Huang X, Hu F (2016) A54 peptide modified and redox-responsive glucolipid conjugate micelles for intracellular delivery of doxorubicin in hepatocarcinoma therapy. ACS Appl Mater Interfaces 8:33148–33156

Lutz KL, Siahaan TJ (1997a) Modulation of the cellular junction protein E-cadherin in bovine brain microvessel endothelial cells by cadherin peptides. Drug Deliv 4:187–193

Lutz KL, Siahaan TJ (1997b) Molecular structure of the apical junction complex and its contribution to the paracellular barrier. J Pharm Sci 86:977–984

Majumdar S, Siahaan TJ (2012) Peptide-mediated targeted drug delivery. Med Res Rev 32:637–658

Majumdar D, Rahman MA, Chen ZG, Shin DM (2016) Anticancer activity of drug conjugates in head and neck cancer cells. Front Biosci 8:358–369

Makagiansar IT, Avery M, Hu Y, Audus KL, Siahaan TJ (2001) Improving the selectivity of HAV-peptides in modulating E-cadherin-E-cadherin interactions in the intercellular junction of MDCK cell monolayers. Pharm Res 18:446–453

Mas-Moruno C, Rechenmacher F, Kessler H (2010) Cilengitide: the first anti-angiogenic small molecule drug candidate design, synthesis and clinical evaluation. Anti Cancer Agents Med Chem 10:753–768

McEligot AJ, Yang S, Meyskens FL Jr (2005) Redox regulation by intrinsic species and extrinsic nutrients in normal and cancer cells. Ann Rev Nutr 25:261–295

Murányi J, Gyulavári P, Varga A, Bökönyi G, Tanai H, Vántus T, Pap D, Ludányi K, Mező G, Kéri G (2016) Synthesis, characterization and systematic comparison of FITC-labelled GnRH-I, –II and -III analogues on various tumour cells. J Pept Sci 22:552–560

Nagy A, Schally AV, Armatis P, Szepeshazi K, Halmos G, Kovacs M, Zarandi M, Groot K, Miyazaki M, Jungwirth A, Horvath J (1996) Cytotoxic analogs of luteinizing hormone-releasing hormone containing doxorubicin or 2-pyrrolinodoxorubicin, a derivative 500–1000 times more potent. Proc Natl Acad Sci U S A 93:7269–7273

Neuwelt EA, Maravilla KR, Frenkel EP, Rapaport SI, Hill SA, Barnett PA (1979) Osmotic blood-brain barrier disruption. Computerized tomographic monitoring of chemotherapeutic agent delivery. J Clin Invest 64:684–688

Neuwelt EA, Hill SA, Frenkel EP (1984) Osmotic blood-brain barrier modification and combination chemotherapy: concurrent tumor regression in areas of barrier opening and progression in brain regions distant to barrier opening. Neurosurgery 15:362–366

Neuwelt EA, Specht HD, Barnett PA, Dahlborg SA, Miley A, Larson SM, Brown P, Eckerman KF, Hellström KE, Hellström I (1987) Increased delivery of tumor-specific monoclonal antibodies to brain after osmotic blood-brain barrier modification in patients with melanoma metastatic to the central nervous system. Neurosurgery 20:885–895

O'Donnell MJ, Maddrell SH (1983) Paracellular and transcellular routes for water and solute movements across insect epithelia. J Exp Biol 106:231–253

On NH, Kiptoo P, Siahaan TJ, Miller DW et al (2014) Modulation of blood-brain barrier permeability in mice using synthetic E-cadherin peptide. Mol Pharm 11:974–981

Páez Pereda M, Ledda MF, Goldberg V, Chervín A, Carrizo G, Molina H, Müller A, Renner U, Podhajcer O, Arzt E, Stalla GK (2000) High levels of matrix metalloproteinases regulate proliferation and hormone secretion in pituitary cells. J Clin Endocrin Metab 85:263–269

Pal D, Audus KL, Siahaan TJ (1997) Modulation of cellular adhesion in bovine brain microvessel endothelial cells by a decapeptide. Brain Res 747:103–113

Pardridge WM (2012) Drug transport across the blood-brain barrier. J Cereb Blood Flow Metab 32:1959–1972

Redko B, Tuchinsky H, Segal T, Tobi D, Luboshits G, Ashur-Fabian O, Pinhasov A, Gerlitz G, Gellerman G (2016) Toward the development of a novel non-RGD cyclic peptide drug conjugate for treatment of human metastatic melanoma. Oncotarget 8:757–768

Regina A, Demeule M, Ché C, Lavallée I, Poirier J, Gabathuler R, Béliveau R, Castaigne JP (2008) Antitumour activity of ANG1005, a conjugate between paclitaxel and the new brain delivery vector Angiopep-2. Br J Pharmacol 155:185–197

Ruoslahti E (1994) Cell adhesion and tumor metastasis. Princess Takamatsu Symp 24:99–105

Ruoslahti E, Pierschbacher MD (1987) New perspectives in cell adhesion: RGD and integrins. Science 238:491–497

Schneeberger EE, Lynch RD (2004) The tight junction: a multifunctional complex. Am J Physiol Cell Physiol 286:C1213–C1228

Sheng Y, Xu J, You Y, Xu F, Chen Y (2015) Acid-sensitive peptide-conjugated doxorubicin mediates the lysosomal pathway of apoptosis and reverses drug resistance in breast cancer. Mol Pharm 12:2217–2228

Sheng Y, You Y, Chen Y (2016) Dual-targeting hybrid peptide-conjugated doxorubicin for drug resistance reversal in breast cancer. Int J Pharm 512:1–13

Sinaga E, Jois SD, Avery M, Makagiansar IT, Tambunan US, Audus KL, Siahaan TJ (2002) Increasing paracellular porosity by E-cadherin peptides: discovery of bulge and groove regions in the EC1-domain of E-cadherin. Pharm Res 19:1170–1179

Smith RAJ, Porteous CM, Gane AM, Murphy MP (2003) Delivery of bioactive molecules to mitochondria *in vivo*. Proc Natl Acad Sci U S A 100:5407–5412

Song H, Zhang J, Wang W, Huang P, Zhang Y, Liu J, Li C, Kong D (2015) Acid-responsive PEGylated doxorubicin prodrug nanoparticles for neuropilin-1 receptor-mediated targeted drug delivery. Colloids Surf B Biointerfaces 136:365–374

Sonoda N, Furuse M, Sasaki H, Yonemura S, Katahira J, Horiguchi Y, Tsukita S (1999) Clostridium perfringens enterotoxin fragment removes specific claudins from tight junction strands: evidence for direct involvement of claudins in tight junction barrier. J Cell Biol 147:195–204

Staat C, Coisne C, Dabrowski S, Stamatovic SM, Andjelkovic AV, Wolburg H, Engelhardt B, Blasig IE (2015) Mode of action of claudin peptidomimetics in the transient opening of cellular tight junction barriers. Biomaterials 54:9–20

Sun W, Li L, Yang QQ, Zhang ZR, Huang Y (2017) Two birds, one stone: dual targeting of the cancer cell surface and subcellular mitochondria by the galectin-3-binding peptide G3-C12. Acta Pharmacol Sin:1–17

Tavelin S, Hashimoto K, Malkinson J, Lazorova L, Toth I, Artursson P (2003) A new principle for tight junction modulation based on occludin peptides. Mol Pharmacol 64:1530–1540

Tella SH, Gallagher JC (2014) Prevention and treatment of postmenopausal osteoporosis. J Steroid Biochem Mol Biol 142:155–170

Thomas FC, Taskar K, Rudraraju V, Goda S, Thorsheim HR, Gaasch JA, Mittapalli RK, Palmieri D, Steeg PS, Lockman PR, Smith QR (2009) Uptake of ANG1005, a novel paclitaxel derivative, through the blood-brain barrier into brain and experimental brain metastases of breast cancer. Pharm Res 26:2486–2494

Uhlig T, Kyprianou T, Martinelli FG, Oppici CA, Heiligers D, Hills D, Calvo XR, Verhaert P (2014) The emergence of peptides in the pharmaceutical business: from exploration to exploitation. EuPA Open Proteom 4:58–69

Ulapane KR, On N, Kiptoo P, Williams TD, Miller DW, Siahaan TJ (2017) Improving brain delivery of biomolecules via BBB modulation in mouse and rat: Detection using MRI, NIRF, and mass spectrometry. Nanotheranostics 1:217–231

Van Itallie CM, Anderson JM (2014) Architecture of tight junctions and principles of molecular composition. Semin Cell Dev Biol 36:157–165

Wang Y, Wang L, Chen G, Gong S (2016) Carboplatin-complexed and cRGD-conjugated unimolecular nanoparticles for targeted ovarian cancer therapy. Macromol Biosci 17:1600292

Wong V, Gumbiner BM (1997) A synthetic peptide corresponding to the extracellular domain of occludin perturbs the tight junction permeability barrier. J Cell Biol 136:399–409

Xie X, Yang Y, Lin W, Liu H, Liu H, Yang Y, Chen Y, Fu X, Deng J (2015) Cell-penetrating peptide-siRNA conjugate loaded YSA-modified nanobubbles for ultrasound triggered siRNA delivery. Colloids Surf B Biointerf 136:641–650

Yusuf-Makagiansar H, Anderson ME, Yakovleva TV, Murray JS, Siahaan TJ (2002) Inhibition of LFA-1/ICAM-1 and VLA-4/VCAM-1 as a therapeutic approach to inflammation and autoimmune diseases. Med Res Rev 22:146–167

Zheng K, Trivedi M, Siahaan TJ (2006) Structure and function of the intercellular junctions: barrier of paracellular drug delivery. Curr Pharm Des 12:2813–2824

Zwanziger D, Hackel D, Staat C, Böcker A, Brack A, Beyermann M, Rittner H, Blasig IE (2012) A peptidomimetic tight junction modulator to improve regional analgesia. Mol Pharm 9:1785–1794

Peptide Lipidation – A Synthetic Strategy to Afford Peptide Based Therapeutics

9

Renata Kowalczyk, Paul W.R. Harris, Geoffrey M. Williams, Sung-Hyun Yang, and Margaret A. Brimble

Abstract

Peptide and protein aberrant lipidation patterns are often involved in many diseases including cancer and neurological disorders. Peptide lipidation is also a promising strategy to improve pharmacokinetic and pharmacodynamic profiles of peptide-based drugs. Self-adjuvanting peptide-based vaccines commonly utilise the powerful TLR2 agonist Pam_nCys lipid to stimulate adjuvant activity. The chemical synthesis of lipidated peptides can be challenging hence efficient, flexible and straightforward synthetic routes to access homogeneous lipid-tagged peptides are in high demand. A new technique coined Cysteine Lipidation on a Peptide or Amino acid (CLipPA) uses a 'thiol-ene' reaction between a cysteine and a vinyl ester and offers great promise due to its simplicity, functional group compatibility and selectivity. Herein a brief review of various synthetic strategies to access lipidated peptides, focusing on synthetic methods to incorporate a Pam_nCys motif into peptides, is provided.

Keywords

Peptide lipidation • PamCys • Self-adjuvanting vaccines • Palmitoylation • Thiol-ene • Vinyl ester

R. Kowalczyk
School of Chemical Sciences, The University of Auckland, 23 Symonds St, Auckland, New Zealand

P.W.R. Harris • G.M. Williams • M.A. Brimble (✉)
School of Chemical Sciences, The University of Auckland, 23 Symonds St, Auckland, New Zealand

School of Biological Sciences, The University of Auckland, 3A Symonds St, Auckland, New Zealand

Maurice Wilkins Centre for Molecular Biodiscovery, University of Auckland,
Private Bag 92019, Auckland 1010, New Zealand
e-mail: m.brimble@auckland.ac.nz

S.-H. Yang
School of Chemical Sciences, The University of Auckland, 23 Symonds St, Auckland, New Zealand

School of Biological Sciences, The University of Auckland, 3A Symonds St, Auckland, New Zealand

© Springer International Publishing AG 2017

A. Sunna et al. (eds.), *Peptides and Peptide-based Biomaterials and their Biomedical Applications*, Advances in Experimental Medicine and Biology 1030, https://doi.org/10.1007/978-3-319-66095-0_9

9.1 Introduction

The market for peptide based therapeutics has been constantly growing since the late 1990s with 140 peptide drugs currently estimated to be undergoing clinical trials and 500 therapeutic peptides in pre-clinical development (Fosgerau and Hoffmann 2015; Kaspar and Reichert 2013; Otvos and Wade 2014). Biologically active peptides are excellent drug candidates due to high receptor selectivity, binding affinity, potency and relatively low toxicity (Fosgerau and Hoffmann 2015; Trabocchi and Guarna 2014). However, the therapeutic potential of peptides can be limited due to their poor chemical and physical stability, short plasma half-life, and low oral bioavailability (Fosgerau and Hoffmann 2015; Trabocchi and Guarna 2014). Peptide drug delivery to the site of action is often challenging and improved technologies to overcome this obstacle are highly desirable (Lewis and Richard 2015). Structural and functional modifications of native peptides using chemical techniques have been used to generate compounds with higher affinity, improved enzymatic stability and/or efficacy compared to the parent peptide (Trabocchi and Guarna 2014). Peptide backbone modifications, cyclization, unnatural amino acid insertion, PEGylation, glycosylation, phosphorylation and lipidation are common techniques to improve the physicochemical and pharmacological profiles of bioactive peptides. (Zhang and Bulaj 2012)

Peptide lipidation is an effective strategy to modify the pharmacokinetic and pharmacodynamic properties of lead peptide therapeutics and has proven to be successful with several marketed peptides including liraglutide (Victoza®) (Jackson et al. 2010; Knudsen et al. 2000) and insulin detemir (Levemir®) (Zhang and Bulaj 2012; Home and Kurtzhals 2006; Le Floch 2010). Incorporation of lipid units onto a peptide backbone can dramatically increase enzymatic stability (Simerska et al. 2011), receptor selectivity and potency (Ward et al. 2013), bioavailability (Hamman et al. 2005; Park et al. 2011; Renukuntla et al. 2013; Karsdal et al. 2015) and drug delivery potential (membrane permeability) (Zhang and Bulaj 2012; Simerska et al. 2011).

This review describes the impact of lipidation on peptide-based drug development and summarises the most recent strategies to incorporate a lipid moiety onto a peptide using chemical techniques. A brief discussion on naturally occuring lipidated proteins and peptides and the potential for lipidation to create bioactive therapeutics is covered. The highlight of this perspective relates to synthetic approaches to incorporate Pam_nCys-based Toll-like receptor 2 (TLR2) lipidated ligands into peptides with the potential to generate self-adjuvanting vaccine constructs.

9.1.1 Protein Lipidation in Nature

Protein lipidation is one of the most important post- and co-translational modifications controlling protein affinity to cellular membranes and influencing protein regulatory and signalling functions (Mejuch and Waldmann 2016; Resh 2013). Altered lipidation patterns are associated with various diseases including cancer, neurological diseases, diabetes, infections (bacterial, fungal and viral) (Resh 2012).

Protein acylation phenomena encompasses a broad range of saturated and unsaturated fatty acids of different length creating proteins with a unique set of functions. Protein-bound lipid types and lipid-protein linkages vary in nature. Covalent attachment of unique fatty acid chains is controlled by the action of specific transferases affording a broad range of lipidated proteins including *N*-myristoylated, *S*- or *N*-palmitoylated, and cholesterol- and isoprenol-enriched moieties (Fig. 9.1). Glycosylphosphatidylinositol (GPI), and phosphatidylethanolamine (PE) conjugation to proteins has also been described (Resh 2013).

Lipid addition occurs at *N*- and *C*- terminal sites of proteins or within the protein sequence directed by specific amino acids such as cysteine, serine, threonine, and lysine (Hannoush 2015). Lipidation can be irreversible when formed *via* an amide bond using an *N*-terminal glycine or cysteine moiety (*N*-myristoylation and *N*-palmitoylation, respectively) or reversible when a thioester bond is formed between the fatty acid the thiol of the cysteine residue

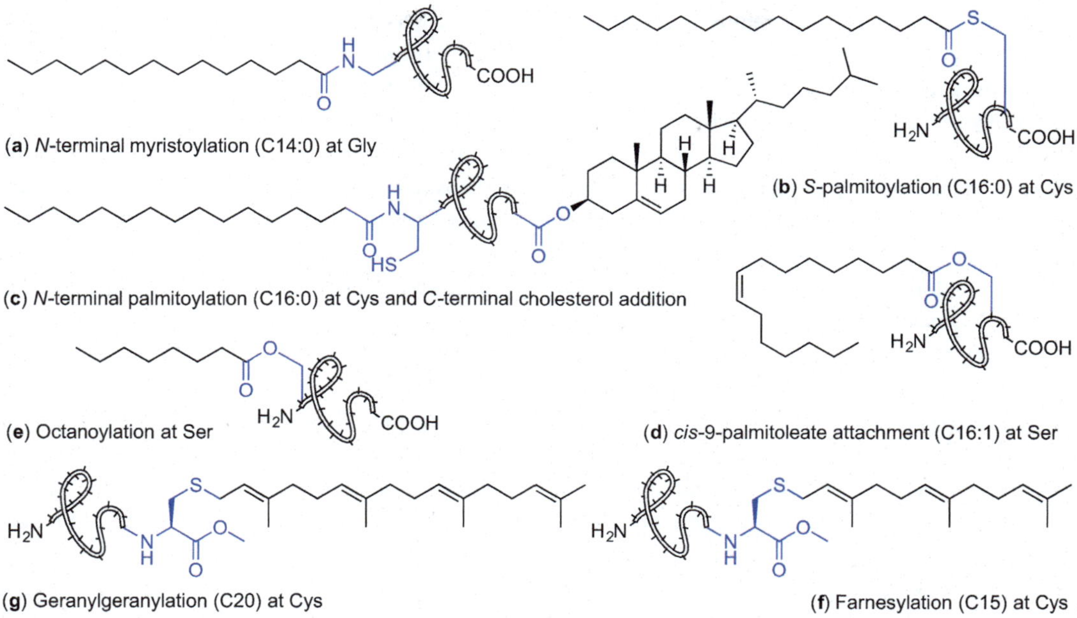

Fig. 9.1 Protein modifications with various lipids found in nature

(*S*-palmitoylation), Fig. 9.1 (Resh 2013; Chamberlain and Shipston 2015).

Proteins can exist in a mono-lipidated state or with multiple-lipid group addition. Membrane proteins such as MARCKS, GPCRs, and K-Ras4B are monolipidated proteins enriched with myristoyl, palmitoyl and farnesyl motifs, respectively. The Hedgehog (Hh) family of proteins which are associated with developmental processes (Lee et al. 2016) are modified with palmitate and cholesterol; similarly, the Src family kinases are myristoylated and palmitoylated and plasma membrane H-Ras and N-Ras proteins are farnesylated and palmitoylated (Resh 2013).

Irreversible protein modification with myristic acid, a 14-carbon fatty acid, is the most prevalent in nature and accounts for 0.5–0.8% of all lipidated eukaryotic proteins. It can occur both co- and post-translationally at the *N*-terminal glycine and is catalysed by *N*-myristoyltransferase (NMT), Fig. 9.1a (Resh 2013; Wright et al. 2010; Resh 2016). *N*-Myristoylation at the *N*ε of lysine was also observed for interleukin 1α (Stevenson et al. 1993) and tumour necrosis factor alpha (TNF) (Stevenson et al. 1992); However, enzymes involved in these acylation processes are yet to be identified (Resh 2016). *N*-Myristoylated proteins such as c-Src, BID, PAK2, or gelsolin play important roles in various biological processes including cellular transformation and effecting protein localization (Hannoush 2015; Wright et al. 2010). *N*-Myristoylation is involved in pathogen survival and altered myristoylation patterns are linked to carcinogenesis (Wright et al. 2010).

S-Palmitoylation is the most common form of protein *S*-acylation affording reversibly-tagged proteins with a 16-carbon palmitic acid unit (Chamberlain and Shipston 2015; Resh 2016). *S*-Palmitoylation can occur at the cysteine moiety located in the proximity of either the *N* -or *C*-terminus of proteins, Fig. 9.1b. Attachment of stearic acid (C18:0) and monounsaturated omega-9 oleic acid (C18:1) *via* the thiol group of a cysteine residue has also been described (Chamberlain and Shipston 2015).

Due to labile nature of the thioester bond used to link a fatty acid with a protein backbone, a dynamic equilibrium between protein *S*-acylation and deacylation with distinct turnover rates occurs that influences intracellular localization, membrane association, and the regulatory functions of a diverse family of proteins. *S*-Acylation of cellular proteins is mediated *via* *S*-acyl transferases from the zDHHC protein family. However, only scant information is available on the *S*-acyl thioesterases involved in protein deacetylation and the dynamic *S*-acylation process (Chamberlain and Shipston 2015). It is proposed that enzymes from the serine hydrolase family including acyl protein thioesterases (APTs) (Davda and Martin 2014), and protein palmitoyl thioesterases (PPTs) (Lin and Conibear 2015) may be involved (Chamberlain and Shipston 2015).

S-Acylation facilitates stable membrane binding of peripheral proteins and mediates protein targeting to specific endoplasmic reticulum (ER) subdomains. Protein *S*-acylation controls trafficking and localization of cellular proteins, and improves protein stability in addition to regulating cellular signalling receptors (Chamberlain and Shipston 2015).

The Hedgehog protein family are critical proteins with roles in embryonic development and tumorigenesis (Resh 2016; Pepinsky et al. 1998). These mature signalling proteins are dually lipidated comprising a palmitate unit which is incorporated through an amide bond at *N*-terminal cysteine (*N*-palmitoylation) *via* the action of hedgehog acyltransferase (Hhat), a member of a membrane-bound *O*-acyltransferases (MBOAT) protein superfamily (Konitsiotis et al. 2015; Matevossian and Resh 2015), and a cholesterol moiety covalently attached to the *C*-terminal glycine *via* its 3β-hydroxyl group, Fig. 9.1c (Resh 2013, 2016). *N*-Palmitoylation is essential for signalling activity of Hh proteins during development while the cholesterol unit aids the signalling functions (Resh 2013, 2016). Aberrant Hh signalling pathways result in birth defects in humans including microencephaly, cyclopia, absent nose or cleft palate. The development of breast, prostate and lung cancer has also been associated with Hh signaling anomalies (Gupta et al. 2010).

Another member of MBOAT superfamily is porcupine (Porcn) transferase which mediates attachment of a monounsaturated *cis*-Δ9-palmitoleate unit *via* a side chain of serine residue to a secreted Wnt glycoprotein family (Resh 2016; Hofmann 2000; Nile and Hannoush 2016; Shindou et al. 2009). This post-translational lipid attachment plays a crucial role in regulating signalling during embryonic development and tissue homeostasis, Fig. 9.1d (Resh 2016; Nile and Hannoush 2016). It has been recently reported that Wnts palmitoylation is reversible; notum hydrolase, which participates in deacylation, affords an inactive form of Wnts with inhibited signalling ability (Resh 2016; Nile and Hannoush 2016; Zhang et al. 2015; Kakugawa et al. 2015). Targeting Wnt signalling pathways using synthetic modulators including small molecules and peptides is therefore a promising tool to inhibit Wnt-driven diseases such as cancer (Nile and Hannoush 2016; Anastas and Moon 2013).

Ghrelin *O*-acyltransferase (GOAT), another MBOAT enzyme, mediates the covalent attachment of octanoic acid onto Ser-3 of the 28-amino acid peptide hormone ghrelin (Fig. 9.1e) (Resh 2016; Yang et al. 2008; Gutierrez et al. 2008; Kojima et al. 1999; Müller et al. 2015). Ghrelin octanoylation is essential for the secretion of insulin and growth hormone, and hormone activity including appetite stimulation, adiposity and cardiovascular functions (Resh 2016; Gutierrez et al. 2008; Müller et al. 2015; Sato et al. 2015). Therefore, ghrelin is an attractive target in novel therapies to treat obesity and diabetes (Müller et al. 2015; Sato et al. 2015).

Protein prenylation refers to a post-translational attachment of isoprenoid lipids. Incorporation of farnesyl (C15) and geranylgeranyl (C20) groups is effected by formation of a thioether bond using a cysteine moiety in the *C*-terminal proximity of the protein *via* protein farnesyltransferase (FT) and geranylgeranyltransferase 1 (GGT 1), Fig. 9.1f, g, respectively (Wang and Casey 2016). The fully processed lipidated protein contains a prenylated cysteine residue with a methylated carboxylic acid moiety, at the protein *C*-terminus. Members from HRAS, KRAS, NRAS, prelamin A, lamin B, and RAS-related GTPases are examples of protein families incorporating isoprenoid lipids within their structures (Wang and Casey 2016). Prenylation controls the oncogenic activity of many proteins including farnesylated RAS proteins that are involved in 30% of human cancers (Wang and Casey 2016).

Another common eukaryotic post-translational lipid modification is the attachment of a complex glycosylphosphatidylinositol anchor to the *C*-terminus of proteins (Paulick and Bertozzi 2008; Ferguson et al. 2009). GPI comprises a phosphoethanolamine linker, a highly conserved glycan core (mannose(α1-2)mannose(α1-6) mannose(α1-4)glucosamine(α1-6)*myo*-inositol) and phospholipid tail which links the GPI anchor to the cell membrane (Paulick and Bertozzi 2008; Ferguson et al. 2009). The sugar-rich domain can be further modified with the addition of various groups including other glycans, sialic acid and phosphoethanolamine moieties affording functionally diverse glycoforms of GPI anchors (Paulick and Bertozzi 2008; Ferguson et al. 2009). The lipid portion of the GPI moiety differs depending on the protein which it is attached to and the organism it originates from. The GPI anchor of human erythrocyte acetylcholinesterase for example, comprises three fatty acids in various states of saturation and lengths ranging from 16 to 22 carbons (Fig. 9.2) (Paulick and Bertozzi 2008; Ferguson et al. 2009; Deeg et al. 1992; Roberts et al. 1988b, a). The exact structure-activity relationship of GPI-anchored proteins is poorly understood due to the complex nature of the GPI anchor structure. GPI-anchored proteins are multifunctional; these proteins have been identified in receptors, hydrolytic enzymes, adhesion and regulatory molecules etc (Paulick and Bertozzi 2008; Ferguson et al. 2009).

Atg8 and LC3 proteins found in yeast and mammals respectively, contain a phospholipid moiety, namely phosphatidylethanolamine (PE) that is post-translationally anchored to a *C*-terminal glycine residue *via* numerous steps of ubiquitination-like reactions catalysed by autophagy-related (Atg) proteins (Resh 2013). It has been reported that increased levels of PE enhance autophagy, a cytoprotective mechanism responsible for degradation of toxic proteins and potentially harmful and damaged organelles (Feng et al. 2014; Rockenfeller et al. 2015). Modulating autophagy can be used for the treatment of human disorders including cancer, diabetes, and Alzheimer's and Parkinsons' disease therefore new autophagy controllers are strongly desirable (Feng et al. 2014; Rockenfeller et al. 2015).

In summary, regulating the action of lipidated proteins may lead to potential therapies to treat infectious disease and human pathologies. Targeting NMT, Hedgehog acyltransferase, FT and GGT 1 inhibitors may play a role in anticancer

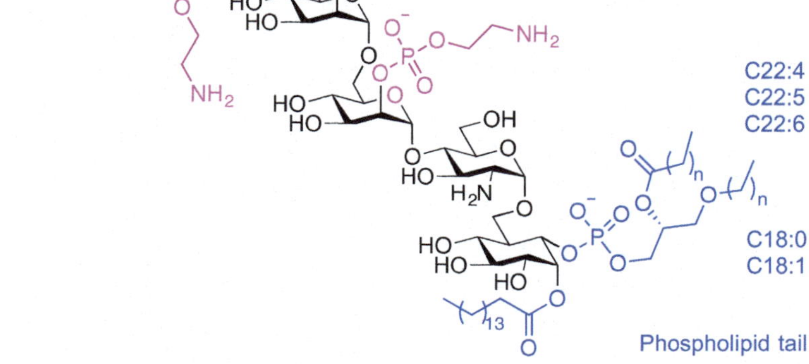

Fig. 9.2 Chemical structure of GPI anchor of human erythrocyte acetylcholinesterase (Paulick and Bertozzi 2008; Ferguson et al. 2009; Deeg et al. 1992; Roberts et al. 1988a, b)

therapies (Wang and Casey 2016; Berndt et al. 2011). Effective techniques to modulate prenylation patterns can be used in hepatitis D and C viruses (HDV and HCV) treatment (Koh et al. 2015; Cory et al. 2015; Ye et al. 2003), premature ageing disorders such as Hutchinson-Gilford progeria syndrome (HGPS) (Gordon et al. 2014; Young et al. 2013) in addition to neurodegenerative pathologies like multiple sclerosis and Alzheimer's disease (Wang and Casey 2016; Gao et al. 2016).

9.1.2 Nature-Derived Lipopeptides with Therapeutic Potential

Lipopeptides isolated from microorganisms such as fungi and bacteria show great therapeutic promise in the development of novel antimicrobial (Cochrane and Vederas 2016), antifungal, antitumor, and anti-inflammatory agents. In case of the plipastatins they can also act as potential therapies for neurological diseases (Dey et al. 2015).

Bacillus and *Paenibacillus* spp. produce lipopeptides of various structures including cyclic cationic and non-cationic lipopeptides where ring formation mostly occurs *via* the ester or amide bond and engages the *C*-terminal carboxylic acid residue (Cochrane and Vederas 2016). The presence of both, D- and L-amino acids together with non-natural amino acids in these lipopeptide sequences is common and improves peptide stability against enzymatic degradation. Branched saturated or unsaturated fatty acids with diverse structures with the main chain varying mostly between C11 to 14 carbons are mostly incorporated into the *N*α-terminal side of the peptides and often feature a β-hydroxyl moiety in their structure (Cochrane and Vederas 2016; Jacques 2011).

Polymyxins, octapeptins, pelgipeptins, and paenibacterins exhibit non-proteinogenic 2,4-diaminobutyric acid (Dab) residues that amplify the cationic character of these peptides (Cochrane and Vederas 2016). Examples of non-cationic cyclic lipopeptides include the iturin-, surfactin-, fengycin-, fusaricidin-, marihysin-, and kurstakin-families (Fig. 9.3) (Cochrane and Vederas 2016).

Linear cationic lipopeptides derived from *Bacillus* and *Paenibacillus* spp. such as cerexins and tridecaptins display promising antibacterial activity against Gram-positive and Gram-negative microbes (Fig. 9.3). A more detailed description of exact structures and biological activities for *Bacillus* and *Pseudomonas* spp. derived lipopeptides has recently been published (Cochrane and Vederas 2016; Jacques 2011; Mnif and Ghribi 2015).

Lipopeptides isolated from *Pseudomonas* spp., which mainly include the viscosins, amphisins and tolaasins in addition to syringomycins, are mostly known for their antiviral and antimicrobial properties (Mnif and Ghribi 2015; Raaijmakers et al. 2006). These structurally diverse cyclic peptides differ in the chain length and comprise 9-25 residues in the form of natural and non-natural amino acids including *allo*-threonine (*allo*-Thr), *allo*-isoleucine (*allo*-Ile), 3-hydroxyaspartic acid, Dab and homoserine (Hse). 4-Chlorothreonine is the amino acid responsible for the antifungal activity of syringomycin (Fig. 9.4) (Grgurina et al. 1994). The fatty acid moiety attached to the *N*-terminus of the peptide chain varies in length and composition and, similar to *Bacillus*-derived peptides, often features the β-hydroxyl unit. The lactone ring is generally formed between the carboxylic acid of the *C*-terminal amino acid and the hydroxyl group of either Ser, Thr or *allo*-Thr present within the peptide chain (Mnif and Ghribi 2015; Raaijmakers et al. 2006).

Other microbial sources of biologically active lipidated peptides with promising therapeutic potential found in nature include strains of *Acremonium*, *Streptomyces*, and *Actinoplanes* (Mnif and Ghribi 2015).

Lipopeptides exhibit a broad spectrum of activities against many pathogens and some naturally-derived compounds, as in the case of daptomycin, polymyxin B or colistin, have already received the Food and Drug Administration (FDA) approval. Daptomycin (Cubicin) isolated from *Streptomyces roseosporus* is a 13-amino acid, cyclic lipopeptide,

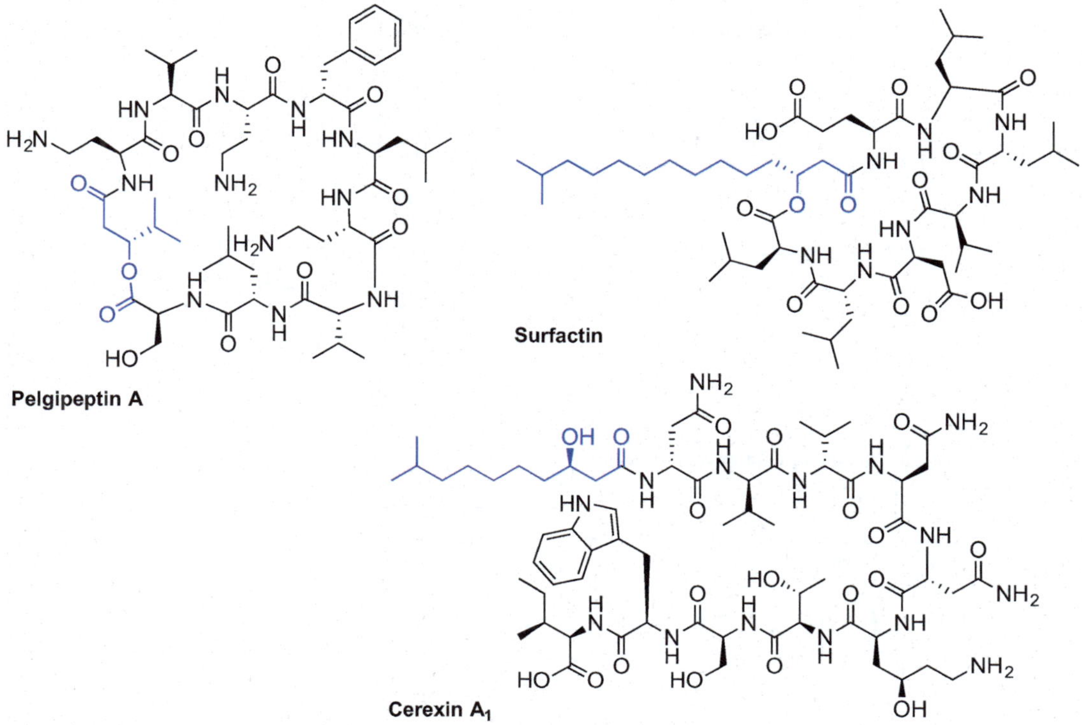

Fig. 9.3 Selected examples of chemical structures of lipopeptides isolated from *Bacillus* and *Paenibacillus* spp (Cochrane and Vederas 2016; Jacques 2011)

Fig. 9.4 Chemical structures of non-natural amino acids found in lipopeptides isolated from *Pseudomonas* spp.

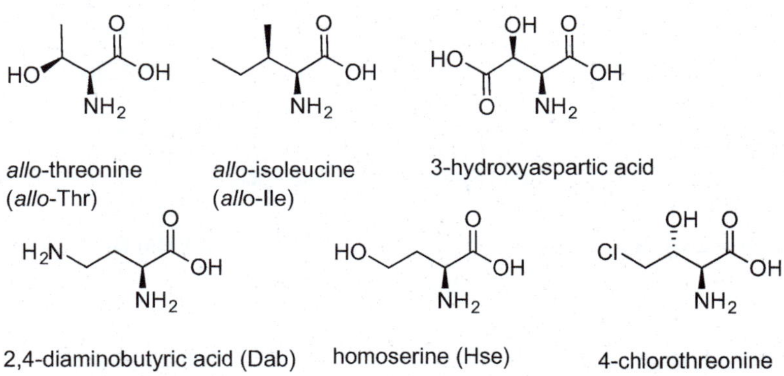

containing decanoic acid at the *N*α-amino group of the *N*-terminal L-tryptophan. Daptomycin exhibits potent activity against Gram-positive pathogens (Fig. 9.5) (Debono et al. 1987; Vilhena and Bettencourt 2012).

Polymyxins are mixed peptide antibiotics produced by *Bacillus polymyxa* and are considered to be the last-line of defence agents against Gram-negative organisms; their use is limited due to concerns with nephrotoxicity (Stansly and Schlosser 1947; Benedict and Langlykke 1947). The general structure of polymyxins comprises a cyclic heptapeptide core attached to a tripeptide unit containing a lipid portion at the *N*α-terminal site of the linear fragment (Velkov et al. 2016). Polymyxins are mixtures of structurally similar peptides. Members of the polymyxin B family mostly differ in the fatty acid component of the

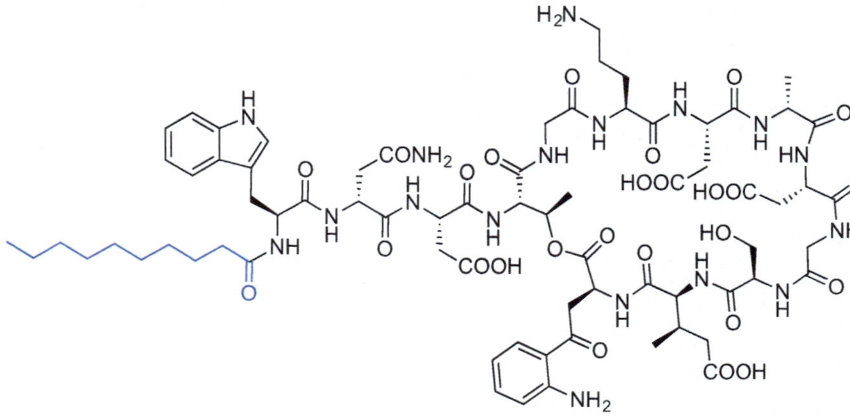

Fig. 9.5 Chemical structure of daptomycin (cubicin)

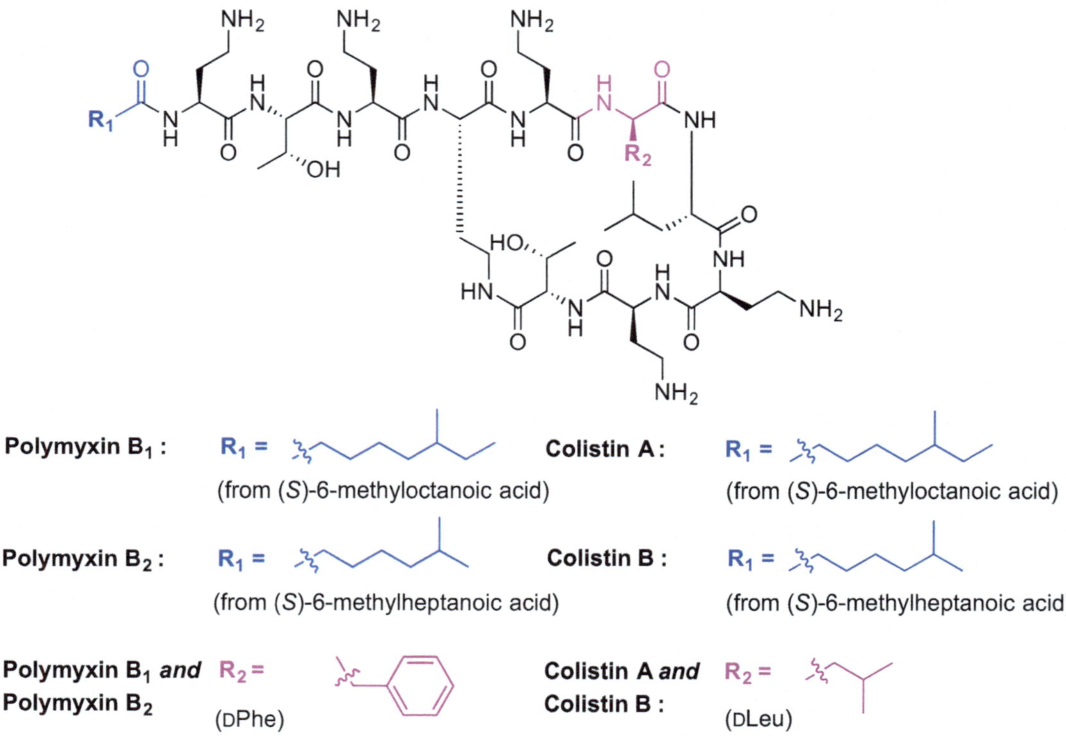

Fig. 9.6 Chemical structure of polymyxin B1, B2 and colistin A and B

antibiotics. Examples include (*S*)-6-methylheptanoic acid for polymyxin B_1 and B_2 respectively (Velkov et al. 2016; Orwa et al. 2001). Colistin A and colistin B are highlighted examples of the polymyxin E family; these antibiotics differ in the substitution of D-phenylalanine to D-leucine at position six of polymyxin B (Fig. 9.6) (Velkov et al. 2016; Brink et al. 2014).

The concept of protein lipidation is clearly not uncommon in nature hence application of this strategy to the therapeutic arena offers enormous potential for the generation of effective peptide-based drug candidates. Therefore, development and synthetic optimisation of naturally derived lipopeptides may afford fine-tuned therapeutics, which are less toxic, more potent and capable of treating multidrug-resistant infections. Interestingly, it has been reported that the attachment of aliphatic chains of various length (C12-C16) can modulate antimicrobial and antifungal activity of otherwise inert short peptides (Makovitzki et al. 2006). Therefore, peptide lipidation can be used as an effective strategy to generate peptide drug leads with clinical potential.

9.1.3 Peptide Lipidation to Generate Peptide-Based Therapeutics

Peptide lipidation can modulate the physicochemical and pharmacological properties of bioactive peptides generating therapeutically useful targets. Increased lipophilicity of peptides due to the presence of fatty acids affects the secondary structure and receptor and membrane binding characteristics of peptides; accordingly lipidation alters absorption, distribution, metabolism, and excretion (ADME) properties and therefore is an attractive tool to convert peptides into drug candidates (Zhang and Bulaj 2012). The most notable examples of clinically relevant lipidated peptides include long-acting insulin detemir (Levemir®) (Home and Kurtzhals 2006; Le Floch 2010) and liraglutide (Victoza®) (Jackson et al. 2010; Knudsen et al. 2000), a glucagon-like peptide-1 (GLP-1) receptor agonist, which are both used to treat diabetes (Fig. 9.7).

The prolonged activity of insulin detemir is due to the presence of C14 myristic acid incorporated into lysine-29 of the B chain of a modified insulin peptide sequence where the threonine-30 residue was removed (Fig. 9.7) (Le Floch 2010; Kurtzhals 2007).

Liraglutide is a long-acting analogue of GLP-1(7-37) where Lys-34 was replaced with Arg and Lys-26 was acylated with a C16 fatty acid attached to γ-glutamic acid as a spacer. The palmitic acid moiety plays a crucial role in delaying liraglutide absorption and extending the half-life of the drug which has been estimated to be 13 hours after subcutaneous injection compared to approximately 2 minutes for the native GLP-1 (Rigato and Fadini 2014; Elbrond et al. 2002). In addition, renal clearance of the drug is reduced due to the shielding effect of the fatty acid moiety; liraglutide binds to plasma albumin *via* the fatty acid group preventing drug degradation by dipeptidyl peptidase-4 (DPP-4) (Malm-Erjefalt et al. 2010; Watson et al. 2010). Lipidation of potent, but unstable GLP-1(7-37), much improved the pharmacokinetic profile of the peptide making it suitable for once-daily administration (Elbrond et al. 2002; Ryan and Hardy 2011). Liraglutide (Saxenda®) has been recently approved by the FDA and the European Medicines Agency (EMA) for adjunctive treatment of obesity (December 2014 and March 2015, respectively) (Iepsen et al. 2015; Bray 2015; Tomlinson et al. 2016).

It has been reported that the type and composition of the fatty acid attached to a bioactive peptide as well as the nature of the spacer between the peptide chain and the fatty acid moiety influences its activity and plasma half-life (Knudsen et al. 2000; Madsen et al. 2007; Lau et al. 2015).

Structure-activity studies of liraglutide analogues revealed the importance of the length, composition, polarity and bulkiness of the fatty acid moiety as well as the type of spacer between the active molecule and the lipid tail on half-life calculations (*in vivo* in pigs) and potency using the cloned human GLP-1 model (Knudsen et al. 2000; Madsen et al. 2007). Linear fatty acids ranging from C10 to C18 (**1**) incorporated into the liraglutide sequence using various linkers including α-, D-γ-glutamic acid, 4-aminobutanoic acid (GABA), β-alanine and triethyleneglycol were evaluated (Fig. 9.8a) (Madsen et al. 2007). Interestingly, prolonged activity increased with the fatty acid chain length starting from 0.8 hours for C10, increasing to 5.1 h (C11), 7.6 (C12), 9 h (C14), 16 h (C16) and 21 h (C18); receptor potency was only affected when the acid chain length was longer than 16 carbons (Madsen et al. 2007). The study underlined the importance of the spacer between the active peptide and the fatty acid and revealed the complete loss of

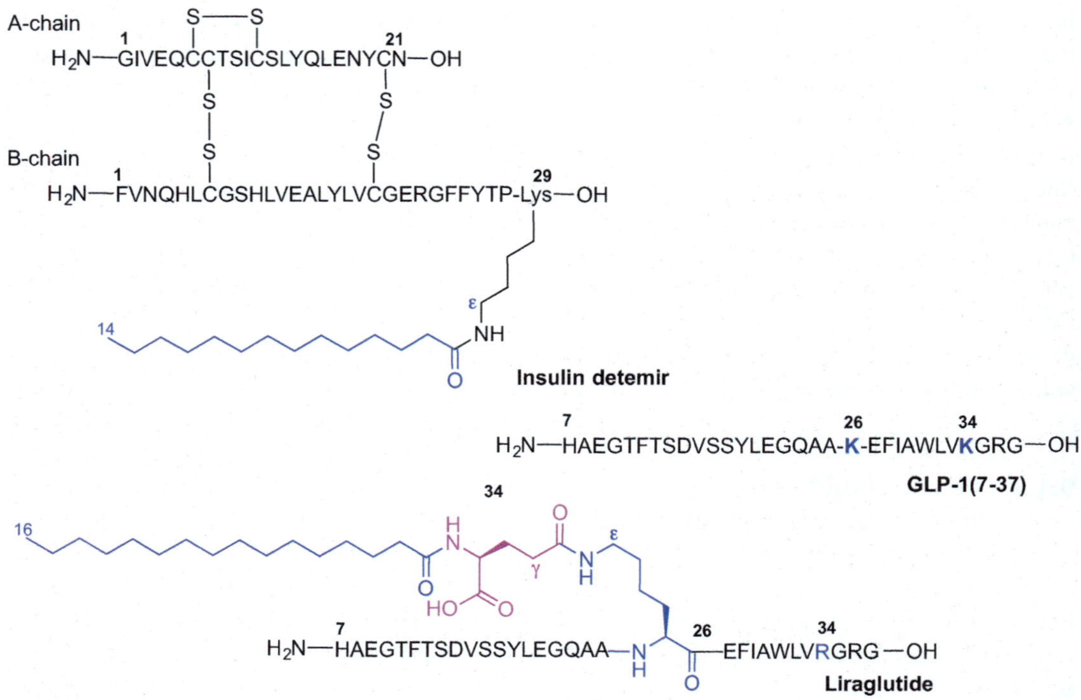

Fig. 9.7 Primary sequence of GLP1(7-37), liraglutide and insulin detemir

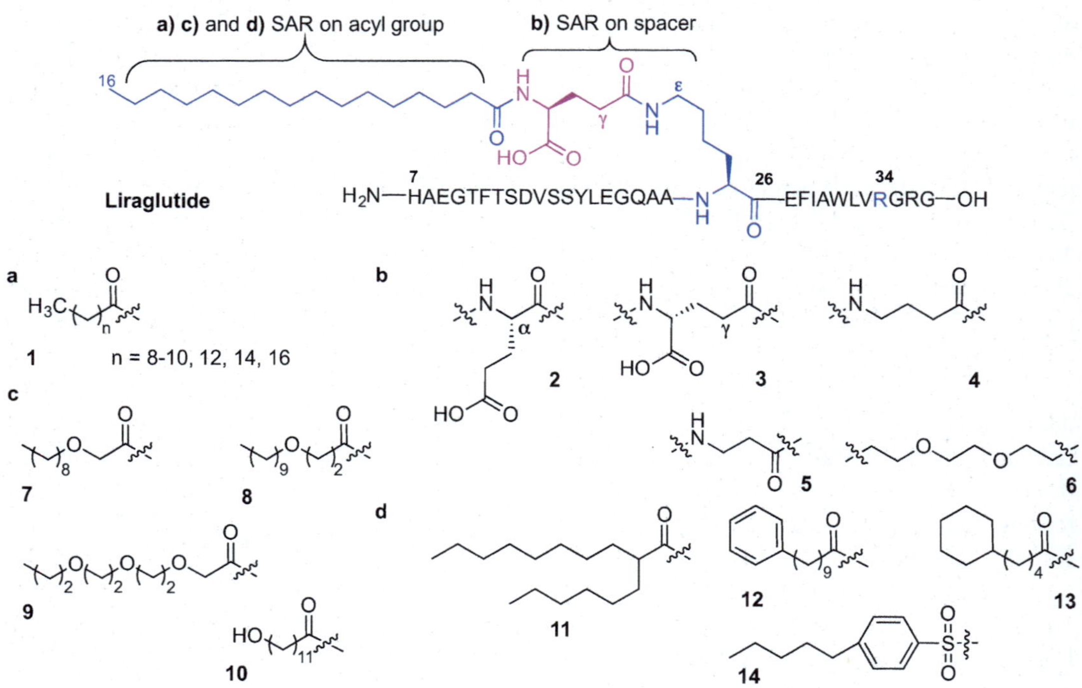

Fig. 9.8 Selected modifications of an acyl component (**a**, **c**, **d**) and spacer (**b**) investigated during the structure-activity study on liraglutide (Madsen et al. 2007)

receptor potency when palmitic acid was directly bound to Lys-26 (Madsen et al. 2007). Liraglutide analogues containing α- or D-γ-glutamic acid (**2** and **3**), GABA (**4**) or β-Ala (**5**), as linkers in place of the native γ-Glu demonstrated similar activities and half-life values to those of liraglutide; unlike the triethylene glycol linker (**6**) which caused a 25-fold decrease in activity (Fig. 9.8b) (Knudsen et al. 2000; Madsen et al. 2007). Increasing the polarity of the fatty acid component by introducing one or more ether groups (**7-9**) or inserting hydroxyl group at the omega terminus (**10**) decreased the protraction of the analogues possibly due to reduced interactions with the fatty acid sites present on albumin, Fig. 9.8c (Madsen et al. 2007). Modification of the C16 palmitic acid in the liraglutide sequence with 2-hexyldecanoyl acid (**11**) which is equivalent to 16 carbon atoms led to slightly improved protraction (18 hours versus 16 h) and a significant decrease in potency of the analogue. Incorporation of more bulky phenyl- and cyclohexyl rings (**12** and **13**, respectively) in place of palmitate, or palmitate replacement with a pentylbenzenesulfonyl group (**14**) was not beneficial in regards to improved potency and half-life values compared to the original molecule (Fig. 9.8d) (Madsen et al. 2007).

Further derivatization of the liraglutide structure resulted in the development of semaglutide (Lau et al. 2015; Nauck et al. 2016). Semaglutide is the once-weekly GLP-1(7-37) analogue currently in phase 3 clinical development for the treatment of type 2 diabetes (Lau et al. 2015; Nauck et al. 2016). Extending the half-life of semaglutide to 165 hours was realised through systematic study of the fatty acid chain type and the spacer attached to liraglutide (Lau et al. 2015). The superior effect of a C18 octadecanedioic acid moiety attached to Lys-26 and a long spacer unit composed of γ-Glu attached to two 8-amino-3,6-dioxaoctanoic acid moieties provided the optimal lead candidate (Fig. 9.9). Non-natural modification of Ala-8 with 2-aminoisobutyric acid (Aib) allowed for additional shielding of the molecule from degradative DPP-4 action (Lau et al. 2015).

The therapeutic potential of peptides as drugs is often hampered by undesirable ADME profiles; peptides are subjected to rapid proteolytic cleavage in the digestive system and are unable to cross the epithelial layer (Karsdal et al. 2015; Di 2015). Oral administration of peptide-based therapeutics is therefore limited. Many strategies to enhance oral delivery of peptides have been described in the literature. Generally, they include attachment of permeation enhancers (such as glycosides, lipids and PEG) and/or targeting proteolytic enzyme inhibitors. Exploration of multifunctional polymers as a polymeric matrix to provide controlled drug release and drug encapsulation in polymeric nanoparticulate systems has also been reported. Using ligand-specific binding and uptake techniques which employ vitamin B12, biotin, folate, and lectins to name a few, as drug carriers was also demonstrated. A more detailed discussion on these topics is covered elsewhere (Park et al. 2011; Karsdal et al. 2011, 2015). A brief discussion of lipidation phenomena affecting oral bioavailability with selected examples of biologically active peptides is described herein.

Chemical modification of the 32-amino acid salmon calcitonin (sCT) with an *N*-palmitoylated cysteine moiety attached to Cys-1 and Cys-7 of sCT *via* disulphide bonds greatly improved the bioavailability of the orally administrated native peptide (Wang et al. 2003). Significant levels of sCT could still be detected in rat plasma up to 12 hours after oral administration of lipidated-sCT compared to undetectable levels after 1 hour when the same dose of native sCT was used (Wang et al. 2003). In this report, a method termed 'reversible aqueous lipidization' (REAL) was used that allows for selective conjugation of a protein to a fatty acid *via* reversible disulphide linkage in aqueous solution using the water soluble *N*-palmitoyl cysteinyl 2-pyridyl disulphide reagent **15** (Scheme 9.1a) and the protein thiol (Ekrami et al. 1995). The REAL technique was applied to the lipidation of other therapeutic peptide drugs including Bowman-Birk protease inhibitor (BBI) (Ekrami et al. 1995), desmopressin (Wang et al. 1999; Wang et al. 2002) and octeotride (Yuan et al. 2005).

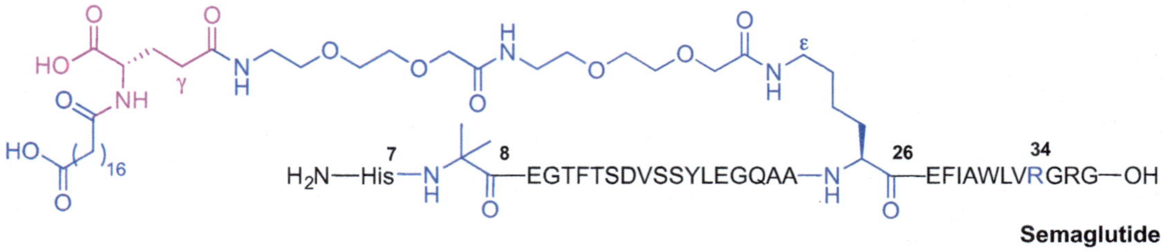

Fig. 9.9 Primary sequence of semaglutide

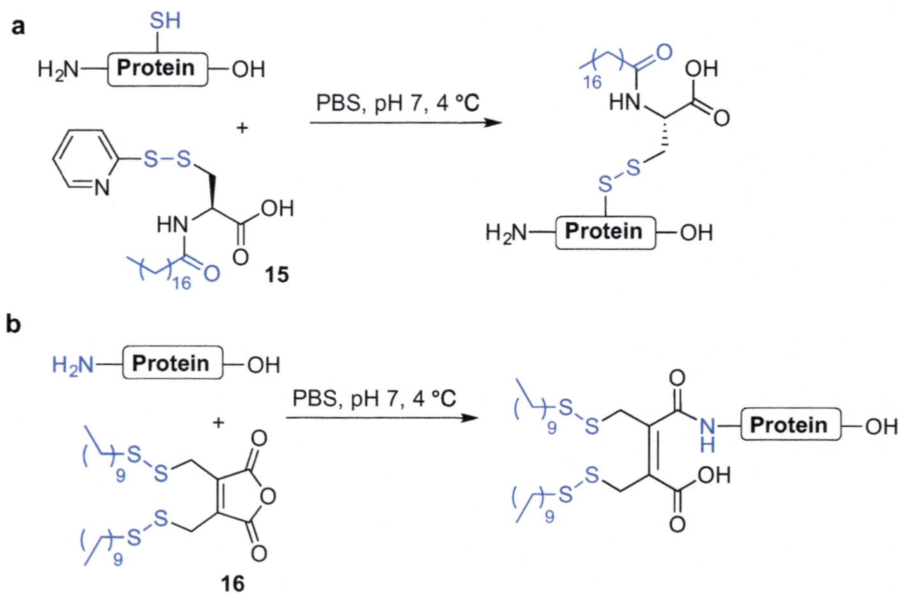

Scheme 9.1 REAL technique to lipidate proteins *via* protein thiol (**a**) and *N*-terminal *N*α-amino group (**b**) (Ekrami et al. 1995; Wang et al. 2006)

Peptide lipidation to improve oral bioavailability was also applied to the endogenous opioid peptide leu-enkephalin (ENK) using a modified REAL technique wherein 3,4-bis(decylthiomethyl)-2,5-furandione **16** was used to introduce a lipophilic moiety onto the *N*α-amino group of the *N*-terminus (Scheme 9.1b) (Wang et al. 2006).

It has been reported that incorporation of lauric acid to the *N*-terminal pyroglutamyl group of thyrotropin-releasing hormone (TRH) significantly improved peptide penetration across the upper small intestine (Muranishi et al. 1991; Tanaka et al. 1996).

There is ongoing interest in developing an insulin formulation that could bypass the requirement for daily subcutaneous insulin injection for the management of diabetes (Wong et al. 2016; Ramesan and Sharma 2014). Promising reports on improved stability of mono- and di-palmitoylated insulin analogues in mucosal tissue homogenates compared to native insulin (Hashimoto et al. 1989; Hashizume et al. 1992) prompted further research into the effects of lipidation on the pharmacokinetic profile of insulin (Asada et al. 1994, 1995). The effect of acylation on the stability and absorption of insulin from the small and large intestines was examined using mono- and di-acylated bovine insulin analogues (Asada et al. 1994, 1995). Mono-acylated analogues were constructed *via* incorporation of caproic (C6), lauric (C12) and palmitic acid (C16) at the *N*α-amino group of Phe-1 of the insulin B chain; di-acylated analogues were prepared by modification of both the *N*α-amino group of Phe-1 and the *N*ε-amino group of Lys-29 of insulin B chain, with two copies of C6, C12, or C16 fatty acids (Fig. 9.10a). Mono-acylated analogues were found to be more stable in small intestinal fluid at 37 °C (Asada et al. 1994) and increased absorption of caproic acid-modified analogues from the large intestine was observed compared to the native compound (Asada et al. 1995).

Acylation of insulin and insulin analogues incorporating arginine residues at various sites within the insulin sequence, with various saturated and unsaturated fatty acids attached to the B chain showed improved solubility at moderately acidic pH inducing long-acting basal control of glucose levels (Flora 2002).

Hexyl-insulin monoconjugate 2 (HIM2) is an insulin analogue that can be administrated as an oral semisolid formulation in hard gelatin capsules (Clement et al. 2002; Still 2002; Kipnes et al. 2003; Clement et al. 2004). HIM2 was created by chemical modification of recombinant insulin by covalent attachment of an amphiphilic oligomer consisting of a lipophilic alkyl unit (C6) and a hydrophilic PEG moiety covalently bound to the *N*ε amino group of the Lys-29 (B chain) (Fig. 9.10b) (Clement et al. 2002, 2004; Still 2002; Kipnes et al. 2003).

Despite various scientific efforts, formulation of orally available insulin and other peptide-based drugs remains a challenging task (Lewis and Richard 2015; Hamman et al. 2005; Renukuntla et al. 2013; Karsdal et al. 2011, 2015).

Peptide lipidation has also been used to mimic the post-translational processes of sterol or lipid attachment facilitating protein association with cell membranes and subsequent initiation of protein activation or deactivation processes (Mejuch and Waldmann 2016; Resh 2013; Avadisian and Gunning 2013). This nature-derived strategy is often designed to generate lipid-anchored drugs including lipidated peptide inhibitors with improved *in vivo* half-life and cell-penetrating potential. The lipid moiety attached to a peptide allows drug anchoring within the cell membrane and enabling action on soluble cytosolic proteins and membrane-bound/associated proteins (Avadisian and Gunning 2013; Rajendran et al.

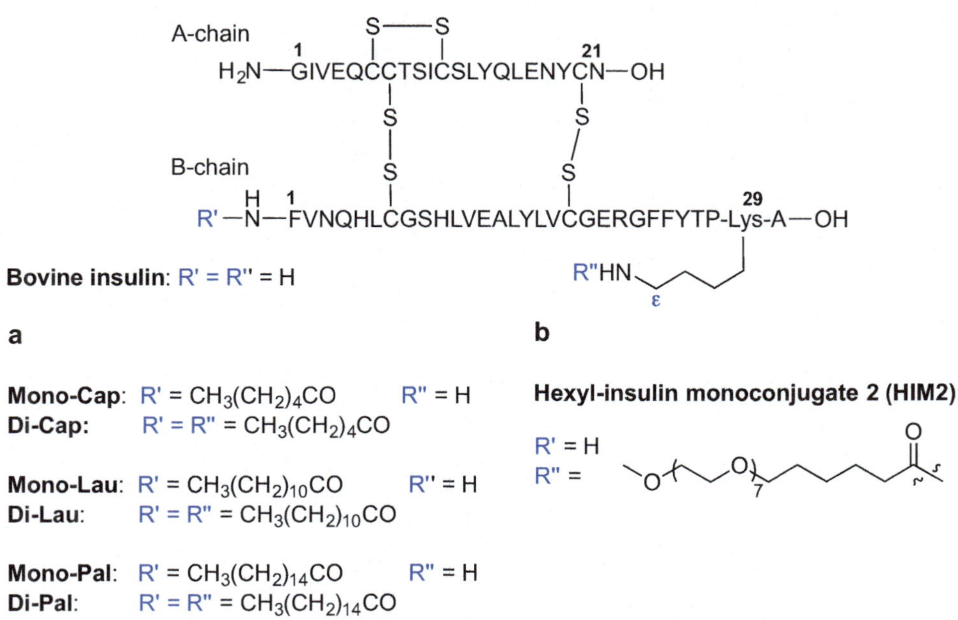

Fig. 9.10 Lipidated insulin analogues created to improve oral bioavailability (Asada et al. 1995; Clement et al. 2002, 2004; Still 2002; Kipnes et al. 2003)

2012). Cholesterol and fatty acids of various chain lengths such as C8-caprylic, C12-lauric, and C16-palmitic are often utilized as lipid motifs that are covalently attached to a peptide inhibitor *via* ester, ether, amide or carbamate bonds (Avadisian and Gunning 2013; Zhao et al. 2012; Wexler-Cohen and Shai 2009; Remsberg et al. 2007; Rajendran et al. 2008a, b; Porotto et al. 2010; Johannessen et al. 2011; Avadisian et al. 2011).

This 'lipid anchoring technique' allowing for subcellular drug delivery by drug conjugation to a lipid *via* a linker, was recently used to effectively inhibit the action of endosomal β-secretase (Rajendran et al. 2008a, b). β-Secretase inhibitors may be useful for the treatment of Alzheimer's disease by blocking the enzyme involved in amyloid formation. The lipid-anchored inhibitors consist of three main parts which include the pharmacophore ('message'), the lipid anchor ('address'), and the linker which conjugates both parts together and allows for optimal flexibility of the pharmacophore within the lipid bilayer to bind with the target (Rajendran et al. 2012). Simons et al. (Rajendran et al. 2008a, b) showed that that conjugation of a sterol to the β-secretase inhibitor (Glu-Val-Asn-statine-Val-Ala-Glu-Phe) *via* a polyglycol linker resulted in greater efficacy; β-secretase cleavage of β-amyloid precursor protein (APP) was decreased resulting in reduced β-amyloid peptide formation (Fig. 9.11). Importantly, the cholesterol-enriched drug was readily internalized into endosomes and cholesterol-sphingolipid domains (rafts) within cellular membranes where β-secretase activity is observed (Rajendran et al. 2008a, b; Hicks et al. 2012; Cordy et al. 2006). Comparison of stearyl-, palmityl-, myristyl-, and oleyl-linked inhibitors revealed cholesterol- and palmitoyl-linked analogues to be superior in terms of raft partitioning ability (Rajendran et al. 2008a).

The lipidation site within the peptide chain is critical as it can determine the pharmacokinetic and pharmacodynamic properties of drug candidates by affecting the solubility and the self-aggregating potential of lipopeptides. Ward et al. (Ward et al. 2013) investigated lipidated glucagon-based peptides to identify acylated co-agonists for the glucagon and glucagon-like peptide 1 receptors (GCGR and GLP-1R, respectively). A number of palmitoylated and *C*-amidated glucagon analogues were prepared where Ser-2 was substituted with an Aib moiety to prevent enzymatic degradation by dipeptidyl peptidase-4. The *N*ε-amino group of Lys-12 or an introduced lysine residue that was used to replace the mid-region moieties of glucagon, namely Tyr-10 or Tyr-13, Leu-14 or Ser-16, Arg-17 or Gln-20, was explored to attach a palmitic acid *via* a γGlu-γGlu dipeptide spacer (Ward et al. 2013). The solubility and aggregate-forming potential of glucagon analogues in phosphate-buffered saline (PBS) (pH 7.4) was variable. Decreased solubility and increased aggregation was observed for the acylated analogue at position 14 which correlated with its reduced *in vivo* activity compared to the other analogues (Ward et al. 2013). Interestingly,

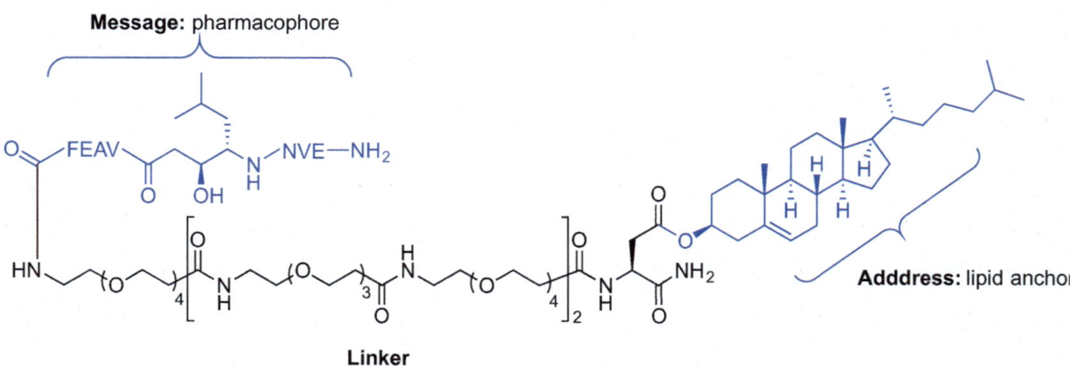

Fig. 9.11 Sterol conjugation to the β-secretase inhibitor using lipid anchoring technique (Rajendran et al. 2012, 2008a, b)

the study also revealed an increased proportion of helical content for all C16 fatty acid-tagged analogues in addition to improved potency at glucagon and GLP-1 receptors for most of the palmitoylated analogues. This is the first indication of enhancing *in vitro* receptor potency through helix stabilization by lipidation (Ward et al. 2013). This finding further reinforced the importance of lipidation in the development of therapeutic peptides (Ward et al. 2013). It was observed that saturated fatty acids with longer chains (>C8) have greater conformation-stabilising potential compared with unsaturated or hydroxyl counterparts due to enhanced hydrophobic interactions with the peptide chains (Zhang and Bulaj 2012). Lipidation was also shown to be an effective tool to induce peptide oligomerization and self-assembly resulting in the formation of micelles, tubules, vesicles, mono- and bilayer structures that can be used in both the drug delivery and tissue engineering fields (Zhang and Bulaj 2012; Hutchinson et al. 2017; Hamley 2015).

Peptide lipidation is an effective strategy to increase the drugable potential of bioactive peptides and has been applied to many other biomolecules not mentioned in this report including angiotensin II (Maletínskâ et al. 1996; Maletinska et al. 1997), BBI (Honeycutt et al. 1996), desmopressin (Wang et al. 1999; Wang et al. 2002), galanin, (Saar et al. 2013; Robertson et al. 2010; Zhang et al. 2009), ghrelin (Bednarek et al. 2000), neuropeptide Y (NPY) (Green et al. 2011; Green et al. 2010), octreotide (Yuan et al. 2005), luteinizing hormone releasing hormone (LHRH) (Toth et al. 1994), tetragastrin (Fujita et al. 1998; Setoh et al. 1995; Yodoya et al. 1994), and more. Further details relating to the above mentioned lipidated analogues can be found in the recent review by Zhang and Bulaj (Zhang and Bulaj 2012).

9.1.4 Pam$_n$Cys Ligand as Adjuvant for Peptide-Based Vaccines

There has been significant interest directed towards the development and synthesis of peptide vaccines as alternatives to conventional vaccines, where potentially toxic, whole live attenuated or killed microorganisms are used to elicit immune responses (Simerska et al. 2011; Moyle and Toth 2008; Li et al. 2014; Brown and Jackson 2005). One of the limitations of peptide-based vaccines is the lack of immunogenicity thus requiring the inclusion of an effective and safe adjuvant (Simerska et al. 2011; Moyle and Toth 2008; Khong and Overwijk 2016).

A less explored class of immune adjuvants are compounds stimulating innate-like T cells, semi-activated T cells with an invariant T cell receptor (TCR) represented by the invariant natural killer T cells (NKT) that recognize glycolipid antigens binding to the lipid antigen-presenting molecule CD1d (Fujii et al. 2003; Hermans et al. 2003). The most well-known CD1d ligand is α-galactosylceramide (α-GalCer, KRN 7000) (Godfrey and Kronenberg 2004) and studies on the use of α-GalCer conjugated to peptide antigens generating potent self-adjuvanting vaccine constructs have been reported (Anderson et al. 2014, 2015 Cavallari et al. 2014).

Toll-like receptors (TLRs) are transmembrane glycoproteins which play an important role in initiating an innate immunity response and developing the adaptive immune response (Gay and Gangloff 2007; Basto and Leitao 2014). Ten members of the human TLR family namely TLR1-TLR10 have been identified. TLR agonists vary and include viral genetic material, microbial nucleic acids and microbial membrane components (Mifsud et al. 2014). Stimulation of TLRs may therefore lead to potent therapies against infectious diseases and many TLR ligands have been evaluated as potential treatments of viral and bacterial infections (Basto and Leitao 2014; Mifsud et al. 2014; Zaman and Toth 2013; Khong and Overwijk 2016).

Lipopeptides derived from bacterial cell wall components including lipoproteins, peptidoglycans, lipoteichoic acid and lipopolysaccharides can activate Toll-like receptor 2 (TLR2) (Basto and Leitao 2014; Zaman and Toth 2013). Conjugation of lipids and liposaccharides to peptide antigens is therefore used to elicit an immune response and plays an important role in self-adjuvanting vaccine development (Simerska et al. 2011; Moyle and Toth 2008; Zaman and Toth 2013).

Common lipidated moieties employed in vaccine design to induce immunogenicity include synthetic analogues of lipoprotein components of *Escherichia coli* (Braun 1975) and *Mycoplasma* (Muhlradt et al. 1998; Muhlradt et al. 1997), namely *S*-[2,3-bis(palmitoyloxy)propyl]-*N*-palmitoyl-L-cysteine (Pam_3Cys) (**17**) and *S*-[2,3-bis(palmitoyloxy)propyl]-L-cysteine (Pam_2Cys) (**18**) (Zeng et al. 2002), respectively (Fig. 9.12) (Khong and Overwijk 2016).

Pam_3Cys and Pam_2Cys have been used as adjuvants in several peptide-based vaccine studies directed towards treating various infectious diseases including, HIV, HBV, hepatitis C (Chua et al. 2008; Chua et al. 2012; Eriksson and Jackson 2007), Lyme disease and influenza (Moyle and Toth 2008; Khong and Overwijk 2016; Zaman and Toth 2013; Chua et al. 2015; Tan et al. 2012) in addition to melanoma (Zom et al. 2014). Better water solubility and similar or improved immunogenicity shown by Pam_2Cys compared to Pam_3Cys (Zaman and Toth 2013; Jackson et al. 2004), makes this motif an even more interesting synthetic target for incorporation into peptide-based vaccines. Structure-activity studies carried out for Pam_2Cys demonstrated enhanced activity by the natural (*R*) configuration at the asymmetric glyceryl carbon, in comparison to the (*S*) isomer, namely *S*-[2(*R*),3-bis(palmitoyloxy)propyl]-L-cysteine [(*R*)-Pam_2Cys], and *S*-[2(*S*),3-bis(palmitoyloxy)propyl]-L-cysteine [(*S*)-Pam_2Cys], respectively (Moyle and Toth 2008; Zaman and Toth 2013; Wu et al. 2010; Takeuchi et al. 2000). Conversely, incorporation of the (*R/S*) diastereoisomer of Pam_3Cys within the MUC1 antitumor vaccine construct elicited immune responses similar to that of the same MUC1 glycopeptide comprising only the (*R*)-enantiomer (Shi et al. 2016).

It has been reported that the Pam_2Cys fatty acid chain length plays a crucial role in determining TLR2 activation; the minimum carbon chain length required for immunogenic activity is C8 and the strength of immune response increases with carbon addition up to C16 (C18=C16>C12>C8) (Moyle and Toth 2008; Zaman and Toth 2013; Buwitt-Beckmann et al. 2005b; Chua et al. 2007). A more soluble derivative of Pam_2Cys, namely Pam_2CysSK_4 showed the most promising activity amongst a range of adjuvants tested in the evaluation of a *Chlamydia trachomatis* vaccine (Cheng et al. 2011; Spohn et al. 2004). It has been reported that the presence of a serine moiety within the Pam_2CysSK_4 motif plays a role in enhanced agonist activity for TLR2 (Wu et al. 2010; Kang et al. 2009).

Further SAR studies on Pam_2CysSK_4 led to identification of a structurally simpler and water soluble monopalmitoylted analogue **19** and its *N*α-amino acetylated variant **20** possessing strong TLR2-agonistic activities, comparable to that of $Pam_2CysSer$, in human (but not murine) blood (Fig. 9.12) (Agnihotri et al. 2011; Salunke et al. 2012). The correct spacing between the ester-linked palmitate and the thioether was found to be crucial for activity of analogue **19** and replacement of the ethyl chain with a propyl chain resulted in loss of activity (Wu et al. 2010; Agnihotri et al. 2011; Salunke et al. 2012).

Replacement of the native amide bond within the Pam_3Cys motif with an urea led to discovery of a novel TLR2 ligand termed UPam; substitution of the native *N*-palmitoyl chain of Pam_3Cys with an *N*-tetradecylcarbamyl moiety afforded a ligand with improved immunostimulatory activity compared to the parent lipopeptide (Fig. 9.12) (Zom et al. 2014, 2016; Willems et al. 2014).

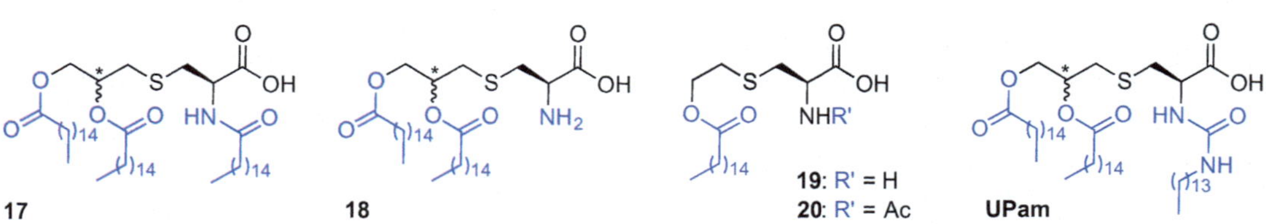

Fig. 9.12 Chemical structure of Pam_3Cys (**17**), Pam_2Cys (**18**), PamCys (**19**), *N*-acetylated PamCys (**20**) and UPam

The use of a cationic lipidated peptide such as R_4Pam_2Cys to elicit T-cell immunity *via* TLR2 stimulation was recently described; the strategy relies on electrostatic attraction of the R_4Pam_2Cys moiety with soluble protein antigens obviating the need for covalent bond generation between the TLR2 ligand and the antigen (Chua et al. 2014).

The use of palmitic acid, lipoamino acids and other lipid-based immunopotentiators, as an alternative to Pam_nCys, covalently bound to synthetic (glyco)peptides to improve the self-adjuvanting effect of vaccine constructs has been reported and is reviewed elsewhere (Moyle and Toth 2008; Khong and Overwijk 2016; Basto and Leitao 2014; Zaman and Toth 2013; McDonald et al. 2015; Steinhagen et al. 2011).

9.1.5 Chemical Approaches for Incorporation of Pam_nCys Ligands

Finding efficient methods to conjugate antigens to lipopeptide adjuvants remains challenging (McDonald et al. 2015). A simple and low-cost synthetic approach for peptide-lipid conjugation to effectively activate TLR2 to afford synthetic material in significant quantities for biological evaluation, is highly desired. A synthetic strategy must be devised using techniques from the chemistry toolbox that are compatible with the presence of lipid, carbohydrate and peptide moieties often required for self-adjuvanting vaccines. Herein, the most recent advances in synthetic techniques used to incorporate TLR2 ligands based on the Pam_nCys moiety into (glyco)peptides are summarized.

A solution phase synthesis of a simple dipeptide by direct condensation of $N\alpha$-9-fluorenylmethoxycarbonyl (Fmoc)-protected *S*-(2,3-bis(hydroxyl)propyl)-L-cysteine with serine where the side chain hydroxyl is protected with a *tert*-butyl (*t*Bu) ether was reported by Jung et al. (Metzger et al. 1991). Subsequent palmitoylation of *S*-glycerylcysteinyl hydroxyls using palmitic acid, *N*,*N'*-diisopropylcarbodiimide (DIC) and 4-(dimethylamino)pyridine (DMAP), followed by *t*Bu protecting group removal from the serine side chain effectively provided Fmoc-Pam_2CysSer (Metzger et al. 1991).

Danishefsky et al. (Kudryashov et al. 2001) employed a solution phase approach to successfully incorporate the Pam_3Cys ligand into a trivalent Lewis Y antigen resulting in antibody production in animal models. However, more common approaches to incorporate the Pam_nCys motif into peptides when designing a synthetic vaccine mostly rely on Fmoc solid phase peptide synthesis (SPPS). In this case, the peptide-based vaccine construct is synthesized first followed by lipid attachment. This approach however may prove problematic when synthesizing long or difficult peptide sequences (Zeng et al. 2011).

Alternatively, a convergent or modular approach can be used requiring initial preparation of vaccine motifs that are later conjugated, mostly *via* a linker, affording a self-adjuvanting vaccine construct (Zeng et al. 1996, 2001, 2002, 2011; Harris et al. 2007; Buwitt-Beckmann et al. 2005a; Metzger et al. 1995). The choice of chemical linkage used for adjuvant-antigen conjugation is very important and may influence the bioactivity of the construct (Zeng et al. 2011).

9.1.5.1 Convergent and Modular Approaches to Self-Adjuvanting Vaccine Constructs

A fully synthetic convergent approach for the preparation of the minimal vaccine construct consisting of the *S*-[2(*R*),3-bis(palmitoyloxy) propyl]-*N*-palmitoyl-L-cysteine [(*R*)-Pam_3Cys)], a helper T cell epitope and T_N antigen (GalNAc) leading to high titres of IgG antibodies in mice was reported by Boons et al. (Buskas et al. 2005). In this example, the resin-bound and side chain protected peptide T cell epitope derived from an outer-membrane protein of *Neisseria meningitides* (Wiertz et al. 1992) was first synthesized using Fmoc SPPS using the extremely acid sensitive 4-(4-hydroxymethyl-3-methoxyphenoxy) butyryl-*p*-methylbenzhydrylamine (HMPB-MBHA) resin affording H_2N-Y(*t*Bu)AFK(Boc) Y(*t*Bu)AR(Pbf)H(Trt)AN(Trt)VGR(Pbf)N(Trt) AFE(O*t*Bu)LFLG-resin (**21**) (Scheme 9.2). To minimize racemization at cysteine, Pam_3Cys was

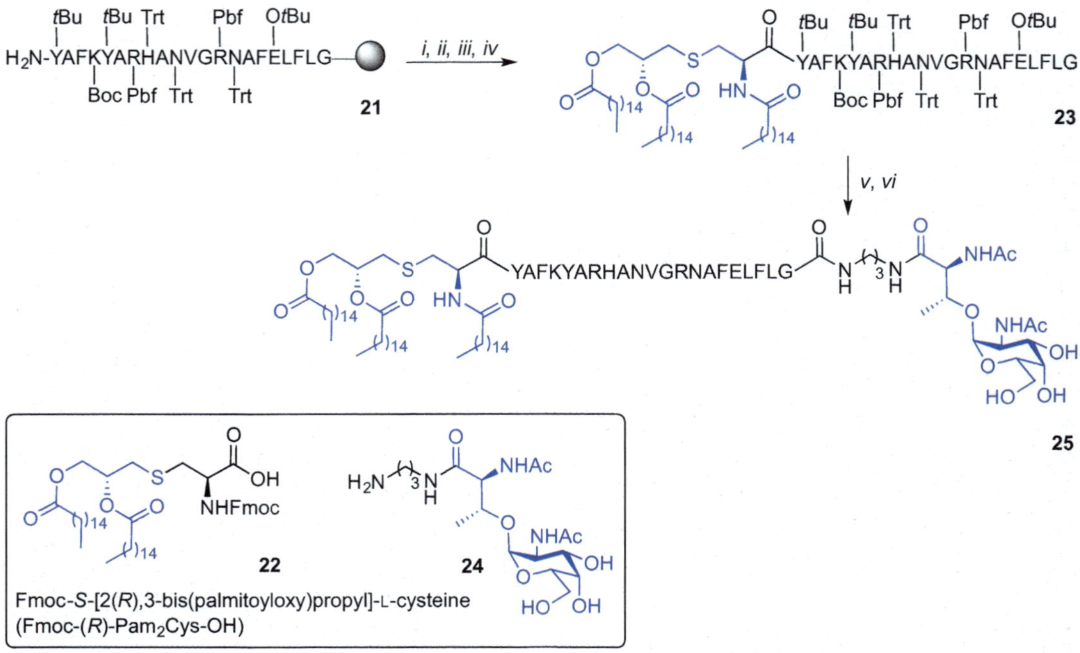

Scheme 9.2 Convergent approach to a self-adjuvanting lipidated vaccine construct incorporating Pam_3Cys TLR2 ligand by Boons et al. (Buskas et al. 2005). Reagents and conditions: (*i*) **22**, PyBOP, HOBt, iPr_2NEt, DMF/CH_2Cl_2 (1:5, v/v); (*ii*) 20% piperidine in DMF; (*iii*) $CH_3(CH_2)_{14}COOH$, PyBOP, HOBt, DMF/CH_2Cl_2 (1:5); (*iv*) 2% TFA in CH_2Cl_2; (*v*) **24**, DIC, HOAt, iPr_2NEt, DMF/CH_2Cl_2 (2:1, v/v), 79%; (*vi*) TFA/H_2O/1,2-ethanedithiol (EDT) (95:2.5:2.5, v/v/v)

introduced into the epitope sequence using the Fmoc-*S*-[2(*R*),3-bis(palmitoyloxy)propyl]-L-cysteine (Fmoc-(*R*)-Pam_2Cys-OH) **22** under the activation of (benzotriazol-1-yloxy)tripyrrolidinophosphonium hexafluorophosphate (PyBOP), 1-hydroxybenzotriazole (HOBt), *N*,*N*-iisopropylethylamine (iPr_2NEt) in a mixture of *N*,*N*-dimethylformamide (DMF) and CH_2Cl_2. Subsequent acylation of the Fmoc-deprotected *N*α-amino group of Cys with palmitic acid and using PyBOP and HOBt, followed by resin cleavage [2% trifluoroacetic acid (TFA) in CH_2Cl_2] gave the side-chain protected Pam_3Cys-tagged lipidated peptide **23**. Finally, condensation of **23** with a spacer containing tumour-associated T_N antigen **24** activated by DIC, 1-hydroxy-7-azabenzotriazole (HOAt) and iPr_2NEt and ultimate side chain protecting group removal using 95% TFA gave the target vaccine construct **25** (Scheme 9.2) (Buskas et al. 2005)

Jackson et al. (Zeng et al. 2011) proposed a modular approach (Zeng et al. 2001) for the preparation of self-adjuvanting vaccine constructs, where standard Fmoc SPPS was used. On-resin incorporation of the Fmoc-Pam_2Cys-OH (Zeng et al. 2002; Metzger et al. 1991; Jones 1975; Hida et al. 1995) *via* a diserine spacer to the *N*ε of an *N*-terminal lysine afforded lipidated $CD4^+$ $T(T_H)$ cell epitope (Zeng et al. 1996, 2002, 2011). The lipid-tagged T_H epitopes were then further *N*-terminally modified to facilitate a chemoselective ligation with complementary functional groups present at the target epitope modules affording oxime-, thioether-, and disulphide bond-linked lipidated vaccine constructs, ready for antibody response studies using animal models (Zeng et al. 2011).

Thus, Fmoc SPPS of T_H epitopes containing *N*-terminal lysine with *N*α-and *N*ε-amino groups orthogonally protected using 1-(4,4-dimethyl-2,6-dioxocyclohexylidene)ethyl (Dde) and Fmoc protecting groups respectively, were prepared affording T_H constructs of general structure **26** (Scheme 9.3). Removal of the Fmoc protecting

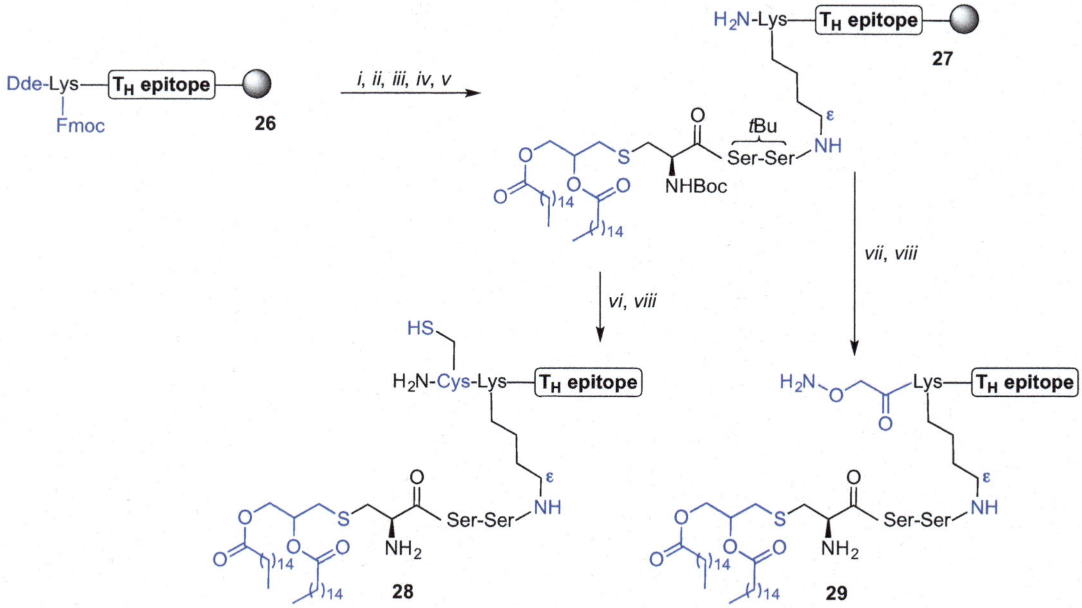

Scheme 9.3 Synthesis of lipidated-T_H epitopes incorporating Pam_2Cys TLR2 ligand by Jackson et al. (Zeng et al. 2011). Reagents and conditions: (*i*) 20% piperidine in DMF; (*ii*) Fmoc-Ser(*t*Bu)-OH, HBTU, HOBt, *i*Pr_2Net, DMF, then (*i*), repeated twice; (*iii*) Fmoc-Pam_2Cys-OH, TBTU, HOBt, *i*Pr_2NEt, CH_2Cl_2, 16 h, then (*i*); (*iv*) Boc_2O, DMF; (*v*) 2% hydrazine hydrate in DMF, 10 min; (*vi*) Boc-Cys(Trt)-OH, HBTU, HOBt, *i*Pr_2Net, DMF; (*vii*) (Boc-aminooxy)acetic acid, DMF; (*viii*) TFA/phenol/H_2O/triisopropylsilane (*i*Pr_3SiH) (88:5:5:2)

group using piperidine then allowed for peptide elongation *via* the exposed *N*ε-amino group to effect incorporation of the diserine spacer. Subsequently, the Fmoc-Pam_2Cys-OH building block was attached using *N*-[(1H-benzotriazol-1-yl)(dimethylamino)methylene]-*N*-methylmethanaminium tetrafluoroborate *N*-oxide (TBTU), HOBt and *i*Pr_2NEt in CH_2Cl_2. The Fmoc protecting group of Pam_2Cys moiety was then exchanged for the *N*-(*tert*-butoxycarbonyl) (Boc) (di-*tert*-butyl dicarbonate, Boc_2O) allowing for orthogonal removal of the Dde from the *N*-terminal amino group of Lys using 2% hydrazine hydrate in DMF, providing lipidated T_H construct **27**. Boc-Cys(Trt)-OH or (Boc-aminooxy) acetic acid were then coupled to the lipidated epitope **27** with subsequent peptide cleavage from the resin using TFA to give Pam_2Cys-tagged T_H epitopes with sulphydryl- (**28**), or aminooxyacetyl-functionality (**29**) at the *N*-terminus, as handles for subsequent elongation with target epitopes (Scheme 9.3).

The target epitopes were separately synthesized using Fmoc SPPS and their *N*-termini acylated with bromoacetic acid or cysteine while still bound to resin. TFA-mediated peptide cleavage from the resin subsequently afforded bromoacetyl-, and thiol-tagged epitopes **30** and **31**, respectively. Alternatively, an additional serine residue was inserted at the *N*-terminus of the peptide sequence allowing for off-resin and sodium periodate-mediated serine oxidation affording an epitope with an *N*-terminal aldehyde handle **32**. Chemoselective ligation between complementary-tagged T_H and target epitopes in buffer solutions, namely **28** and **30** (*aq* buffer, pH 8), **28** and **31** (2,2′-dipyridyl disulphide), **29** and **32** (*aq* buffer, pH 3) gave thioether-, disulphide, and oxime-bond linked self-adjuvanting peptide-based vaccine constructs **33-35**, ready for further bioanalysis (Zeng et al. 2011) (Scheme 9.4).

This modular approach (Zeng et al. 2001) ensured that attachment of the Pam_2Cys motif at the *N*ε-amino group of Lys "in between" both

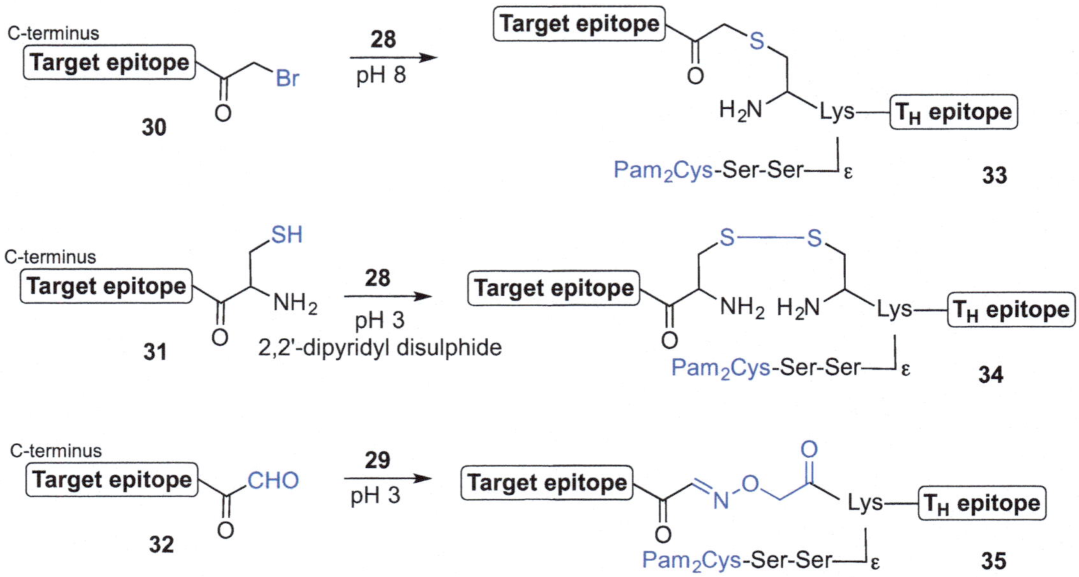

Scheme 9.4 Schematic representation of modular approach to a self-adjuvanting lipidated vaccine constructs **33**-**35** incorporating Pam_2Cys TLR2 ligand and different epitopes linked *via* thioether- (**33**), disulphide- (**34**), and oxime bond (**35**) by Jackson et al. (Zeng et al. 2011)

epitopes orientating the vaccine constructs in a branched configuration. The Pam_2Cys motif can also be incorporated at the *N*α-amino group at the *N*-terminus of a vaccine construct; however, decreased immunogenic activity resulted following linear assembly, partially due to reduced solubility, compared to the branched vaccine counterparts (Zeng et al. 2002).

A new thioether ligation strategy to create self-adjuvanting peptide vaccine constructs using the Pam_3CysSK_4 moiety has been recently reported (Cai et al. 2013). This approach takes advantage of the complementary modified Pam_3CysSK_4 motif with a bromo-handle and thiol-containing antigen that are subsequently linked together *via* a thioether bond. Key to this approach was the initial preparation of an active intermediate Pam_3CysSK_4-K($COCH_2Br$)-OH **36** that was accessed by microwave-enhanced (MW) Fmoc SPPS. Herein, a Wang-resin was initially preloaded with lysine orthogonally protected with Fmoc at *N*α and with a 1-(4,4-dimethyl-2,6-dioxo-cyclohexylidene)-3-methyl-butyl (ivDde) at *N*ε. Subsequent peptide chain elongation *via* the *N*α-amino group followed by lipidation using Pam_3Cys-pentafluorophenyl (Pfp) ester [HOBt in *N*-methyl-2-pyrrolidone (NMP) for 45 min at 50 °C] afforded resin-bound and side-chain protected $Pam_3CysS(OtBu)[K(Boc)]_4$-K(ivDde). The ivDde protecting group was then removed using hydrazine, and the *N*ε-amino group acylated with pentafluorophenyl bromoacetate. Subsequent TFA-mediated peptide cleavage gave Pam_3CysSK_4-K($COCH_2Br$)-OH **36**. The key intermediate **36** was then converted into an active iodo-acetyl derivative using potassium iodide (KI) in urea/sodium acetate (NaOAc) mixture affording **37** (Scheme 9.5a). The iodo-acetyl moiety **37** was then ligated with several peptide epitopes that incorporated a thiol-terminated PEG spacer at their *N*-terminus. For example construct **38** was treated with **37** and trimethylamine (Et_3N) in DMF at 40 °C affording construct **39** (Scheme 9.5b). The authors successfully applied this strategy for conjugation of a Pam_3CysSK_4 motif *via* a thioether linkage to B- and T-cell epitopes affording various self-adjuvanting vaccine constructs (Cai et al. 2013). The three-component construct **39** comprising P4 tetanus toxoid T cell epitope (Demotz et al. 1989; Monji and Pious 1997), linked *via* a PEG spacer with MUC1 glycopeptide comprising T_N antigen,

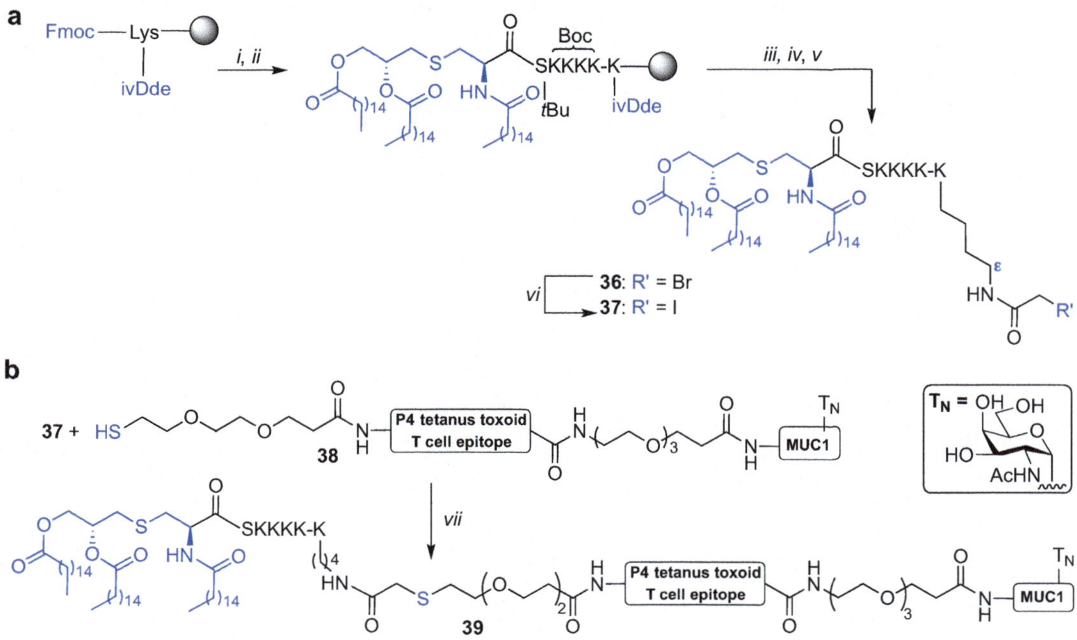

Scheme 9.5 Exemplified synthesis of three component synthetic vaccine incorporating Pam_3Cys TLR2 ligand using thioether ligation strategy by Kunz et al. (Cai et al. 2013). Reagents and conditions: (*i*) MW Fmoc SPPS; (*ii*) (*R*)-Pam_3Cys-OPfp, HOBt, NMP, 45 min, 50 °C; (*iii*) 2% hydrazine hydrate in DMF, 5 min, rt (repeated 3 x); (*iv*) $BrCH_2COOPfp$, HOBt, 4 h, rt; (*v*) TFA/iPr_3SiH/H_2O (15:0.9:0.9, v/v/v); (*vi*) KI, 8 M urea/0.1M NaOAc, 30 min; (*vii*) NEt_3, DMF, 40 °C

and a conjugated Pam_3CysSK_4 *via* a thioether linkage proved most efficacious (Cai et al. 2013).

9.1.5.2 Native Chemical Ligation Approach to Self-Adjuvanting Vaccine Constructs

Native Chemical Ligation (NCL) (Dawson et al. 1994) enables synthetic access to long peptides and large biomolecules and has been used by our research group in numerous studies (Yang et al. 2013; Harris and Brimble 2015; Medini et al. 2015; Harris et al. 2015; Harris and Brimble 2013; Medini et al. 2016; Lee et al. 2011; Brimble et al. 2015; Son et al. 2014; Harris and Brimble 2010). NCL conjugates two synthetic partners containing complementary reactive sites, namely an *N*-terminal cysteine and a *C*-terminal thioester moiety *via* a thiol-catalysed chemoselective reaction affording a thioester-linked product; subsequent S→N transfer ensures the formation of a native peptide bond (Dawson et al. 1994). Brimble et al. (Harris et al. 2007) explored synthetic pathways to access Pam_2Cys-linked thioester moiety that could be later incorporated into a long peptide *via* NCL. The initial effort to synthesise a more soluble derivative of Pam_2Cys, namely Pam_2CysSK_4G thioester using *tert*-butyloxycarbonyl (Boc) SPPS resulted in unexpected cleavage of the palmitoyl esters during the final hydrofluoric acid (HF)-mediated peptide removal from the resin (Zeng et al. 2011). Successful synthesis of Pam_2CysSK_4G thioester was however completed using an alternative Fmoc SPPS strategy employing a sulfonamide 'safety catch linker' (Backes and Ellman 1999; Ingenito et al. 1999) and Fmoc-*S*-[2(*S*),3-bis(palmitoyloxy)propyl]-L-cysteine (Fmoc-(*S*)-Pam_2Cys-OH) (**40**) as the building block (Scheme 9.6) (Harris et al. 2007). Loading of 4-sulfamylbutyryl aminomethyl polystyrene resin with Fmoc-Gly-OH was initially performed [DIC, *N*-methylimidazole (*N*-MeIm) in DMF/CH_2Cl_2 mixture] followed by standard Fmoc SPPS affording side chain protected peptidyl-

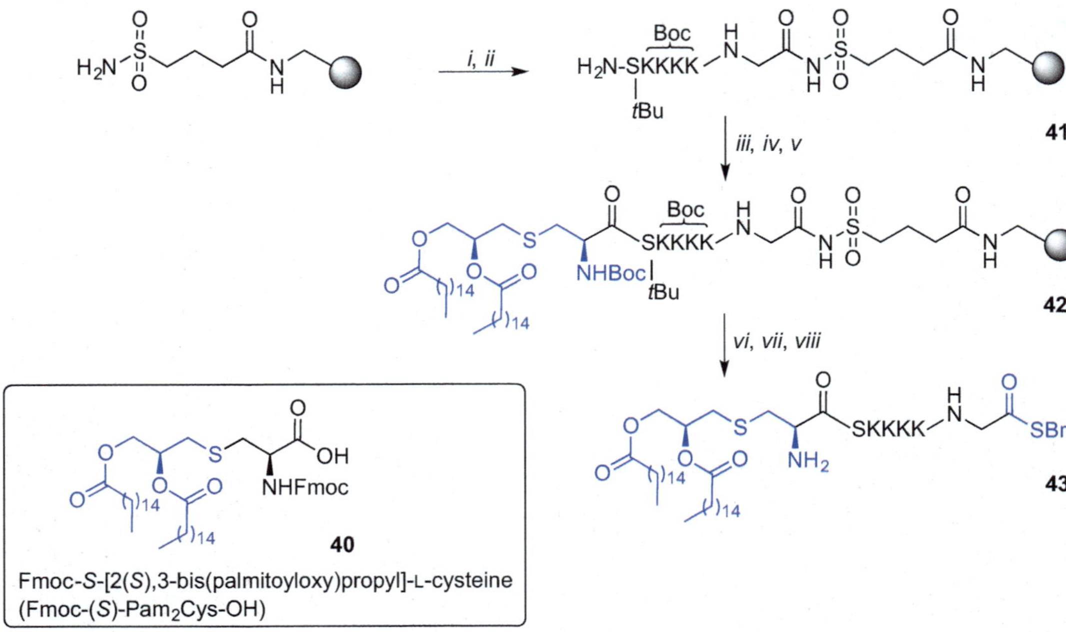

Scheme 9.6 Synthesis of a *C*-terminal thioester derivative of the Pam_2CysSKKKK using Fmoc SPPS by Brimble et al. (Harris et al. 2007). Reagents and conditions: (*i*) Fmoc-Gly-OH, DIC, *N*-MeIm, CH_2Cl_2, DMF; (*ii*) Fmoc SPPS; (*iii*) **40**, PyBOP, HOBt, CH_2Cl_2; (*iv*) 20% piperidine in DMF; (*v*) Boc_2O, CH_2Cl_2, DMF; (*vi*) ICH_2CN, NMP; (*vii*) BnSH, DMF; (*viii*) TFA/phenol/*i*Pr_3SiH/H_2O (88:5:2:5, v/v/v/v)

resin **41**. Subsequent coupling of lipidated building block **40** (Metzger et al. 1991; Hida et al. 1995) was effected (PyBOP/HOBt) and the Fmoc protecting group was exchanged to Boc (Boc_2O in DMF/CH_2Cl_2 mixture) to provide **42**. Resin-bound **42** was then activated with iodoacetonitrile in NMP, with subsequent cleavage from resin using benzyl thiol (BnSH). Finally side chain protecting groups removal using TFA afforded the desired Pam_2CysSK_4G thioester **43** (Harris et al. 2007).

Boons et al. (Ingale et al. 2006) were the first to demonstrate a successful synthesis of a three-component glycolipidated peptide vaccine by sequential NCL of the suitably prepared ligation fragments; Fmoc SPPS was employed to synthesise the T-cell epitope C(Acm) YAFKYARHANVGRNAFELFLG-thioester (**44**), the tumour-associated glycopeptide fragment derived from MUC-1 CTSAPDT(GalNAc) RPAP (**45**), and the TLR2 ligand Pam_3CysSK_4G-thioester (**46**). Due to limited success when ligation of **44** with **45** was undertaken using standard NCL conditions (phosphate buffer containing 6 M guanidinium hydrochloride, thiophenol, 37 °C), new methodology involving incorporation of **44** with **45** into liposomes to aid solubility was used. A film of dodecylphosphocholine (DPC), thioester **44** and thiol **45** were hydrated *via* incubating in a phosphate buffer (pH 7.5) for 4 h at 37 °C in the presence of tris(2-carboxyethyl)phosphine (TCEP) and ethylenediaminetetraacetic acid (EDTA) to suppress disulphide bond formation. The mixture was then sonicated and the resulting peptide/lipid suspension formed uniform 1 μm vesicles. Sodium 2-mercaptoethane sulfonate (MESNA) was subsequently added and ligation completed after 2 h at 37 °C affording **47** in high 78% yield after reversed-phase high-performance column chromatography (RP HPLC) purification (Ingale et al. 2006). Ligation of Pam_3CysSK_4G-thioester **46** with thiol **48**, accessed by removal of the acetamidomethyl (Acm) protecting group from **47** [$Hg(OAc)_2$], using liposome-mediated NCL afforded a three-component vaccine construct **49** in 83% yield

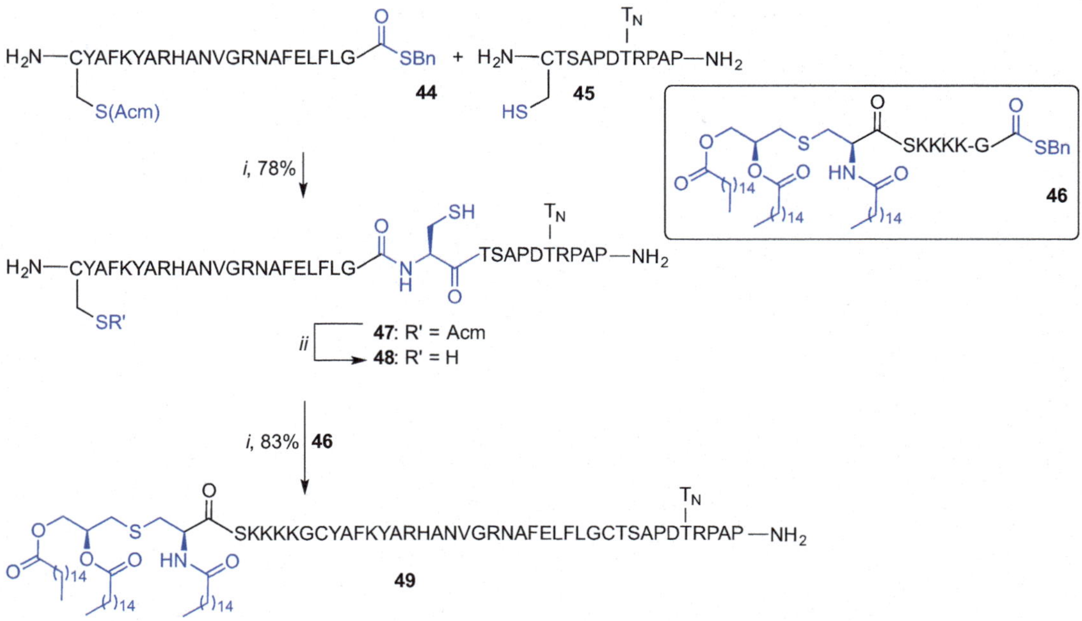

Scheme 9.7 Liposome-mediated NCL to the synthesis of three-component vaccine construct incorporating Pam_3Cys TLR2 ligand by Boons et al. (Ingale et al. 2006). Reagents and conditions: (*i*) 200 mM sodium phosphate buffer (pH 7.5), DPC, TCEP, EDTA, sonication, extrusion, and then MESNA; (*ii*) $Hg(OAc)_2$, 10% *aq* HOAc, 50 mM DL-dithiothreitol (DTT), 89%

after purification by chromatography (Scheme 9.7). The scope of this technique was later demonstrated by the synthesis of other self-adjuvanting vaccine constructs that differ in the composition of the (glyco)peptide and lipid component; some of the constructs proved highly immunogenic when tested in mice models (Ingale et al. 2006; Ingale et al. 2007; Lakshminarayanan et al. 2012; Abdel-Aal et al. 2014; Ingale et al. 2009).

The liposome-mediated NCL approach allowed for the generation of a native amide linkage between each of the required vaccine modules. However, the use of dodecylphosphocholine liposomes in ligation buffers can be limiting owing to the need for RP HPLC purification after each ligation step to isolate the product (McDonald et al. 2015; Ingale et al. 2006).

9.1.5.3 Fragment Condensation Approach to Self-Adjuvanting Vaccine Constructs

Kunz et al. (Kaiser et al. 2010) and Payne et al. (Wilkinson et al. 2010) described a fragment condensation approach to incorporate a Pam_3Cys TLR2 ligand into mono- and per-glycosylated MUC1 glycopeptides respectively, using a PEG-based spacer to access fully synthetic vaccine constructs.

The Kunz approach involved initial synthesis of the lipidated, side-chain protected and the *C*-terminal carboxylic acid $Pam_3CysS(tBu)$ K(Boc)K(Boc)K(Boc)K(Boc) (**50**) unit using Fmoc SPPS. The MUC1 glycopeptides *N*-terminally modified with PEG linker, namely $H_2N(CH_2CH_2O)_3CH_2CH_2CONH$-PAH-GVT(sugar)-SAP-DTR-PAP-GST-AP-OH, comprising either T_N- (**51**), T- (**52**) or 2,6-sialyl-T-antigen (**53**) at the singly glycosylated Thr-6 were then accessed *via* Fmoc SPPS. The fragment condensation was subsequently effected in solution and using *N*-[(dimethylamino)-1*H*-1,2,3-triazolo[4,5-*b*] pyridin-1-ylmethylene]-*N*-methylmethanaminium hexafluorophosphate *N*-oxide (HATU)/HOAt and 4-methylmorpholine (NMM) in DMF which was followed by TFA-

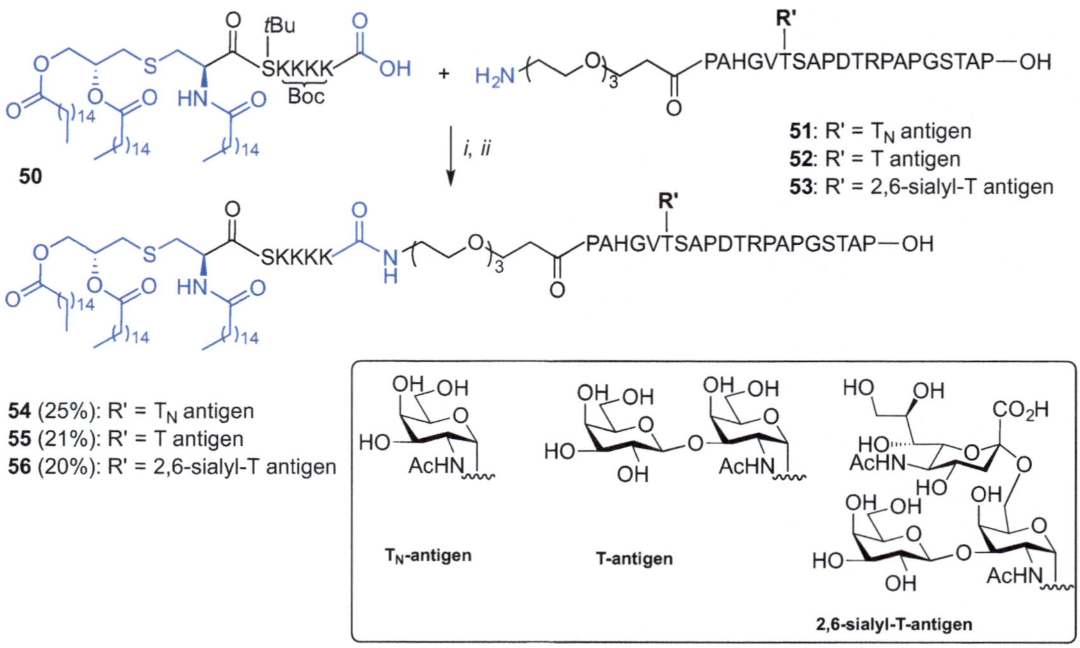

Scheme 9.8 Fragment condensation for the synthesis of the vaccine construct incorporating Pam_3Cys TLR2 ligand by Kunz et al. (Kaiser et al. 2010). Reagents and conditions: (*i*) HATU, HOAt, NMM, DMF; (*ii*) TFA/*i*Pr_3SiH/H_2O (10:1:1, v/v/v), 1.5 h

mediated protecting group removal and purification affording three novel vaccine constructs, **54**, **55** and **56** in 25%, 21% and 20% yield, respectively (Scheme 9.8). Importantly, bio-assessment of TLR2 ligand-MUC1 assembly comprising T-antigen **55** showed the ability to elicit humoral immune response in mice (Kaiser et al. 2010).

The Payne group employed the lipopeptide component with a PEG-like spacer at *C*-terminus, namely Pam_3CysS(*t*Bu)-CONH$(CH_2CH_2O)_2$ CH_2COOH (**57**), and per-glycosylated full copies of the MUC1 VNTR domain epitope (GVT(sugar)-S(sugar)-APDT(sugar)-RPAPGS(sugar)T(sugar) APPAH), incorporating no copies (**58**) or multiple-copies of either T_N- (**59**) or T-antigen (**60**), for convergent conjugation. All peptide fragments **57-60** were synthesized using Fmoc SPPS. The free carboxylic acid of the lipid partner **57** was pre-activated using pentafluorophenyl ester with ensuing fragment condensation with the requisite MUC1 epitopes **58**, **59** or **60** using HOBt and *i*Pr_2NEt in DMF affording desired MUC1-Pam_3Cys chimeras with no sugars **60**, or containing five copies of either T_N-or T-antigen, **62** and **63**, respectively (Scheme 9.9) (Wilkinson et al. 2010). This fragment condensation approach was also used in other studies by the Payne group to synthesise multiple-component vaccine constructs incorporating Pam_3Cys (Wilkinson et al. 2012; McDonald et al. 2014; Wilkinson et al. 2011). The fragment condensation strategy is a good alternative to the liposome-mediated NCL approach reported by Boons et al. (Ingale et al. 2006, 2007; Lakshminarayanan et al. 2012) with no requirements for solubilizing agents.

9.1.5.4 Linear Approach to Self-Adjuvanting Vaccine Construct

The Boons group has recently reported a linear synthesis to access a three-component cancer vaccine composed of a B-cell epitope glycosylated with a sialyl-T_N moiety, a T_H epitope derived from polio virus (Leclerc et al. 1991) and a Pam_3CysSK_4 ligand. The key strategies employed by the Boons group included the use of microwave-enahanced Fmoc SPPS and on resin incorporation of the Fmoc-(*S*)-Pam_2Cys-OH (**40**) building block onto the free *N*α-amino group of

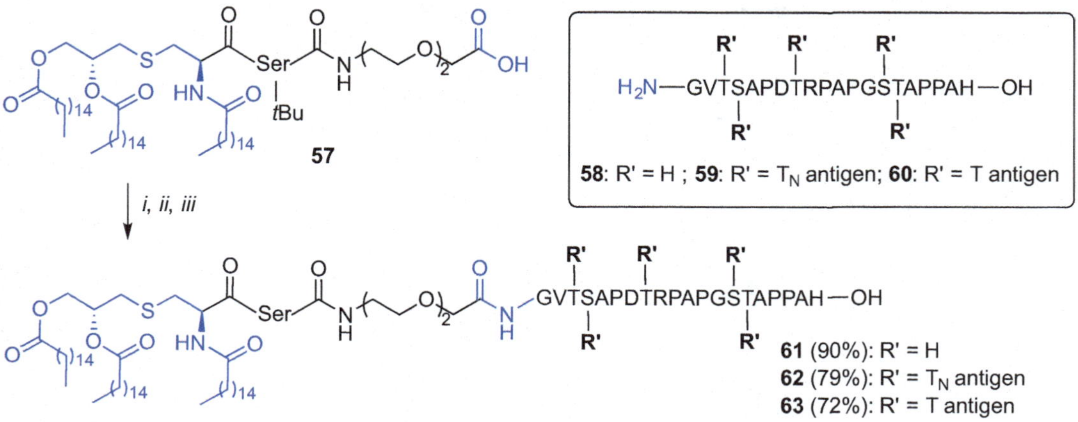

Scheme 9.9 Fragment condensation approach for the synthesis of vaccine constructs incorporating Pam_3Cys by Payne et al. (Wilkinson et al. 2010). Reagents and conditions: (*i*) pentafluorophenol, DIC, CH_2Cl_2;(*ii*) **58** or **59** or **60**, HOBt, *i*Pr_2NEt, DMF; (*iii*) TFA/*i*Pr_3SiH (1:1, v/v)

the pre-synthesised glycopeptide construct containing deprotected hydroxyl groups of the sugar moiety (Thompson et al. 2015). Fmoc protecting group removal from the Fmoc-(*S*)-Pam_2Cys-tagged vaccine construct (piperidine) could be then followed by *N*α-amino group palmitoylation using palmitic acid, HATU, HOAt and *i*Pr_2NEt in DMF. Finally, TFA treatment afforded fully synthetic vaccine construct **64** incorporating the Pam_3Cys TLR2 ligand (Fig. 9.13). Biological evaluation demonstrated induction of potent humoral and cellular immune responses in transgenic mice (Thompson et al. 2015).

A three-component vaccine construct similar to that described above, but incorporating the unnatural T_N moiety, namely α-*O*-GalNAc-α-methylserine in place of threonine, within the MUC1 epitope was recently accessed using the MW-enhanced Fmoc SPPS strategy previously reported by Boons et al. (Thompson et al. 2015; Martinez-Saez et al. 2016). This novel vaccine construct however, showed only comparable efficacy to that reported for the assembly containing native threonine.

As shown above, a linear approach for the synthesis of complex multi-component lipidated peptides containing only natural peptide bonds demonstrates the efficiency of the microwave-assisted Fmoc SPPS technique. However, longer and/or more hydrophobic lipopeptide constructs may still be difficult to access when using a linear SPPS and alternative synthetic routes for lipid incorporation are in demand.

9.1.5.5 TLR2 Ligand Conjugation Using Copper(I)-Catalysed Huisgen 1,3-Dipolar Cycloaddition

The need for large quantities of Fmoc-Pam_2Cys building block required for SPPS conjugation poses a considerable obstacle due to the difficulty and cost involved in its synthesis. An alternative conjugation approach to incorporate the Pam_2Cys moiety into a peptide could mitigate this conundrum. The copper(I)-catalysed Huisgen 1,3-dipolar cycloaddition of alkynes and azides to afford a 1,2,3-triazole conjugate (CuAAC 'click chemistry') offered promise for the conjugation of Pam_2Cys with a peptide due to its tolerance of various functional groups and its complete regioselectivity to form 1,4-disubstituted products (Tornoe et al. 2002; Rostovtsev et al. 2002). The Brimble group therefore designed a Pam_2Cys click building block containing an azide handle in place of the *N*α-amino group of the cysteine residue which could be then clicked to a peptide functionalized with a propargyl moiety (Yeung et al. 2012). However, initial attempts to directly introduce an azide onto a free *N*α-amino group of Pam_2Cys using a diazotransfer reaction (Goddard-Borger and Stick 2007) proved unsuccessful, potentially due to obstruction of the reactive sites

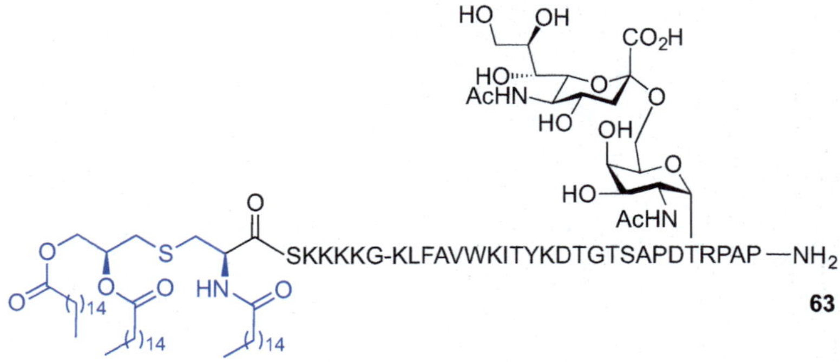

63

Fig. 9.13 A fully synthetic vaccine construct prepared using MW-enhanced linear Fmoc SPPS and incorporating Pam_3Cys TLR2 ligand by Boons et al. (Thompson et al. 2015)

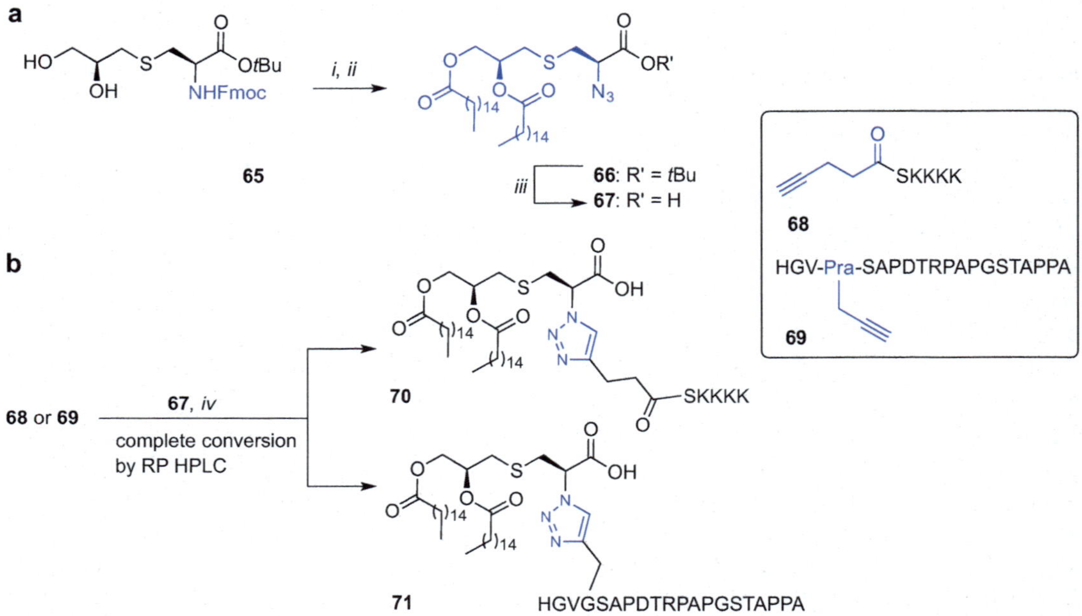

Scheme 9.10 Synthesis of Pam_2Cys azide **67** and Cu(I) 'click' conjugation of **67** with alkyne-modified peptides to get lipidated **70** and **71** by Brimble et al. (Yeung et al. 2012). Reagents and conditions: (*i*) piperidine, CH_2Cl_2, then imidazole-1-sulfonyl azide·HCl, K_2CO_3, $CuSO_4$, MeOH, 50% over 2 steps;(*ii*) $CH_3(CH_2)_{14}COOH$, DIC, DMAP, tetrahydrofuran (THF), 74%;(*iii*) TFA, 84%; (*iv*) CuI·P(OEt)$_3$, *i*Pr$_2$NEt, DMF, 30 min

by the long palmitate groups (Yeung et al. 2012). A revised strategy was developed starting from an *S*-glyceryl cysteine intermediate **65** (Metzger et al. 1991; Pattabiraman et al. 2008) which was subjected to *N*α-amino group deprotection (piperidine in CH_2Cl_2) to reveal the amino group for the ensuing diazotranfer reaction using imidazole-1-sulfonylazide·HCl, K_2CO_3 and $CuSO_4 \cdot 5H_2O$ in MeOH (Goddard-Borger and Stick 2007) affording an azide-diol in 50% yield over 2 steps (Scheme 9.10a). Subsequent palmitoylation of the azide-diol [palmitic acid, DIC, and catalytic 4-(dimethylamino)pyridine (DMAP)] provided *t*Bu-protected Pam_2Cys azide **66** in 74% yield. Subsequent TFA treatment to remove the carboxyl protecting group gave the desired lipidated and azide-tagged 'click' ligation partner **67** in 84% yield (Yeung et al. 2012).

The synthesis of the alkyne-containing peptides for subsequent Cu(I) conjugation with **67** was undertaken using Fmoc SPPS (Yeung et al. 2012). Pentynoyl acid was coupled to the *N*-terminus affording **68** and propargylglycine (Pra) was used as an alkyne handle within the modified MUC1 peptide sequence, namely HGV-Pra-SAPDTRPAPGSTAPPA **69**. The 'click' reaction of both alkyne-enriched peptides **68** and **69** using azide **67** was completed within 30 min as evidenced by RP HPLC using CuI·P(OEt)$_3$ and *i*Pr$_2$NEt in DMF affording 1,2,3-triazole-linked Pam$_2$Cys peptides **70** and **71**, respectively. The amenability of the Pam$_2$Cys azide to direct conjugation onto suitably modified peptides using the 'click' technique was successfully demonstrated (Scheme 9.10b) (Yeung et al. 2012). However, construct **70** was immunologically inactive possibly due to difference in the distance between the serine and the Pam$_2$Cys (unpublished data). It has been reported that the exact length and geometry around the Cys-Ser unit is critical for activity of the Pam$_2$CysSK$_4$ motif (Wu et al. 2010; Kang et al. 2009).

Kunz et al. were the first to report CuAAC-assisted ligation of Pam$_3$CysSK$_4$ to a MUC1 glycopeptide to synthesise mono-, di- and tetra-valent MUC1 tandem repeat glycopeptide constructs to prepare of fully synthetic antitumour vaccines (Cai et al. 2011). The Kunz approach for the synthesis of monovalent MUC1 derivatives used Fmoc SPPS of MUC1 glycopeptide in which the *N*-terminal *N*α-amino group was acylated with a PEG linker suitably modified with an azide handle affording construct **72**. The 'click' synthetic partner **73** incorporated an alkyne group *via* a PEG spacer linking with the Pam$_3$CysSK$_4$ ligand by the *N*ε-amino group of the additional *C*-terminal lysine residue. The copper(I)-mediated reaction of the suitably prepared 'click' partners was then performed using copper acetate and Na ascorbate in H_2O at 40 °C affording the monovalent vaccine construct **74** with >70% yield (Scheme 9.11) (Cai et al. 2011).

The *N*ε-amino group of the *C*-terminal lysine linked to the Pam$_3$CysSK$_4$ moiety was later used as a point of attachment of additional lysine groups forming a multibranched lysine core which terminated with two or four copies of PEG-alkyne handles. Subsequent Cu(I) 'click' using the Pam$_3$CysSK$_4$ ligand incorporating two- or four alkyne groups and azide construct **72** afforded the desired di- (**75**) and tetra-valent (**76**) assemblies, respectively (Scheme 9.11) (Cai et al. 2011). Importantly, the tetravalent construct of general structure **76** synthesized using this strategy that incorporated the ST$_N$ glycoside within the MUC1 sequence proved effective in inducing strong immune responses in mice including stimulation of killer cells (Cai et al. 2014).

The Sucheck group has reported the 1,2,3-triazole-mediated conjugation of a Pam$_3$Cys ligand equipped with a *C*-terminal alkyne, with a 20-amino acid azide-tagged tandem repeat of MUC1 incorporating the T$_N$ unit (Sarkar et al. 2013). The alkyne-containing 'click' partner was available from Fmoc-Pam$_2$Cys(O*t*Bu) **77** by *tert*-butyl protection removal (TFA) followed by coupling with propargyl amine in the presence of PyBOP, HOBt and *i*Pr$_2$NEt in CH_2Cl_2. Fmoc protecting group removal and acylation of the revealed *N*α-amino group with palmitic acid using PyBOP, HOBt, and *i*Pr$_2$NEt gave the alkyne functionalized Pam$_3$Cys **78**, Scheme 9.12a. The glycopeptide-azide was prepared *via* Fmoc SPPS on Fmoc-Ala-WANG resin using DIC/HOBt as coupling reagent and piperidine in DMF for Fmoc removal, affording resin-bound **79**. The azido group was installed on-resin by coupling 6-azidohexanoic acid to the *N*-terminal proline residue of MUC1 followed by TFA-mediated peptide cleavage from the resin and acetyl deprotection of the T$_N$ hydroxyls (sodium methoxide in MeOH) to provide azide-containing MUC1 epitope **80** (Scheme 9.12b). The 'click' conjugation of both constructs, alkyne-functionalized Pam$_3$Cys **78** and the azide-MUC1 component **80** was undertaken with $CuSO_4{\cdot}5H_2O$, Na ascorbate with the aid of a Cu(I) stabilizing agent tris[(1-benzyl-1*H*-1,2,3-triazol-4-yn)methyl]amine (TBTA) in water/MeOH/THF mixture affording Pam$_3$Cys-MUC1 conjugate **81** quantitatively (Scheme 9.12c) (Sarkar et al. 2013).

The Brimble group recently employed a Pam$_2$CysSK$_4$ motif for the synthesis of a series of lipopeptide-based TLR2 agonists using 'click' chemistry (Wright et al. 2013c). Incorporation of an acetylene handle at the *C*-terminal end of the Pam$_2$CysSK$_4$ construct would allowing for the chemoselective 'click' conjugation with an azide-tagged epitope. Unlike the previous study by the Brimble group (Yeung et al. 2012) this approach

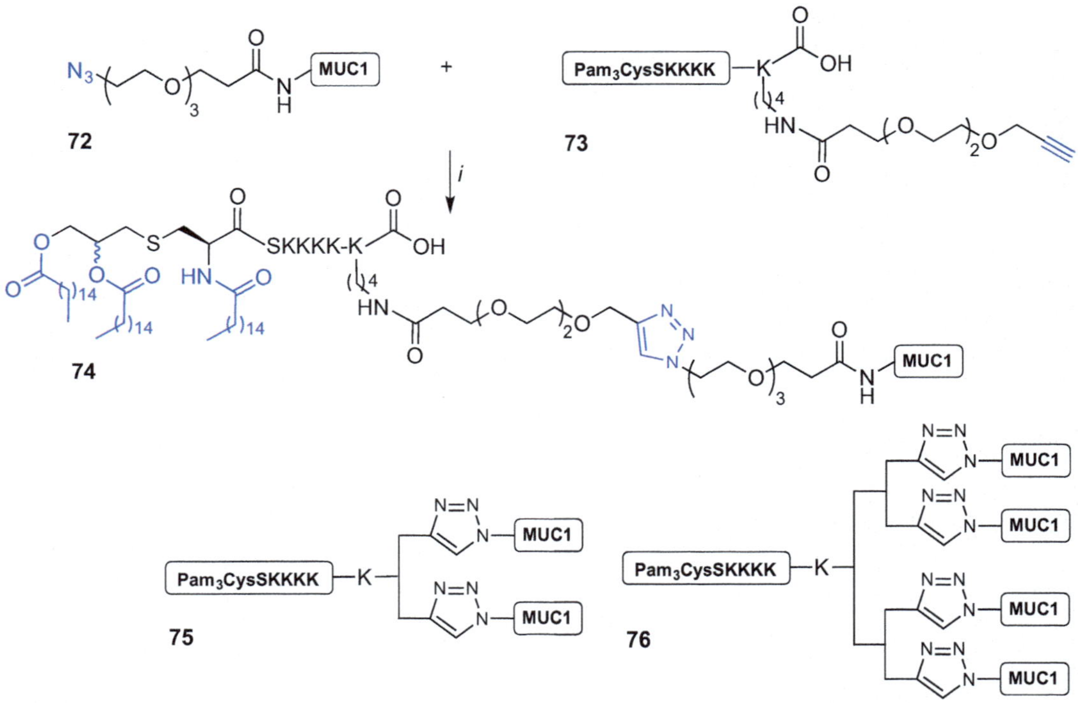

Scheme 9.11 Synthesis of mono-valent 'click' construct **74** from azide-modified MUC1 antigen **72** and alkyne-modified Pam_3Cys **73** and graphical representation of di- and tetra-valent vaccine constructs **75** and **76** by Kuntz et al. (Cai et al. 2011, 2014). Reagents and conditions: (*i*) copper acetate, Na ascorbate, H_2O, 40 °C, >70%

maintained the critical atomic distance between the Pam_2Cys and adjacent serine moiety[159]. Additionally, both the self-adjuvanting lipopeptide construct and the epitope were directly conjugated *via* a 1,2,3-triazole unit in contrast to approach by Kunz et al. where a PEG linker spaced these units apart (Cai et al. 2011).

It has been reported that the immunogenicity of the antigen incorporated to a vaccine construct may be suppressed by the presence of a linker (Buskas et al. 2004). We were also interested if the location of the triazole between the antigen and the Pam_2CysSK_4 moiety affects the TLR2-mediated stimulation of innate immunity; antigen conjugation with the lipid at either the *N*- or *C*-terminus of the peptide antigen was therefore investigated. It has been reported that acetylation of the *N*α-amino group of the monoacyl PamCys moiety improved TLR2 activity (Salunke et al. 2012) hence the effects of this modification were also evaluated in this study (Wright et al. 2013c).

A lipidated and *C*-propargylated 'click' partner **82**, in addition to the *N*-acetylated analogue **83** were first synthesized (Scheme 9.13a). Synthesis began by the *N*-terminal coupling of the Fmoc-(*S*)-Pam_2Cys-OH building block **40** prepared from L-cysteine (Zeng et al. 2002; Metzger et al. 1991; Jones 1975; Hida et al. 1995), to the resin-bound *C*-terminal propargylated H_2N-S(tBu)K(Boc)K(Boc)K(Boc)K(Boc)-Pra-resin peptide synthesized using standard Fmoc SPPS (Wright et al. 2013c). The peptide was lipidated using **40** and conditions adapted from Albericio et al. [benzotriazol-1-yloxytris(dimethylamino)phosphonium hexafluorophosphate (BOP), 2,4,6-collidine, CH_2Cl_2/DCM (1:1)] (Han et al. 1996). Subsequent Fmoc-deprotection, followed by TFA-mediated resin cleavage and RP HPLC purification afforded the desired construct **82**. Acylation of the *N*α-amino group of cysteine to give **83** was performed using a mixture of acetic anhydride and iPr_2NEt in

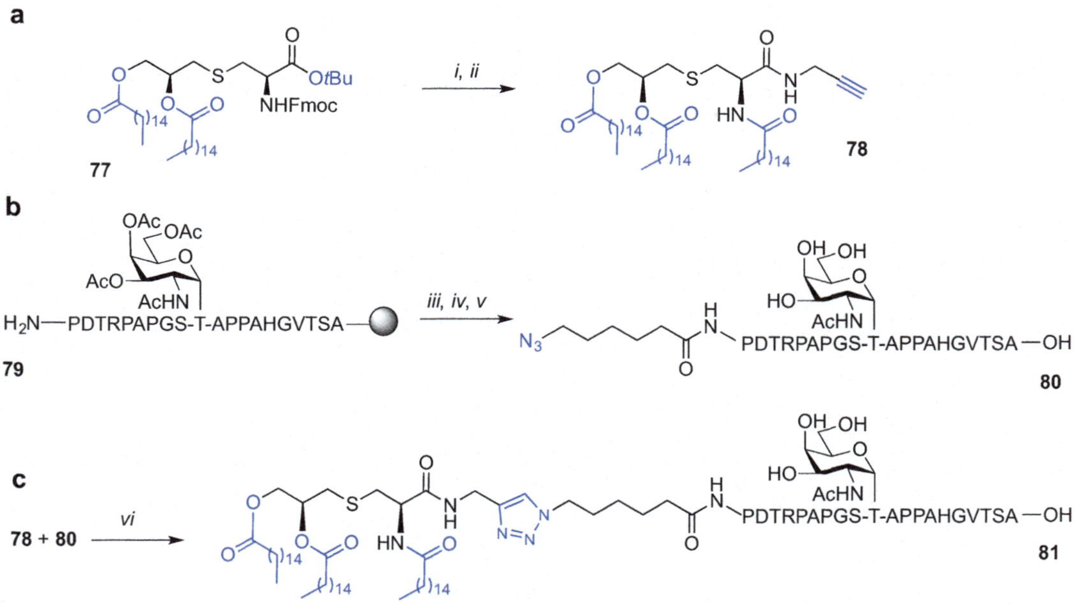

Scheme 9.12 Synthesis of Pam_3Cys-MUC1-T_N conjugate using 'click' chemistry by Sucheck et al. (Sarkar et al. 2013). Reagents and conditions: (*i*) TFA, 1 h, rt, then propargyl amine, PyBOP, HOBt, *i*Pr_2NEt, 4 Å molecular sieves, CH_2Cl_2, 4 h, rt, 66% over two steps; (*ii*) CH_3CN/CH_2Cl_2/Et_2NH (2:1:2), 2 h, rt, then $CH_3(CH_2)_{14}COOH$, PyBOP, HOBt, *i*Pr_2NEt, 4 Å molecular sieves, CH_2Cl_2, 4 h, rt, 80% over two steps; (*iii*) 6-azidohexanoic acid, DIC, HOBt, NMP; (*iv*) TFA/thioanisole/EDT/H_2O phenol (88:3:5:2:2, v/v/v/v/v); (*v*) NaOMe, MeOH, 2 h, rt, 100%; (*vi*) $CuSO_4{\cdot}5H_2O$, Na ascorbate, TBTA, H_2O/MeOH/THF (1:1:2), 40 h, rt, 100%

DMF, prior to peptide cleavage and purification (Wright et al. 2013c). A truncated fragment of ppUL83 protein, namely NLVPMVATV, derived from the cytomegalovirus (CMV) known to stimulate $CD8^+$ cytotoxic T-cells (Kopycinski et al. 2010) was chosen as a model epitope for the 'click' reaction. Synthesis of two NLVPMVATV analogues incorporating an azide handle at either the *N*- or *C*-terminus was also required. For the preparation of an azide-tagged antigen at the *N*-terminal site of the peptide **84**, Fmoc SPPS was employed starting from 4-(hydroxymethyl) phenoxypropanoic acid (HMPP) resin and coupling of 2-azidoacetic acid to the *N*-terminal Asn at the last step of the SPPS. Subsequent acid-mediated peptide cleavage from the resin followed by RP HPLC purification afforded the desired 'click' partner **84** (Scheme 9.13b).

For the synthesis of the NLVPMVATV analogue with the *C*-terminally-tagged azide moiety 2-azidoacetic acid was incorporated *via* the *N*ε-amino group of an inserted lysine moiety at the *C*-terminus of the peptide. An orthogonally protected lysine residue [Dde-Lys(Fmoc)] was coupled to the Rink-amide resin, followed by the selective *N*ε-Fmoc protecting group removal (20% piperidine in DMF) and coupling of the 2-azidoacetic acid moiety. Subsequent hydrazine hydrate-mediated Dde group deprotection allowed for the iterative peptide chain elongation using Fmoc SPPS through readily unmasked *N*α-amino group of the lysine residue, affording construct **85** (Scheme 9.13b). Chemoselective conjugation of propargylated-, or propargylated and *N*-acetylated- Pam_2CysSK_4 motives **82** and **83**, respectively with azidopeptides **84** and **85** under activation with $CuSO_4$ and Na ascorbate in DMSO, gave 1,2,3-triazole-linked constructs **86-89** in good yields (30-40%) and high purities (>95% by RP HPLC) (Scheme 9.13c) (Wright et al. 2013c).

Biological evaluation of **82** and *N*-acetylated analogue **83** using fresh human blood and measuring the level of CD80 surface expression compared to commercially sourced Pam_3CSK_4 interestingly revealed no major difference in

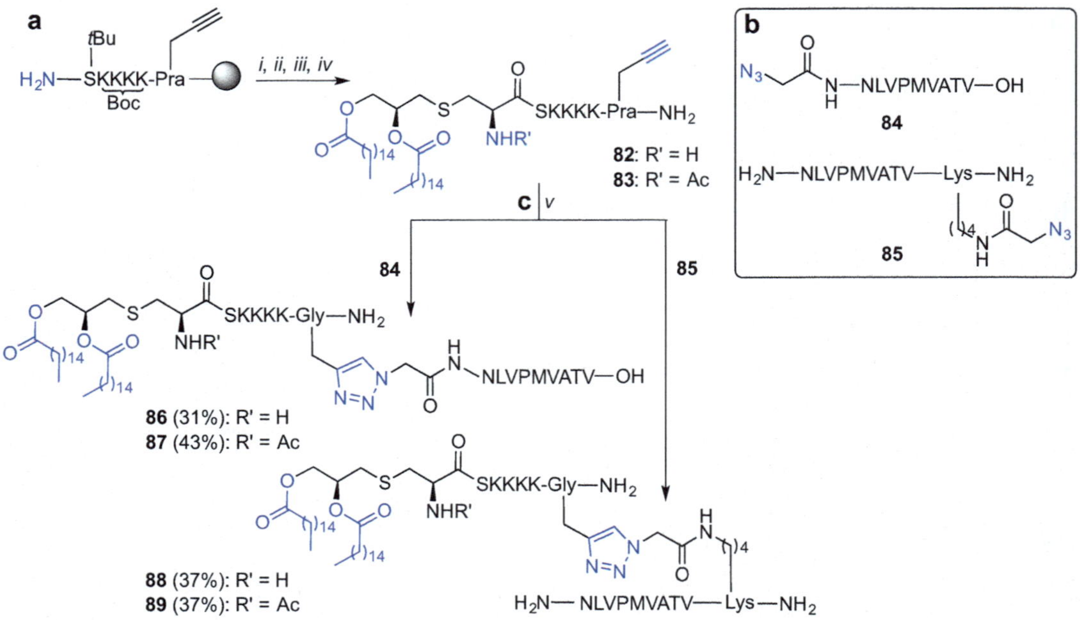

Scheme 9.13 Synthesis of Pam_2Cys lipopeptide-based TLR2 agonists using 'click' chemistry by Brimble et al. (Wright et al. 2013c). Reagents and conditions: (*i*) **40**, BOP, 2,4,6-collidine, CH_2Cl_2/DMF (1:1); (*ii*) 20% piperidine in DMF; (*iii*) Ac_2O, $i Pr_2NEt$, DMF (only for **83**); (*iv*) TFA/iPr_3SiH/DODT/H_2O (94:1:2.5:2.5, v/v/v/v), 4 h, rt; (*v*) $CuSO_4$, Na ascorbate, DMSO, 10 min, 70 °C

CD80 expression between both propargylated Pam_2Cys analogues with free- (**82**) and *N*-acetylated-*N*α-amine (**83**) in contrast to published reports (Salunke et al. 2012). Importantly, there were no preferences regarding the *N*- or *C*-terminus for the antigen conjugation with lipidated adjuvant *via* 1,2,3-triazole and similar CD80 expression levels were observed for both 'clicked' analogues **86** and **87** and activity of 'clicked' lipopetides was comparable with the activity of commercially available Pam_3CysSK_4 (Wright et al. 2013c). This efficient procedure can therefore be generally applied for rapid generation of lipopeptides providing access to vaccine constructs (Wright et al. 2013c).

9.1.5.6 Cysteine Lipidation on a Peptide or Amino acid (CLipPA)

The 'thiol-ene' reaction, a radical-promoted alkylation of a thiol with an alkene has been gaining in popularity in polymer and material science (Lowe 2010; Lowe 2014) as well as providing an effective strategy for bioconjugation and for site-selective modification of protein and organic molecules (Dondoni and Marra 2012; Hoyle and Bowman 2010; Liu and Li 2012; Krall et al. 2016; Madder and Gunnoo 2016). The Brimble group have recently applied for the first time, a single step 'thiol-ene' coupling to synthesise monoacyl lipopeptides that showed self-adjuvanting antigenic activity with potency comparable to that of the synthetically challenging Pam_3Cys moiety (Wright et al. 2013a, b; Brimble et al. 2014). We envisaged lipid attachment *via* the 'post-translational' route where the desired peptide constructs incorporating a cysteine at the *N*-terminus are first synthesized followed by *S*-lipidation with inexpensive and commercially available vinyl palmitate using the 'thiol-ene' reaction. The viability of the transformation was first tested by preparation of the *S*-palmitoylated, *N*α-Fmoc protected cysteine, starting from commercially available Fmoc-Cys(Trt)-OH which thiol protecting group was removed (TFA) affording Fmoc-Cys-OH (**90**). This was followed by hydrothiolation of vinyl palmitate **91** using UV light at 365 nm and 2,2-dimethoxy-2-phenylacetophenone (DMPA)

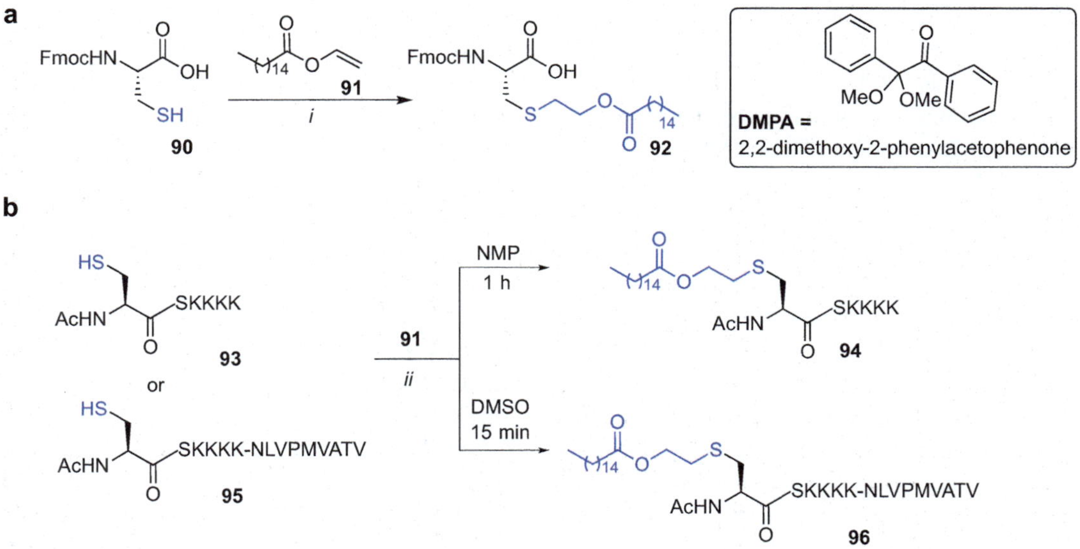

Scheme 9.14 (a) Model 'thiol-ene' reaction of Fmoc-Cys-OH (**90**) with vinyl palmitate (**91**). (b) Direct *S*-palmitoylation of **93** and antigenic peptide **95** using vinyl palmitate **91** and 'thiol-ene' reaction by Brimble et al. (Wright et al. 2013a; Wright et al. 2013b). Reagents and conditions: (*i*) **91** (2 equiv) DMPA (0.2 equiv), CH_2Cl_2, 1 h, *hν* 365 nm, 44%; (*ii*) **91** (5 equiv), DMPA (0.4 equiv), DTT (3 equiv), *hν* 365 nm

as photoinitiator in CH_2Cl_2 for 60 min. The *S*-palmitoylated, *N*α-Fmoc protected cysteine **92** was obtained in satisfactory yield (44%) (Scheme 9.14a) (Wright et al. 2013a, b). Subsequent direct lipidation of short, unprotected peptides $CysSK_4$ and *N*α-acetylated $CysSK_4$ using **91**, DMPA and photoinitiation (365 nm), was examined. The study revealed the need for extraneous thiols to obviate problems of vinyl palmitate telomerization and mixed disulphide formation. The choice of solvent also proved critical for a successful reaction. The optimized 'thiol-ene' conditions (DMPA, DTT as thiol additive, NMP, h*ν* 365 nm) were then used to directly lipidate *N*α-acetylated $CysSK_4$ **93** using vinyl palmitate (**91**) with high conversion (>90%) to the *S*-palmitoylated peptide **94** (Scheme 9.14b).

The utility of direct lipidation was explored using more structurally complex antigenic peptide substrate derived from the cytomegalovirus ppUL85 protein (Kopycinski et al. 2010) comprising an *N*-terminally $CysSK_4$, motif Ac-CSKKKK-NLVPMVATV (**95**). Pleasingly, good conversion of **95** to *S*-palmitoylated peptide antigen **96** using the photoinitiated 'thiol-ene' reaction, **91** and optimized conditions (DMPA, DTT, DMSO) was observed as judged by RP HPLC profile (Scheme 9.14b) (Wright et al. 2013a; Wright et al. 2013b). We therefore coined the term 'Cysteine Lipidation on a Peptide or Amino acid (CLipPA)' to describe this efficient transformation allowing for one step lipidation of *N*α-protected cysteine derivatives using vinyl palmitate.

We subsequently focused on a detailed study to optimise conditions for highly selective and effective mono-*S*-palmitoylation of peptides using CLipPA technology (Yang et al. 2016). Our first goal was to provide optimal conditions for the synthesis of a lipidated *N*α-protected cysteine building block that could be used directly in SPPS. The *N*α-protecting group, radical initiator and activation method were revised. Treatment of *N*α-protected Fmoc, Boc or *N*α-acetylated cysteine with an excess of vinyl palmitate in the presence of DMPA or 2,2-azo-bis(2-methylpropioniytile (AIBN) as radical initiator in either CH_2Cl_2 or 1,2-dichloroethane as solvent and under thermal heating (reflux at 90 °C), microwave irradiation (100 W, 70 °C) or UV light (365 nm) was studied.

The *S*-palmitoylated products obtained were readily purified by silica gel chromatography without the need for RP HPLC.

An optimal conversion of *N*α-protected with Fmoc- or Boc cysteine **90** and **96** was observed under UV light activation, using excess DMPA (1 equiv) for 1 h in CH_2Cl_2 affording **92** and **97** in 85% yield (Scheme 9.15a). Heating, either conventional or using microwave, gave lower yields due to the premature cleavage of Fmoc protecting group and the instability of the Boc group to high temperatures. Conversely, lipidation of *N*α-Ac cysteine **98** appeared to be straightforward under all conditions tested giving good to excellent yields of the expected *N*α-Ac and *S*-palmitoylated product **99**. However the most effective conversion was when CH_2Cl_2 and AIBN were used under microwave heating (100 W, 70 °C) for 80 min leading to quantitative formation of desired product **99** (Scheme 9.15b) (Yang et al. 2016).

The choice of *N*α-protecting group may influence the degree of racemization during the coupling step when SPPS is performed (Zhang et al. 2012). Therefore, the coupling of *S*-palmitoylated, *N*α–protected building blocks, **92** or **99** to a model peptide sequence was evaluated (Kopycinski et al. 2010). The Met residue of NLVPMVATV was substituted with Cys(*t*Bu) to demonstrate applicability of the 'thiol-ene' reaction conditions to a suitably protected cysteine thiol. The resin-bound and side-chain protected peptide H_2N-S(*t*Bu)K(Boc)K(Boc)K(Boc)K(Boc)-N(Trt)LVPC(*t*Bu)VAT(*t*Bu)V-resin (**100**) was prepared using Fmoc SPPS at room temperature with HATU/*i*Pr_2NEt and piperidine as coupling and Fmoc deprotection reagents and acylation with the lipidated building blocks, **92** or **99** was undertaken using

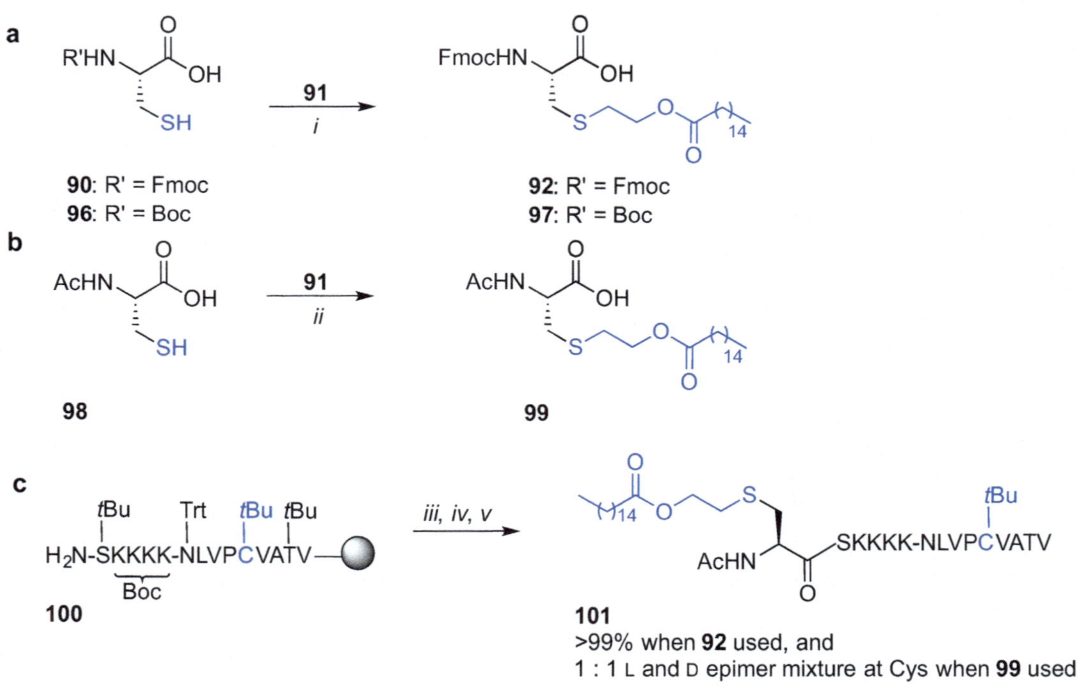

Scheme 9.15 (**a**) Lipidation of Fmoc-Cys-OH (**90**) and Boc-Cys-OH (**96**) with vinyl palmitate (**91**) using optimized conditions for CLipPA; (**b**) Lipidation of Ac-Cys-OH (**98**) with vinyl palmitate (**91**) using optimized conditions for CLipPA; (**c**) On-resin lipidation of antigen **100** using *N*α-protected *S*-palmitoylated cysteine building blocks **92** and **99** (Wright et al. 2013a; Wright et al. 2013b; Yang et al. 2016). Reagents and conditions: (*i*) **91** (1.5 equiv), DMPA (1 equiv), CH_2Cl_2, 1 h, *hν* 365 nm, 85%; (*ii*) **91** (1.5 equiv), AIBN (1 equiv), CH_2Cl_2, 80 min, MW, 100 W, 70 °C, 99%; (*iii*) **92** or **99**, PyBOP, 2,4,6-trimethylcollidine, CH_2Cl_2/DMF (1:1), 1 h, rt; (*iv*) (for building block **92**): 20% piperidine in DMF, then 20% Ac_2O in DMF; (*v*) TFA/3,6-dioxa-1,8-octane-dithiol (DODT)/H_2O/*i*Pr_3SiH (94:2.5:2.5:1, v/v/v/v)

racemization-suppressing conditions (PyBOP, 2,4,6-collidine, room temperature) (Zhang et al. 2012; Carpino et al. 1994; Carpino and El-Faham 1994). In the case of *N*α-Fmoc-protected **92**, the Fmoc protecting group was removed after coupling and subsequently exchanged for an acetyl group before TFA-mediated peptide cleavage was performed affording **101** (Scheme 9.15c). This allowed for a direct comparison of RP HPLC profiles to assess the degree of racemization. The RP HPLC chromatogram investigation of crude **101**, obtained by using either **92** or **99** building block revealed that 1:1 ratio of epimers was formed when acetamide protecting group was used for lipidated cysteine incorporation. No detectable epimerization was however observed when *N*α-Fmoc-protected **92** was used for lipid incorporation. The type of *N*α-protecting group clearly influenced the degree of racemization during the study indicating the preferred choice of Fmoc-protected building block **92** for Fmoc SPPS-mediated peptide lipidation.

We then focused on reaction conditions that would allow direct lipidation of a thiol-containing peptide affording an *S*-palmitoylated construct **101** in a convergent-like approach.

The construct **102**, derived from resin-bound **100**, incorporated two cysteine residues; an *N*-terminal Cys with a sulfhydryl group ready for 'thiol-ene' conjugation and the side chain of the second, internally located cysteine was masked with *t*Bu. Subsequent photoinitiated lipidation at 365 nm of **102** using vinyl palmitate **91** (7 equiv) and previously reported conditions [DMPA (0.5 equiv), DTT (3 equiv) in NMP for 60 min] afforded *S*-palmitoylated peptide **101** albeit in variable yields (Scheme 9.16a) (Wright et al. 2013a; Wright et al. 2013b). A careful examination of LC-MS profiles of the 'thiol-ene' reaction leading to desired conjugate **101** identified formation of unwanted by-products such as DTT-adducts and bis-palmitoylated peptide **104**. The competitive formation of **104** by-product was found to increase with increasing levels of vinyl palmitate in the reaction mixture. Substitution of DTT with the more bulky mercaptan *tert*-butyl thiol (*t*BuSH) proved superior in suppressing an unwanted addition of the thiol scavenger to the

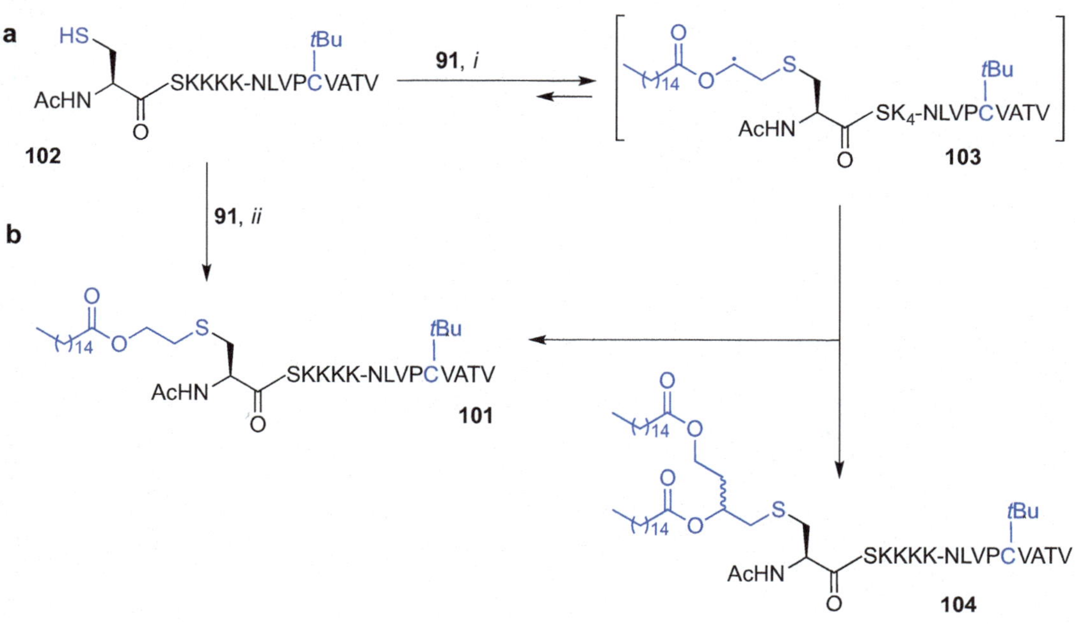

Scheme 9.16 CLipPA direct conjugation of vinyl palmitate (**91**) and semiprotected peptide **102** under unoptimised conditions (**a**) and optimized conditions (**b**) (Yang et al. 2016). Reagents and conditions: (*i*) **91** (7 equiv), DMPA (0.5 equiv), DTT (3 equiv) NMP, 1 h, *hν* 365 nm; (*ii*) **91** (70 equiv), DMPA (0.5 equiv), *t*BuSH (80 equiv), *i*Pr_3SiH (80 equiv), TFA/NMP (5:95, v/v), 30 min, *hν* 365 nm

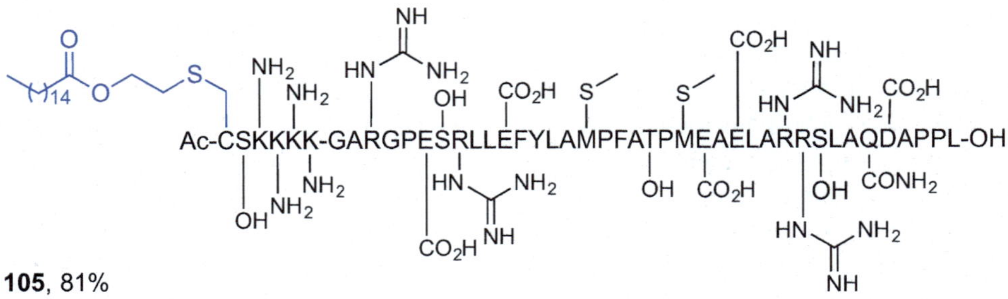

105, 81%

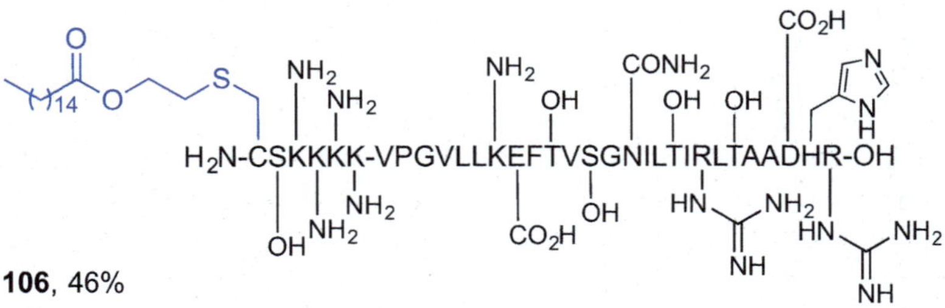

106, 46%

Fig. 9.14 *S*-Palmitoylated long peptide products accessed using CLipPA technology (Yang et al. 2016)

carbon-centered radical **103**. Formation of undesired bis-palmitoylated adduct **104** was also diminished by including an organosilane-based coreductant (*i*Pr$_3$SiH) that facilitated hydrogen transfer to the radical intermediate **103**. Furthermore, decreasing the pH of reaction mixture with TFA led to a cleaner reaction profile, presumably a result of protonation of electron-rich amine residues. Moreover, a large excess of vinyl palmitate (**91**), *tert*-butyl mercaptan and *i*Pr$_3$SiH were also needed to maximise conversion of **102** to the desired **101**. Although a large excess of vinyl palmitate was used in the optimized, photoinitiated (*hν* 365 nm) conditions [**91** (70 equiv), DMPA (0.5 equiv), *t*BuSH (80 equiv), *i*Pr$_3$SiH (80 equiv), TFA (5% v/v) in NMP for 30 min], a now quantitative conversion of peptide **102** to the *S*-monopalmitoylated construct **101** (95%, based on the corresponding peak integration on the RP HPLC profile) was observed with negligible levels of bis-adduct **104** formed (Scheme 9.16b).

The optimized CLipPA technology could be used to effect direct *S*-monopalmitoylation of complex, unprotected peptide substrates as demonstrated for long peptides including Ac-CSKKKK-GARGPESRLLEFYLAMPFATPMEAELARRSLAQDAPPL-OH and H$_2$N-CSKKKK-VPGVLLKEFTVSGNILTIRLTAADHR-OH, derived from NY-ESO-1(79-116) and NY-ESO-1(118-143), respectively. An excellent conversion to the desired lipidated peptide **105** (81%) and good 46% conversion to **106**, based on RP HPLC profiles, demonstrated the power of this new strategy (Fig. 9.14) (Yang et al. 2016).

The CLipPA technology offers a feasible one-step approach to lipidated peptide constructs containing all-natural bonds. We believe that this technique has strong potential to play a key role in self-adjuvanting peptide-based vaccine development in the future. The use of CLipPA eliminates the need for complex, multi-step and timeconsuming solution-phase synthesis of lipidated building blocks that are not readily available in all research laboratories. Depending on the vaccine construct requirements, either a stepwise SPPS approach, or a direct, convergent-like substrate lipidation can be executed using the 'thiol-ene' reaction and the optimized CLipPA conditions to afford S-palmitoylated assemblies in excellent yields with high selectivity.

9.2 Conclusions

Lipidation of peptides and proteins plays an important role in improving pharmacokinetic and pharmacodynamic profiles of peptides which may lead to potent analogues with clinical potential. Lipidated peptides activating TLR2 are crucial for peptide-based self-adjuvanting vaccine development. A simple, efficient and low-cost synthetic approach for incorporation of lipid motifs into peptides for subsequent bioevaluation is required. Synthesis of lipidated peptides *via* a standard SPPS technique using orthogonal protecting group strategy poses a challenge due to decreased solubility of lipopeptides. Novel synthetic advances such as the atom economical and functional group compatible CLipPA technique provides a useful approach to access *S*-palmitoylated peptides with a range of applications including vaccine design.

References

Abdel-Aal AB, Lakshminarayanan V, Thompson P, Supekar N, Bradley JM, Wolfert MA, Cohen PA, Gendler SJ, Boons GJ (2014) Immune and anticancer responses elicited by fully synthetic aberrantly glycosylated MUC1 tripartite vaccines modified by a TLR2 or TLR9 agonist. ChemBioChem 15(10):1508–1513

Agnihotri G, Crall BM, Lewis TC, Day TP, Balakrishna R, Warshakoon HJ, Malladi SS, David SA (2011) Structure-activity relationships in toll-like receptor 2-agonists leading to simplified monoacyl lipopeptides. J Med Chem 54(23):8148–8160

Anastas JN, Moon RT (2013) WNT signalling pathways as therapeutic targets in cancer. Nat Rev Cancer 13(1):11–26

Anderson RJ, Tang CW, Daniels NJ, Compton BJ, Hayman CM, Johnston KA, Knight DA, Gasser O, Poyntz HC, Ferguson PM, Larsen DS, Ronchese F, Painter GF, Hermans IF (2014) A self-adjuvanting vaccine induces cytotoxic T lymphocytes that suppress allergy. Nat Chem Biol 10(11):943–949

Anderson RJ, Compton BJ, Tang C-W, Authier-Hall A, Hayman CM, Swinerd GW, Kowalczyk R, Harris P, Brimble MA, Larsen DS, Gasser O, Weinkove R, Hermans IF, Painter GF (2015) NKT cell-dependent glycolipid-peptide vaccines with potent anti-tumour activity. Chem Sci 6(9):5120–5127

Asada H, Douen T, Mizokoshi Y, Fujita T, Murakami M, Yamamoto A, Muranishi S (1994) Stability of acyl derivatives of insulin in the small intestine: relative importance of insulin association characteristics in aqueous solution. Pharm Res 11(8):1115–1120

Asada H, Douen T, Waki M, Adachi S, Fujita T, Yamamoto A, Muranishi S (1995) Absorption characteristics of chemically modified-insulin derivatives with various fatty acids in the small and large intestine. J Pharm Sci 84(6):682–687

Avadisian M, Gunning PT (2013) Extolling the benefits of molecular therapeutic lipidation. Mol Biosyst 9(9):2179–2188

Avadisian M, Fletcher S, Liu B, Zhao W, Yue P, Badali D, Xu W, Schimmer AD, Turkson J, Gradinaru CC, Gunning PT (2011) Artificially induced protein–membrane anchorage with cholesterol-based recognition agents as a new therapeutic concept. Angew Chem Int Ed Engl 50(28):6248–6253

Backes BJ, Ellman JA (1999) An alkanesulfonamide "safety-catch" linker for solid-phase synthesis. J Org Chem 64(7):2322–2330

Basto AP, Leitao A (2014) Targeting TLR2 for vaccine development. J Immunol Res 2014:619410

Bednarek MA, Feighner SD, Pong SS, McKee KK, Hreniuk DL, Silva MV, Warren VA, Howard AD, Van Der Ploeg LH, Heck JV (2000) Structure-function studies on the new growth hormone-releasing peptide, ghrelin: minimal sequence of ghrelin necessary for activation of growth hormone secretagogue receptor 1a. J Med Chem 43(23):4370–4376

Benedict RG, Langlykke AF (1947) Antibiotic activity of Bacillus polymyxa. J Bacteriol 54(1):24

Berndt N, Hamilton AD, Sebti SM (2011) Targeting protein prenylation for cancer therapy. Nat Rev Cancer 11(11):775–791

Braun V (1975) Covalent lipoprotein from the outer membrane of escherichia coli. Biochim Biophys Acta Rev Biomemr 415(3):335–377

Bray GA (2015) Obesity: liraglutide[mdash]another weapon in the war against obesity? Nat Rev Endocrinol 11(10):569–570

Brimble MA, Wright TH, Dunbar RP, Williams GM (2014) Amino acid and peptide conjugates and conjugation process. WO2014207708 A3

Brimble MA, Edwards PJ, Harris PW, Norris GE, Patchett ML, Wright TH, Yang SH, Carley SE (2015) Synthesis of the antimicrobial S-linked glycopeptide, glycocin F. Chemistry 21(9):3556–3561

Brink AJ, Richards GA, Colombo G, Bortolotti F, Colombo P, Jehl F (2014) Multicomponent antibiotic substances produced by fermentation: implications for regulatory authorities, critically ill patients and generics. Int J Antimicrob Agents 43(1):1–6

Brown LE, Jackson DC (2005) Lipid-based self-adjuvanting vaccines. Curr Drug Deliv 2(4):383–393

Buskas T, Li Y, Boons GJ (2004) The immunogenicity of the tumor-associated antigen Lewis(y) may be suppressed by a bifunctional cross-linker required for coupling to a carrier protein. Chemistry 10(14):3517–3524

Buskas T, Ingale S, Boons GJ (2005) Towards a fully synthetic carbohydrate-based anticancer vaccine: syn-

thesis and Immunological evaluation of a lipidated glycopeptide containing the tumor-associated Tn antigen. Angew Chem Int Ed Engl 44(37):5985–5988
Buwitt-Beckmann U, Heine H, Wiesmuller KH, Jung G, Brock R, Akira S, Ulmer AJ (2005a) Toll-like receptor 6-independent signaling by diacylated lipopeptides. Eur J Immunol 35(1):282–289
Buwitt-Beckmann U, Heine H, Wiesmuller KH, Jung G, Brock R, Ulmer AJ (2005b) Lipopeptide structure determines TLR2 dependent cell activation level. FEBS J 272(24):6354–6364
Cai H, Huang ZH, Shi L, Zhao YF, Kunz H, Li YM (2011) Towards a fully synthetic MUC1-based anticancer vaccine: efficient conjugation of glycopeptides with mono-, di-, and tetravalent lipopeptides using click chemistry. Chemistry 17(23):6396–6406
Cai H, Sun ZY, Huang ZH, Shi L, Zhao YF, Kunz H, Li YM (2013) Fully synthetic self-adjuvanting thioether-conjugated glycopeptide-lipopeptide antitumor vaccines for the induction of complement-dependent cytotoxicity against tumor cells. Chemistry 19(6):1962–1970
Cai H, Sun ZY, Chen MS, Zhao YF, Kunz H, Li YM (2014) Synthetic multivalent glycopeptide-lipopeptide antitumor vaccines: impact of the cluster effect on the killing of tumor cells. Angew Chem Int Ed Engl 53(6):1699–1703
Carpino LA, El-Faham A (1994) Effect of tertiary bases on O-benzotriazolyluronium salt-induced peptide segment coupling. J Org Chem 59(4):695–698
Carpino LA, El-Faham A, Albericio F (1994) Racemization studies during solid-phase peptide synthesis using azabenzotriazole-based coupling reagents. Tetrahedron Lett 35(15):2279–2282
Cavallari M, Stallforth P, Kalinichenko A, Rathwell DC, Gronewold TM, Adibekian A, Mori L, Landmann R, Seeberger PH, De Libero G (2014) A semisynthetic carbohydrate-lipid vaccine that protects against pneumoniae in mice. Nat Chem Biol 10(11):950–956
Chamberlain LH, Shipston MJ (2015) The physiology of protein S-acylation. Physiol Rev 95(2):341–376
Cheng C, Jain P, Bettahi I, Pal S, Tifrea D, de la Maza LM (2011) A TLR2 agonist is a more effective adjuvant for a Chlamydia major outer membrane protein vaccine than ligands to other TLR and NOD receptors. Vaccine 29(38):6641–6649
Chua BY, Zeng W, Lau YF, Jackson DC (2007) Comparison of lipopeptide-based immunocontraceptive vaccines containing different lipid groups. Vaccine 25(1):92–101
Chua BY, Eriksson EM, Brown LE, Zeng W, Gowans EJ, Torresi J, Jackson DC (2008) A self-adjuvanting lipopeptide-based vaccine candidate for the treatment of hepatitis C virus infection. Vaccine 26(37):4866–4875
Chua BY, Johnson D, Tan A, Earnest-Silveira L, Sekiya T, Chin R, Torresi J, Jackson DC (2012) Hepatitis C VLPs delivered to dendritic cells by a TLR2 targeting lipopeptide results in enhanced antibody and cell-mediated responses. PLoS One 7(10):e47492
Chua BY, Olson MR, Bedoui S, Sekiya T, Wong CY, Turner SJ, Jackson DC (2014) The use of a TLR2 agonist-based adjuvant for enhancing effector and memory CD8 T-cell responses. Immunol Cell Biol 92(4):377–383
Chua BY, Wong CY, Mifsud EJ, Edenborough KM, Sekiya T, Tan AC, Mercuri F, Rockman S, Chen W, Turner SJ, Doherty PC, Kelso A, Brown LE, Jackson DC (2015) Inactivated influenza vaccine that provides rapid, innate-immune-system-mediated protection and subsequent long-term adaptive immunity. MBio 6(6):e01024–e01015
Clement S, Still JG, Kosutic G, McAllister RG (2002) Oral insulin product hexyl-insulin monoconjugate 2 (HIM2) in type 1 diabetes mellitus: the glucose stabilization effects of HIM2. Diabetes Technol Ther 4(4):459–466
Clement S, Dandona P, Still JG, Kosutic G (2004) Oral modified insulin (HIM2) in patients with type 1 diabetes mellitus: results from a phase I/II clinical trial. Metabolism 53(1):54–58
Cochrane SA, Vederas JC (2016) Lipopeptides from Bacillus and Paenibacillus spp.: a gold mine of antibiotic candidates. Med Res Rev 36(1):4–31
Cordy JM, Hooper NM, Turner AJ (2006) The involvement of lipid rafts in Alzheimer's disease. Mol Membr Biol 23(1):111–122
Cory D, Choong I, Glenn JS (2015) Treatment of hepatitis delta virus (HDV) infection with combination of lonafarnib and ritonavir. WO 2015168648:A1
Davda D, Martin BR (2014) Acyl protein thioesterase inhibitors as probes of dynamic S-palmitoylation. MedChemComm 5(3):268–276
Dawson P, Muir T, Clark-Lewis I, Kent S (1994) Synthesis of proteins by native chemical ligation. Science 266(5186):776–779
Debono M, Barnhart M, Carrell CB, Hoffmann JA, Occolowitz JL, Abbott BJ, Fukuda DS, Hamill RL, Biemann K, Herlihy WC (1987) A21978C, a complex of new acidic peptide antibiotics: isolation, chemistry, and mass spectral structure elucidation. J Antibiot 40(6):761–777
Deeg MA, Humphrey DR, Yang SH, Ferguson TR, Reinhold VN, Rosenberry TL (1992) Glycan components in the glycoinositol phospholipid anchor of human erythrocyte acetylcholinesterase. Novel fragments produced by trifluoroacetic acid. J Biol Chem 267(26):18573–18580
Demotz S, Lanzavecchia A, Eisel U, Niemann H, Widmann C, Corradin G (1989) Delineation of several DR-restricted tetanus toxin T cell epitopes. J Immunol 142(2):394–402
Dey G, Bharti R, Sen R, Mandal M (2015) Microbial amphiphiles: a class of promising new-generation anticancer agents. Drug Discov Today 20(1):136–146
Di L (2015) Strategic approaches to optimizing peptide ADME properties. AAPS J 17(1):134–143

Dondoni A, Marra A (2012) Recent applications of thiol-ene coupling as a click process for glycoconjugation. Chem Soc Rev 41(2):573–586
Ekrami HM, Kennedy AR, Shen WC (1995) Water-soluble fatty acid derivatives as acylating agents for reversible lipidization of polypeptides. FEBS Lett 371(3):283–286
Elbrond B, Jakobsen G, Larsen S, Agerso H, Jensen LB, Rolan P, Sturis J, Hatorp V, Zdravkovic M (2002) Pharmacokinetics, pharmacodynamics, safety, and tolerability of a single-dose of NN2211, a long-acting glucagon-like peptide 1 derivative, in healthy male subjects. Diabetes care 25(8):1398–1404
Eriksson EM, Jackson DC (2007) Recent advances with TLR2-targeting lipopeptide-based vaccines. Curr Protein Pept Sci 8(4):412–417
Feng Y, He D, Yao Z, Klionsky DJ (2014) The machinery of macroautophagy. Cell Res 24(1):24–41
Ferguson MAJ, Kinoshita T, Hart GW (2009) Glycosylphosphatidylinositol anchors. In: Varki A, Cummings RD, Esko JD, Freeze HH, Bertozzi CR, Hart GW, Etzler ME (eds) Essentials of glycobiology, 2nd edn. Cold Spring Harbor Laboratory Press, New York, pp 143–161
Flora DB 2002. Fatty acid-acylated insulin analogs. US6444641 B1.
Fosgerau K, Hoffmann T (2015) Peptide therapeutics: current status and future directions. Drug Discov Today 20(1):122–128
Fujii S, Shimizu K, Smith C, Bonifaz L, Steinman RM (2003) Activation of natural killer T cells by alpha-galactosylceramide rapidly induces the full maturation of dendritic cells in vivo and thereby acts as an adjuvant for combined CD4 and CD8 T cell immunity to a coadministered protein. J Exp Med 198(2):267–279
Fujita T, Kawahara I, Y-s Q, Hattori K, Takenaka K, Muranishi S, Yamamoto A (1998) Permeability characteristics of tetragastrins across intestinal membranes using the Caco-2 monolayer system: comparison between acylation and application of protease inhibitors. Pharm Res 15(9):1387–1392
Gao S, Yu R, Zhou X (2016) The role of geranylgeranyltransferase I-mediated protein prenylation in the brain. Mol Neurobiol 53(10):6925–6937
Gay NJ, Gangloff M (2007) Structure and function of Toll receptors and their ligands. Annu Rev Biochem 76:141–165
Goddard-Borger ED, Stick RV (2007) An efficient, inexpensive, and shelf-stable diazotransfer reagent: imidazole-1-sulfonyl azide hydrochloride. Org Lett 9(19):3797–3800
Godfrey DI, Kronenberg M (2004) Going both ways: immune regulation via CD1d-dependent NKT cells. J Clin Invest 114(10):1379–1388
Gordon LB, Rothman FG, López-Otín C, Misteli T (2014) Progeria: a paradigm for translational medicine. Cell 156(3):400–407
Green BR, White KL, McDougle DR, Zhang L, Klein B, Scholl EA, Pruess TH, White HS, Bulaj G (2010) Introduction of lipidization-cationization motifs affords systemically bioavailable neuropeptide Y and neurotensin analogs with anticonvulsant activities. J Pept Sci 16(9):486–495
Green BR, Smith M, White KL, White HS, Bulaj G (2011) Analgesic neuropeptide W suppresses seizures in the brain revealed by rational repositioning and peptide engineering. ACS Chem Neurosci 2(1):51–56
Grgurina I, Barca A, Cervigni S, Gallo M, Scaloni A, Pucci P (1994) Relevance of chlorine-substituent for the antifungal activity of syringomycin and syringotoxin, metabolites of the phytopathogenic bacterium Pseudomonas syringae pv. syringae. Experientia 50(2):130–133
Gupta S, Takebe N, LoRusso P (2010) Review: targeting the hedgehog pathway in cancer. Ther Adv Med Oncol 2(4):237–250
Gutierrez JA, Solenberg PJ, Perkins DR, Willency JA, Knierman MD, Jin Z, Witcher DR, Luo S, Onyia JE, Hale JE (2008) Ghrelin octanoylation mediated by an orphan lipid transferase. Proc Natl Acad Sci U S A 105(17):6320–6325
Hamley IW (2015) Lipopeptides: from self-assembly to bioactivity. Chem. Commun 51(41):8574–8583
Hamman JH, Enslin GM, Kotze AF (2005) Oral delivery of peptide drugs: barriers and developments. BioDrugs 19(3):165–177
Han Y, Bontems SL, Hegyes P, Munson MC, Minor CA, Kates SA, Albericio F, Barany G (1996) Preparation and applications of xanthenylamide (XAL) handles for solid-phase synthesis of C-terminal peptide amides under particularly mild conditions. J Org Chem 61(18):6326–6339
Hannoush RN (2015) Synthetic protein lipidation. Curr Opin Chem Biol 28:39–46
Harris PW, Brimble MA (2010) Toward the total chemical synthesis of the cancer protein NY-ESO-1. Biopolymers 94(4):542–550
Harris PW, Brimble MA (2013) A comparison of Boc and Fmoc SPPS strategies for the preparation of C-terminal peptide alpha-thiolesters: NY-ESO-1 (3)(9)Cys-(6)(8) Ala-COSR. Biopolymers 100(4):356–365
Harris PW, Brimble MA (2015) Chemical synthesis of a polypeptide backbone derived from the primary sequence of the cancer protein NY-ESO-1 enabled by kinetically controlled ligation and pseudoprolines. Biopolymers 104(2):116–127
Harris PWR, Brimble MA, Dunbar R, Kent SBH (2007) Synthesis of a C-terminal thioester derivative of the lipopeptide Pam2CSKKKKG using Fmoc SPPS. Synlett 2007(5):0713–0716
Harris PW, Squire C, Young PG, Brimble MA (2015) Chemical synthesis of gamma-secretase activating protein using pseudoglutamines as ligation sites. Biopolymers 104(1):37–45
Hashimoto M, Takada K, Kiso Y, Muranishi S (1989) Synthesis of palmitoyl derivatives of insulin and their biological activities. Pharm Res 6(2):171–176
Hashizume M, Douen T, Murakami M, Yamamoto A, Takada K, Muranishi S (1992) Improvement of large

intestinal absorption of insulin by chemical modification with palmitic acid in rats. J Pharm Pharmacol 44(7):555–559
Hermans IF, Silk JD, Gileadi U, Salio M, Mathew B, Ritter G, Schmidt R, Harris AL, Old L, Cerundolo V (2003) NKT cells enhance CD4+ and CD8+ T cell responses to soluble antigen in vivo through direct interaction with dendritic cells. J Immunol 171(10):5140–5147
Hicks DA, Nalivaeva NN, Turner AJ (2012) Lipid rafts and Alzheimer's disease: protein-lipid interactions and perturbation of signaling. Front Physiol 3:189
Hida T, Hayashi K, Yukishige K, Tanida S, Kawamura N, Harada S (1995) Synthesis and biological activities of TAN-1511 analogues. J Antibiot 48(7):589–603
Hofmann K (2000) A superfamily of membrane-bound O-acyltransferases with implications for Wnt signaling. Trends Biochem Sci 25(3):111–112
Home P, Kurtzhals P (2006) Insulin detemir: from concept to clinical experience. Expert Opin Pharmacother 7(3):325–343
Honeycutt L, Wang J, Ekrami H, Shen WC (1996) Comparison of pharmacokinetic parameters of a polypeptide, the Bowman-Birk protease inhibitor (BBI), and its palmitic acid conjugate. Pharm Res 13(9):1373–1377
Hoyle CE, Bowman CN (2010) Thiol-ene click chemistry. Angew Chem Int Ed Engl 49(9):1540–1573
Hutchinson JA, Burholt S, Hamley IW (2017) Peptide hormones and lipopeptides: from self-assembly to therapeutic applications. J Pept Sci 23(2):82–94
Iepsen EW, Torekov SS, Holst JJ (2015) Liraglutide for Type 2 diabetes and obesity: a 2015 update. Expert Rev Cardiovasc Ther 13(7):753–767
Ingale S, Buskas T, Boons GJ (2006) Synthesis of glyco(lipo)peptides by liposome-mediated native chemical ligation. Org Lett 8(25):5785–5788
Ingale S, Wolfert MA, Gaekwad J, Buskas T, Boons GJ (2007) Robust immune responses elicited by a fully synthetic three-component vaccine. Nat Chem Biol 3(10):663–667
Ingale S, Wolfert MA, Buskas T, Boons GJ (2009) Increasing the antigenicity of synthetic tumor-associated carbohydrate antigens by targeting Toll-like receptors. ChemBioChem 10(3):455–463
Ingenito R, Bianchi E, Fattori D, Pessi A (1999) Solid phase synthesis of peptide C-terminal thioesters by Fmoc/t-Bu chemistry. J Am Chem Soc 121(49):11369–11374
Jackson DC, Lau YF, Le T, Suhrbier A, Deliyannis G, Cheers C, Smith C, Zeng W, Brown LE (2004) A totally synthetic vaccine of generic structure that targets Toll-like receptor 2 on dendritic cells and promotes antibody or cytotoxic T cell responses. Proc Natl Acad Sci U S A 101(43):15440–15445
Jackson SH, Martin TS, Jones JD, Seal D, Emanuel F (2010) Liraglutide (Victoza): the first once-daily incretin mimetic injection for type-2 diabetes. P T 35(9):498–529
Jacques P (2011) Surfactin and other lipopeptides from *Bacillus* spp. In: Soberón-Chávez G (ed) Biosurfactants, Microbiology monographs, vol 20. Springer, Berlin/Heidelberg, pp 57–91
Johannessen L, Remsberg J, Gaponenko V, Adams KM, Barchi JJ, Tarasov SG, Jiang S, Tarasova NI (2011) Peptide structure stabilization by membrane anchoring and its general applicability to the development of potent cell-permeable inhibitors. ChemBioChem 12(6):914–921
Jones AR (1975) The metabolism of 3-chloro,- 3-bromo- and 3-iodoprpan-1,2-diol in rats and mice. Xenobiotica 5(3):155–165
Kaiser A, Gaidzik N, Becker T, Menge C, Groh K, Cai H, Li YM, Gerlitzki B, Schmitt E, Kunz H (2010) Fully synthetic vaccines consisting of tumor-associated MUC1 glycopeptides and a lipopeptide ligand of the Toll-like receptor 2. Angew Chem Int Ed Engl 49(21):3688–3692
Kakugawa S, Langton PF, Zebisch M, Howell S, Chang T-H, Liu Y, Feizi T, Bineva G, O'Reilly N, Snijders AP, Jones EY, Vincent J-P (2015) Notum deacylates Wnts to suppress signalling activity. Nature 519(7542):187–192
Kang JY, Nan X, Jin MS, Youn SJ, Ryu YH, Mah S, Han SH, Lee H, Paik SG, Lee JO (2009) Recognition of lipopeptide patterns by Toll-like receptor 2-Toll-like receptor 6 heterodimer. Immunity 31(6):873–884
Karsdal MA, Henriksen K, Bay-Jensen AC, Molloy B, Arnold M, John MR, Byrjalsen I, Azria M, Riis BJ, Qvist P, Christiansen C (2011) Lessons learned from the development of oral calcitonin: the first tablet formulation of a protein in phase III clinical trials. J Clin Pharmacol 51(4):460–471
Karsdal MA, Riis BJ, Mehta N, Stern W, Arbit E, Christiansen C, Henriksen K (2015) Lessons learned from the clinical development of oral peptides. Br J Clin Pharmacol 79(5):720–732
Kaspar AA, Reichert JM (2013) Future directions for peptide therapeutics development. Drug Discov Today 18(17–18):807–817
Khong H, Overwijk WW (2016) Adjuvants for peptide-based cancer vaccines. J Immunother Cancer 4:1–11
Kipnes M, Dandona P, Tripathy D, Still JG, Kosutic G (2003) Control of postprandial plasma glucose by an oral insulin product (HIM2) in patients with type 2 diabetes. Diabetes care 26(2):421–426
Knudsen LB, Nielsen PF, Huusfeldt PO, Johansen NL, Madsen K, Pedersen FZ, Thogersen H, Wilken M, Agerso H (2000) Potent derivatives of glucagon-like peptide-1 with pharmacokinetic properties suitable for once daily administration. J Med Chem 43(9):1664–1669
Koh C, Canini L, Dahari H, Zhao X, Uprichard SL, Haynes-Williams V, Winters MA, Subramanya G, Cooper SL, Pinto P, Wolff EF, Bishop R, Ai Thanda Han M, Cotler SJ, Kleiner DE, Keskin O, Idilman R, Yurdaydin C, Glenn JS, Heller T (2015) Oral prenylation inhibition with lonafarnib in chronic hepatitis

D infection: a proof-of-concept randomised, double-blind, placebo-controlled phase 2A trial. Lancet Infect Dis 15(10):1167–1174

Kojima M, Hosoda H, Date Y, Nakazato M, Matsuo H, Kangawa K (1999) Ghrelin is a growth-hormone-releasing acylated peptide from stomach. Nature 402(6762):656–660

Konitsiotis AD, Jovanović B, Ciepla P, Spitaler M, Lanyon-Hogg T, Tate EW, Magee AI (2015) Topological analysis of hedgehog acyltransferase, a multipalmitoylated transmembrane protein. J Biol Chem 290(6):3293–3307

Kopycinski J, Osman M, Griffiths PD, Emery VC (2010) Sequence flexibility of the immunodominant HLA A*0201 restricted ppUL83 CD8 T-cell epitope of human cytomegalovirus. J Med Virol 82(1):94-103

Krall N, da Cruz FP, Boutureira O, Bernardes GJL (2016) Site-selective protein-modification chemistry for basic biology and drug development. Nat Chem 8(2):103–113

Kudryashov V, Glunz PW, Williams LJ, Hintermann S, Danishefsky SJ, Lloyd KO (2001) Toward optimized carbohydrate-based anticancer vaccines: epitope clustering, carrier structure, and adjuvant all influence antibody responses to Lewis(y) conjugates in mice. Proc Natl Acad Sci U S A 98(6):3264–3269

Kurtzhals P (2007) Pharmacology of insulin detemir. Endocrinol Metab Clin North Am 36(1):14–20

Lakshminarayanan V, Thompson P, Wolfert MA, Buskas T, Bradley JM, Pathangey LB, Madsen CS, Cohen PA, Gendler SJ, Boons GJ (2012) Immune recognition of tumor-associated mucin MUC1 is achieved by a fully synthetic aberrantly glycosylated MUC1 tripartite vaccine. Proc Natl Acad Sci U S A 109(1):261–266

Lau J, Bloch P, Schaffer L, Pettersson I, Spetzler J, Kofoed J, Madsen K, Knudsen LB, McGuire J, Steensgaard DB, Strauss HM, Gram DX, Knudsen SM, Nielsen FS, Thygesen P, Reedtz-Runge S, Kruse T (2015) Discovery of the once-weekly glucagon-like peptide-1 (GLP-1) analogue semaglutide. J Med Chem 58(18):7370–7380

Le Floch J-P (2010) Critical appraisal of the safety and efficacy of insulin detemir in glycemic control and cardiovascular risk management in diabetics. Diabetes Metab Syndr Obes 3:197–213

Leclerc C, Deriaud E, Mimic V, van der Werf S (1991) Identification of a T-cell epitope adjacent to neutralization antigenic site 1 of poliovirus type 1. J Virol 65(2):711–718

Lee DJ, Harris PW, Brimble MA (2011) Synthesis of MUC1 Neoglycopeptides using efficient microwave-enhanced chaotrope-assisted click chemistry. Org Biomol Chem 9(5):1621–1626

Lee RTH, Zhao Z, Ingham PW (2016) Hedgehog signalling. Development 143(3):367–372

Lewis AL, Richard J (2015) Challenges in the delivery of peptide drugs: an industry perspective. Ther Deliv 6(2):149–163

Li W, Joshi MD, Singhania S, Ramsey KH, Murthy AK (2014) Peptide vaccine: progress and challenges. Vaccines 2(3):515–536

Lin David TS, Conibear E (2015) Enzymatic protein depalmitoylation by acyl protein thioesterases. Biochem Soc Trans 43(2):193

Liu C-F, Li F (2012) Method for modification of organic molecules. WO 2012/158122 A1

Lowe AB (2010) Thiol-ene "click" reactions and recent applications in polymer and materials synthesis. Polym Chem 1(1):17–36

Lowe AB (2014) Thiol-ene "click" reactions and recent applications in polymer and materials synthesis: a first update. Polym Chem 5(17):4820–4870

Madder A, Gunnoo SB (2016) Chemical protein modification via cysteine. ChemBioChem 17(7):529–553

Madsen K, Knudsen LB, Agersoe H, Nielsen PF, Thøgersen H, Wilken M, Johansen NL (2007) Structure–activity and protraction relationship of long-acting glucagon-like peptide-1 derivatives: importance of fatty acid length, polarity, and bulkiness. J Med Chem 50(24):6126–6132

Makovitzki A, Avrahami D, Shai Y (2006) Ultrashort antibacterial and antifungal lipopeptides. Proc Natl Acad Sci U S A 103(43):15997–16002

Maletínskâ L, Neugebauer W, Parê M-C, Pêrodin J, Pham D, Escher E (1996) Lipid masking and reactivation of angiotensin analogues. Helv Chim Acta 79(7):2023–2034

Maletinska L, Neugebauer W, Perodin J, Lefebvre M, Escher E (1997) Angiotensin analogues palmitoylated in positions 1 and 4. J Med Chem 40(20):3271–3279

Malm-Erjefalt M, Bjornsdottir I, Vanggaard J, Helleberg H, Larsen U, Oosterhuis B, van Lier JJ, Zdravkovic M, Olsen AK (2010) Metabolism and excretion of the once-daily human glucagon-like peptide-1 analog liraglutide in healthy male subjects and its in vitro degradation by dipeptidyl peptidase IV and neutral endopeptidase. Drug Metab Dispos 38(11):1944–1953

Martinez-Saez N, Supekar NT, Wolfert MA, Bermejo IA, Hurtado-Guerrero R, Asensio JL, Jimenez-Barbero J, Busto JH, Avenoza A, Boons GJ, Peregrina JM, Corzana F (2016) Mucin architecture behind the immune response: design, evaluation and conformational analysis of an antitumor vaccine derived from an unnatural MUC1 fragment. Chem Sci 7(3):2294–2301

Matevossian A, Resh MD (2015) Membrane topology of hedgehog acyltransferase. J Biol Chem 290(4):2235–2243

McDonald DM, Wilkinson BL, Corcilius L, Thaysen-Andersen M, Byrne SN, Payne RJ (2014) Synthesis and immunological evaluation of self-adjuvanting MUC1-macrophage activating lipopeptide 2 conjugate vaccine candidates. Chem Commun 50(71):10273–10276

McDonald DM, Byrne SN, Payne RJ (2015) Synthetic self-adjuvanting glycopeptide cancer vaccines. Front Chem 3:60

Medini K, Harris PW, Hards K, Dingley AJ, Cook GM, Brimble MA (2015) Chemical synthesis of a pore-forming antimicrobial protein, caenopore-5, by using native chemical ligation at a Glu-Cys site. ChemBioChem 16(2):328–336

Medini K, Harris PWR, Menorca A, Hards K, Cook GM, Brimble MA (2016) Synthesis and activity of a diselenide bond mimetic of the antimicrobial protein caenopore-5. Chem Sci 7(3):2005–2010

Mejuch T, Waldmann H (2016) Synthesis of lipidated proteins. Bioconjug Chem 27(8):1771–1783

Metzger JW, Wiesmuller KH, Jung G (1991) Synthesis of N alpha-Fmoc protected derivatives of S-(2,3-dihydroxypropyl)-cysteine and their application in peptide synthesis. Int J Pept Protein Res 38(6):545–554

Metzger JW, Beck-Sickinger AG, Loleit M, Eckert M, Bessler WG, Jung G (1995) Synthetic S-(2,3-dihydroxypropyl)-cysteinyl peptides derived from the N-terminus of the cytochrome subunit of the photoreaction centre of Rhodopseudomonas viridis enhance murine splenocyte proliferation. J Pept Sci 1(3):184–190

Mifsud EJ, Tan AC, Jackson DC (2014) TLR agonists as modulators of the innate immune response and their potential as agents against infectious disease. Front Immunol 5:79

Mnif I, Ghribi D (2015) Review lipopeptides biosurfactants: mean classes and new insights for industrial, biomedical, and environmental applications. Biopolymers 104(3):129–147

Monji T, Pious D (1997) Exogenously provided peptides fail to complex with intracellular class II molecules for presentation by antigen-presenting cells. J Immunol 158(7):3155–3164

Moyle PM, Toth I (2008) Self-adjuvanting lipopeptide vaccines. Curr Med Chem 15(5):506–516

Muhlradt PF, Kiess M, Meyer H, Sussmuth R, Jung G (1997) Isolation, structure elucidation, and synthesis of a macrophage stimulatory lipopeptide from Mycoplasma fermentans acting at picomolar concentration. J Exp Med 185(11):1951–1958

Muhlradt PF, Kiess M, Meyer H, Sussmuth R, Jung G (1998) Structure and specific activity of macrophage-stimulating lipopeptides from Mycoplasma hyorhinis. Infect Immun 66(10):4804–4810

Müller TD, Nogueiras R, Andermann ML, Andrews ZB, Anker SD, Argente J, Batterham RL, Benoit SC, Bowers CY, Broglio F, Casanueva FF, D'Alessio D, Depoortere I, Geliebter A, Ghigo E, Cole PA, Cowley M, Cummings DE, Dagher A, Diano S, Dickson SL, Diéguez C, Granata R, Grill HJ, Grove K, Habegger KM, Heppner K, Heiman ML, Holsen L, Holst B, Inui A, Jansson JO, Kirchner H, Korbonits M, Laferrère B, LeRoux CW, Lopez M, Morin S, Nakazato M, Nass R, Perez-Tilve D, Pfluger PT, Schwartz TW, Seeley RJ, Sleeman M, Sun Y, Sussel L, Tong J, Thorner MO, van der Lely AJ, van der Ploeg LHT, Zigman JM, Kojima M, Kangawa K, Smith RG, Horvath T, Tschöp MH (2015) Ghrelin. Mol Metab 4(6):437–460

Muranishi S, Sakai A, Yamada K, Murakami M, Takada K, Kiso Y (1991) Lipophilic peptides: synthesis of lauroyl thyrotropin-releasing hormone and its biological activity. Pharm Res 8(5):649–652

Nauck MA, Petrie JR, Sesti G, Mannucci E, Courrèges J-P, Lindegaard ML, Jensen CB, Atkin SL (2016) A phase 2, randomized, dose-finding study of the novel once-weekly human GLP-1 analog, semaglutide, compared with placebo and open-label liraglutide in patients with type 2 diabetes. Diabetes care 39(2):231–241

Nile AH, Hannoush RN (2016) Fatty acylation of Wnt proteins. Nat Chem Biol 12(2):60–69

Orwa JA, Govaerts C, Busson R, Roets E, Van Schepdael A, Hoogmartens J (2001) Isolation and structural characterization of polymyxin B components. J Chromatogr A 912(2):369–373

Otvos L, Wade JD (2014) Current challenges in peptide-based drug discovery. Front Chem 2:62

Park K, Kwon IC, Park K (2011) Oral protein delivery: current status and future prospect. React Funct Polym 71(3):280–287

Pattabiraman VR, McKinnie SM, Vederas JC (2008) Solid-supported synthesis and biological evaluation of the lantibiotic peptide bis(desmethyl) lacticin 3147 A2. Angew Chem Int Ed Engl 47(49):9472–9475

Paulick MG, Bertozzi CR (2008) The glycosylphosphatidylinositol anchor: a complex membrane-anchoring structure for proteins. Biochemistry 47(27):6991–7000

Pepinsky RB, Zeng C, Wen D, Rayhorn P, Baker DP, Williams KP, Bixler SA, Ambrose CM, Garber EA, Miatkowski K, Taylor FR, Wang EA, Galdes A (1998) Identification of a palmitic acid-modified form of human sonic hedgehog. J Biol Chem 273(22):14037–14045

Porotto M, Yokoyama CC, Palermo LM, Mungall B, Aljofan M, Cortese R, Pessi A, Moscona A (2010) Viral entry inhibitors targeted to the membrane site of action. J Virol 84(13):6760–6768

Raaijmakers JM, de Bruijn I, de Kock MJD (2006) Cyclic lipopeptide production by plant-associated Pseudomonas spp.: diversity, activity, biosynthesis, and regulation. Mol Plant Microbe Interact 19(7):699–710

Rajendran L, Schneider A, Schlechtingen G, Weidlich S, Ries J, Braxmeier T, Schwille P, Schulz JB, Schroeder C, Simons M, Jennings G, Knölker H-J, Simons K (2008a) Efficient inhibition of the Alzheimer's disease β-secretase by membrane targeting. Science 320(5875):520–523

Rajendran L, Schneider A, Schlechtingen G, Weidlich S, Ries J, Braxmeier T, Schwille P, Schulz JB, Schroeder C, Simons M, Jennings G, Knölker H-J, Simons K (2008b) Corrections and clarifications. Science 321(5891):912–912

Rajendran L, Udayar V, Goodger ZV (2012) Lipid-anchored drugs for delivery into subcellular compartments. Trends Pharmacol Sci 33(4):215–222

Ramesan RM, Sharma CP (2014) Recent advances in the oral delivery of insulin. Recent Pat Drug Deliv Formul 8(2):155–159

Remsberg JR, Lou H, Tarasov SG, Dean M, Tarasova NI (2007) Structural analogues of smoothened intracellular loops as potent inhibitors of hedgehog pathway and cancer cell growth. J Med Chem 50(18):4534–4538

Renukuntla J, Vadlapudi AD, Patel A, Boddu SHS, Mitra AK (2013) Approaches for enhancing oral bioavailability of peptides and proteins. Int J Pharm 447(0):75-93

Resh MD (2012) Targeting protein lipidation in disease. Trends Mol Med 18(4):206–214

Resh MD (2013) Covalent lipid modifications of proteins. Curr Biol 23(10):R431–R435

Resh MD (2016) Fatty acylation of proteins: the long and the short of it. Prog Lipid Res 63:120-131

Rigato M, Fadini GP (2014) Comparative effectiveness of liraglutide in the treatment of type 2 diabetes. Diabetes Metab Syndr Obes 7:107–120

Roberts WL, Myher JJ, Kuksis A, Low MG, Rosenberry TL (1988a) Lipid analysis of the glycoinositol phospholipid membrane anchor of human erythrocyte acetylcholinesterase. Palmitoylation of inositol results in resistance to phosphatidylinositol-specific phospholipase C. J Biol Chem 263(35):18766–18775

Roberts WL, Santikarn S, Reinhold VN, Rosenberry TL (1988b) Structural characterization of the glycoinositol phospholipid membrane anchor of human erythrocyte acetylcholinesterase by fast atom bombardment mass spectrometry. J Biol Chem 263(35):18776–18784

Robertson CR, Scholl EA, Pruess TH, Green BR, White HS, Bulaj G (2010) Engineering galanin analogues that discriminate between GalR1 and GalR2 receptor subtypes and exhibit anticonvulsant activity following systemic delivery. J Med Chem 53(4):1871–1875

Rockenfeller P, Koska M, Pietrocola F, Minois N, Knittelfelder O, Sica V, Franz J, Carmona-Gutierrez D, Kroemer G, Madeo F (2015) Phosphatidylethanolamine positively regulates autophagy and longevity. Cell Death Differ 22(3):499–508

Rostovtsev VV, Green LG, Fokin VV, Sharpless KB (2002) A stepwise Huisgen cycloaddition process: Copper(I)-catalyzed regioselective "ligation" of azides and terminal alkynes. Angew Chem Int Ed Engl 41(14):2596–2599

Ryan GJ, Hardy Y (2011) Liraglutide: once-daily GLP-1 agonist for the treatment of type 2 diabetes. J Clin Pharm Ther 36(3):260–274

Saar I, Lahe J, Langel K, Runesson J, Webling K, Järv J, Rytkönen J, Närvänen A, Bartfai T, Kurrikoff K, Langel Ü (2013) Novel systemically active galanin receptor 2 ligands in depression-like behavior. J Neurochem 127(1):114–123

Salunke DB, Shukla NM, Yoo E, Crall BM, Balakrishna R, Malladi SS, David SA (2012) Structure–activity relationships in human Toll-like Receptor 2-specific monoacyl lipopeptides. J Med Chem 55(7):3353–3363

Sarkar S, Salyer AC, Wall KA, Sucheck SJ (2013) Synthesis and immunological evaluation of a MUC1 glycopeptide incorporated into l-rhamnose displaying liposomes. Bioconjug Chem 24(3):363–375

Sato T, Oishi K, Ida T, Kojima M (2015) Physiological functions and pathology of ghrelin. Am J Life Sci 3(3-2):8–16

Setoh K, Murakami M, Araki N, Fujita T, Yamamoto A, Muranishi S (1995) Improvement of transdermal delivery of tetragastrin by lipophilic modification with fatty acids. J Pharm Pharmacol 47(10):808–811

Shi L, Cai H, Huang ZH, Sun ZY, Chen YX, Zhao YF, Kunz H, Li YM (2016) Synthetic MUC1 antitumor vaccine candidates with varied glycosylation pattern bearing R/S-configured Pam3CysSerLys4. ChemBioChem 17(15):1412–1415

Shindou H, Eto M, Morimoto R, Shimizu T (2009) Identification of membrane O-acyltransferase family motifs. Biochem Biophys Res Commun 383(3):320–325

Simerska P, Moyle PM, Toth I (2011) Modern lipid-, carbohydrate-, and peptide-based delivery systems for peptide, vaccine, and gene products. Med Res Rev 31(4):520–547

Son SJ, Harris PW, Squire CJ, Baker EN, Kent SB, Brimble MA (2014) Total chemical synthesis of an orf virus protein, ORFV002, an inhibitor of the master gene regulator NF-kappaB. Biopolymers 102(2):137–144

Spohn R, Buwitt-Beckmann U, Brock R, Jung G, Ulmer AJ, Wiesmuller KH (2004) Synthetic lipopeptide adjuvants and Toll-like receptor 2-structure-activity relationships. Vaccine 22(19):2494–2499

Stansly PG, Schlosser ME (1947) Studies on polymyxin: isolation and identification of Bacillus polymyxa and differentiation of polymyxin from certain known antibiotics. J Bacteriol 54(5):549–556

Steinhagen F, Kinjo T, Bode C, Klinman DM (2011) TLR-based immune adjuvants. Vaccine 29(17):3341–3355

Stevenson FT, Bursten SL, Locksley RM, Lovett DH (1992) Myristyl acylation of the tumor necrosis factor alpha precursor on specific lysine residues. J Exp Med 176(4):1053–1062

Stevenson FT, Bursten SL, Fanton C, Locksley RM, Lovett DH (1993) The 31-kDa precursor of interleukin 1 alpha is myristoylated on specific lysines within the 16-kDa N-terminal propiece. Proc Natl Acad Sci U S A 90(15):7245–7249

Still GJ (2002) Development of oral insulin: progress and current status. Diabetes Metab Res Rev 18(S1):S29–S37

Takeuchi O, Kaufmann A, Grote K, Kawai T, Hoshino K, Morr M, Muhlradt PF, Akira S (2000) Cutting edge: preferentially the R-stereoisomer of the mycoplasmal lipopeptide macrophage-activating lipopeptide-2 activates immune cells through a toll-like receptor 2- and MyD88-dependent signaling pathway. J Immunol 164(2):554–557

Tan AC, Mifsud EJ, Zeng W, Edenborough K, McVernon J, Brown LE, Jackson DC (2012) Intranasal administration of the TLR2 agonist Pam2Cys provides rapid protection against influenza in mice. Mol Pharm 9(9):2710–2718

Tanaka K, Fujita T, Yamamoto Y, Murakami M, Yamamoto A, Muranishi S (1996) Enhancement of intestinal transport of thyrotropin-releasing hormone via a carrier-mediated transport system by chemical modification with lauric acid. Biochim Biophys Acta 1283(1):119–126

Thompson P, Lakshminarayanan V, Supekar NT, Bradley JM, Cohen PA, Wolfert MA, Gendler SJ, Boons GJ (2015) Linear synthesis and immunological properties of a fully synthetic vaccine candidate containing a sialylated MUC1 glycopeptide. Chem Commun 51(50):10214–10217

Tomlinson B, Hu M, Zhang Y, Chan P, Liu Z-M (2016) Investigational glucagon-like peptide-1 agonists for the treatment of obesity. Expert Opin Investig Drugs 25(10):1167–1179

Tornoe CW, Christensen C, Meldal M (2002) Peptidotriazoles on solid phase: [1,2,3]-triazoles by regiospecific copper(I)-catalyzed 1,3-dipolar cycloadditions of terminal alkynes to azides. J Org Chem 67(9):3057–3064

Toth I, Flinn N, Hillery A, Gibbons WA, Artursson P (1994) Lipidic conjugates of luteinizing hormone releasing hormone (LHRH)+ and thyrotropin releasing hormone (TRH)+ that release and protect the native hormones in homogenates of human intestinal epithelial (Caco-2) cells. Int J Pharm 105(3):241–247

Trabocchi A, Guarna A (2014) The basics of peptidomimetics. In: Peptidomimetics in organic and medicinal chemistry. Wiley, Chichester, pp 1–17

Velkov T, Roberts KD, Thompson PE, Li J (2016) Polymyxins: a new hope in combating Gram-negative superbugs? Future Med Chem 8(10):1017–1025

Vilhena C, Bettencourt A (2012) Daptomycin: a review of properties, clinical use, drug delivery and resistance. Mini Rev Med Chem 12(3):202–209

Wang M, Casey PJ (2016) Protein prenylation: unique fats make their mark on biology. Nat Rev Mol Cell Biol 17(2):110–122

Wang J, Shen D, Shen WC (1999) Preparation, purification, and characterization of a reversibly lipidized desmopressin with potentiated anti-diuretic activity. Pharm Res 16(11):1674–1679

Wang J, Wu D, Shen WC (2002) Structure-activity relationship of reversibly lipidized peptides: studies of fatty acid-desmopressin conjugates. Pharm Res 19(5):609–614

Wang J, Chow D, Heiati H, Shen W-C (2003) Reversible lipidization for the oral delivery of salmon calcitonin. J Control Release 88(3):369–380

Wang J, Hogenkamp DJ, Tran M, Li W-Y, Yoshimura RF, Johnstone TBC, Shen W-C, Gee KW (2006) Reversible lipidization for the oral delivery of leu-enkephalin. J Drug Target 14(3):127–136

Ward BP, Ottaway NL, Perez-Tilve D, Ma D, Gelfanov VM, Tschöp MH, DiMarchi RD (2013) Peptide lipidation stabilizes structure to enhance biological function. Mol Metab 2(4):468–479

Watson E, Jonker DM, Jacobsen LV, Ingwersen SH (2010) Population pharmacokinetics of liraglutide, a once-daily human glucagon-like peptide-1 analog, in healthy volunteers and subjects with type 2 diabetes, and comparison to twice-daily exenatide. J Clin Pharmacol 50(8):886–894

Wexler-Cohen Y, Shai Y (2009) Membrane-anchored HIV-1 N-heptad repeat peptides are highly potent cell fusion inhibitors via an altered mode of action. PLOS Pathog 5(7):e1000509

Wiertz EJ, van Gaans-van den Brink JA, Gausepohl H, Prochnicka-Chalufour A, Hoogerhout P, Poolman JT (1992) Identification of T cell epitopes occurring in a meningococcal class 1 outer membrane protein using overlapping peptides assembled with simultaneous multiple peptide synthesis. J Exp Med 176(1):79–88

Wilkinson BL, Malins LR, Chun CK, Payne RJ (2010) Synthesis of MUC1-lipopeptide chimeras. Chem Commun 46(34):6249–6251

Wilkinson BL, Day S, Malins LR, Apostolopoulos V, Payne RJ (2011) Self-adjuvanting multicomponent cancer vaccine candidates combining per-glycosylated MUC1 glycopeptides and the Toll-like receptor 2 agonist Pam3CysSer. Angew Chem Int Ed Engl 50(7):1635–1639

Wilkinson BL, Day S, Chapman R, Perrier S, Apostolopoulos V, Payne RJ (2012) Synthesis and immunological evaluation of self-assembling and self-adjuvanting tricomponent glycopeptide cancer-vaccine candidates. Chemistry 18(51):16540–16548

Willems MMJHP, Zom GG, Khan S, Meeuwenoord N, Melief CJM, van der Stelt M, Overkleeft HS, Codée JDC, van der Marel GA, Ossendorp F, Filippov DV (2014) N-Tetradecylcarbamyl lipopeptides as novel agonists for Toll-like Receptor 2. J Med Chem 57(15):6873–6878

Wong CY, Martinez J, Dass CR (2016) Oral delivery of insulin for treatment of diabetes: status quo, challenges and opportunities. J Pharm Pharmacol 68(9):1093–1108

Wright MH, Heal WP, Mann DJ, Tate EW (2010) Protein myristoylation in health and disease. J Chem Biol 3(1):19–35

Wright TH, Brooks AES, Didsbury AJ, Williams GM, Harris PWR, Dunbar PR, Brimble MA (2013a) Direct peptide lipidation through thiol–ene coupling enables rapid synthesis and evaluation of self-adjuvanting vaccine candidates. Angew Chem Int Ed Engl 52(40):10616–10619

Wright TH, Brooks AES, Didsbury AJ, Williams GM, Harris PWR, Dunbar PR, Brimble MA (2013b) Corrigendum: direct peptide lipidation through thiol–ene coupling enables rapid synthesis and evaluation of self-adjuvanting vaccine candidates. Angew Chem Int Ed Engl 52(45):11686–11686

Wright TH, Brooks AES, Didsbury AJ, McIntosh JD, Burkert K, Yeung H, Williams GM, Dunbar PR,

Brimble MA (2013c) An improved method for the synthesis of lipopeptide TLR2-agonists using click chemistry. Synlett 24(14):1835–1841

Wu W, Li R, Malladi SS, Warshakoon HJ, Kimbrell MR, Amolins MW, Ukani R, Datta A, David SA (2010) Structure-activity relationships in toll-like receptor-2 agonistic diacylthioglycerol lipopeptides. J Med Chem 53(8):3198–3213

Yang J, Brown MS, Liang G, Grishin NV, Goldstein JL (2008) Identification of the acyltransferase that octanoylates ghrelin, an appetite-stimulating peptide hormone. Cell 132(3):387–396

Yang SH, Wojnar JM, Harris PW, DeVries AL, Evans CW, Brimble MA (2013) Chemical synthesis of a masked analogue of the fish antifreeze potentiating protein (AFPP). Org Biomol Chem 11(30):4935–4942

Yang SH, Harris PWR, Williams GM, Brimble MA (2016) Lipidation of cysteine or cysteine-containing peptides using the thiol-ene reaction (CLipPA). Eur J Org Chem 2016(15):2608–2616

Ye J, Wang C, Sumpter R, Brown MS, Goldstein JL, Gale M (2003) Disruption of hepatitis C virus RNA replication through inhibition of host protein geranylgeranylation. Proc Natl Acad Sci U S A 100(26):15865–15870

Yeung H, Lee DJ, Williams GM, Harris PWR, Dunbar RP, Brimble MA (2012) A method for the generation of Pam2Cys-based lipopeptide mimics via CuAAC click chemistry. Synlett 23(11):1617–1620

Yodoya E, Uemura K, Tenma T, Fujita T, Murakami M, Yamamoto A, Muranishi S (1994) Enhanced permeability of tetragastrin across the rat intestinal membrane and its reduced degradation by acylation with various fatty acids. J Pharmacol Exp Ther 271(3):1509–1513

Young SG, Yang SH, Davies BSJ, Jung H-J, Fong LG (2013) Targeting protein prenylation in progeria. Sci Transl Med 5(171):171ps173-171ps173

Yuan L, Wang J, Shen WC (2005) Reversible lipidization prolongs the pharmacological effect, plasma duration, and liver retention of octreotide. Pharm Res 22(2):220–227

Zaman M, Toth I (2013) Immunostimulation by synthetic lipopeptide-based vaccine candidates: structure-activity relationships. Front Immunol 4:318

Zeng W, Jackson DC, Rose K (1996) Synthesis of a new template with a built-in adjuvant and its use in constructing peptide vaccine candidates through polyoxime chemistry. J Pept Sci 2(1):66–72

Zeng W, Ghosh S, Macris M, Pagnon J, Jackson DC (2001) Assembly of synthetic peptide vaccines by chemoselective ligation of epitopes: influence of different chemical linkages and epitope orientations on biological activity. Vaccine 19(28-29):3843–3852

Zeng W, Ghosh S, Lau YF, Brown LE, Jackson DC (2002) Highly immunogenic and totally synthetic lipopeptides as self-adjuvanting immunocontraceptive vaccines. J Immunol 169(9):4905–4912

Zeng W, Horrocks KJ, Robevska G, Wong CY, Azzopardi K, Tauschek M, Robins-Browne RM, Jackson DC (2011) A modular approach to assembly of totally synthetic self-adjuvanting lipopeptide-based vaccines allows conformational epitope building. J Biol Chem 286(15):12944–12951

Zhang L, Bulaj G (2012) Converting peptides into drug Leads by lipidation. Curr Med Chem 19(11):1602–1618

Zhang L, Robertson CR, Green BR, Pruess TH, White HS, Bulaj G (2009) Structural requirements for a lipoamino acid in modulating the anticonvulsant activities of systemically active galanin analogues. J Med Chem 52(5):1310–1316

Zhang Y, Muthana SM, Farnsworth D, Ludek O, Adams K, Barchi JJ Jr, Gildersleeve JC (2012) Enhanced epimerization of glycosylated amino acids during solid-phase peptide synthesis. J Am Chem Soc 134(14):6316–6325

Zhang X, Cheong S-M, Amado NG, Reis AH, MacDonald BT, Zebisch M, Jones EY, Abreu JG, He X (2015) Notum is required for neural and head induction via Wnt deacylation, oxidation, and inactivation. Dev Cell 32(6):719–730

Zhao L, Tong P, Chen Y-X, Hu Z-W, Wang K, Zhang Y-N, Zhao D-S, Cai L-F, Liu K-L, Zhao Y-F, Li Y-M (2012) A multi-functional peptide as an HIV-1 entry inhibitor based on self-concentration, recognition, and covalent attachment. Org Biomol Chem 10(32):6512–6520

Zom GG, Khan S, Britten CM, Sommandas V, Camps MG, Loof NM, Budden CF, Meeuwenoord NJ, Filippov DV, van der Marel GA, Overkleeft HS, Melief CJ, Ossendorp F (2014) Efficient induction of antitumor immunity by synthetic toll-like receptor ligand-peptide conjugates. Cancer Immunol Res 2(8):756–764

Zom GG, Welters MJ, Loof NM, Goedemans R, Lougheed S, Valentijn RR, Zandvliet ML, Meeuwenoord NJ, Melief CJ, de Gruijl TD, Van der Marel GA, Filippov DV, Ossendorp F, Van der Burg SH (2016) TLR2 ligand-synthetic long peptide conjugates effectively stimulate tumor-draining lymph node T cells of cervical cancer patients. Oncotarget 7(41):67087–67100

Molecular Engineering of *Conus* Peptides as Therapeutic Leads

10

James T. Daniel and Richard J. Clark

Abstract

The venom from carnivorous marine snails of the *Conus* genus is a cocktail of peptides, proteins and small molecules that is used by the snail to capture prey. The peptides within this venom have been the focus of many drug design efforts as they exhibit potent and selective targeting of therapeutically important receptors, transporters and channels, particularly in relation to the treatment of chronic pain. The most well studied class of *Conus* peptides are the conotoxins, which are disulfide-rich and typically have well-defined three dimensional structures that are important for both biological activity and stability. In this chapter we discuss the molecular engineering approaches that have been used to modify these conotoxins to improve their pharmacological properties, including potency, selectivity, stability, and minimisation of the bioactive pharmacophore. These engineering strategies include sidechain modifications, disulfide substitution and deletion, backbone cyclisation, and truncations. Several of these re-engineered conotoxins have progressed to pre-clinical or clinical studies, which demonstrates the promise of using these molecular engineering techniques for the development of therapeutic leads.

Keywords

Peptide engineering • Conotoxins • Cone snails • Drug design • Chronic pain • Structure/activity relationships • Stabilisation

10.1 Introduction

Animal venoms are increasingly recognised as a valuable source of novel therapeutic peptides. The pharmacological diversity and potency of venom constituents toward physiological targets provides researchers an almost inexhaustible

J.T. Daniel • R.J. Clark (✉)
School of Biomedical Sciences, The University of Queensland, Brisbane 4072, Australia
e-mail: richard.clark@uq.edu.au

© Springer International Publishing AG 2017

A. Sunna et al. (eds.), *Peptides and Peptide-based Biomaterials and their Biomedical Applications*, Advances in Experimental Medicine and Biology 1030, https://doi.org/10.1007/978-3-319-66095-0_10

reservoir of compounds with potential applications as biomedical research tools and pharmaceutical lead molecules. Among the most successful venomous fauna are carnivorous marine gastropods of the *Conus* genus, or cone snails, which populate tropical waters and prey upon fish, worms and molluscs. In order to defend against or capture their comparatively faster prey, cone snails have evolved potent and fast-acting paralytic venoms, which are injected hypodermically through a harpoon-like tooth (Olivera et al. 2015). Early studies discovered that short bioactive peptide molecules from the venom were responsible for the rapid inhibition of neuromuscular currents in animal tissues (Endean et al. 1974; Gray et al. 1981; Olivera 1985). It is now recognised that *Conus* venoms comprise an astoundingly diverse cocktail of peptide toxins, termed conopeptides, many of which are potent and highly selective modulators of important neurophysiological ion-channels, G-protein coupled receptors (GPCRs) and membrane transporters (Lewis et al. 2012; Akondi et al. 2014a, b). Proteomic analysis of venom extracts by liquid chromatography and mass spectrometry has yielded current estimates in excess of 1000 mature conopeptides per species (Biass et al. 2009; Davis et al. 2009). With >700 species of cone snails thought to exist in the wild, *Conus* venoms represent an enormous and virtually untapped source of potential lead molecules for drug discovery and biomedical research.

Conopeptides bind with high affinity and specificity for their protein targets, which often play crucial roles in transduction of neurological stimuli, particularly ligand- and voltage-gated ion channels. Furthermore, although *Conus* venoms are potentially lethal in humans, small doses of individual conopeptides may produce analgesia *in vivo*. These appealing pharmacological properties have stimulated research into the design and development of conopeptides as novel analgesic drug leads for treatment of neuropathic pain: a chronic pain condition that occurs following pathological damage to somatosensory neural pathways involved in pain. Current first line pain therapeutics such as small molecule anticonvulsants and opioids have displayed only limited clinical efficacy due to poor specificity for their targets, dependence and limited routes of administration (Baron et al. 2010), thus there is demand for improved pharmacological approaches. The therapeutic potential of conopeptides is exemplified by the success of ω-conotoxin MVIIA, a novel voltage-gated calcium channel blocker, which was approved by the FDA in 2004 for treatment of chronic neuropathic pain. However, like other peptide and protein based drug leads, conopeptides suffer from poor reductive and proteolytic stability, low membrane penetration and rapid clearance *in vivo*, thus limiting bioavailability. Chemical modifications of conopeptide drug leads, with the purpose of increasing biological stability and activity, have therefore become invaluable techniques in the preclinical development of these compounds (Craik and Adams 2007; Clark and Craik 2010; Clark et al. 2012; Brady et al. 2013a). Here we outline various synthetic approaches that have been successfully used to engineer conotoxin analogues with improved structural and pharmacological properties.

10.2 Conotoxin Structure and Pharmacology

10.2.1 Discovery and Classification

Historically, discovery and characterisation of conopeptides has been primarily achieved through venom extraction, liquid chromatographic fractionation and mass spectral analysis of *Conus* venom constituents. Three-dimensional structural analysis of conopeptide molecules is routinely performed by nuclear magnetic resonance (NMR) spectroscopy and, in combination with *in vitro* and *in vivo* pharmacological screens, allows detailed characterisation of their structure-activity relationships (SARs). Together, these approaches have led to identification of distinct conopeptide classes with highly conserved cysteine frameworks, three-dimensional structures and pharmacological targets, and are categorised accordingly (Table 10.1). There are several families of conopeptides which lack disulfide bonds, however the majority, referred to as conotoxins, are cysteine-rich and form well defined disulfide-stabilised structures. The molecular engineering of these disulfide-rich conotoxins will be the focus of this review. To date, some 1700 conotoxins have been characterised and detailed structural and pharmacological data is compiled in an online database known as ConoServer (Kaas et al. 2010, 2012), although this number is estimated to represent less than 0.1% of existing conotoxin sequences (Akondi et al. 2014b). Notable examples include the α-conotoxins, which inhibit nicotinic acetylcholine receptors (nAChR), ω-conotoxins that are voltage-gated calcium channel antagonists, μ-conotoxins that inhibit voltage-gated sodium channels (Na_Vs), and the χ-conotoxins that block the noradrenaline transporter (NET). Examples of analgesic conotoxins and their classifications are summarised in Table 10.1.

Recently, the predominant form of conotoxin discovery has shifted towards next-generation transcriptomic sequencing of *Conus* venom ducts and generation of cDNA libraries. Conotoxins are encoded as precursor genes possessing an endoplasmic reticulum (ER) signal sequence, a propeptide region and a hypervariable mature peptide region and thus genomic data can be mined for novel peptide sequences. The continuously increasing availability of genomic data from *Conus* species has revealed distinct but highly conserved ER signal peptide sequences among conotoxin precursors, leading to classification based on gene superfamilies (Robinson and Norton 2014). Before being secreted into the venom ducts, conotoxins undergo maturation steps in the form of post-translational modifications (PTMs) including side chain modifications (e.g. hydroxylation of prolines and carboxylation of glutamic acid) (Espiritu et al. 2014) and disulfide formation (folding) assisted by an expanded library of protein disulfide isomerases (Safavi-Hemami et al. 2016). Therefore, cDNA libraries provide valuable information for predicting mature conotoxins sequences and synthesis of possible bioactive peptides without requiring direct access to venom samples. However, nucleic acid sequence data does not detect PTMs or disulfide connectivity and thus a combinatorial approach utilising both transcriptomic and proteomic analysis ("venomics") has proved most effective for investigating conotoxin diversity and discovering novel bioactive peptides (Lluisma et al. 2012; Robinson et al. 2014; Kaas and Craik 2015). Interestingly, *Conus* venomics have revealed only a limited number of conotoxin precursor genes (~100) responsible for the almost 10-fold greater number of mature conotoxins found in the venom. This diversity is thought to result from "messy" transcriptional regulation producing sequences with truncations/elongations, alternate propeptide cleavage sites and variable incorporation of post-translational modi-

Table 10.1 Classification of analgesic conotoxins

Gene superfamily	Cys Framework [Connectivity]	Pharmacological family	Pain target	Examples (# amino acids)
O1	Type VI/VII	ω	$Ca_V2.2$ (direct)	MVIIA (25)
	C-C-CC-C-C			GVIA (27)
	[I–III, II–IV]			CVID (27)
		μO	$Na_V1.8$	MrVIB (31)
A	Type I	α	nAChR and $Ca_V2.2$ (indirect)	Vc1.1 (16)
	CC-C-C			RgIA (13)
	[I–III, II–IV]			AuIB (15)
T	Type X	χ	Norepinephrine Transporter	MrIA (13)
	CC-C-[PO]-C			
	[I–IV, II–III]			
M	Type III	μ	$Na_V1.2$	KIIIA (16)
	CC-C-C-CC			
	[I–IV, II–V, III–VI]			

fications, producing a greatly expanded library of mature peptides and contributing accelerated evolution of conotoxins (Dutertre et al. 2013; Jin et al. 2013; Robinson et al. 2014).

10.2.2 Cysteine Framework

Conotoxins owe their diverse structural and pharmacological properties to variations in both sequence and fold. Due to their disulfide-rich nature, the number and distribution of cysteine residues, referred to as the cysteine framework, is a primary determinant of three-dimensional conotoxin structure. Over 20 distinct frameworks have been identified (Kaas et al. 2010) and novel frameworks are being continuously discovered (Bernaldez et al. 2013; Kancherla et al. 2015). As in larger proteins, formation of intramolecular disulfide bonds in conotoxins confers structural integrity by stabilising secondary structure motifs such as α-helices, turns and β-strands and enabling specific three-dimensional interaction between the peptide pharmacophore and its respective receptor interface (Marx et al. 2006). In addition, these folded structures help to increase biological stability of the peptides by burying cleavage sites and limiting access of proteases (Gongora-Benitez et al. 2014). Clearly, with the presence of multiple cysteine residues, there is the possibility of distinct disulfide connectivities, or isomers, for example a conotoxin with four cysteines can form three disulfide isomers, six cysteine residues can form 15 possible connectivities and the presence of eight cysteines results in 105 possible isomers (and so on). Commonly, only one isomer exhibits bioactivity with the alternate arrangements generally leading to structurally unstable and pharmacologically inactive analogues (Gehrman et al. 1998; Townsend et al. 2009). However, non-native disulfide isomers of both α-conotoxins AuIB and Vc1.1 exhibit similar activity compared to their native isomers (Dutton et al. 2002; Carstens et al. 2016a).

Conotoxins with identical cysteine framework, disulfide connectivities and pharmacological families may exhibit large variations in intercysteine residues, commonly referred to as "loops". Variations in loop size and composition can confer differential affinity, potency and selectivity towards their pharmacological targets. A striking example is the α-conotoxin family of nAChR inhibitors, which have been extensively researched for their therapeutic potential (Armishaw 2010). They possess a type I cysteine framework (CC-C-C) with a native CysI-III, II-IV (globular) connectivity and have loop sizes varying from 1–5 and 2–8 residues for loops 1 and 2, respectively. Certain members of this family discriminate between different neuronal nAChR receptor subtypes, including α7 selective ImI (Quiram and Sine 1998), α3β4 selective AuIB (Luo et al. 1998), and α9α10 selective Vc1.1 (Sandall et al. 2003; Vincler et al. 2006). In terms of drug design, subtype selectivity of conotoxins is an important attribute since the expression and composition of many conotoxin-targeted ion-channel subunits is variable throughout the healthy and pathological CNS. This selectivity allows targeting of subtypes that are dysfunctional in certain pathologies and therefore conotoxin-based drugs may have an advantage over notoriously promiscuous small molecule drugs. Similarly, their target specificity has also allowed researchers to utilise conotoxins in the interrogation of the structure and function of receptor subtypes to improve understanding of their differential roles *in vivo* (Janes 2005; Hannon and Atchison 2013).

10.2.3 Synthesis and Structure-Activity Relationships

Although naturally extracted *Conus* venoms provide enough biological material for analysis with sensitive mass spectral and genomic techniques, a far greater quantity is required for examining their structure and biological activity than is generally accessible from the cone snail. Fortunately, the chain length of conotoxins makes them amenable to production using chemical synthesis which to date has been predominantly achieved by solid-phase peptide synthesis (SPPS) (Merrifield 1963; Nishiuchi and Sakakibara 1982; Gray et al. 1983).

A functionalised resin is used to sequentially couple N-terminal and side chain-protected amino acids to enable C-to-N peptide elongation. Two main types of SPPS strategies are used based on the N-protecting groups, namely *tert*-butyloxycarbonyl (Boc) and 9-fluorenylmethyloxycarbonyl (Fmoc), which are either acid or base labile, respectively. Subsequent cleavage of the peptide from the resin and purification using chromatography (typically reverse-phase high performance liquid chromatography (RP-HPLC)) results in production of the linear reduced conotoxin. Not only does SPPS offer the advantage of both scalability and automation, but importantly can also be used to incorporate substitutions and modifications to the peptide sequence, which has been instrumental in understanding conotoxin structure-activity relationships (SARs) and engineering analogues with improved biopharmaceutical properties.

Following SPPS, the synthetic conotoxin must be folded into the desired disulfide isomer, which is generally achieved by dissolving the linear precursor in aqueous oxidative buffer (pH ~8) or aqueous/organic buffer. However, from a synthetic standpoint, the possibility of multiple disulfide isomers presents a challenge for the folding of conotoxins since random oxidation can produce isomeric mixtures. Optimisation of oxidation buffers to favour formation of the native isomer has produced surprising improvements in relative yield of correctly folded peptide (Gyanda et al. 2013). However, a more effective strategy has been to incorporate cysteine residues with orthogonal thiol-protecting groups during SPPS, most commonly the S-acetamidomethyl (Acm) group (Veber et al. 1972). The Acm group is stable to acidic cleavage and oxidative folding conditions, but can be readily oxidised in the presence of excess I_2 in acetate, enabling regioselective disulfide formation. This directed approach is particularly effective in conotoxins with two disulfide bonds and can be used to selectively generate all three disulfide isomers of α-conotoxins (Gehrman et al. 1998; Townsend et al. 2009; Armishaw et al. 2010). Similarly, incorporation of additional orthogonal protecting groups such as 4-methylbenzyl (Meb) enables regioselective folding of conotoxins with three or even four disulfides (Cuthbertson and Indervoll 2003; Vetter et al. 2012).

The three-dimensional shape of a conotoxin is crucial in defining its bioactivity and as such, investigating their biophysical properties is essential to understanding and developing conotoxin-based drugs. Based on their small size and conformational stability, the predominant method for interrogating conotoxin structure has been two-dimensional NMR spectroscopy (Daly et al. 2011). This powerful technique is used to measure resonance frequencies in response to a magnetic field and determine chemical shift values of individual Hα and NH protons along the peptide chain. Chemical shifts of structured conotoxins are referenced to random coil values for an unstructured peptide to generate secondary shifts that reflect the global fold of the peptide (Wishart et al. 1995). Comparison of secondary shifts between native conotoxins and their synthetic analogues also enables identification of relative structural perturbations. Furthermore, information regarding atom distances and torsional angles obtained using two-dimensional NMR spectroscopy provides parameters for computational determination and visualisation of three-dimensional conotoxin structures such as those shown in Fig. 10.1. Structural data from NMR in combination with synthetic mutagenesis and concentration-response or binding data, allows identification of key SARs and understanding ligand-receptor interactions. The study of conotoxin SARs has been extensively reviewed previously (Lewis et al. 2012; Akondi et al. 2014b) and is not the focus of this chapter, however data from these types of studies can be invaluable in peptide engineering efforts to improve conotoxin stability and bioactivity and guide the structure-based development of conotoxin therapeutics.

10.2.4 Conotoxin Analgesia

The sensation of pain begins with activation of somatosensory neurons originating in the dorsal root ganglion (DRG) known as nociceptors (Woolf and Ma 2007). These afferent nerves project throughout the body and express a diverse

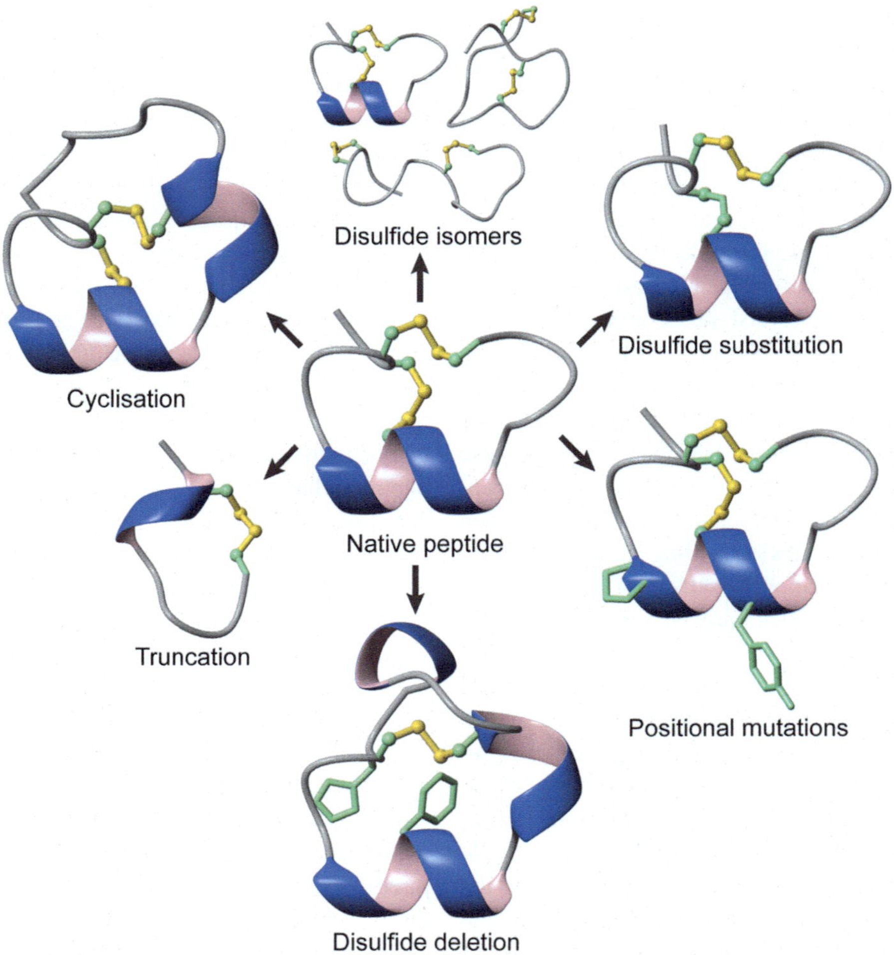

Fig. 10.1 Strategies for engineering conotoxins. Three-dimensional NMR structures of native Vc1.1 (*centre*) and various structurally modified derivatives reported in literature are displayed as examples. The two disulfide bonds in Vc1.1 may rearrange to form three different disulfide isomers (*globular, ribbon, beads*) with differential pharmacological properties (*top*). Disulfide substitution with non-reducible linkages, such as a [2–8]dicarba bond in Vc1.1 (*top right*), can improve biological stability. Positional mutations may influence structure and pharmacology, e.g. the post-translationally modified vc1a (*bottom right*) which loses calcium channel inhibition. Disulfide deletion (*bottom*; [C2H,C8F]cVc1.1) can prevent disulfide scrambling while maintaining bioactivity. Truncated conotoxins that retain pharmacological activity, such as [Ser^3]Vc1.1(1–8) (*bottom left*), offer minimised lead molecules with reduced production time and costs. Cyclisation of conotoxins by joining termini reduces proteolytic degradation and improves biological stability, as reported for cyclic Vc1.1 (*top left*). Structural data was obtained from Protein Data Bank entries and three-dimensional ribbon diagrams were generated using MOLMOL software

collection of membrane channels and receptors that activate differentially in response to noxious stimuli and initiate membrane depolarisation (Gold and Gebhart 2010). Various nociceptor-specific voltage-gated ion channels subtypes propagate the resulting membrane potential toward the DRG activating pre-synaptic voltage-gated calcium channels (Ca_Vs) leading to neurotransmitter exocytosis and communication to the central nervous system (CNS). The signal continues to the brain where it is processed by pain centres, including the thalamus and brainstem, to produce an aversive physical and emotional response in the organism (Basbaum et al.

2009). Although pain is a crucial survival mechanism, severe injury or diseases such as diabetes mellitus and cancer can result in the sustained presence of aberrant biochemical signals, particularly inflammatory mediators, which can profoundly influence the surface expression and activation kinetics of nociceptive proteins (White et al. 2007; Gold and Gebhart 2010). Often this leads to pathological hypersensitivity in the underlying nociceptive circuitry and ectopic generation of pain signals, which manifests as a chronic or 'neuropathic' pain state *in vivo* that persists long after the initial stimulus. Epidemiological studies estimate up to 6–10% of the population suffer from moderate to severe symptoms (Smith and Torrance 2012; van Hecke et al. 2014). Depending on the severity, neuropathic pain conditions can prevent normal daily activities and may require specialist medical care, therefore it is associated with a substantial personal and economic burden.

Current therapeutic approaches to alleviate pathological pain include tricyclic antidepressants, $Ca_V2.2$ inhibitors and opioids (Baron et al. 2010). However, these approaches are ineffective in 40–60% of neuropathic pain patients and their clinical efficacy is hindered by severe adverse effects, problems with dependence and limited routes of administration (Dworkin et al. 2007). As such, there is considerable demand for pharmacological agents to improve quality of life of those suffering from neuropathic pain. Incidentally, conotoxins offer an exciting new class of potential analgesic peptides for the treatment of neuropathic pain. Characteristics such as structural and conformational stability, low micromolar or nanomolar potency and selectivity for receptor subtypes are desirable pharmaceutical qualities and for this reason, significant efforts have been made to develop conotoxins as therapeutic agents. Furthermore, since many conotoxins selectively inhibit ion channels and receptors that are implicated in pain transmission (Lewis et al. 2012), they are ideal candidates for development of novel pharmaceuticals to reduce pathological hyperexcitability associated with neuropathic pain.

N-type Ca_V channels ($Ca_V2.2$) are widely expressed in the central and peripheral nervous system and consist of a pore-forming α_1 subunit and auxiliary β, α_2, δ, and γ subunits (Murakami et al. 2004). Upon membrane depolarisation, these channels open to allow Ca^{2+} influx leading to neurotransmitter exocytosis and propagation of membrane potential across synapses (Hannon and Atchison 2013). Members of the ω-conotoxin family, such as MVIIA, fold into an inhibitory cysteine knot (ICK) conformation (Kohno et al. 1995) and irreversibly inhibit $Ca_V2.2$ currents (Lew et al. 1997). Accordingly, intrathecal injection of MVIIA produces anti-nociceptive effects in rodent pain models (Malmberg and Yaksh 1995; Bowersox et al. 1996). These properties led to the development of ziconotide, a synthetic version of MVIIA, which is currently the only conotoxin clinically approved for neuropathic pain treatment (Hannon and Atchison 2013). Although clinical trials demonstrated analgesic efficacy in many patients, ziconotide exhibits quite substantial neurological side effects (Staats et al. 2004; Rauck et al. 2006). Other ω-conotoxins such as CVID-F that exhibit reversible binding to $Ca_V2.2$ have shown decreased toxicity but similar analgesic activity in rodents compared to MVIIA and may offer alternatives with improved therapeutic windows (Scott et al. 2002; Berecki et al. 2010).

Another family of analgesic conotoxins that possess three disulfides but do not form an ICK structure (Type III framework) are the μ-conotoxins from the M superfamily, which act as potent inhibitors of muscle or neuronal Na_Vs (Knapp et al. 2012). The role of Na_Vs in conducting membrane potentials is well understood and several subtypes including the tetrodotoxin-sensitive (TTX-S) $Na_V1.3$ (He et al. 2010) and $Na_v1.7$ (Schmalhofer et al. 2008), and TTX-resistant $Na_v1.8$ and $Na_v1.9$ (Leo et al. 2010) have been specifically implicated in pain transmission. Moreover, $Na_v1.3$, $Na_v1.8$ and $Na_v1.9$ subtypes appear to be upregulated in response to nerve damage (Siqueira et al. 2009; He et al. 2010). μ-Conotoxins bind to Na_Vs primarily through electrostatic interactions with positively charged residues of the peptide and the Na_v pore-forming α-subunit and sterically hinder ion flow through the cavity (Hui et al. 2002). Residues which lie in the α-helix are also important for activity (Zhang et al. 2007). Although most μ-conotoxins preferentially inhibit the TTX-S Na_V subtypes, several members such as KIIIA also show inhibitory activity at the TTX-R Na_Vs which are the more desirable targets for selective pain relief (Bulaj et al. 2005). Notably, KIIIA exhibited analgesic activity in rodent model of inflammatory pain with an ED_{50} of 0.1 mg/kg (Zhang et al. 2007). A separate class of Na_V inhibitors from the O1 superfamily has also been discovered, termed μO-conotoxins, which exhibit little sequence homology to their μ-conotoxin counterparts (McIntosh et al. 1995) but are more closely related to ω-conotoxins and also adopt an ICK conformation (Daly et al. 2004). One example is MrVIB, which was found to preferentially inhibit the $Na_V1.8$ subtype (Bulaj et al. 2006) but unlike μ-conotoxins binds to a distinct site which inhibits the activation of the voltage-sensor (Leipold et al. 2007). Accordingly, MrVIB elicits long-lasting analgesia (Bulaj et al. 2006) with substantially fewer motor deficits compared to non-selective Na_V antagonists at analgesic doses (Ekberg et al. 2006) and therefore may be advantageous over the less selective μ-conotoxins. However, despite extensive research and development, to date no μ-conotoxins have entered clinical trials.

The α-conotoxin family are 12–19 amino acids long with two disulfides in a native globular connectivity (CysI-III, II-IV) forming a characteristic α-helix, and are potent antagonists of nAChRs (Lebbe et al. 2014). These receptors are pentameric ion-channels which consist of α_1-α_{10}, β_2-β_4, γ, δ or ε subunits, resulting in numerous subtypes expressed differentially throughout the skeletomuscular and nervous systems where they mediate fast synaptic transmission (Jensen et al. 2005). While the early-discovered α-conotoxins such as GI target neuromuscular subtypes (Gehrman et al. 1998), more recently discovered members of this family are highly selective for neuronal nAChR subtypes such as α3β2, α3β4 and α9α10 (Vincler et al. 2006; Clark et al. 2006; Armishaw 2010). The neuronal subtypes are appealing therapeutic targets since they have been implicated in various neurological disorders such as Parkinson's, Alzheimer's, addiction, depression and pain (Gotti and Clementi 2004). Indeed, several α-conotoxins including Vc1.1, RgIA and AuIB display potent analgesic effects in rodent pain models (Satkunanathan et al. 2005; Vincler et al. 2006; Klimis et al. 2011) and contribute to neuroprotection and functional recovery (Mannelli et al. 2014) with no obvious cognitive side-effects (Napier et al. 2012). Vc1.1 entered clinical trials (ACV1) but was discontinued after phase two due to differential activity at the human and rat α9α10 nAChR subtypes (Metabolic Pharmaceuticals 2007). More recently, a subset of α-conotoxins, including Vc1.1, RgIA and AuIB, were found to also inhibit $Ca_V2.2$ indirectly through a novel $GABA_BR$- and src kinase-dependent pathway and has been proposed as an alternative analgesic mechanism (Callaghan et al. 2008; Klimis et al. 2011; Berecki et al. 2014; Huynh et al. 2015). However, little is known regarding the structural and functional determinants that underpin this alternate interaction. Despite their controversial mechanism of action and failure in clinical trials, α-conotoxins have been subjected to extensive preclinical development as pharmaceuticals. These studies have led to engineered analogues which may be more suitable as leads for neuropathic pain treatment, such as the orally active cyclic Vc1.1 (Clark et al. 2010) and dicarba-bridged Vc1.1 analogues that are selective for nAChR or $Ca_V2.2$ inhibition (van Lierop et al. 2013).

The χ-conotoxin MrIA is 13-amino acids in length with two disulfides that acts as a potent, allosteric noradrenaline transporter (NET) inhibitor, which has been implicated as a therapeutic target for neurological disorders including pain (Sharpe et al. 2001; Sharpe et al. 2003). Unlike the globular conformation of α-conotoxins, native MrIA adopts a ribbon disulfide connectivity, giving rise to a β-hairpin turn, which presents the pharmacophore comprising residues Gly6, Tyr7, Lys8 and Leu9 to the receptor-binding interface (Brust et al. 2009). A more chemically stable analogue, denoted Xen2174, displayed analgesic activity in the rodent chronic constriction injury (CCI) model when administered by intrathecal injection (Nielsen et al. 2005) and subsequently entered clinical trials but was discontinued at phase II.

10.3 Conotoxin Engineering

Although the therapeutic potential of conotoxins as analgesics has been well documented, several challenges are faced on their journey toward clinical use. Their peptidic nature imparts a high susceptibility to attack by endogenous proteases and the presence of reducing agents such as the antioxidant glutathione can result in disulfide rearrangement or "scrambling" to form inactive isomers. This biological instability translates to poor bioavailability and *in vivo* efficacy, particularly when delivered orally (Olivera 1985) and improving these characteristics might aid their therapeutic development. In this sense, introduction of non-native modifications to create analogues with more desirable biopharmaceutical characteristics is a principal strategy for the preclinical development of conotoxins (Craik and Adams 2007; Brady et al. 2013a). Some of the major techniques for engineering conotoxins are outlined in Fig. 10.1, which displays α-conotoxin Vc1.1 and its modified analogues as examples. However, based on the similarities among families and folds, the methods described here can likely be applied to many known and undiscovered conotoxin drug leads.

10.3.1 Non-disulfide Modifications

Post-translational modifications are widely incorporated into conotoxins during their biosynthesis, with the most common, other than the disulfide bonds, being an amidated C-terminus, hydroxyproline (Hyp) and γ-carboxyglutamate (Gla), and can influence the structure and function of the peptide (Espiritu et al. 2014). For example, in the case of Vc1.1, which was originally discovered from a cDNA screen, the corresponding venom peptide contained PTMs at postions 6 (Hyp) and 14 (Gla) (Townsend et al. 2009). When compared to Vc1.1, vc1a was equipotent at the α9α10 nAChR but did not block $Ca_V2.2$ and was unable to reverse mechanical allodynia in a rodent pain model (Nevin et al. 2007). This suggests that rationally designed synthetic modifications might have potential for creating conotoxin analogues with more favourable biopharmaceutical properties, such as increased chemical and structural stability.

The N-terminal asparagine in native MrIA readily undergoes aspartamide formation producing undesirable α- and β-aspartyl analogues (Brust et al. 2009). In order to improve the chemical stability of MrIA, Brust et al. (2009) synthesised a library of Asn1 substitutions that were stable toward aspartamide formation. When Asn1 was replaced with pyroglutamtic acid the resulting analogue, Xen2174, retained identical backbone structure and displayed slightly increased potency for NET inhibition compared to MrIA (Brust et al. 2009). Furthermore, intrathecal injection of Xen2174 reversed signs of allodynia in both CCI and spinal nerve ligation rodent models of pain (Nielsen et al. 2005). The more stable Xen2174 subsequently progressed through phase I clinical trials for treatment of chronic and post-operative pain, but trials were cancelled at phase II (Lewis 2015).

The conserved proline in loop 1 of most α-conotoxins has been established as a key structural and functional residue which provides a backbone constraint that induces the characteristic 3_{10} α-helix (Gehrmann et al. 1999; Clark et al. 2006) and also contributes favourable hydrophobic interactions with nAChRs (Dutertre et al. 2005). Consequently, non-conservative substitutions at this position result in perturbations to backbone structure and abrogate pharmacological activity in many α-conotoxins (Nevin et al. 2007; Ellison et al. 2008; Halai et al. 2009; Grishin et al. 2013). In contrast, strengthening favourable hydrophobic interactions with nAChRs by introducing artificial proline derivatives presents a strategy for improving affinity and potency. To this end, Armishaw et al. (2009) synthesised a series of ImI and PnIA[A10L] analogues in which Pro6 was replaced with a range of 4- or 5-substituted proline derivatives. Reductions in affinity and IC_{50} at the α7 nAChR subtype were observed for all analogues except ImI[P6/5-(R)-Ph] which showed marked improvements in both affinity and potency. Homology modelling and docking experiments suggested that the newly introduced phenyl group in ImI[P6/5-(R)-Ph] increased hydrophobic contacts with Tyr93 side chain in the nAChR binding pocket.

A major caveat of many hydrophilic peptide-based drug leads is their low lipid solubility, which can translate to poor membrane permeability and low bioavailability, particularly when delivered orally. One technique used to increase the lipid content of peptides has been to couple lipophilic fatty acid chains to increase their membrane penetration and improve biodistribution (Zhang and Bulaj 2012). Lipophilic analogues of α-conotoxin MII have been synthesised by incorporating 2-amino-L-D-dodecanoic (Laa) at either the N-terminus (LaaMII) or in place of Asn5 (5LaaMII) (Blanchfield et al. 2003). LaaMII retained wild-type activity at the α3β4 receptor whereas the structure and activity of 5LaaMII were impaired, consistent with the role of Asn5 in receptor binding (Everhart et al. 2004). Similarly, addition of a lipophilic N-terminal amino-tetratdecanoic acid (Adta) to MrIA was reported to improve NET inhibition with 3-fold increase in potency over wild-type MrIA (Dekan et al. 2007). In the Caco-2 transepithelial permeability assay, LaaMII was significantly more permeable than native MII (Blanchfield et al. 2003). However, when the oral absorption and distribution of tritium labelled LaaMII were investigated, similar levels of overall biodistribution were observed compared to native MII (7.86% vs 6.22%) (Blanchfield et al. 2007). Although lipidation has been reported to increase oral bioavailability of calcitonin and Leu-enkephalin (Wang et al. 2003; Wang et al. 2006), similar improvements were not observed for MII, suggesting this approach may be less viable for the disulfide-rich α-conotoxins.

10.3.2 Disulfide Substitutions

The presence of internal disulfide bonds is a feature common to many bioactive peptides and is crucial for defining the three-dimensional structures adopted by these molecules. Despite this, disulfides are inherently unstable *in vivo* and highly susceptible to reduction in thiol-rich environments, which results in loss of bioactivity and reduced *in vivo* half-life (Gongora-Benitez et al. 2014). To circumvent such pharmaceutically undesirable properties, a variety of synthetic disulfide substitutions have been introduced into conotoxins with the aim of improving structural and biological stability. The most common synthetic approach is to replace cysteine pairs during SPPS with non-natural amino acids possessing side chains designed to enable regioselective formation of covalent linkages with increased resistance to reduction or degradation. Many synthetic amino acids are now commercially available and reagent costs continue to decrease, thus disulfide substitutions an attractive avenue for conotoxin drug design. Table 10.2 outlines the structure of several synthetic disulfide substitutions that have been applied to conotoxins.

10.3.2.1 Diselenide Bonds

Selenocysteine (Sec) is a naturally occurring amino acid, which is able to form diselenide bridges through oxidation of its selenol side chain, with similar bond length to disulfide bonds. However, in contrast to disulfides, diselenides exhibit superior reductive stability as evidenced by a substantially lower redox potential (−180 mV vs. −381 mV) (Besse et al. 1997), making them appealing isosteric surrogates to ameliorate disulfide scrambling. Moreover, the pKa of Sec (5.73) is significantly lower than that of the Cys thiol groups (8.53) and thus is able to react selectively at lower pH, a property which has been exploited to allow the selective folding of several disulfide-rich peptides including endothelin (Pegoraro et al. 1998) and apamin (Fiori et al. 2000). The applicability of this approach toward conotoxins was first demonstrated using ImI whereby one or both cysteine pairs were substituted with Sec residues during synthesis allow-

Table 10.2 Conotoxin disulfide substitutions

Disulfide substitution	Chemical structure	Examples	Reduces scrambling?	Improves stability?	References
Diselenide	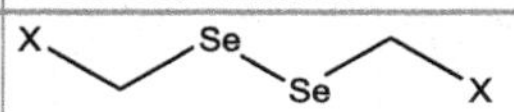	ImI MI AuIB PnIA SIIIA MrVIB GVIA	✓	×	Armishaw et al. (2006), Walewska et al. (2009), Muttenthaler et al. (2010), de Araujo et al. (2011) and Gowd et al. (2010)
Lactam	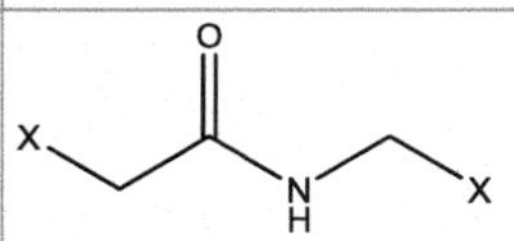	SI	✓	N.D.	Hargittai et al. (2000)
Thioether	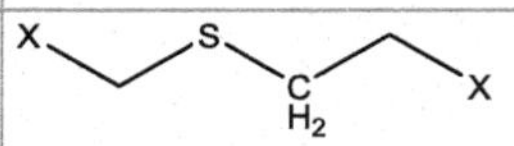	GI ImI	✓	N.D.	Bondebjerg et al. (2003) and Dekan et al. (2011)
Dicarba	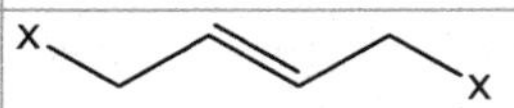	ImI Vc1.1 RgIA	✓	✓	MacRaild et al. (2009), van Lierop et al. (2013) and Chhabra et al. (2014)
Triazole	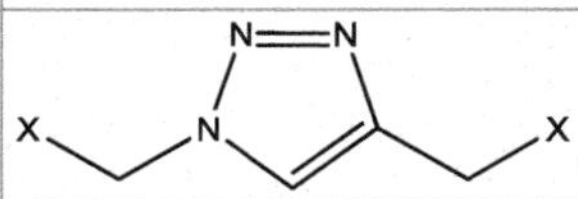	MrIA	✓	✓	Gori et al. (2015)

ing selective folding of these analogues (Armishaw et al. 2006). The resulting peptides exhibited highly conserved NMR secondary shifts to the native peptide and retained IC_{50} values similar to wild-type ImI. Importantly, the ImI diselenide analogues were highly resistant toward reduction and displayed significantly decreased disulfide scrambling in reductive environments (Armishaw et al. 2006). Expanding on this work, Muttenthaler et al. (2010) employed selective diselenide formation in a number of other α-conotoxins including MI, AuIB, [A10L]PnIA, and also observed the conservation of secondary NMR structure, diselenide bond lengths and dihedral angles compared to the native peptides. Remarkably, analogues of MI, AuIB and [A10L] PnIA all displayed improved inhibitory potency at their respective nAChR subtypes, suggested to be a result of increase hydrophobicity surrounding the CysII-IV disulfide.

Diselenide substitution has also been utilised in conotoxins containing three disulfides, where the oxidative folding problem is far more apparent considering the 15 possible disulfide isomers that can potentially form. Selenocysteine substitution of any of the three native cysteine pairs in μ-conotoxin SIIIA, and initial regioselective formation of diselenide bonds, significantly improved overall yields of correctly folded peptide following subsequent thiol oxidation, without compromising activity at $Na_v1.2$ channels (Walewska et al. 2009). Interestingly, when this strategy was applied the $Na_v1.8$ selective μ-conotoxin, MrVIB, which is notoriously difficult to fold due to its high hydrophobic amino acid content, a 4-fold increase in overall yield of correctly folded peptide and similar levels of analgesia *in vivo* compared to native MrVIB were observed (de Araujo et al. 2011). Consistent improvements in yield were observed in all three diselenide-substituted analogues of ω-conotoxin GVIA, also containing an ICK-motif (Gowd et al. 2010), and later used to reveal the importance of the C1-C16 disulfide which, although non-crucial for activity, assists formation of the remaining disulfides into the native fold (Gowd et al. 2012). Although the utility of synthetic diselenide incorporation for engineering isomorphous conotoxin analogues with improved folding kinetics, reduced disulfide scrambling and potentially increased potency is well established, no improvements in stability toward degradation in rat plasma were observed (Muttenthaler et al. 2010), highlighting the requirement for alternative substitution strategies to reduce susceptibility to proteolytic cleavage and improve bioavailability *in vivo*.

10.3.2.2 Side-Chain Lactam Bridge

One of the earliest described structural modifications to mimic conotoxin disulfides involved replacing either cysteine pair in α-conotoxin SI with a Glu/Lys pair, which can be readily coupled using standard SPPS *in situ* neutralisation to form side chain lactam bridges (Hargittai et al. 2000). Replacing the Cys[2–7] disulfide completely abolished affinity for the nAChR, however the [3–13]-lactam analogue in either orientation retained meaningful affinity, with the [Glu3-Lys13] orientation exhibiting ~70 fold increase in affinity over native SI despite the comparatively larger distance between the α-carbons in lactam analogues (8 vs 4 atoms). Although no structural studies were performed, the loss of affinity in [2–7]-lactam analogues might be explained by a greater sensitivity to synthetic modifications in this region. Cys7 lies central to the α-helix, which comprises the key α-conotoxin binding motif thought to interact with the nAChR binding pocket (Ulens et al. 2006). This suggests that substitutions of the alternate CysII-IV disulfide may be more suitable for maintaining structural integrity and receptor affinity in α-conotoxins.

10.3.2.3 Thioether Bonds

In order to closely mimic the disulfide bond length and maintain three-dimensional structure, a substitution strategy involving the regioselective formation of a non-reducible cystathionine thioether bridge, formed by a reaction between the cysteine thiol and a γ-chlorinated side chain, has been applied to α-conotoxins GI (Bondebjerg et al. 2003) and ImI (Dekan et al. 2011). Although both α-conotoxins saw modest decreases in IC_{50} when both disulfides were replaced with thioether surrogates, a single [3–12]cystathionine bridge in ImI retained full wild-type activity. Furthermore, structural analysis of ImI by NMR spectroscopy revealed a high degree of overlap between the lowest energy structures of all three possible analogues, highlighting the effectiveness of this substitution for maintaining the topological conformation of conotoxins.

10.3.2.4 Dithiol Derivatives

Chen et al. (2014) recently developed a cysteine derivative that contains γ- and δ-thiol groups (Dtaa) to examine its potential as a substitute for adjacent cysteines. As α-conotoxins possess a type I framework containing adjacent CysI and CysII residues, ImI was used to investigate the effect of a CysI-CysII → Dtaa-Ala substitution on structure and activity. One resulting analogue, in which the δ-sulfur and γ-sulfur were bonded to Cys8 and Cys12 respectively (ImI1), displayed nearly eight-fold increased inhibitory activity at the α7 nAChR whereas the alternate connectivity, as well as a CysI-CysII → Dtaa substitution, were equipotent compared to wild-type ImI. Since both arrangements retain their inhibitory activity, shuffling may be less likely to impair their analgesic effects and offer an advantage over the native disulfide configurations. Additionally, NMR structural data indicated that the tertiary structure of ImI1 was highly similar to native ImI, highlighting the utility of Dtaas as conservative Cys-Cys substitutions.

10.3.2.5 Dicarba Linkages

The discovery of ruthenium-based (Grubbs) catalysts has enabled an unprecedented efficiency of ring-closing olefin metathesis (RCM) and the facile formation of C = C bonds (Dias et al. 1997). RCM has since become a widely used technique for introducing conformational constraints in peptides and stabilising secondary structure (Gleeson et al. 2016), and is thus an appealing strategy to engineer more stable conotoxin analogues. Indeed, the incorporation of olefin side chains as cysteine substitutes and regioselective formation of dicarba bridges by RCM has been successfully applied to α-conotoxins ImI, Vc1.1 and RgIA (MacRaild et al. 2009; van Lierop et al. 2013, Chhabra et al. 2014). Although dicarba bond lengths are shorter than disulfides (2.02 and 1.34 Å, respectively), all dicarba analogues displayed highly similar backbone structures with only minor perturbations surrounding substituted regions. However, for Vc1.1 and RgIA, which both belong to a subset of indirect $Ca_V2.2$ inhibitors, these slight conformational changes were significant enough to profoundly alter their selectivity for either α9α10 nAChR or $Ca_V2.2$ inhibition. In both cases, [I-III]-dicarba analogues were unable to inhibit ACh-evoked currents in *Xenopus* oocytes. This loss of affinity is consistent with involvement of CysI-III in maintaining the peptide fold and it forming favourable interactions with the nAChR (Ulens et al. 2006). Interestingly however, when tested for activity at high-voltage activated $Ca_V2.2$ currents in rat DRG neurons at 100 nM no loss of inhibition was observed compared to the native peptides. In contrast, the alternate [II-IV]-dicarba analogues did not inhibit $Ca_V2.2$ currents, but wild-type α9α10 inhibitory activity was retained. Since the precise molecular mechanisms underlying the analgesic effects of this subset of α-conotoxins is still controversial, these analogues with differential receptor preference may provide useful tools for further interrogation of their analgesic targets.

10.3.2.6 1,2,3-Triazole

In recent years, the development of the copper-catalysed alkyne-azide cycloaddition (CuAAC) reaction has enabled the rapid and selective formation of 1,4-disubstituted-1,2,3-triazole bridges between alkyne and azide moieties and has been adapted as a technique for various forms of peptide conjugation (Rostovtsev et al. 2002; Tornøe et al. 2002). Moreover, incorporation of modified amino acids containing alkyne and azide side chains and formation of intramolecular triazole bridges was successfully used to substitute both disulfide-bonds in the antimicrobial peptide Tachyplesin-1 (TP-1) without affecting structure or function (Holland-Nell and Meldal 2011). To examine the potential of intramolecular triazoles to mimic disulfides in conotoxins, Gori et al. (2015) applied this strategy to MrIA by replacing either disulfide with triazole bridges. MrIA analogues with substitutions by modified amino acids containing azide or alkyne side chains at position 4 and 13 respectively, exhibited highly conserved structures and increased NET IC_{50} compared to native MrIA, whereas the [5–10]-triazole analogues were significantly less active due to structural aberrations surrounding the sensitive Gly6-Leu9 pharmacophore. In addition, the [4–13]-triazole analogues produced levels of *in vivo* analgesia similar to wild-type MrIA in the PNL neuropathic pain model. Importantly, the reductive stability of these analogues in glutathione was superior to the native peptide, which correlated with profound improvements in stability in rat plasma over a period of 48 hours, with 80–90% remaining as opposed to complete breakdown of MrIA. Although this technique has not been investigated for peptides with globular folds such as the α-conotoxins, CuAAC offers a cheap and efficient method for engineering more stable disulfide mimetics and, through optimisation of linkage length and selective replacement of less functionally important disulfides, can likely be applied to a range of conotoxin families.

10.3.2.7 Disulfide Deletion

Another approach used to reduce conotoxin complexity is the deletion of disulfides by mutating orthogonal pairs of Cys residues. In the case of ICK conotoxins, GVIA and KIIIA, the CysI-IV disulfide has been identified as non-critical for activity (Flinn et al. 1999; Khoo et al. 2009). Guided by these observations, Stevens et al. (2012) synthesised ten analogues of Cys1–9 disulfide-deleted KIIIA and assessed their structure and activity. The [S4R,C9A]KIIIA(2–16) analogue, denoted R2-Midi, displayed nanomolar potency at the neuronal $Na_V1.2$ subtype (34.1 nM) and similar subtype selectivity to native KIIIA. Interestingly however, unlike KIIIA, the three-dimensional NMR structure of R2-midi did not exhibit the native backbone α-helix. More recently, a cyclic CysI-III disulfide-deleted analogue of Vc1.1 (hcVc1.1) was designed in which cysteines were substituted with His2 and Phe8 to mimic the hydrophobic core of Vc1.1 (Yu et al. 2015). The hcVc1.1 analogue showed little difference in Hα chemical shifts, three-dimensional solution structures and

$Ca_V2.2$ inhibition compared to native globular Vc1.1. However, hcVc1.1 was about 40-fold less potent at α9α10 nAChR and showed a 30% reduction in serum stability over 24 hours. Therefore, while disulfide-deleted analogues offer the advantage of simplified folding and decreased scrambling, in the case of Vc1.1, the cost was not only reduced activity at one of its targets, but also reduced proteolytic stability.

10.3.3 Cyclisation

Cyclisation of conotoxins and other peptide lead molecules is one of the most routinely used techniques for improving biological stability (Clark and Craik 2012; Clark et al. 2012). The rationale for this approach comes from the discovery and characterisation of a range of naturally existing disulfide-rich peptides with N-to-C-terminal amide bonds that form a cyclic backbone structure, including the plant cyclotides (Craik et al. 1999), mammalian theta-defensins (Tang et al. 1999) and sunflower trypsin inhibitors (Luckett et al. 1999). A notable example is the 29 amino acid cyclotide kalata-B1, which displays exceptional stability in harsh denaturing environments including 6 M guanidine hydrochloride, 8 M urea and 0.5 M HCl at 80 °C and in the presence of various proteases including trypsin, endoGlu-C, pepsin, and thermolysin (Colgrave and Craik 2004). The exquisite structural and biological stability of cyclic peptides are highly desirable pharmaceutical properties and suggests their potential as templates in peptide-based drug design.

10.3.3.1 Design and Synthesis

The similarities in size and fold between conotoxins and natural cyclic peptides, make them ideal candidates for cyclisation and indeed, the last decade has seen the development of a range of cyclic conotoxin analogues from α-, χ-, ω- and κ-conotoxin families. To date, the majority of cyclic conotoxin analogues have been synthesised using native-chemical ligation (NCL) which is suitable for both Boc- and Fmoc-based SPPS (Dawson et al. 1994; Blanco-Canosa and Dawson 2008; Zheng et al. 2013). The mechanism involves the reaction between an N-terminal cysteine thiol and C-terminal thioester, which undergoes thioester exchange, followed by an intramolecular rearrangement resulting in formation of a native amide bond. Although originally used to synthesise larger proteins by joining linear fragments (Dawson et al. 1994), it has since been adapted as a means of generating cyclic peptides via an intramolecular N- to C-terminal ligation (Tam and Lu 1997; Craik et al. 1999; Clark et al. 2010). An advantage of disulfide-rich peptides like conotoxins in this approach is that native cysteines can be used as the N-terminal Cys for NCL without the requirement for incorporation of non-native cysteines or a desulfurisation reaction (Clark and Craik 2010).

The biological activity of conotoxins relies heavily on their fold, thus an important consideration when designing cyclic conotoxins is the preservation of secondary structure. This has traditionally been achieved through the incorporation of inert peptidic (Gly-Ala) linkers, which maintain distances between native termini and minimise structural perturbations. For example, the globular 4/7 α-conotoxins Vc1.1 and MII whose termini are ~11–12 Å apart require a 6–7 residue linker (Clark et al. 2005, 2010; Halai et al. 2011) whereas conotoxins with closer termini such as AuIB and MrIA, 2-3 residue linkers are more appropriate (Lovelace et al. 2006; Armishaw et al. 2011). In fact, there is a linear trend between termini distance and number of linker residues required for cyclisation, as shown in Fig. 10.2b (Clark et al. 2010).

10.3.3.2 Conotoxin Cyclisation Using NCL

The α-conotoxins are particularly amenable to cyclisation based on the close proximity of their termini, small and well-characterised structures and relatively poor *in vivo* stability. The principle was first demonstrated using the 4/7 α-conotoxin MII in which three analogues containing a 5-, 6- or 7-residue spacer were cyclised by NCL (Clark et al. 2005). Compared to the native peptide, cMII-6 and cMII-7 retained identical backbone structures and nAChR inhibitory activity, whereas

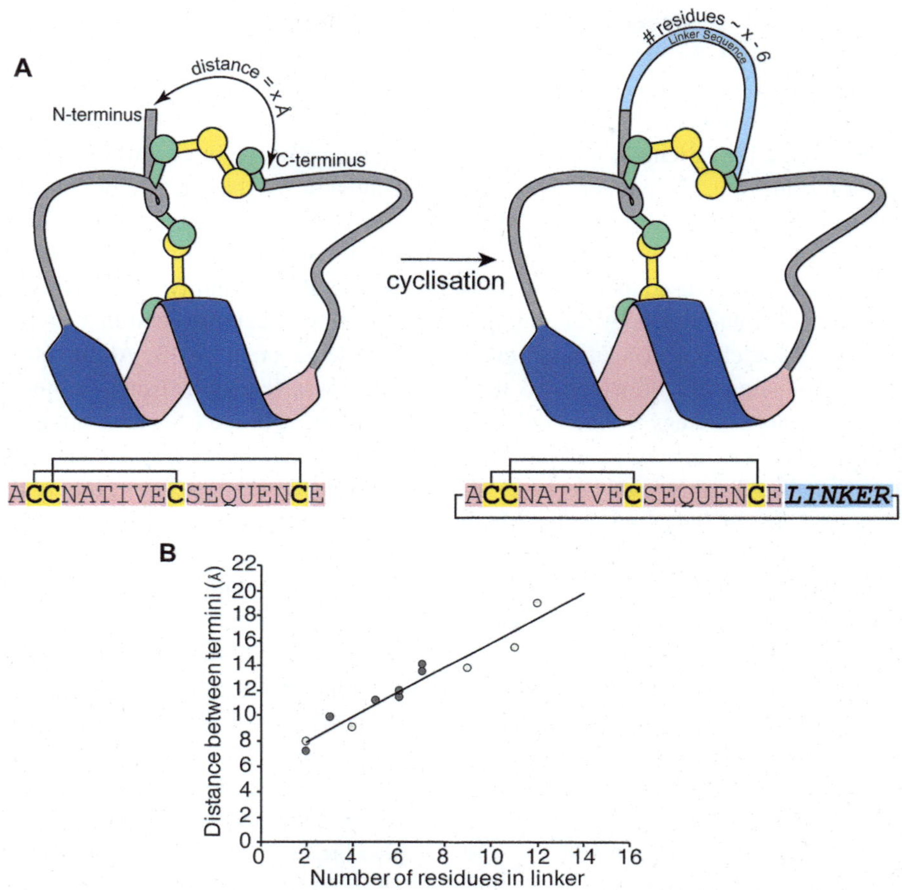

Fig. 10.2 **Strategy for design of cyclic conotoxins.** (**a**) Linker regions (*light blue*) generally consisting of inert Gly and Ala residues are incorporated into the native sequence (*highlighted in pink*) during peptide synthesis. N- and C-termini are joined to form a cyclic backbone typically via native chemical ligation or enzyme-mediated ligation. A generic α-conotoxin structure and its cyclic analogue are shown as an example. (**b**) Linear relationship between the optimal number of linker residues to maintain conotoxin structure and function to the distance between termini. Circles represent cyclic peptide analogues from literature reported to maintain native structure and pharmacology and shaded circles indicate conotoxins (adapted from Clark et al. 2010). The correlation between the N-to-C distance and the number of residues required to span it indicates that a linker length containing approximately x-6 residues is required, where x is the distance between the termini in Ångstroms

cMII-5 was pharmacologically inactive due to major structural perturbations. Furthermore, cyclisation conferred a 20% reduction in degradation in human plasma over 24 hours, highlighting its utility for improving conotoxin stability. Consistent improvements in proteolytic stability without sacrificing secondary structure or nAChR activity were observed for several other cyclic α-conotoxins including ImI (Armishaw et al. 2009), Vc1.1 (Clark et al. 2010), AuIB (Lovelace et al. 2011), and RgIA (Halai et al. 2011) as well as the NET-inhibitor χ-conotoxin MrIA (Lovelace et al. 2006). Perhaps the most significant finding to come from this research is the development of an orally active cyclic analogue of the analgesic conotoxin Vc1.1 (Clark et al. 2010). Not only did cVc1.1 exhibit increased stability towards disulfide scrambling in simulated *in vivo* environments including gastric fluid, intestinal fluid and human serum, but remarkably, when dosed orally in a rat CCI model of neuropathic pain produced analgesia at 3 mg/kg comparable to 30 mg/kg of first line analgesic gabapentin. Oral activity is a highly desirable property for pharmaceuticals as it allows ease of administration and improves patient comfort and compliance.

In addition to the linker length, the composition of linker residues can also influence the bioactivity of cyclic conotoxins. Originally, linkers consisting of relatively inert amino acids such as Gly and Ala residues were used to avoid altering peptide structure and pharmacology. In principle however, altering the chemical nature of linker regions by introducing functional amino acid sequences represents a strategy for manipulation of conotoxin structure, activity and physicochemical properties. To this end, Dekan et al. (2012) designed an analogue of cyclic MrIA (cMrIA) with a bioactive RGD linker in place of the AGN sequence originally used to produce cMrIA (Lovelace et al. 2006). The RGD sequence is a binding motif, present in extracellular matrix (ECM) proteins and disintegrin peptides, which interacts with the integrin family of transmembrane receptors that mediate cellular processes such as adhesion and migration. The cMrIA(RGD) analogue displayed a backbone structure which closely matched cMrIA, and retained NET inhibitory activity as well as inhibition of integrin receptors and platelet aggregation. The RGD linker also imparted cMrIA with greater stability towards degradation in rat plasma compared to the GA linker, however, this did not correlate with any notable improvements in pharmacokinetic parameters following intravenous injection in Sprague-Dawley rats. A more recent study by Carstens et al. (2016b) explored the effect of altering the charge and hydrophobicity of the cVc1.1 linker sequence on its pharmacological properties. The GGAAGG linker in cVc1.1 was replaced with GKGKGK (L1), GVGVGV (L2) and GEGEGE (L3) linkers (underlined residues are N-methylated). The secondary structures of all analogues were similar to that of cVc1.1. The cVc1.1-L2 analogue exhibited modest improvements in Caco-2 membrane permeability assay, attributed to its increased hydrophobicity. However, despite structural similarities, all three analogues were unable to inhibit Ca_V currents at 500 nM in rat DRG neurons suggesting that these linkers are unfavourable for receptor interaction.

Fewer studies have focused on cyclisation of larger conotoxins with three disulfides since they are comparatively more difficult to synthesise and already exhibit greater biological stability than α-conotoxins. Three ω-conotoxins, MVIIA from the O1-superfamily, and gm9a and bru9a from the P-superfamily, which all form ICK motifs, have been successfully cyclised using NCL (Hemu et al. 2013; Akcan et al. 2015). MVIIA was cyclised using a novel Fmoc-compatible approach incorporating a GGPG linker and N-methyl Cys as a thioester surrogate. The presence of the cyclic peptide in the correct fold was confirmed by mass spectrometry and partial reduction/alkylation; however, no structure, activity or stability data were reported for the cyclic analogue (Hemu et al. 2013). In the case of gm9a and bru9a, direct cyclisation led to misfolding of the peptide, but inclusion of a cyclotide-like GLP linker engendered a native backbone structure (Akcan et al. 2015). Cyclic gm9a (cgm9a-GLP) inhibited Ca_V currents in rat DRG neurons at a concentration of 1 μM, whereas neither bru9a nor its cyclic analogue were active. In an *in vivo* test using *Drosophila*, cgm9a-GLP produced significantly greater inhibition of tergotrochanteral muscle response to stimulation of the giant fibre compared to native gm9a. Both cyclic analogues maintained wild-type stability in serum with >90% remaining after 24 h.

10.3.3.3 Enzymatic Conotoxin Cyclisation

Enzymatic approaches have recently been introduced as an alternative to NCL for the production of cyclic peptides (Antos et al. 2009). Sortase A (SrtA) is a transpeptidase expressed in Gram-positive bacteria, which attaches proteins to the cell wall by cleaving at a C-terminal LPXT/G motif and forming a thioester-enzyme intermediate that subsequently undergoes nucleophilic attack by glycine-based nucleophiles (Ton-That et al. 1999). This mechanism has been exploited to generate cyclic peptides by incorporating a C-terminal LPXTG motif and an N-terminal polyglycine tail into the native linear peptides

(Bolscher et al. 2011; Maximillian et al. 2011). Using this approach, Jia et al. (2014) aimed to examine its applicability toward cyclisation of disulfide-rich peptides, including Vc1.1. The resulting cyclo-[G]Vc1.1[GLPET] analogue showed close overlap of Hα secondary shifts with both native and cyclic Vc1.1, indicating successful SrtA-mediated cyclisation. However, the pharmacological activity of cyclo-[G] Vc1.1[GLPET] was not examined in this study. A cyclic analogue of the κ-conotoxin PVIIA, which like many cyclotides and conotoxins contains an ICK motif, has also been synthesised using SrtA (Kwon et al. 2016). Interestingly, although cyc-PVIIA retained a highly similar structure to the native PVIIA, it displayed only partial inhibitory activity at *Shaker* K_V channels, which was attributed to the loss of electrostatic interaction between the N-terminus and channel pore. Furthermore, the serum stability of cyc-PVIIA was substantially less than that of native PVIIA (50% vs >95% remaining after 12 h).

NCL requires strong oxidative conditions for thioester exchange to occur so folding is generally carried out post-cyclisation. However, in some cases such as ImI and AuIB this pre-constrained backbone can drastically reduce the yield of native conformations (Armishaw et al. 2010; Lovelace et al. 2011). In contrast, SrtA cyclisation is more versatile and can be carried out both before folding, which is more suitable for Vc1.1, or after folding, which is more favourable for kB1 (Jia et al. 2014). Despite its versatility, a major disadvantage of SrtA-mediated cyclisation is the requirement for a high enzyme-to-peptide ratio (1:3) and overnight incubation, which is unfavourable for rapid large-scale production of cyclic peptides. The biosynthetic pathway of naturally existing cyclic plant peptides, or cyclotides, is thought to involve asparaginyl endoproteases (AEPs) which cleave and cyclise the linear precursors, suggesting these enzymes may have enhanced catalytic efficiency for peptide cyclisation over bacterial transpeptidases. Indeed, the recently characterised AEP butelase-1, isolated from *Clitoria ternatea,* is able to cyclise diverse peptide substrates including conotoxin MrIA with 95% efficiency in <12 min at an enzyme-to-peptide ratio of 1:400, demonstrating a remarkable improvement over SrtA (Nguyen et al. 2014; Nguyen et al. 2015). It is likely that continued discovery and optimisation of enzyme-catalysed cyclisation will drastically reduce costs and work-up times associated with cyclic peptide synthesis and enable high-throughput production of cyclic conotoxin libraries.

10.3.4 Structural Minimisation

From a drug design perspective, removal of non-critical residues presents a strategy for engineering smaller conotoxin analogues that are not only synthesised more cheaply and efficiently, but are also less susceptible to inherent instability of larger peptides. One approach has been to truncate non-essential residues from intercysteine loops. For example, the loop 2 in α-conotoxins varies greatly in size and composition between members (2–8 residues) which led Jin et al. (2008) to reason that the larger 4/7 α-conotoxins might be compatible with deletions in this loop. To investigate the effect of loop 2 deletions, a series of sequentially truncated analogues of [A10L]PnIA were designed and synthesised. Electrophysiological recordings of α7 nAChR revealed that truncations down to 4/4 did not significantly reduce inhibitory activity, however, further truncation to the 4/3 analogue completely abolished the activity. Interestingly, truncations beyond 4/6 caused a considerable increase in disulfide scrambling. Moreover, the affinity for AChBP was also significantly reduced due to loss of hydrogen bonding between loop 2 residues and the receptor. More recently, Carstens et al. (2016a) used a similar approach to reduce the size of α-conotoxins that inhibit $Ca_V2.2$ channels. In contrast to loop 2, the loop 1 *SXPX* motif in α-conotoxins is highly conserved and plays crucial roles in peptide structure and activity. In particular, the α9α10/$Ca_V2.2$ inhibitors Vc1.1, RgIA, as well as the α7-selective ImI, all share identical residues 1–8. To examine whether this loop 1 motif was sufficient for activity at these targets, an analogue consisting of only residues 1–8 braced by the Cys2–8 disulfide, referred to as

[Ser[3]]Vc1.1(1–8) was synthesised. Interestingly, this analogue retained wild-type inhibition of $Ca_V2.2$ currents in rodent DRG neurons at 1 μM, and also inhibited α7 nAChR currents, but lost its ability to inhibit the α9α10 nAChR subtype. [Ser[3]]Vc1.1(1–8) was tested in an *in vivo* model of chronic visceral pain and remarkably was able to significantly inhibit the visceromotor response following colorectal distension when administered intracolonically. Thus, this minimal functional α-conotoxin motif offers an exciting new analgesic lead molecule, with the advantage of faster and more efficient synthesis compared to full-length Vc1.1.

The effects of truncations on the structure and function of three-disulfide μ-conotoxins have also been explored. The conserved α-helix from Lys7-Asp11 of analgesic μ-conotoxin KIIIA has been identified as a key requirement for inhibition of Na_Vs (Zhang et al. 2007). Additionally, removal of the Cys1–9 disulfide does not perturb this α-helix and it retains similar potency to wild-type KIIIA (Khoo et al. 2009). Based on these observations, Khoo *et al.* (2011) designed a truncated KIIIA(7–14) analogue comprising the α-helical residues and the functionally important Arg14. Furthermore, a side chain lactam bond, which has previously been shown to stabilise α-helices, was introduced between positions 9 and 13 substitutions. The resulting analogues maintained their native α-helical structure and similar selectivity for inhibition of Na_V subtypes expressed in *Xenopus* oocytes to wild-type KIIIA. Although the consequence of this truncation was significantly reduced potency, it was still within the low micromolar range, which is a promising result for such a substantial reduction in peptide size.

The logical next step to minimising conotoxin structures is to mimic the peptide pharmacophore by positioning side chain functional groups in the correct spatial orientation using non-peptidic scaffolds. This strategy has been applied to several conotoxins in an attempt to create non-peptidic or "peptidomimetic" analogues that retain biological activity but exhibit improved pharmacokinetic properties. A peptidomimetic of the ω-conotoxin MVIIA was designed using a "dendroid" approach that utilised a branched aromatic core that projected Arg10, Leu11 and Tyr13 side chains to mimic the native peptide (Menzler et al. 1998). Despite similar orientation of the pharmacophore, this analogue was virtually inactive at $Ca_V2.2$ (Menzler et al. 2000). However, shortening the dendroid scaffold branches to reduce overall flexibility of the molecule produced an analogue with low micromolar potency for $Ca_V2.2$ inhibition. GVIA peptidomimetics have also been designed and synthesised by positioning the functionally important Lys2, Tyr13 and Arg17 side chains around either a benzothiazole or anthranilimide scaffold (Baell et al. 2001). In subsequent studies, both analogues showed good overlap of side chain orientations with the corresponding three-dimensional parent peptide structure (Baell et al. 2004, 2006). Radioligand displacement of ^{125}I-GVIA from $Ca_V2.2$ in rat brain preparations yielded inhibitory constants (K_i) for both scaffolds in the 1–5 μM range and maintained specificity for $Ca_V2.2$ over the P/Q-type Ca_V ($Ca_V2.1$). Further reductions to the molecular size (−193 g/mol) of the benzothiazole-based mimetic by removal of the Lys2 moiety surprisingly maintained $Ca_V2.2$ inhibition equipotent its tripeptide parent molecule (Duggan et al. 2009). A peptidomimetic analogue of μ-conotoxin KIIIA that mimics the bond vectors of Lys7, Trp8 His12 has also been synthesised but exhibited minimal inhibitory activity at $Na_V1.7$ in xenopus oocytes (Brady et al. 2013b). Despite major reductions in bioactivity for all reported conotoxin peptidomimetics compared to their parent peptides, their functionally minimised structures offer ease of synthesis and their non-peptidic composition may afford greater stability *in vivo*. Furthermore, peptidomimetics are more amenable to medicinal chemistry optimisation of molecular structure which may engender future analogues with improved receptor affinity and potency.

10.3.5 Dendrimers

Renal clearance of short peptides is a primary elimination pathway and contributes to their low bioavailability. Synthesis of multimeric ligands or "dendrimers" by attaching multiple copies of a peptide to a polylysine core is a recent approach for increasing molecular size. As a result, dendrimers are expected to have reduced renal clearance and may also exhibit a multivalency effect that increases receptor affinity over monomeric ligands. Dendrimers of both ImI (Wan et al. 2015) and MrIA (Wan et al. 2016) have been constructed by coupling 5-hexynoic acid to polylysine dendrons and subsequently attaching azido-functionalised conotoxin monomers using CuAAC. The dendrimer design also included polyethylene glycol (PEG) spacers separating the N-termini and azide moiety of the peptide. In the case of ImI, PEG(9) divalent and tetravalent dendrimers were synthesised, whereas PEG(9) or PEG(24) tetravalent, and PEG(24) octavalent dendrimers of MrIA were constructed. All dendrimeric analogues of ImI and MrIA closely matched the Hα secondary shift values of their parent peptide indicating conserved secondary structure. Multivalent effects were observed for both dimeric and tetrameric ImI, which displayed 5–10-fold greater affinity for the α7 nAChR subtype and 60–100-fold increased inhibitory potency over wild-type ImI, whereas MrIA dendrimers only exhibited wild-type levels of NET inhibition. The MrIA dendrimers were also tested *in vivo* using a PNL rat pain model but unexpectedly, were unable to produce any analgesic effect following intrathecal injection. The loss of *in vivo* activity, which was thought to result from an inability to diffuse through the spinal cord, therefore suggests that dendrimerisation not a viable approach for improving MrIA bioactivity. However, since MrIA is the only example tested *in vivo*, it is unclear whether dendrimers of conotoxins that elicit analgesia through alternate mechanisms would also lose activity.

10.4 Conclusions

In recent decades, the therapeutic potential of conotoxins has been extensively explored as they possess a number of desirable pharmaceutical properties including high potency and selectivity for neurophysiological targets that are implicated in various neurological disorders, defined and rigid three-dimensional structures, and relative ease of synthesis and modifications. However, like other peptide drug leads they are susceptible to degradation *in vivo* by proteolysis or reduction, which impairs their bioavailability and presents a major hurdle toward the development of conotoxin-based drugs. For this reason, the diverse synthetic strategies described herein, such as disulfide substitution, cyclisation and structural minimisation, have been implemented to engineer conotoxins with improved biopharmaceutical properties. In many cases, engineered conotoxins display similar or even improved potency to the native parent peptides and preservation of three-dimensional structure, while exhibiting improved biological stability. Although the engineering techniques described here have only been applied to a handful of conotoxins, the high conservation of structure and function among families and folds suggest that these methods will be applicable to known and novel conotoxin drug leads. Furthermore, the continuous discovery and development of new chemical and enzymatic approaches will likely simplify and accelerate the production of engineered conotoxin analogues. Such improvements are expected to facilitate the therapeutic development of conotoxins with the hope that a greater number of conotoxin-based drugs will reach the clinic.

References

Akcan M, Clark RJ, Daly NL, Conibear AC, de Faoite A, Heghinian MD, Sahil T, Adams DJ, Marí F, Craik DJ (2015) Transforming conotoxins into cyclotides: backbone cyclization of P-superfamily conotoxins. Pept Sci 104(6):682–692

Akondi KB, Lewis RJ, Alewood PF (2014a) Re-engineering the mu-conotoxin SIIIA scaffold. Biopolymers 101(4):347–354

Akondi KB, Muttenthaler M, Dutertre S, Kaas Q, Craik DJ, Lewis RJ, Alewood PF (2014b) Discovery, synthesis, and structure-activity relationships of conotoxins. Chem Rev 114(11):5815–5847

Antos JM, Popp MW-L, Ernst R, Chew G-L, Spooner E, Ploegh HL (2009) A straight path to circular proteins. J Biol Chem 284(23):16028

Armishaw C, Jensen A, Balle L, Scott K, Sorensen L, Stramgaard K (2011) Improving the stability of alpha-conotoxin AuIB through N-to-C cyclization: the effect of linker length on stability and activity at nicotinic acetylcholine receptors. Antioxid Redox Signal 14(1):65–76

Armishaw C, Jensen A, Balle T, Clark R, Harpsoe K, Skonberg C, Liljefors T, Stromgaard K (2009) Rational design of alpha-Conotoxin analogues targeting alpha7 nicotinic acetylcholine receptors: improved antagonistic activity by incorporation of proline derivatives. J Biol Chem 284(14):9498–9512

Armishaw CJ (2010) Synthetic alpha-conotoxin mutants as probes for studying nicotinic acetylcholine receptors and in the development of novel drug leads. Toxins 2(6):1471

Armishaw CJ, Daly N, Nevin S, Adams D, Craik D, Alewood P (2006) Alpha-selenoconotoxins, a new class of potent alpha7 neuronal nicotinic receptor antagonists. J Biol Chem 281(20):14136–14143

Armishaw CJ, Dutton JL, Craik DJ, Alewood PF (2010) Establishing regiocontrol of disulfide bond isomers of alpha-conotoxin ImI via the synthesis of N-to-C cyclic analogs. Biopolymers 94(3):307–313

Baell JB, Duggan PJ, Forsyth SA, Lewis RJ, Lok YP, Schroeder CI, Shepherd NE (2006) Synthesis and biological evaluation of anthranilamide-based non-peptide mimetics of omega-conotoxin GVIA. Tetrahedron 62(31):7284–7292

Baell JB, Duggan PJ, Forsyth SA, Lewis RJ, Phei Lok Y, Schroeder CI (2004) Synthesis and biological evaluation of nonpeptide mimetics of omega-conotoxin GVIA. Biorg Med Chem 12(15):4025–4037

Baell JB, Forsyth SA, Gable RW, Norton RS, Mulder RJ (2001) Design and synthesis of type-III mimetics of omega-conotoxin GVIA. J Comput Aided Mol Des 15(12):1119–1136

Baron R, Binder A, Wasner G (2010) Neuropathic pain: diagnosis, pathophysiological mechanisms, and treatment. Lancet Neurol 9(8):807–819

Basbaum AI, Bautista DM, Scherrer G, Julius D (2009) Cellular and molecular mechanisms of pain. Cell 139(2):267–284

Berecki G, McArthur JR, Cuny H, Clark RJ, Adams DJ (2014) Differential Cav2.1 and Cav2.3 channel inhibition by baclofen and alpha-conotoxin Vc1.1 via GABAB receptor activation. J Gen Physiol 143(4):465–479

Berecki G, Motin L, Haythornthwaite A, Vink S, Bansal P, Drinkwater R, Wang CI, Moretta M, Lewis RJ, Alewood PF, Christie MJ, Adams DJ (2010) Analgesic omega-conotoxins CVIE and CVIF selectively and voltage-dependently block recombinant and native N-type calcium channels. Mol Pharmacol 77(2):139

Bernaldez J, Román-Gonźalez SA, Martińez O, Jimeńez S, Vivas O, Arenas I, Corzo G, Arreguiń R, García DE, Possani LD, Licea A (2013) A Conus regularis conotoxin with a novel eight-cysteine framework inhibits Cav2.2 channels and displays an anti-nociceptive activity. Mar Drugs 11(4):1188–1202

Besse D, Siedler F, Diercks T, Kessler H, Moroder L (1997) The redox potential of selenocystine in unconstrained cyclic peptides. Angew Chem Int Ed Engl 36(8):883–885

Biass D, Dutertre S, Gerbault A, Menou J-L, Offord R, Favreau P, Stöcklin R (2009) Comparative proteomic study of the venom of the piscivorous cone snail Conus Consors. J Proteome 72(2):210–218

Blanchfield J, Dutton J, Hogg R, Gallagher O, Craik D, Jones A, Adams D, Lewis R, Alewood P, Toth I (2003) Synthesis, structure elucidation, in vitro biological activity, toxicity, and Caco-2 cell permeability of lipophilic analogues of alpha-Conotoxin MII. J Med Chem 46(7):1266–1272

Blanchfield JT, Gallagher OP, Cros C, Lewis RJ, Alewood PF, Toth I (2007) Oral absorption and in vivo biodistribution of alpha-conotoxin MII and a lipidic analogue. Biochem Biophys Res Commun 361(1):97–102

Blanco-Canosa JB, Dawson PE (2008) An efficient Fmoc-SPPS approach for the generation of thioester peptide precursors for use in native chemical ligation. Angew Chem Int Ed Engl 47(36):6851–6855

Bolscher JGM, Oudhoff MJ, Nazmi K, Antos JM, Guimaraes CP, Spooner E, Haney EF, Garcia Vallejo JJ, Vogel HJ, van't Hof W, Ploegh HL, Veerman ECI (2011) Sortase A as a tool for high-yield histatin cyclization. FASEB J 25(8):2650–2658

Bondebjerg J, Grunnet M, Jespersen T, Meldal M (2003) Solid-phase synthesis and biological activity of a thioether analogue of conotoxin G1. Chembiochem 4(2–3):186–194

Bowersox SS, Gadbois T, Singh T, Pettus M, Wang YX, Luther RR (1996) Selective N-type neuronal voltage-sensitive calcium channel blocker, SNX-111, produces spinal antinociception in rat models of acute, persistent and neuropathic pain. J Pharmacol Exp Ther 279(3):1243

Brady MR, Baell BJ, Norton SR (2013a) Strategies for the development of conotoxins as new therapeutic leads. Mar Drugs 11(7):2293–2313

Brady RM, Zhang M, Gable R, Norton RS, Baell JB (2013b) De novo design and synthesis of a μ-conotoxin KIIIA peptidomimetic. Bioorg Med Chem Lett 23(17):4892–4895

Brust A, Palant E, Croker DE, Colless B, Drinkwater R, Patterson B, Schroeder CI, Wilson D, Nielsen CK, Smith MT, Alewood D, Alewood PF, Lewis RJ (2009) Chi-Conopeptide pharmacophore development: toward a novel class of norepinephrine transporter inhibitor (Xen2174) for pain. J Med Chem 52(22):6991–7002

Bulaj G, West PJ, Garrett JE, Watkins M, Zhang M-M, Norton RS, Smith BJ, Yoshikami D, Olivera BM (2005) Novel conotoxins from Conus Striatus and Conus Kinoshitai selectively block TTX-resistant sodium channels. Biochemistry 44(19):7259–7265

Bulaj G, Zhang M-M, Green BR, Fiedler B, Layer RT, Wei S, Nielsen JS, Low SJ, Klein BD, Wagstaff JD, Chicoine L, Harty TP, Terlau H, Yoshikami D, Olivera BM (2006) Synthetic muO-conotoxin MrVIB blocks TTX-resistant sodium channel NaV1.8 and has a long-lasting analgesic activity. Biochemistry 45(23):7404–7414

Callaghan B, Haythornthwaite A, Berecki G, Clark RJ, Craik DJ, Adams DJ (2008) Analgesic alpha-conotoxins Vc1.1 and Rg1A inhibit N-type calcium channels in rat sensory neurons via GABAB receptor activation. J Neurosci 28(43):10943–10951

Carstens BB, Berecki G, Daniel JT, Lee HS, Jackson KAV, Tae HS, Sadeghi M, Castro J, O'Donnell T, Deiteren A, Brierley SM, Craik DJ, Adams DJ, Clark RJ (2016a) Structure–activity studies of cysteine-rich alpha-conotoxins that inhibit high-voltage-activated calcium channels via GABAB receptor activation reveal a minimal functional motif. Angew Chem Int Ed 55(15):4692–4696

Carstens BB, Swedberg J, Berecki G, Adams DJ, Craik DJ, Clark RJ (2016b) Effects of linker sequence modifications on the structure, stability, and biological activity of a cyclic alpha-conotoxin. Pept Sci 106(6):864–875

Chen S, Gopalakrishnan R, Schaer T, Marger F, Hovius R, Bertrand D, Pojer F, Heinis C (2014) Dithiol amino acids can structurally shape and enhance the ligand-binding properties of polypeptides. Nat Chem 6(11):1009–1016

Chhabra S, Belgi A, Bartels P, van Lierop BJ, Robinson SD, Kompella SN, Hung A, Callaghan BP, Adams DJ, Robinson AJ, Norton RS (2014) Dicarba analogues of alpha-conotoxin RgIA. Structure, stability, and activity at potential pain targets. J Med Chem 57(23):9933

Clark R, Craik D (2012) Engineering cyclic peptide toxins. Methods Enzymol 503:57–74

Clark RJ, Akcan M, Kaas Q, Daly NL, Craik DJ (2012) Cyclization of conotoxins to improve their biopharmaceutical properties. Toxicon 59(4):446–455

Clark RJ, Craik DJ (2010) Invited review native chemical ligation applied to the synthesis and bioengineering of circular peptides and proteins. Biopolymers 94(4):414–422

Clark RJ, Fischer H, Dempster L, Daly NL, Rosengren KJ, Nevin ST, Meunier FA, Adams DJ, Craik DJ (2005) Engineering stable peptide toxins by means of backbone cyclization: stabilization of the alpha-conotoxin MII. Proc Natl Acad Sci U S A 102(39):13767–13772

Clark RJ, Fischer H, Nevin ST, Adams DJ, Craik DJ (2006) The synthesis, structural characterization, and receptor specificity of the alpha-conotoxin Vc1.1. J Biol Chem 281(32):23254–23263

Clark RJ, Jensen J, Nevin ST, Callaghan BP, Adams DJ, Craik DJ (2010) The engineering of an orally active conotoxin for the treatment of neuropathic pain. Angew Chem Int Ed Engl 49(37):6545–6548

Colgrave ML, Craik DJ (2004) Thermal, chemical, and enzymatic stability of the cyclotide kalata B1: the importance of the cyclic cystine knot. Biochemistry 43(20):5965

Craik DJ, Adams DJ (2007) Chemical modification of conotoxins to improve stability and activity. ACS Chem Biol 2(7):457–468

Craik DJ, Daly NL, Bond T, Waine C (1999) Plant cyclotides: a unique family of cyclic and knotted proteins that defines the cyclic cystine knot structural motif. J Mol Biol 294(5):1327–1336

Cuthbertson A, Indrevoll B (2003) Regioselective formation, using orthogonal cysteine protection, of an alpha-conotoxin dimer peptide containing four disulfide bonds. Org Lett 5(16):2955

Daly N, Ekberg J, Thomas L, Adams D, Lewis R, Craik D (2004) Structures of muO-conotoxins from Conus marmoreus: inhibitors of tetrodotosin (TTX)-sensitive and TTX-resistant sodium channels in mammalian sensory neurons. J Biol Chem 279(24):25774–25782

Daly NL, Rosengren KJ, Henriques ST, Craik DJ (2011) NMR and protein structure in drug design: application to cyclotides and conotoxins. Eur Biophys J 40(4):359–370

Davis J, Jones A, Lewis RJ (2009) Remarkable inter- and intra-species complexity of conotoxins revealed by LC/MS. Peptides 30(7):1222–1227

Dawson PE, Muir TW, Clark-Lewis I (1994) Synthesis of proteins by native chemical ligation. Science 266(5186):776–779

de Araujo AD, Callaghan B, Nevin ST, Daly NL, Craik DJ, Moretta M, Hopping G, Christie MJ, Adams DJ, Alewood PF (2011) Total synthesis of the analgesic conotoxin MrVIB through selenocysteine-assisted folding. Angew Chem 50(29):6527

Dekan Z, Paczkowski FA, Lewis R, Alewood P (2007) Synthesis and in vitro biological activity of cyclic lipophilic chi-conotoxin MrIA analogues. Int J Pept Res Ther 13(1–2):307–312

Dekan Z, Vetter I, Daly NL, Craik DJ, Lewis RJ, Alewood PF (2011) Alpha-Conotoxin ImI incorporating stable cystathionine bridges maintains full potency and identical three-dimensional structure. J Am Chem Soc 133(40):15866–15869

Dekan Z, Wang C-iA, Andrews RK, Lewis RJ, Alewood PF (2012) Conotoxin engineering: dual pharmacophoric noradrenaline transport inhibitor/integrin binding peptide with improved stability. Org Biomol Chem 10(30):5791–5794

Dias EL, Nguyen ST, Grubbs RH (1997) Well-defined ruthenium olefin metathesis catalysts: mechanism and activity. J Am Chem Soc 119(17):3887–3897

Duggan PJ, Lewis RJ, Phei Lok Y, Lumsden NG, Tuck KL, Yang A (2009) Low molecular weight non-peptide mimics of omega-conotoxin GVIA. Bioorg Med Chem Lett 19(10):2763–2765

Dutertre S, Jin AH, Kaas Q, Jones A, Alewood PF, Lewis RJ (2013) Deep venomics reveals the mechanism for expanded peptide diversity in cone snail venom. Mol Cell Proteomics 12(2):312–329

Dutertre S, Nicke A, Lewis RJ (2005) Beta2 subunit contribution to 4/7 alpha-conotoxin binding to the nicotinic acetylcholine receptor. J Biol Chem 280(34):30460–30468

Dutton J, Bansal P, Hogg R, Adams D, Alewood P, Craik D (2002) A new level of conotoxin diversity, a non-native disulfide bond connectivity in alpha-conotoxin AuIB reduces structural definition but increases biological activity. J Biol Chem 277(50):48849–48857

Dworkin RH, O'Connor AB, Backonja M, Farrar JT, Finnerup NB, Jensen TS, Kalso EA, Loeser JD, Miaskowski C, Nurmikko TJ, Portenoy RK, Rice ASC, Stacey BR, Treede R-D, Turk DC, Wallace MS (2007) Pharmacologic management of neuropathic pain: evidence-based recommendations. Pain 132(3):237–251

Ekberg J, Jayamanne A, Vaughan CW, Aslan S, Thomas L, Mould J, Drinkwater R, Baker MD, Abrahamsen B, Wood JN, Adams DJ, Christie MJ, Lewis RJ (2006) muO-Conotoxin MrVIB selectively blocks Nav1.8 sensory neuron specific sodium channels and chronic pain behavior without motor deficits. Proc Natl Acad Sci USA 103(45):17030

Ellison M, Feng Z-P, Park AJ, Zhang X, Olivera BM, McIntosh JM, Norton RS (2008) Alpha-RgIA, a novel conotoxin that blocks the alpha9alpha10 nAChR: structure and identification of key receptor-binding residues. J Mol Biol 377(4):1216

Endean R, Parish G, Gyr P (1974) Pharmacology of the venom of conus geographus. Toxicon 12(2):131–138

Espiritu MJ, Cabalteja CC, Sugai CK, Bingham J-P (2014) Incorporation of post-translational modified amino acids as an approach to increase both chemical and biological diversity of conotoxins and conopeptides. Amino Acids 46(1):125–151

Everhart D, Cartier G, Malhotra A, Gomes A, McIntosh J, Luetje C (2004) Determinants of potency on alpha-conotoxin MII, a peptide antagonist of neuronal nicotinic receptors. Biochemistry 43(10):2732–2737

Fiori S, Pegoraro S, Rudolph-Böhner S, Cramer J, Moroder L (2000) Synthesis and conformational analysis of apamin analogues with natural and non-natural cystine/selenocystine connectivities. Biopolymers 53(7):550

Flinn JP, Pallaghy PK, Lew MJ, Murphy R, Angus JA, Norton RS (1999) Role of disulfide bridges in the folding, structure and biological activity of omega-conotoxin GVIA. BBA Prot Struct Molec Enzym 1434(1):177–190

Gehrmann J, Alewood PF, Craik DJ (1998) Structure determination of the three disulfide bond isomers of alpha-conotoxin GI: a model for the role of disulfide bonds in structural stability. J Mol Biol 278(2):401–415

Gehrmann J, Daly NL, Alewood PF, Craik DJ (1999) Solution structure of alpha-conotoxin ImI by 1H nuclear magnetic resonance. J Med Chem 42(13):2364

Gleeson EC, Jackson WR, Robinson AJ (2016) Ring-closing metathesis in peptides. Tetrahedron Lett 57(39):4325–4333

Gold MS, Gebhart GF (2010) Nociceptor sensitization in pain pathogenesis. Nat Med 16(11):1248–1257

Gongora-Benitez M, Tulla-Puche J, Albericio F (2014) Multifaceted roles of disulfide bonds. Peptides as therapeutics. Chem Rev 114(2):901–926

Gori A, Wang CIA, Harvey PJ, Rosengren KJ (2015) Stabilization of the cysteine-rich conotoxin MrIA by using a 1,2,3-triazole as a disulfide bond mimetic. Angew Chem 54(4):1361–1364

Gotti C, Clementi F (2004) Neuronal nicotinic receptors: from structure to pathology. Prog Neurobiol 74(6):363–396

Gowd KH, Blais KD, Elmslie KS, Steiner AM, Olivera BM, Bulaj G (2012) Dissecting a role of evolutionary-conserved but noncritical disulfide bridges in cysteine-rich peptides using omega-conotoxin GVIA and its selenocysteine analogs. Biopolymers 98(3):212

Gowd KH, Yarotskyy V, Elmslie KS, Skalicky JJ, Olivera BM, Bulaj G (2010) Site-specific effects of diselenide bridges on the oxidative folding of a cystine knot peptide, omega-selenoconotoxin GVIA. Biochemistry 49(12):2741–2752

Gray WR, Luque A, Olivera BM, Barrett J, Cruz LJ (1981) Peptide toxins from conus geographus venom. J Biol Chem 256(10):4734

Gray WR, Rivier JE, Galyean R, Cruz LJ, Olivera BM (1983) Conotoxin MI. Disulfide bonding and conformational states. J Biol Chem 258(20):12247–12251

Grishin AA, Cuny H, Hung A, Clark RJ, Brust A, Akondi K, Alewood PF, Craik DJ, Adams DJ (2013) Identifying key amino acid residues that affect alpha-conotoxin AuIB inhibition of alpha3beta4 nicotinic acetylcholine receptors. J Biol Chem 288(48):34428–34442

Gyanda R, Banerjee J, Chang Y-P, Phillips AM, Toll L, Armishaw CJ (2013) Oxidative folding and preparation of alpha-conotoxins for use in high-throughput structure–activity relationship studies. J Pept Sci 19(1):16–24

Halai R, Callaghan B, Daly NL, Clark RJ, Adams DJ, Craik DJ (2011) Effects of cyclization on stability, structure, and activity of alpha-conotoxin RgIA at the alpha9alpha10 nicotinic acetylcholine receptor and GABA(B) receptor. J Med Chem 54(19):6984–6992

Halai R, Clark RJ, Nevin ST, Jensen JE, Adams DJ, Craik DJ (2009) Scanning mutagenesis of alpha-conotoxin Vc1.1 reveals residues crucial for activity at the alpha9alpha10 nicotinic acetylcholine receptor. J Biol Chem 284(30):20275–20284

Hannon HE, Atchison WD (2013) Omega-conotoxins as experimental tools and therapeutics in pain management. Mar Drugs 11(3):680–699

Hargittai B, Sole N, Groebe S, Abramson G, Barany B (2000) Chemical syntheses and biological activities of lactam analogues of alpha-conotoxin SI. J Med Chem 43(25):4787–4792

He X-H, Zang Y, Chen X, Pang R-P, Xu J-T, Zhou X, Wei X-H, Li Y-Y, Xin W-J, Qin Z-H, Liu X-G (2010) TNF-alpha contributes to up-regulation of Nav1.3 and

Nav1.8 in DRG neurons following motor fiber injury. Pain 151(2):266–279

Hemu X, Taichi M, Qiu Y, Dx L, Tam JP (2013) Biomimetic synthesis of cyclic peptides using novel thioester surrogates. Pept Sci 100(5):492–501

Holland-Nell K, Meldal M (2011) Maintaining biological activity by using triazoles as disufide bond mimetics. Angew Chem 123(22):5310–5312

Hui K, Lipkind G, Fozzard HA, French RJ (2002) Electrostatic and steric contributions to block of the skeletal muscle sodium channel by mu-conotoxin. J Gen Physiol 119(1):45–54

Huynh TG, Cuny H, Slesinger PA, Adams DJ (2015) Novel mechanism of voltage-gated N-type (Cav2.2) calcium channel inhibition revealed through alpha-conotoxin Vc1.1 activation of the GABA(B) receptor. Mol Pharmacol 87(2):240–250

Janes R (2005) Alpha-Conotoxins as selective probes for nicotinic acetylcholine receptor subclasses. Curr Opin Pharmacol 5:280–292

Jensen AA, Frølund B, Liljefors T, Krogsgaard-Larsen P (2005) Neuronal nicotinic acetylcholine receptors: structural revelations, target identifications, and therapeutic inspirations. J Med Chem 48(15):4705–4745

Jia X, Kwon S, Wang C-IA, Huang Y-H, Chan LY, Tan CC, Rosengren KJ, Mulvenna JP, Schroeder CI, Craik DJ (2014) Semienzymatic cyclization of disulfide-rich peptides using Sortase A. J Biol Chem 289(10):6627

Jin A, Dutertre S, Kaas Q, Lavergne V, Kubala P, Lewis R, Alewood P (2013) Transcriptomic messiness in the venom duct of Conus miles contributes to conotoxin diversity. Mol Cell Proteomics 12(12):3824–3833

Jin A-H, Daly NL, Nevin ST, Wang C-IA, Dutertre S, Lewis RJ, Adams DJ, Craik DJ, Alewood PF (2008) Molecular engineering of conotoxins: the importance of loop size to alpha-conotoxin structure and function. J Med Chem 51(18):5575–5584

Kaas Q, Craik DJ (2015) Bioinformatics-aided venomics. Toxins 7(6):2159–2187

Kaas Q, Westermann J-C, Craik DJ (2010) Conopeptide characterization and classifications: an analysis using ConoServer. Toxicon 55(8):1491–1509

Kaas Q, Yu R, Jin A-H, Dutertre S, Craik DJ (2012) ConoServer: updated content, knowledge, and discovery tools in the conopeptide database. Nucleic Acids Res 40:D325

Kancherla AK, Meesala S, Jorwal P, Palanisamy R, Sikdar SK, Sarma SP (2015) A disulfide stabilized beta-sandwich defines the structure of a new cysteine framework M-superfamily conotoxin. ACS Chem Biol 10(8):1847–1860

Khoo KK, Feng ZP, Smith BJ, Zhang MM, Yoshikami D, Olivera BM, Bulaj G, Norton RS (2009) Structure of the analgesic mu-conotoxin KIIIA and effects on the structure and function of disulfide deletion. Biochemistry 48(6):1210–1219

Khoo KK, Wilson MJ, Smith BJ, Zhang M-M, Gulyas J, Yoshikami D, Rivier JE, Bulaj G, Norton RS (2011) Lactam-stabilized helical analogues of the analgesic mu-conotoxin KIIIA. J Med Chem 54(21):7558

Klimis H, Adams DJ, Callaghan B, Nevin S, Alewood PF, Vaughan CW, Mozar CA, Christie MJ (2011) A novel mechanism of inhibition of high-voltage activated calcium channels by alpha-conotoxins contributes to relief of nerve injury-induced neuropathic pain. Pain 152(2):259–266

Knapp O, McArthur JR, Adams DJ (2012) Conotoxins targeting neuronal voltage-gated sodium channel subtypes: potential analgesics? Toxins 4(12):1236–1260

Kohno T, Kim JI, Kobayashi K, Kodera Y, Maeda T, Sato K (1995) Three-dimensional structure in solution of the calcium channel blocker omega-conotoxin MVIIA. Biochemistry 34(32):10256–10265

Kwon S, Bosmans F, Kaas Q, Cheneval O, Conibear AC, Rosengren KJ, Wang CK, Schroeder CI, Craik DJ (2016) Efficient enzymatic cyclization of an inhibitory cystine knot-containing peptide. Biotechnol Bioeng 113(10):2202–2212

Lebbe KE, Peigneur S, Wijesekara I, Tytgat J (2014) Conotoxins targeting nicotinic acetylcholine receptors: an overview. Mar Drugs 12(5):2970

Leipold E, DeBie H, Zorn S, Adolfo B, Olivera BM, Terlau H, Heinemann SH (2007) muO-Conotoxins inhibit Nav channels by interfering with their voltage sensors in domain-2. Channels 1(4):253–262

Leo S, D'Hooge R, Meert T (2010) Exploring the role of nociceptor-specific sodium channels in pain transmission using Nav1.8 and Nav1.9 knockout mice. Behav Brain Res 208(1):149–157

Lew MJ, Flinn JP, Pallaghy PK, Murphy R, Whorlow SL, Wright CE, Norton RS, Angus JA (1997) Structure-function relationships of omega-conotoxin GVIA. Synthesis, structure, calcium channel binding, and functional assay of alanine-substituted analogues. J Biol Chem 272(18):12014

Lewis RJ (2012) Discovery and development of the chi-conopeptide class of analgesic peptides. Toxicon 59(4):524–528

Lewis RJ (2015) CHAPTER 9 case study 1: development of the analgesic drugs prialt® and xen2174 from cone snail venoms. In: venoms to drugs: venom as a source for the development of human therapeutics. R Soc Chem 245–254

Lewis RJ, Dutertre S, Vetter I, Christie MJ (2012) Conus venom peptide pharmacology. Pharmacol Rev 64(2):259

Lluisma AO, Milash BA, Moore B, Olivera BM, Bandyopadhyay PK (2012) Novel venom peptides from the cone snail Conus pulicarius discovered through next-generation sequencing of its venom duct transcriptome. Mar Genomics 5:43–51

Lovelace ES, Armishaw CJ, Colgrave ML, Wahlstrom ME, Alewood PF, Daly NL, Craik DJ (2006) Cyclic MrIA: a stable and potent cyclic conotoxin with a novel topological fold that targets the norepinephrine transporter. J Med Chem 49(22):6561

Lovelace ES, Gunasekera S, Alvarmo C, Clark RJ, Nevin ST, Grishin AA, Adams DJ, Craik DJ, Daly NL (2011) Stabilization of alpha-conotoxin AuIB: influences of disulfide connectivity and backbone cyclization. Antioxid Redox Signal 14(1):87

Luckett S, Garcia RS, Barker JJ, Konarev AV, Shewry PR, Clarke AR, Brady RL (1999) High-resolution structure of a potent, cyclic proteinase inhibitor from sunflower seeds. J Mol Biol 290(2):525–533

Luo S, Kulak JM, Cartier GE, Jacobsen RB, Yoshikami D, Olivera BM, McIntosh JM (1998) Alpha-Conotoxin AuIB selectively blocks alpha3beta4 nicotinic acetylcholine receptors and nicotine-evoked norepinephrine release. J Neurosci 18(21):8571

MacRaild CA, Illesinghe J, BJV L, Townsend AL, Chebib M, Livett BG, Robinson AJ, Norton RS (2009) Structure and activity of (2,8)-dicarba-(3,12)-cystino alpha-IMI, an alpha-conotoxin containing a nonreducible cystine analogue. J Med Chem 52(3):755–762

Malmberg AB, Yaksh TL (1995) Effect of continuous intrathecal infusion of omega-conopeptides, N-type calcium-channel blockers, on behavior and antinociception in the formalin and hot-plate tests in rats. Pain 60(1):83–90

Mannelli LDC, Cinci L, Micheli L, Zanardelli M, Pacini A, McIntosh JM, Ghelardini C (2014) Alpha-conotoxin RgIA protects against the development of nerve injury-induced chronic pain and prevents both neuronal and glial derangement. Pain 155(10):1986–1995

Marx UC, Daly NL, Craik DJ (2006) NMR of conotoxins: structural features and an analysis of chemical shifts of post-translationally modified amino acids. Magn Reson Chem 44:S41

Maximilian WP, Stephanie KD, Tzu-Ying C, Eric S, Hidde LP (2011) Sortase-catalyzed transformations that improve the properties of cytokines. Proc Natl Acad Sci U S A 108(8):3169

McIntosh JM, Hasson A, Spira ME, Gray WR, Li W, Marsh M, Hillyard DR, Olivera BM (1995) A new family of conotoxins that blocks voltage-gated sodium channels. J Biol Chem 270(28):16796–16802

Menzler S, Bikker JA, Horwell DC (1998) Synthesis of a non-peptide analogue of omega-conotoxin MVIIA. Tetrahedron Lett 39(41):7619–7622

Menzler S, Bikker JA, Suman-Chauhan N, Horwell DC (2000) Design and biological evaluation of non-peptide analogues of omega-conotoxin MVIIA. Bioorg Med Chem Lett 10(4):345–347

Merrifield RB (1963) Solid phase peptide synthesis. I. The synthesis of a tetrapeptide. J Am Chem Soc 85(14):2149–2154

Metabolic Pharmaceuticals (2007) Metabolic discontinues clinical trial programme for neuropathic pain drug, ACV1. ASX Announcement

Murakami M, Nakagawasai O, Suzuki T, Mobarakeh II, Sakurada Y, Murata A, Yamadera F, Miyoshi I, Yanai K, Tan K, Sasano H, Tadano T, Iijima T (2004) Antinociceptive effect of different types of calcium channel inhibitors and the distribution of various calcium channel alpha(1) subunits in the dorsal horn of spinal cord in mice. Brain Res 1024(1-2):122–129

Muttenthaler M, Nevin ST, Grishin AA, Ngo ST, Choy PT, Daly NL, Hu S-H, Armishaw CJ, Wang C-IA, Lewis RJ, Martin JL, Noakes PG, Craik DJ, Adams DJ, Alewood PF (2010) Solving the alpha-conotoxin folding problem: efficient selenium-directed on-resin generation of more potent and stable nicotinic acetylcholine receptor antagonists. J Am Chem Soc 132(10):3514

Napier IA, Klimis H, Rycroft BK, Jin AH, Alewood PF, Motin L, Adams DJ, Christie MJ (2012) Intrathecal alpha-conotoxins Vc1.1, AuIB and MII acting on distinct nicotinic receptor subtypes reverse signs of neuropathic pain. Neuropharmacology 62(7):2202

Nevin ST, Clark RJ, Klimis H, Christie MJ, Craik DJ, Adams DJ (2007) Are alpha9alpha10 nicotinic acetylcholine receptors a pain target for alpha-conotoxins? Mol Pharmacol 72(6):1406–1410

Nguyen GK, Wang S, Qiu Y, Hemu X, Lian Y, Tam JP (2014) Butelase 1 is an Asx-specific ligase enabling peptide macrocyclization and synthesis. Nat Chem Biol 10(9):732–738

Nguyen GKT, Kam A, Loo S, Jansson AE, Pan LX, Tam JP (2015) Butelase 1: a versatile ligase for peptide and protein macrocyclization. J Am Chem Soc 137(49):15398

Nielsen CK, Lewis RJ, Alewood D, Drinkwater R, Palant E, Patterson M, Yaksh TL, McCumber D, Smith MT (2005) Anti-allodynic efficacy of the chi-conopeptide, Xen2174, in rats with neuropathic pain. Pain 118(1):112–124

Nishiuchi Y, Sakakibara S (1982) Primary and secondary structure of conotoxin GI, a neurotoxic tridecapeptide from a marine snail. FEBS Lett 148(2):260–262

Olivera BM (1985) Peptide neurotoxins from fish-hunting cone snails. Science 230:1338

Olivera BM, Seger J, Horvath MP, Fedosov AE (2015) Prey-capture strategies of fish-hunting cone snails: behavior, neurobiology and evolution. Brain Behav Evol 86(1):58–74

Pegoraro S, Fiori S, Rudolph-Böhner S, Watanabe TX, Moroder L (1998) Isomorphous replacement of cystine with selenocystine in endothelin: oxidative refolding, biological and conformational properties of [Sec3,Sec11,Nle7]-endothelin-11. J Mol Biol 284(3):779–792

Quiram PA, Sine SM (1998) Identification of residues in the neuronal alpha7 acetylcholine receptor that confer selectivity for conotoxin ImI. J Biol Chem 273(18):11001

Rauck RL, Wallace MS, Leong MS, Minehart M, Webster LR, Charapata SG, Abraham JE, Buffington DE, Ellis D, Kartzinel R, the Ziconotide 301 Study Group R (2006) A randomized, double-blind, placebo-controlled study of intrathecal ziconotide in adults with severe chronic pain. J Pain Symptom Manag 31(5):393–406

Robinson DS, Norton SR (2014) Conotoxin gene superfamilies. Mar Drugs 12(12):6058–6101

Robinson SD, Safavi-Hemami H, McIntosh LD, Purcell AW, Norton RS, Papenfuss AT (2014) Diversity of conotoxin gene superfamilies in the venomous snail, Conus victoriae. PLoS One 9(2):e87648

Rostovtsev VV, Green LG, Fokin VV, Sharpless KB (2002) A stepwise huisgen cycloaddition process: copper(I)-catalyzed regioselective "ligation" of azides and terminal alkynes. Angew Chem Int Ed 41(14):2596–2599

Safavi-Hemami H, Li Q, Jackson RL, Song AS, Boomsma W, Bandyopadhyay PK, Gruber CW, Purcell AW, Yandell M, Olivera BM, Ellgaard L (2016) Rapid expansion of the protein disulfide isomerase gene family facilitates the folding of venom peptides. Proc Natl Acad Sci U S A 113(12):3227–3232

Sandall DW, Satkunanathan N, Keays DA, Polidano MA, Liping X, Pham V, Down JG, Khalil Z, Livett BG, Gayler KR (2003) A novel alpha-conotoxin identified by gene sequencing is active in suppressing the vascular response to selective stimulation of sensory nerves in vivo. Biochemistry 42:6904–6911

Satkunanathan N, Livett B, Gayler K, Sandall D, Down J, Khalil Z (2005) Alpha-conotoxin Vc1.1 alleviates neuropathic pain and accelerates functional recovery of injured neurones. Brain Res 1059(2):149–158

Schmalhofer WA, Calhoun J, Burrows R, Bailey T, Kohler MG, Weinglass AB, Kaczorowski GJ, Garcia ML, Koltzenburg M, Priest BT (2008) ProTx-II, a selective inhibitor of Nav1.7 sodium channels, blocks action potential propagation in nociceptors. Mol Pharmacol 74(5):1476–1484

Scott DA, Wright CE, Angus JA (2002) Actions of intrathecal omega-conotoxins CVID, GVIA, MVIIA, and morphine in acute and neuropathic pain in the rat. Eur J Pharmacol 451(3):279–286

Sharpe I, Palant E, Schroeder C, Kaye D, Adams D, Alewood P, Lewis R (2003) Inhibition of the norepinephrine transporter by the venom peptide chi-MrIA: site of action, Na+ dependence, and structure-activity relationship. J Biol Chem 278(41):40317–40323

Sharpe IA, Gehrmann J, Loughnan ML, Thomas L, Adams DA, Atkins A, Palant E, Craik DJ, Adams DJ, Alewood PF, Lewis RJ (2001) Two new classes of conopeptides inhibit the alpha1-adrenoceptor and noradrenaline transporter. Nat Neurosci 4(9):902

Siqueira SRDT, Alves B, Malpartida HMG, Teixeira MJ, Siqueira JTT (2009) Abnormal expression of voltage-gated sodium channels Nav1.7, Nav1.3 and Nav1.8 in trigeminal neuralgia. Neuroscience 164(2):573–577

Smith BH, Torrance N (2012) Epidemiology of neuropathic pain and its impact on quality of life. Curr Pain Headache Rep 16(3):191–198

Staats PS, Yearwood T, Charapata SG, Presley RW, Wallace MS, Byas-Smith M, Fisher R, Bryce DA, Mangieri EA, Luther RR, Mayo M, McGuire D, Ellis D (2004) Intrathecal Ziconotide in the treatment of refractory pain in patients with cancer or AIDS: a randomized controlled trial. JAMA 291(1):63–70

Stevens M, Peigneur S, Dyubankova N, Lescrinier E, Herdewijn P, Tytgat J (2012) Design of bioactive peptides from naturally occurring mu-conotoxin structures. J Biol Chem 287(37):31382–31392

Tam JP, Lu Y-A (1997) Synthesis of large cyclic cystine-knot peptide by orthogonal coupling strategy using unprotected peptide precursor. Tetrahedron Lett 38(32):5599–5602

Tang Y-Q, Yuan J, Ösapay G, Ösapay K, Tran D, Miller CJ, Ouellette AJ, Selsted ME (1999) A cyclic antimicrobial peptide produced in primate leukocytes by the ligation of two truncated alpha-defensins. Science 286(5439):498–502

Ton-That H, Liu G, Mazmanian SK, Faull KF, Schneewind O (1999) Purification and characterization of sortase, the transpeptidase that cleaves surface proteins of Staphylococcus Aureus at the LPXTG motif. Proc Natl Acad Sci U S A 96(22):12424–12429

Tornøe CW, Christensen C, Meldal M (2002) Peptidotriazoles on solid phase: [1,2,3]-triazoles by regiospecific copper(I)-catalyzed 1,3-dipolar cycloadditions of terminal alkynes to azides. J Org Chem 67(9):3057–3064

Townsend A, Livett BG, Bingham JP, Truong HT, Karas JA, O'Donnell P, Williamson NA, Purcell AW, Scanlon D (2009) Mass spectral identification of vc1.1 and differential distribution of conopeptides in the venom duct of Conus Victoriae. Effect of post-translational modifications and disulfide isomerisation on bioactivity. Int J Pept Res Ther 15(3):195–203

Ulens C, Hogg RC, Celie PH, Bertrand D, Tsetlin V, Smit AB, Sixma TK (2006) Structural determinants of selective alpha-conotoxin binding to a nicotinic acetylcholine receptor homolog AChBP. Proc Natl Acad Sci U S A 103(10):3615–3620

van Hecke O, Austin SK, Khan RA, Smith BH, Torrance N (2014) Neuropathic pain in the general population: a systematic review of epidemiological studies. Pain 155(4):654–662

van Lierop BJ, Robinson SD, Kompella SN, Belgi A, McArthur JR, Hung A, MacRaild CA, Adams DJ, Norton RS, Robinson AJ (2013) Dicarba alpha-conotoxin Vc1.1 analogues with differential selectivity for nicotinic acetylcholine and GABAB receptors. ACS Chem Biol 8(8):1815–1821

Veber D, Milkowski J, Varga S, Denkewalter R, Hirschmann R (1972) Acetamidomethyl. A novel thiol protecting group for cysteine. J Am Chem Soc 94(15):5456–5461

Vetter I, Dekan Z, Knapp O, Adams DJ, Alewood PF, Lewis RJ (2012) Isolation, characterization and total regioselective synthesis of the novel muO-conotoxin MfVIA from Conus magnificus that targets voltage-gated sodium channels. Biochem Pharmacol 84(4):540–548

Vincler M, Wittenauer S, Parker R, Ellison M, Olivera BM, McIntosh JM (2006) Molecular mechanism for analgesia involving specific antagonism of alpha9alpha10 nicotinic acetylcholine receptors. Proc Natl Acad Sci U S A 103(47):17880–17884

Walewska A, Mm Z, Skalicky JJ, Yoshikami D, Olivera BM, Bulaj G (2009) Integrated oxidative folding of cysteine/selenocysteine containing peptides: improving chemical synthesis of conotoxins. Angew Chem 121(12):2255–2258

Wan J, Brust A, Bhola RF, Jha P, Mobli M, Lewis RJ, Christie MJ, Alewood PF (2016) Inhibition of the norepinephrine transporter by chi-conotoxin dendrimers. J Pept Sci 22(5):280–289

Wan J, Cooper MA, Paul B, Abraham N, Tae H-S, Lawson J, Mobli M, Vetter I, Adams DJ, Lewis RJ, Huang JX, Alewood PF (2015) Alpha-conotoxin dendrimers have enhanced potency and selectivity for homomeric nicotinic acetylcholine receptors. J Am Chem Soc 137(9):3209–3212

Wang J, Chow D, Heiati H, Shen W-C (2003) Reversible lipidization for the oral delivery of salmon calcitonin. J Control Release 88(3):369–380

Wang J, Hogenkamp DJ, Tran M, Li WY, Yoshimura RF, Johnstone TB, Shen WC, Gee KW (2006) Reversible lipidization for the oral delivery of leu-enkephalin. J Drug Target 14(3):127–136

White FA, Jung H, Miller RJ (2007) Chemokines and the pathophysiology of neuropathic pain. Proc Natl Acad Sci U S A 104(51):20151–20158

Wishart DS, Bigam CG, Holm A, Hodges RS, Sykes BD (1995) H-1, C-13 and N-15 random coil NMR chemical-shifts of the common amino-acids 1. Investigations of nearest-neighbor effects. J Biomol NMR 5(1):67–81

Woolf CJ, Ma Q (2007) Nociceptors--noxious stimulus detectors. Neuron 55(3):353–364

Yu R, Seymour VAL, Berecki G, Jia X, Akcan M, Adams DJ, Kaas Q, Craik DJ (2015) Less is more: design of a highly stable disulfide-deleted mutant of analgesic cyclic alpha-conotoxin Vc1.1. Sci Rep 5:13264

Zhang L, Bulaj G (2012) Converting peptides into drug leads by lipidation. Curr Med Chem 19(11):1602–1618

Zhang M-M, Green BR, Catlin P, Fiedler B, Azam L, Chadwick A, Terlau H, McArthur JR, French RJ, Gulyas J, Rivier JE, Smith BJ, Norton RS, Olivera BM, Yoshikami D, Bulaj G (2007) Structure/function characterization of mu-conotoxin KIIIA, an analgesic, nearly irreversible blocker of mammalian neuronal sodium channels. J Biol Chem 282(42):30699–30706

Zheng J-S, Tang S, Qi Y-K, Wang Z-P, Liu L (2013) Chemical synthesis of proteins using peptide hydrazides as thioester surrogates. Nat Protocols 8(12):2483–2495

Uptake Mechanism of Cell-Penetrating Peptides

11

Maxime Gestin, Moataz Dowaidar, and Ülo Langel

Abstract

Cell-penetrating peptides have been extensively used since their discovery for delivering cargoes unable to cross the cell membrane. Among other transported cargoes, they have shown very efficient delivery for oligonucleotides making cell-penetrating peptides a powerful tool for gene therapy. Numerous cell-penetrating peptides have now been discovered offering a wide library of structures and mechanisms of actions. Nevertheless, if it is known that different pathways are available for particles to be taken up, most mechanisms by which these particles enter cells are still to be characterized more precisely. Indeed it is admitted that cell-penetrating peptides are taken up either by direct translocation or by endocytosis but classes of cell-penetrating peptides are usually not related to specific entrance mechanisms. Actually, for most particles, different pathways can be detected during their uptake which makes the literature sometimes contradictory. Recent studies have nevertheless shown convergent uptake patterns for individual structures. Acetylated cell-penetrating peptides complexed with oligonucleotides have been shown to interact to scavenger receptor class A to induce caveolae-mediated endocytosis whereas antimicrobial peptides create pores in the cell membrane for direct translocation. Arginine-rich peptides have presented concentration-dependent mechanisms, being taken up either by membrane destabilization

Maxime Gestin and Moataz Dowaidar contributed equally to this work.

M. Gestin (✉) • M. Dowaidar
Department of Neurochemistry, The Arrhenius Laboratories for Natural Sciences, Stockholm University, Svante Arrhenius väg 16B, SE-10691 Stockholm, Sweden
e-mail: maxime.gestin@neurochem.su.se

Ü. Langel
Department of Neurochemistry, The Arrhenius Laboratories for Natural Sciences, Stockholm University, Svante Arrhenius väg 16B, SE-10691 Stockholm, Sweden

Laboratory of Molecular Biotechnology, Institute of Technology, University of Tartu, Nooruse, 50411 Tartu, Estonia

© Springer International Publishing AG 2017

A. Sunna et al. (eds.), *Peptides and Peptide-based Biomaterials and their Biomedical Applications*, Advances in Experimental Medicine and Biology 1030, https://doi.org/10.1007/978-3-319-66095-0_11

or clathrin-mediated endocytosis. Relating the structure of cell-penetrating peptides or their particles to distinct mechanisms would allow this delivery platform to become even more specific by using rational design to fit to the desired uptake pathway.

Keywords
Cell-penetrating peptide • Uptake mechanism • Endocytosis • Pore formation • Delivery platform • Gene therapy

Abbreviations

CPP	Cell-penetrating peptides
GAG	Glycosaminoglycan
HS	Heparan sulfate
HSPG	Heparan sulfate proteoglycan
LDL	Low-density lipoprotein
LPS	Lipopolysaccharide
NRP1	Neuropilin-1
SCARA	Scavenger receptor class A

11.1 Introduction

Cell-penetrating peptides (CPPs) are peptides able to cross cell membranes and to carry bioactive cargo (small drug molecules, oligonucleotides, proteins) into the cytoplasm (Kristensen and Nielsen 2016). Although CPPs have been extensively employed to transport cargo molecules the uptake mechanism of the particles formed by CPPs and their cargo is poorly understood as it seems to depends on lots of different factors such as membrane structure, peptide's primary and secondary structures, cargos nature, concentration of particle (Pan et al. 2016). Nevertheless, two kinds of the mechanism have been extensively reported for CPPs uptake, namely independent energy pathway and endocytic pathway. CPPs seem to be able to either interact with membranes for direct translocation either to use endocytic pathways to be internalized in cells and further release through endosomal escape (Durzyńska et al. 2015). Relating those uptake mechanisms to CPP's family is needed to fully understand the effect of this treatment that has been used for the delivery of numerous cargoes including genetic material (Dixon et al. 2016).

Uptake mechanisms have been extensively studied during the past years, and several of them have been described and sorted between energy-dependent and energy-independent pathways. The energy-independent pathway involves membrane interaction that allows a direct penetration into the cytosol. This penetration can be due to pore formation during the membrane interactions in the same scheme as antimicrobial peptides (Murray et al. 2016; Milani et al. 2009). Energy-dependent pathways are usually related to endocytic mechanisms, and several of them have been detected subsequently to CPP administration. Macropinocytosis has been shown to be able to incorporate nanoparticles including CPPs, and their complexes CPPs seems to be able to interact with different receptor on the surface of the membrane depending on their structure (Liu et al. 2015) but also clathrin-mediated and caveolae-mediated endocytosis (Kawaguchi et al. 2016; Matsubara et al. 2017). The involvement of anionic receptor such as neuropilin-1 and heparan sulfate proteoglycans have also been shown in several studies (Christianson and Belting 2014; Pang et al. 2015).

As the knowledge about the different pathways increases concerning receptor involvement, membrane interactions, different endocytic mechanism, it becomes now a challenge to detect which pathway is involved in various kind of particle structure. Even though the literature remains contradictory on this subject, recent studies show convergent uptake mechanism for several types of CPPs, namely acetylated CPPs (Lindberg et al. 2015), Arginine-rich CPPs (Brock 2014) and antimicrobial peptide (Murray et al. 2016).

11.2 Mechanisms

11.2.1 The Role of Heparan Sulfate Proteoglycan and Neuropilin-1

The cell membrane proteoglycans, in special, heparan sulfate proteoglycans (HSPGs), perform critical functions in synergy with peptides on the plasma membrane and possibly operating as significant CPP interacting receptor (Chen et al. 2015). They are composed of a central protein-glycosaminoglycan (GAG) of heparan sulfate (HS), linear-chains polysaccharides constituted of varying N-sulfated glucosamine or N-acetylated and uronic acids (Bishop et al. 2007; Chen et al. 2015). HS and some sulfated GAGs are among the conventional extremely anionic biopolymers (Sarrazin et al. 2011). Hence, they could attract positively charged CPPs by electrostatic force and perform as primary binding sites to enhance their cargo uptake efficacy (Naik et al. 2013). The uptake of CPPs is then a synergy of the cationic peptides with anionic elements of the cellular membrane such as HSPGs (Mandani et al. 2011). The cationic amino acids include the core to the cell-penetrating efficacy of positively charged CPPs and numerous bioactive proteins. Durable electrostatic synergies among the cationic residues and the negatively charged HS chains of proteoglycans have usually been considered as the fundamental mechanism in inducing the uptake of CPPs and their complexes including macromolecular cargoes via endocytosis or energy-independent pathways (Farkhani et al. 2014; Durzyńska et al. 2015).

Neuropilin-1 (NRP1) has also been shown to participate in the uptake of CPPs across the cell membrane (Pang et al. 2015). NRP1 binding mechanism demands that the peptide owns a C-terminal arginine (CendR motif) and the native L-conformation of its amino acids. Furthermore, the carboxyl group of the arginine should be clear of the substituent (Teesalu et al. 2009). It has been demonstrated that NRP1 and HSPG can work in cooperation to induce a macropinocytotic pathway. Indeed, using electronic microscopy, different nanoparticles interacting with either NRP1 or HSPG have been detected in the same vesicle when administrated together (Pang et al. 2015) (Fig. 11.1).

11.2.2 Energy-Independent Pathways

Energy-independent pathways involve membrane interaction or disturbance to induce direct translocation of the particle into the cytosol. Antimicrobial peptides have the ability to interact with microbe's membrane to disrupt it. This property has been used in several studies by fusing five antimicrobial peptides with cell-penetrating peptides (Reyes-Cortes et al. 2016; Brezden et al. 2016). The knowledge about antimicrobial peptide's mechanism of action is therefore required to adopt a rational design of those fused peptides. The action of antimicrobial has been described for a few years now (Brogden 2005; Murray et al. 2016) and different models have been proposed: The "barrel-stave" pathway, the "carpet" model the "toroidal-pore" pathway (Fig. 11.1).

An experimental platform has been introduced, based on the Whole-cell voltage-clamp investigation of currents proceeding through the plasma membrane of exterior rod fragments separated from frog retina, wherever endogenous conductance could be entirely hindered by intense light (Vedovato and Rispoli 2007; Madan et al. 2007). By applying this method, the membrane permeabilization features of several artificial or natural peptides were examined, and an explanation was purposed to discriminate within "toroidal", "barrel-stave," and "carpet" model.

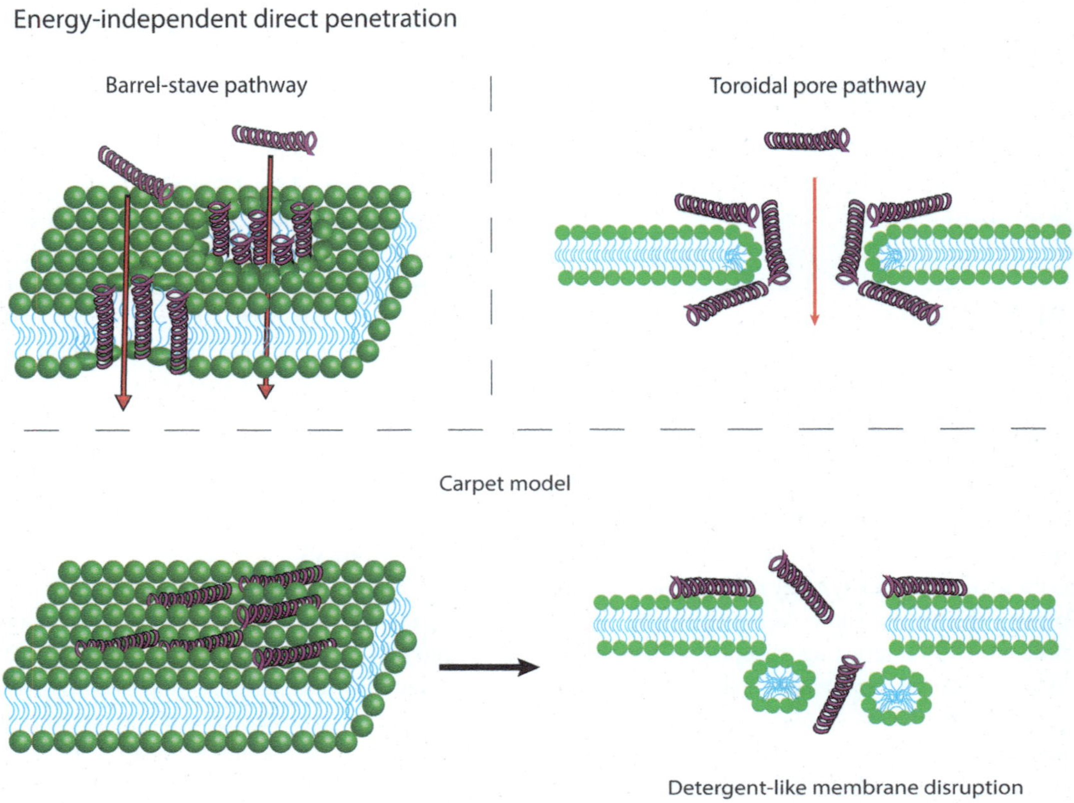

Fig. 11.1 Schematic representation of pore-formation pathways

The barrel-stave pathway demands that a particular quantity of monomers adhere collectively in the cell membrane to establish a conductive ion pore. Whenever a number of monomers implanted in the plasma membrane are inadequate, it is presumed that the pores appear and repeatedly disaggregate, providing maintained single channel effects. The use of higher peptide concentrations is supposed to generate perceptible currents that retrieve to zero after extracellular peptide relocation since the synergy within the peptides and the cell membrane is not anymore adequate to preserve the peptides implanted in the cell membrane when the extracellular accumulation is terminated. This is estimated to be a signature of the barrel-stave permeabilization manner (Vedovato and Rispoli 2007).

In the state of the toroidal path, the attractive molecular forces among the hydrophilic head groups of phospholipids and the peptides force the lipid tails into shifting up and creating pores whose surfaces are composed of both peptide and hydrophilic head groups of phospholipids monomers. This interaction does provide a prompt pore generation or disaggregation. The administration of peptide is supposed to generate noticeable currents beside a "delay" and higher time constants of recovery and current activation than in the state of the barrel-stave pathway (Madan et al. 2007). Thus if the peptide administration is brief and at average concentration, the monomers omitted on the membrane following peptide relocation do not have a level sufficient to supply inflation to conductive pores: those peptides should still contribute to producing extra pores once the peptide is implemented over the cellular membrane. Consequently, reformed peptide treatments are assumed to excite currents that mainly had retrieved to the zero level simultaneously to peptide relocation, and next to induce reduction of delay and a continuous rise of the steady-state current amplitude. This mechanism would ultimately drive the generation of stationary conductive channels that would generate growing higher background currents simultaneous peptide translocation. This is accordingly regarded as a signature of that permeabilization manner. (Vedovato et al. 2007).

In the carpet model, the dynamic electrostatic interaction among the phospholipid head groups and the peptides drive to a peptide-induced membrane 'carpeting' consequence. Ultimately, the peptides can produce micelles, therefore commencing to bilayer disintegration under a detergent-like mode (Durzyńska et al. 2015).

11.2.3 Endocytosis

11.2.3.1 Caveolae-Mediated Endocytosis

Caveolae are lipid raft invaginations of the cell membrane. They are usually coated with receptors able to bind to lipid chain such as HSPGs and scavenger receptor (Fig. 11.2a). Scavenger receptors (SCARA) are a class of cell membrane glycoproteins initially identified to bind transformed low-density lipoproteins (LDL) like acetylated and oxidized LDLs (Brown and Goldstein 1983). Initially, those receptors were considered only to scavenge acetylated LDL within macrophages, but the definition has expanded to include receptors able to bind a various set of ligands, e.g., Lipoproteins and endogenous modified proteins, bacterial lipopolysaccharides (LPS), and more recently silver nanoparticles (Wang et al. 2012) and acetylated CPP (Lindberg et al. 2015). As they recognize so many pathogen-associated molecular patterns, it is compatible that the receptors are profoundly expressed on macrophages, where they were formerly distinguished (Canton et al. 2013). Next, they were discovered in several diverse cell types, wherever subtypes were recognized in various cells as smooth muscle cells, endothelial and epithelial cells, fibroblasts and splenic dendritic cells. The number of cell membrane receptors named as scavenger receptors is continuously increasing from the initial scheme of classes A-F to incorporate 19 receptors that are classified into class A to I. The different classes of the receptor are similar in the capacity to bind negatively charged ligands. Scavenger receptors are also commonly acknowledged as a subclass of the membrane-bound guide perception receptors (Canton et al. 2013).

11.2.3.2 Clathrin-Coated Vesicles

Clathrin-coated vesicles are one of the most used receptor-dependent transport mechanism for mammalian cells (Bitsikas et al. 2014). The mechanism of their formation have been well studied over the years (Mousavi et al. 2004; Lampe et al. 2016). Clathrin-coated vesicles' main functions are the recycling the internalization of a membrane receptor. Thus they form on the site with a high density of receptor where adaptor proteins AP-180 and epsins recruit clathrins. Clathrin is a triskelion-shaped protein formed with 3 heavy chains and 3 light chains in three-legged structure (Robinson 2015). This recruitment starts to induce a curvation of the membrane. The particular structure of clathrins allows them to interact together to form a polyhedral coat around the membrane that has been deformed into a pit shape. Dynamin2 is recruited to the nascent pit and creates a collar-like structure before GTP-hydrolysis changes its conformation and lead to membrane fusion and release of the clathrin-coated vesicle into the cytosol (Mulcahy et al. 2014). The vesicle is then uncoated to form an endosome (Fig. 11.2b). The high density of receptor on the surface of the pits made clathrin-coated a major way of entering cells, and several CPPs have been reported to be transported in such vesicles (Kawaguchi et al. 2016).

11.2.3.3 Macropinocytosis

Macropinocytosis is a lipid-raft-dependent, a receptor-independent endocytic pathway where actin projections form membrane ruffles which wrap in on the cell and transport the enclosing medium inside the cell (Fig. 11.2c). Its mechanism is still not fully understood, but some of its characteristics have been elucidated. Macropinocytosis requires cholesterol, actin, rac1 and Na^+/K^+

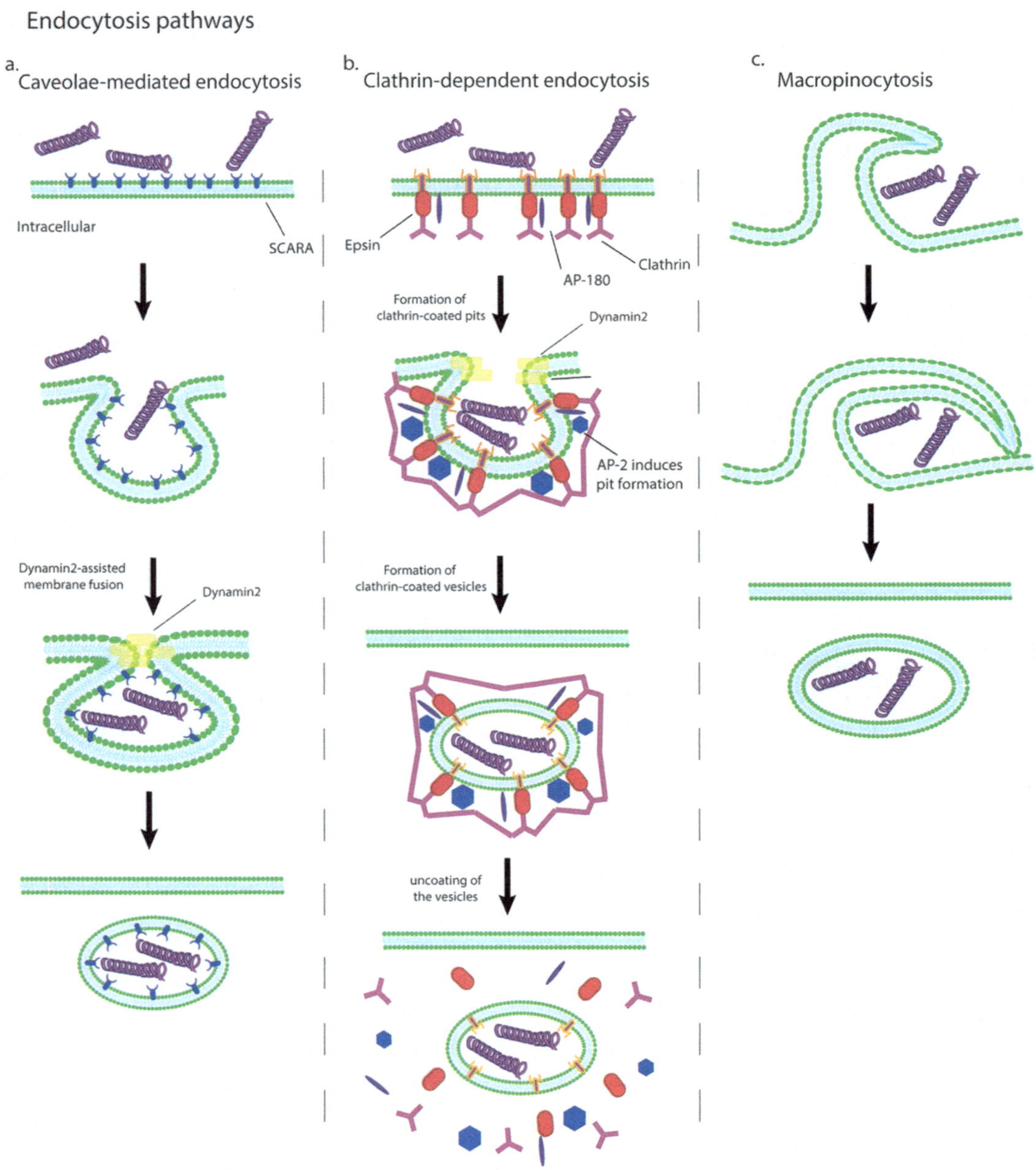

Fig. 11.2 Schematic representation of endocytic pathways

exchanger capacities. Cholesterol is needed to recruit rac1, a GTPase which plays a significant role in macropinocytosis. Besides blocking the Na^+/K^+ exchangers hindered uptake by macropinocytosis (Mulcahy et al. 2014). Macropinocytosis is believed to happen in every cell kinds, and due to the dimensions of macropinosomes, big extracellular particles are delivered up inside the cells.

11.3 CPPs and Mechanisms Associated

11.3.1 Arginine-Rich CPPss

The arginine-rich CPP Tat has been extensively studied, and several schemes have been proposed for its uptake. It has been shown that nanoparticles functionalized with Tat interacted with NRP1 due to its CendR motif. Indeed cells lacking NRP1 knocked down the uptake of Tat nanoparticles (Pang et al. 2015).

An arginine-rich CPPs resembling the native TAT protein, penetratin have been observed to associate with the plasma membrane HSPGs with a significant association (Naik et al. 2013). Furthermore, Åmand et al. showed that an enhanced arginine amount promotes proteoglycan-based uptake and other results present proof that Syndecan-4, a class I transmembrane proteins HSPGs, predicaments and arbitrates delivery of arginine-rich CPPs via the cellular membrane within the cells through clathrin-mediated endocytosis (Kawaguchi et al. 2016). However, it has also been shown that macropinocytosis was also involved as a significant uptake mechanism of the arginine-rich peptide. Indeed other studies show that when macropinocytosis is blocked with different inhibitor, the uptake is abolished (Kaplan et al. 2005; Wadia et al. 2004).

These two kinds of the receptor can induce both energy-independent and endocytic pathways, and it has been shown that Arginine-rich CPPs and Tat, in particular, can be taken up with both pathways. Nevertheless, the conditions for inducing each pathway have been elucidated. At a concentration below 2 μM, the uptake is energy-independent and relies on membrane interactions and pore formation. Between 2 μM and 20 μM, an endocytic pathway was detected, and below 20 μM the uptake mechanism changed back to an energy-independent pathway with rapid uptake from restricted membrane areas. Furthermore, the inhibition of endocytosis lowered the concentration threshold to reach this rapid energy-independent pathway (Brock 2014). The author of this study explains this phenomenon by the accumulation of peptide on the extracellular middle creating membrane disturbance. Other studies have also shown these membrane interaction leading to direct penetration with arginine-rich CPPs and polyarginine (Chakrabarti et al. 2014; Herce et al. 2014; Zeller et al. 2015).

11.3.2 Antimicrobial Peptide Conjugated CPPs

An antimicrobial peptide CM_{18} was conjugated to Tat_{11} was employed to assist the cytosolic uptake of a broad series of co-administered cargoes (GFP, calcein, dextrans) (Salomone et al. 2012), and to provide adequate transfection for DNA plasmids complexed (Salomone et al. 2013, 2014). CM_{18}-Tat_{11} was proved to grasp both Tat_{11} capacities to penetrate eukaryotic cells plus CM_{18} membrane-perturbing features. The pathway by which this chimeric peptide remodels endosomal-membrane integrity within eukaryotic cells is nevertheless not entirely explained but considering CM_{18}-Tat_{11} includes an adequate antimicrobial peptide (CM_{18}), so it is proposed that it uses a pore-formation pathway to enter cells.

Three different functionalized lactoferrin with antimicrobial properties have been administrated to bacteria cells, and their uptake was followed by electronic microscopy. The results showed membrane disruption starting from pore formation, and the peptide showed a diffuse pattern of distribution in the cytoplasm (Reyes-Cortes et al. 2016).

11.3.3 Acetylated CPPs

PepFect 14, an acetylated amphipathic CPPs, form nanoparticles with plasmid DNA (PF14/pDNA) and these particles showed a negative ζ-potential that allowed interactions with scavenger receptor class A (Ezzat et al. 2012). The silencing of specific SCARA subtypes notably reduced the gene expression of the plasmid cargo. That proposes the latter receptor intermediates the delivery of PF14/pDNA nanoparticles in association with caveolae. Indeed SCARAs are uptaken through caveolae-mediated endocytosis in macrophages. Furthermore, PF14/pDNA nanoparticles apparently provoked partial interruption of the endosomal membranes in some endosomes and inducing recruitment of SCARA from intracellular membranes to the extra cellular plasma membrane (Juks et al. 2017). It has been shown that scavenger receptor classes A 3 and 5 (SCARA3 and SCARA5) as partly responsible for the cellular penetration of acetylated CPPs by applying both several pharmacological inhibitors and by silencing the receptor employing siRNA (Ezzat et al. 2011; Ezzat et al. 2012)

11.4 Conclusions

As seen before, the mechanisms of the different uptake mechanisms are more and more characterized (Durzyńska et al. 2015; Liu et al. 2015; Madani and Gräslund 2015) due to extensive studies of the abilities of CPPs. However, until recent years, research has been more focused on the development of new CPPs and particles than in the broad investigation of their methods of action leading to a contradictory literature. The challenge for the CPP platform lies now in sorting the different class of CPPs to characterize standard schemes in the uptake. This knowledge is required to understand which mechanism is used depending on the biological, physical and chemical characteristics of the CPP and its complex with the cargo as well as on the cell membrane constitution. Studies are now turning in that way, and the recent results are elucidating uptake mechanisms for arginine-rich CPPs, antimicrobial peptides and acetylated CPPs (Lindberg et al. 2015; Brock 2014; Murray et al. 2016). This way of studying CPPs could allow a more rational design of the peptide to induce a particular pathway and reach a more efficient treatment that would push the CPP delivery platform to the front plan of the delivery system.

References

Bishop JR, Schuksz M, Esko JD (2007) Heparan sulphate proteoglycans fine-tune mammalian physiology. Nature 446:1030–1037. https://doi.org/10.1038/nature05817

Bitsikas V, Corrêa IR, Nichols BJ (2014) Clathrin-independent pathways do not contribute significantly to endocytic flux. elife 3. https://doi.org/10.7554/eLife.03970

Brezden A, Mohamed MF, Nepal M, Harwood JS, Kuriakose J, Seleem MN, Chmielewski J (2016) Dual targeting of intracellular pathogenic bacteria with a cleavable conjugate of kanamycin and an antibacterial cell-penetrating peptide. J Am Chem Soc 138:10945–10949. https://doi.org/10.1021/jacs.6b04831

Brock R (2014) The uptake of arginine-rich cell-penetrating peptides: putting the puzzle together. Bioconjug Chem 25:863–868. https://doi.org/10.1021/bc500017t

Brogden KA (2005) Antimicrobial peptides: pore formers or metabolic inhibitors in bacteria? Nat Rev Microbiol 3:238–250. https://doi.org/10.1038/nrmicro1098

Brown MS, Goldstein JL (1983) Lipoprotein metabolism in the macrophage: implications for cholesterol deposition in atherosclerosis. Annu Rev Biochem 52:223–261. https://doi.org/10.1146/annurev.bi.52.070183.001255

Canton J, Neculai D, Grinstein S (2013) Scavenger receptors in homeostasis and immunity. Nat Rev Immunol 13:621–634. https://doi.org/10.1038/nri3515

Chakrabarti A, Witsenburg JJ, Sinzinger MD, Richter M, Wallbrecher R, Cluitmans JC, Verdurmen WPR, Tanis S, Adjobo-Hermans MJW, Rademann J, Brock R (2014) Multivalent presentation of the cell-penetrating peptide nona-arginine on a linear scaffold strongly increases its membrane-perturbing capacity. Biochim Biophys Acta BBA Biomembr 1838:3097–3106. https://doi.org/10.1016/j.bbamem.2014.08.001

Chen C-J, Tsai K-C, Kuo P-H, Chang P-L, Wang W-C, Chuang Y-J, Chang MD-T (2015) A Heparan sulfate-binding cell penetrating peptide for tumor targeting and migration inhibition. Biomed Res Int 2015:1. https://doi.org/10.1155/2015/237969

Christianson HC, Belting M (2014) Heparan sulfate proteoglycan as a cell-surface endocytosis receptor. Matrix Biol 35:51–55. https://doi.org/10.1016/j.matbio.2013.10.004

Dixon JE, Osman G, Morris GE, Markides H, Rotherham M, Bayoussef Z, El Haj AJ, Denning C, Shakesheff KM (2016) Highly efficient delivery of functional cargoes by the synergistic effect of GAG binding motifs and cell-penetrating peptides. Proc Natl Acad Sci U S A 113:E291–E299. https://doi.org/10.1073/pnas.1518634113

Durzyńska J, Przysiecka Ł, Nawrot R, Barylski J, Nowicki G, Warowicka A, Musidlak O, Goździcka-Józefiak A (2015) Viral and other cell-penetrating peptides as vectors of therapeutic agents in medicine. J Pharmacol Exp Ther 354:32–42. https://doi.org/10.1124/jpet.115.223305

Ezzat K, EL Andaloussi S, Zaghloul EM, Lehto T, Lindberg S, Moreno PMD, Viola JR, Magdy T, Abdo R, Guterstam P, Sillard R, Hammond SM, Wood MJA, Arzumanov AA, Gait MJ, Smith CIE, Hällbrink M, Langel Ü (2011) PepFect 14, a novel cell-penetrating peptide for oligonucleotide delivery in solution and as solid formulation. Nucleic Acids Res 39:5284–5298. https://doi.org/10.1093/nar/gkr072

Ezzat K, Helmfors H, Tudoran O, Juks C, Lindberg S, Padari K, El-Andaloussi S, Pooga M, Langel Ü (2012) Scavenger receptor-mediated uptake of cell-penetrating peptide nanocomplexes with oligonucleotides. FASEB J 26:1172–1180. https://doi.org/10.1096/fj.11-191536

Farkhani SM, Valizadeh A, Karami H, Mohammadi S, Sohrabi N, Badrzadeh F (2014) Cell penetrating peptides: efficient vectors for delivery of nanoparticles, nanocarriers, therapeutic and diagnostic molecules. Peptides 57:78–94. https://doi.org/10.1016/j.peptides.2014.04.015

Herce HD, Garcia AE, Cardoso MC (2014) Fundamental molecular mechanism for the cellular uptake of Guanidinium-rich molecules. J Am Chem Soc 136:17459–17467. https://doi.org/10.1021/ja507790z

Juks C, Lorents A, Arukuusk P, Langel Ü, Pooga M (2017) Cell-penetrating peptides recruit type A scavenger receptors to the plasma membrane for cellular delivery of nucleic acids. FASEB J Off Publ Fed Am Soc Exp Biol 31:975–988. doi: 10.1096/fj.201600811R

Kaplan IM, Wadia JS, Dowdy SF (2005) Cationic TAT peptide transduction domain enters cells by macropinocytosis. J Control Release 102:247–253. https://doi.org/10.1016/j.jconrel.2004.10.018

Kawaguchi Y, Takeuchi T, Kuwata K, Chiba J, Hatanaka Y, Nakase I, Futaki S (2016) Syndecan-4 is a receptor for Clathrin-mediated endocytosis of arginine-rich cell-penetrating peptides. Bioconjug Chem 27:1119–1130. https://doi.org/10.1021/acs.bioconjchem.6b00082

Kristensen M, Nielsen HM (2016) Cell-penetrating peptides as tools to enhance non-injectable delivery of biopharmaceuticals. Tissue Barriers 4:e1178369. https://doi.org/10.1080/21688370.2016.1178369

Lampe M, Vassilopoulos S, Merrifield C (2016) Clathrin coated pits, plaques and adhesion. J Struct Biol 196:48–56. https://doi.org/10.1016/j.jsb.2016.07.009

Lindberg S, Regberg J, Eriksson J, Helmfors H, Muñoz-Alarcón A, Srimanee A, Figueroa RA, Hallberg E, Ezzat K, Langel Ü (2015) A convergent uptake route for peptide- and polymer-based nucleotide delivery systems. J Control Release 206:58–66. https://doi.org/10.1016/j.jconrel.2015.03.009

Liu BR, Huang Y-W, Aronstam RS, Lee H-J (2015) Comparative mechanisms of protein transduction mediated by cell-penetrating peptides in prokaryotes. J Membr Biol 248:355–368. https://doi.org/10.1007/s00232-015-9777-x

Madan V, Sánchez-Martínez S, Vedovato N, Rispoli G, Carrasco L, Nieva JL (2007) Plasma membrane-porating domain in poliovirus 2B protein. A Short Peptide Mimics Viroporin Activity J Mol Biol 374:951–964. https://doi.org/10.1016/j.jmb.2007.09.058

Mandani F, Lindberg S, Langel L, Futaki S, Gräslund A (2011) Mechanisms of cellular uptake of cell-penetrating peptides. J Biophys 2011:1. https://doi.org/10.1155/2011/414729

Madani F, Gräslund A (2015) Investigating membrane interactions and structures of CPPs. In: Langel Ü (ed) cell-penetrating peptides. Springer New York, pp 73–87

Matsubara T, Otani R, Yamashita M, Maeno H, Nodono H, Sato T (2017) Selective intracellular delivery of ganglioside GM3-binding peptide through Caveolae/raft-mediated endocytosis. Biomacromolecules 18:355–362. https://doi.org/10.1021/acs.biomac.6b01262

Milani A, Benedusi M, Aquila M, Rispoli G (2009) Pore forming properties of Cecropin-Melittin hybrid peptide in a natural membrane. Molecules 14:5179–5188. https://doi.org/10.3390/molecules14125179

Mousavi SA, Malerød L, Berg T, Kjeken R (2004) Clathrin-dependent endocytosis. Biochem J 377:1–16. https://doi.org/10.1042/BJ20031000

Mulcahy LA, Pink RC, Carter DRF (2014) Routes and mechanisms of extracellular vesicle uptake. J Extracell Vesicles 3:24641. https://doi.org/10.3402/jev.v3.24641

Murray B, Pearson CS, Aranjo A, Cherupalla D, Belfort G (2016) Mechanism of four de novo designed antimicrobial peptides. J Biol Chem 291:25706–25715. https://doi.org/10.1074/jbc.M116.733816

Naik RJ, Chatterjee A, Ganguli M (2013) Different roles of cell surface and exogenous glycosaminoglycans in controlling gene delivery by arginine-rich peptides with varied distribution of arginines. Biochim Biophys Acta BBA - Biomembr 1828:1484–1493. https://doi.org/10.1016/j.bbamem.2013.02.010

Pan R, Xu W, Ding Y, Lu S, Chen P (2016) Uptake mechanism and direct translocation of a new CPP for siRNA delivery. Mol Pharm 13:1366–1374. https://doi.org/10.1021/acs.molpharmaceut.6b00030

Pang H-B, Braun GB, Ruoslahti E (2015) Neuropilin-1 and heparan sulfate proteoglycans cooperate in cellular uptake of nanoparticles functionalized by cationic cell-penetrating peptides. Sci Adv 1:e1500821. https://doi.org/10.1126/sciadv.1500821

Reyes-Cortes R, Acosta-Smith E, Mondragón-Flores R, Nazmi K, Bolscher JGM, Canizalez-Roman A, Leon-Sicairos N (2016) Antibacterial and cell penetrating effects of LFcin17–30, LFampin265–284, and LF chimera on enteroaggregative Escherichia Coli. Biochem Cell Biol 95:76–81. https://doi.org/10.1139/bcb-2016-0088

Robinson MS (2015) Forty years of Clathrin-coated vesicles. Traffic 16:1210–1238. https://doi.org/10.1111/tra.12335

Salomone F, Breton M, Leray I, Cardarelli F, Boccardi C, Bonhenry D, Tarek M, Mir LM, Beltram F (2014) High-yield nontoxic gene transfer through conjugation of the CM18-Tat11 chimeric peptide with nanosecond electric pulses. Mol Pharm 11:2466–2474. https://doi.org/10.1021/mp500223t

Salomone F, Cardarelli F, Di Luca M, Boccardi C, Nifosì R, Bardi G, Di Bari L, Serresi M, Beltram F (2012) A novel chimeric cell-penetrating peptide with membrane-disruptive properties for efficient endosomal escape. J Control Release 163:293–303. https://doi.org/10.1016/j.jconrel.2012.09.019

Salomone F, Cardarelli F, Signore G, Boccardi C, Beltram F (2013) In vitro efficient transfection by CM18-Tat11 hybrid peptide: a new tool for gene-delivery applications. PLoS One 8:e70108. https://doi.org/10.1371/journal.pone.0070108

Sarrazin S, Lamanna WC, Esko JD (2011) Heparan sulfate proteoglycans. Cold Spring Harb Perspect Biol 3:a004952. https://doi.org/10.1101/cshperspect.a004952

Teesalu T, Sugahara KN, Kotamraju VR, Ruoslahti E (2009) C-end rule peptides mediate neuropilin-1-dependent cell, vascular, and tissue penetration. Proc Natl Acad Sci U S A 106:16157–16162. https://doi.org/10.1073/pnas.0908201106

Vedovato N, Baldini C, Toniolo C, Rispoli G (2007) Pore-forming properties of Alamethicin F50/5 inserted in a biological membrane. Chem Biodivers 4:1338–1346. https://doi.org/10.1002/cbdv.200790114

Vedovato N, Rispoli G (2007) A novel technique to study pore-forming peptides in a natural membrane. Eur Biophys J 36:771–778. https://doi.org/10.1007/s00249-007-0152-4

Wadia JS, Stan RV, Dowdy SF (2004) Transducible TAT-HA fusogenic peptide enhances escape of TAT-fusion proteins after lipid raft macropinocytosis. Nat Med 10:310–315. https://doi.org/10.1038/nm996

Wang H, Wu L, Reinhard BM (2012) Scavenger receptor mediated endocytosis of silver nanoparticles into J774A.1 macrophages is heterogeneous. ACS Nano 6:7122–7132. https://doi.org/10.1021/nn302186n

Zeller S, Choi C, Uchil PD, Ban H, Siefert A, Fahmy TM, Mothes W, Lee SK, Kumar P (2015) Attachment of cell-binding ligands to arginine-rich cell penetrating peptides enables cytosolic translocation of complexed siRNA. Chem Biol 22:50–62. https://doi.org/10.1016/j.chembiol.2014.11.009

Empowering the Potential of Cell-Penetrating Peptides for Targeted Intracellular Delivery via Molecular Self-Assembly

12

Yejiao Shi, João Conde, and Helena S. Azevedo

Abstract

Cell-penetrating peptides (CPPs) have been widely explored as an effective tool to deliver a variety of molecules and nanoparticles into cells due to their intrinsic property to translocate across cell membranes. CPPs are easier to synthesize and functionalize, and their incorporation into delivery vehicles could be achieved by both non-covalent and covalent methods. Recent advances in molecular self-assembly have demonstrated the possibility to fabricate various nanostructures with precise control over the shape, size and presentation of diverse functionalities. Through rational design, CPPs could be used as a building block for the nanostructure formation via self-assembly, while providing the functionality for intracellular delivery. In this book chapter, we will describe strategies to design self-assembling CPP conjugates and illustrate how their self-assembled nanostructures are manipulated for effective intracellular delivery. Fundamental knowledge on CPPs and molecular self-assembly will also be described.

Keywords

Self-assembly • Cell-penetrating peptides • Multifunctional • Stimuli-responsive • Targeted delivery • Intracellular delivery

12.1 Introduction

Targeted delivery of therapeutic and diagnostic agents to desired locations in the human body to improve their efficacy and prevent undesirable side effects on normal cells, organs and tissues, has been a major research effort in the past few decades. With this purpose, a variety of delivery carriers, including liposomes, polymeric nanoparticles and inorganic nanoparticles, have

Y. Shi • J. Conde • H.S. Azevedo (✉)
School of Engineering and Materials Science, Institute of Bioengineering, Queen Mary, University of London, London E1 4NS, UK
e-mail: h.azevedo@qmul.ac.uk

© Springer International Publishing AG 2017

A. Sunna et al. (eds.), *Peptides and Peptide-based Biomaterials and their Biomedical Applications*, Advances in Experimental Medicine and Biology 1030, https://doi.org/10.1007/978-3-319-66095-0_12

been developed. These delivery carriers have been designed and modified based on both passive and active targeting strategies. The passive delivery is achieved by increasing the blood circulation time of the delivery carriers and enhancing their permeability and retention (EPR) effect at the target site. On the other hand, active delivery is achieved through specific interactions between ligands (monoclonal antibodies, peptides, aptamers or small molecules) bound on the surface of the delivery carriers and up-regulated antigens or receptors on the surface of target cells (Koren and Torchilin 2012). In addition, many diseases are characterized by alterations of physiological conditions, including acidic extracellular pH, increased temperature, hypoxia and overexpression of specific analytes. These variations in physiological parameters provide an opportunity to design stimuli-responsive delivery carriers for targeted delivery (Torchilin 2014).

Although the accumulation of therapeutic and diagnostic agents has been significantly improved at the target site following the strategies mentioned above, their insufficient cellular internalization, especially macromolecular drugs (proteins and nucleic acids) still remain an important challenge. Furthermore, many therapeutic and diagnostic targets have been found located within cells, such as nuclei, mitochondria, and lysosomes. To enhance the intracellular delivery and target specific intracellular organelles, cell-penetrating peptides (CPPs), with the ability to enter cells, have received great deal of attention.

Since being discovered, CPPs have been widely used for the intracellular delivery of a variety of cargos. However, in most of the cases, CPPs are bound to the cargos by either non-covalent or covalent methods, to provide cell-penetrating ability (Koren and Torchilin 2012). However, despite the enormous potential of CPPs, they still suffer from non-selectivity. Thus, several strategies, based on the rational modification of CPPs, or their controlled presentation using different intrinsic and external triggers at the target site, have been exploited and reviewed (MacEwan and Chilkoti 2013). Nonetheless, these modifications require complex and multiple functionalization steps. In contrast, in molecular self-assembly, multiple functionalities could be integrated into a single self-assembling molecule, enabling the multivalent presentation of chemically defined ligands into one nanostructure. Employing molecular self-assembly to develop nanostructures, in which CPPs could be used as both structural and functional elements to fabricate nanostructures via self-assembly, offer a simple approach to develop multifunctional and stimuli-responsive intercellular delivery carriers with cell penetrating ability. These approaches require, however, a good understanding on the structure features of CPPs and principles of molecular self-assembly, in order to design stable multifunctional and stimuli-responsive delivery carriers. Thus, in this book chapter, we will first describe basic information on the CPP properties and the self-assembly principles. This knowledge is essential for the rational integration of CPPs into well-known self-assembling blocks, which will be further discussed along the chapter, as well as their applications in biomedicine.

12.1.1 Cell-Penetrating Peptides (CPPs)

CPPs, also known as protein transduction domains (PTDs), are a class of short peptides that comprise 5–40 amino acids and are noteworthy for their ability to translocate across cell membranes (Raucher and Ryu 2015). The first CPP, HIV transactivator of transcription protein (Tat), was discovered in 1988 by Frankel and Pabo, who showed evidence that Tat could be effectively internalized by cells *in vitro,* resulting in transactivation of the viral promoter (Frankel and Pabo 1988). Since the initial discovery, research on CPPs has evolved rapidly over the past 30 years and more than 100 protein-derived and synthetic peptide sequences have been described with membrane-crossing properties (Milletti 2012).

With more and more CPPs emerging, they have been divided into different subgroups, according to their origins, sequence characteristics and cellular uptake mechanisms (Milletti 2012).

However, the best-known classification of CPPs is based on their physicochemical properties, in which CPPs have been divided into three major classes: cationic, amphipathic and hydrophobic. Cationic CPPs are typically non-specific for cells and tissues and their membrane-cross ability relies primarily on the electrostatic interactions between their positively charged amino acids and the negatively charged phospholipids, proteoglycans or glycosaminoglycans (GAGs) on the cell surface (Copolovici et al. 2014). Studies have shown that the number of positive charges is critical for efficient translocation and effective cellular uptake of most cationic CPPs, suggesting a minimum of eight positively charged residues (Futaki et al. 2001). However, when comparing arginine with lysine residues, both with positively charged side chains, arginine residues have shown to be more effective in terms of cell penetrating ability. Their guanidine head group could form hydrogen bonding with the negatively charged phosphates and sulfates on the surface of cellular membranes (Tünnemann et al. 2008). Amphipathic CPPs, including transportan, model amphipathic peptide (MAP) and Pep-1, comprise both polar and non-polar amino acids and form the largest class (44%) of CPPs (Milletti 2012). Compared to cationic and amphipathic CPPs, only few (15%) hydrophobic CPPs, which contain only non-polar amino acids and a low net charge, have been discovered (Milletti 2012).

Depending on the different physicochemical properties of CPPs, their cellular uptake mechanisms vary considerably. In addition to the direct translocation pathway that primarily relies on the electrostatic interactions and hydrogen bonding between CPPs and cellular membranes, energy-dependent endocytosis pathways, including macropinotcytosis, clathrin- and caveolea-mediated endocytosis and clathrin- and caveolea-independent endocytosis, also exist and may even coexist in the cellular entry process of CPPs (Copolovici et al. 2014).

Although the cellular uptake mechanism of CPPs is still not fully understood, a vast number of studies have reported the CPP-assisted intracellular delivery of peptides and proteins (Mae and Langel 2006; Fonseca et al. 2009), nucleic acids (Said Hassane et al. 2010; Nakase et al. 2012), chemotherapeutic drugs (Reissmann 2014), imaging agents (Juliano et al. 2009; Wang et al. 2014) and also nanoparticles (Koren and Torchilin 2012; Farkhani et al. 2014), by both covalent and non-covalent modification approaches (Koren and Torchilin 2012), with improved delivery efficiency into various tissues, cells and subcellular compartments. The origins, sequences, structures and proposed internalization mechanisms of the most commonly used CPPs have been summarized in Table 12.1.

Regardless of the effective cell membrane translocation ability of CPPs, their non-specificity has also raised concerns when intended for targeted delivery. Therefore, great efforts have been devoted to "tame" CPPs through their rational modification or controlled presentation of CPPs using different intrinsic and external triggers at the target site (Zhang et al. 2015). These efforts have resulted in the development of multifunctional and stimuli-responsive delivery systems with improved performance (Tu and Zhu 2015). In contrast to the CPP modification strategies, the use of CPPs as building blocks to create supramolecular nanostructures as intracellular delivery vectors has received attention only recently and has not yet been fully exploited.

Molecular self-assembly offers a simple approach to develop multifunctional and stimuli-responsive delivery systems as multiple functionalities can be easily incorporated into a single self-assembling molecule. CPPs, as functionality for effective cell internalization, could also be used as a building block for the self-assembly of multifunctional and stimuli-responsive delivery carriers. Considering the peptide nature of CPPs, they could be integrated into known classes of self-assembling peptides (e.g. peptide amphiphiles, β-sheet forming peptides) or formulated into new designs.

12.1.2 Molecular Self-Assembly

Molecular self-assembly is a bottom-up approach for producing well-defined nanostructures resulting from the spontaneous organization of

Table 12.1 List of commonly used CPPs, their origins, sequences, secondary structures and mechanisms of cellular internalization

Name	Origin	Sequence (one-letter code)	Class and Secondary Structure	Proposed Internalization Mechanism	References
$Tat_{(48–60)}$	HIV-1 transcriptional activator	GRKKRRQRRRPPQ	Cationic; random coil/polyproline II helix	Direct penetration, pore formation	Frankel and Pabo (1988) and Green and Loewenstein (1988)
Penetratin	Antennapedia *Drosophila melanogaster*	RQIKIWFQNRRMKWKK	Amphipathic; α–helical/β-sheet	Direct penetration, endocytosis	Derossi et al. (1994)
Pep-1	Synthetic/chimeric (nuclear localization sequence of SV40 large T antigen)	KETWWETWWTEWSQPKKKRKV	Amphipathic; α–helical	Direct penetration, pore formation	Chaloin et al. (1998)
MAP	Model amphipathic peptide	KLALKLALKALKAALKLA	Amphipathic; α–helical	Multiple mechanisms	Oehlke et al. (1998)
Transportan	Galanin-mastoparan	GWTLNS/AGYLLGKINLKALAALAKKIL	Amphipathic; α–helical	Endocytosis, direct penetration	Pooga et al. (1998)
Polyarginines (Arg_4 to Arg_8)	Synthetic	RRRRRRRR	Cationic; random coil, α–helical	Direct penetration, endocytosis	Futaki et al. (2001)
pVEC	Murine vascular endothelial cadherin	LLIILRRRIRKQAHAHSK	Amphipathic; β-sheet	Direct penetration, transporter-mediated	Saalik et al. (2004)
CADY	Chimeric	GLWRALWRLLRSLWRLLWRA	Amphipathic; random coil; α–helical	Direct penetration, endocytosis	Crombez et al. (2009)

rationally designed molecules (Cui et al. 2010). This self-organization occurs as a consequence of specific molecular interactions, including hydrogen bonding, hydrophobic and electrostatic interactions, π-π stacking and van der waals forces (Mendes et al. 2013). These interactions could act alone or in concert to generate precisely self-assembled nanostructures. Compared to top-down approach, self-assembly fabrication offers several advantages, such as three-dimensional assembly, versatility, cost-effective large-scale production and achievement of near-atomic resolution (Toksöz and Guler 2009). It has become an attractive strategy to fabricate diverse functional nanomaterials, ranging from electronic and detection materials to biomaterials, due to its simplicity and precise control over both the morphology and function of the assembled nanostructures (Cui et al. 2010).

A variety of building blocks, from both synthetic and biological origin, have been used in molecular self-assembly, allowing the fabrication of nanostructures with defined shape and size for diverse biomedical applications (Mendes et al. 2013). Among all the building blocks, peptides are particularly attractive and have been widely used due to their chemical versatility, biodegradability, biocompatibility and inherent chemical functionality for establishing supramolecular interactions, such as hydrogen bonding, electrostatic and hydrophobic interactions as well as aromatic stacking (Habibi et al. 2016). Naturally occurring peptide motifs with α–helix and β-sheet secondary structures could be used to drive the self-assembly process (Lim et al. 2009). Amphiphilic peptides, consisting of a hydrophilic bioactive headgroup and a hydrophobic segment, have also been shown with the ability to self-assemble (Lim et al. 2009). Therefore, peptide sequences could both direct the nanostructure formation and provide functionalities for efficient intracellular delivery of the nanostructures. The self-assembled peptidic nanocarrier could not only act as therapeutic itself, but could also enable the integration of additional functionalities, including targeting, cell-penetrating, endosomal escape and stimuli-responsive peptide sequences. The ionic or hydrophobic nature of the peptide sequences could further allow the binding of therapeutic and diagnostic payloads. The incorporation of all functionalities into a nanocarrier, being generated in a single step, is an energy- and cost-effective process (Habibi et al. 2016). This unique synergy between molecular self-assembly and inherent peptide functionality provides a simple approach for design of multi-functional and stimuli-responsive nanostructures. Herein, strategies to design self-assembling CPP-conjugates for the formation of nanostructures will be summarized and examples will be given on different types of CPP-conjugates.

12.2 Design of Self-Assembling Cell-Penetrating Peptide (CPP) Conjugates

While research on CPPs has witnessed considerable progress, the use of CPPs into self-assembled nanostructures for intracellular delivery has received attention only recently. Generally, there are two strategies to design CPP conjugates with the ability of self-assembly into nanostructures.

The first strategy is based on hydrophobic interactions to drive the self-assembly process. Amphiphilic block molecule, in which one of its blocks is hydrophilic and the other is hydrophobic, tends to aggregate above a critical concentration in aqueous solution, with the hydrophobic tails forming the inner core, while the hydrophilic head groups are presented on the outer surface, facing the aqueous solution. Based on the fact that most CPPs are hydrophilic, self-assembling CPP conjugates could be designed by attaching appropriate hydrophobic segments (lipid tails, hydrophobic drugs, *etc.*) to the CPP sequence.

The second strategy relies on the formation of hydrogen bonding and π-π stacking to drive the self-assembly process. Helical structures, including α–helix, polyproline helix, collagen triple helix, 3_{10}-helix and π-helix, together with β-sheet structures, are commonly found in biological systems. These structural motifs are stabilized by hydrogen bonding and steric effects, resulting in their self-assembly behavior, and could be conjugated to CPPs to drive the nanostructure formation.

In this section, we will discuss design rules for creating self-assembling CPP-conjugates, based on the two strategies mentioned above, and provide examples of different types of CPP-conjugates.

12.2.1 CPP-Lipid Tail Conjugates

CPP-lipid tail conjugates have been designed following the first strategy by taking advantage of the hydrophobicity of lipid tails. Considering the results from earlier research, that amphiphilic molecules could adopt various morphologies depending on the relative volume fraction between the hydrophilic and hydrophobic segments, various studies were conducted to establish rules to control the size, shape and even cellular uptake mechanism of the assembled nanostructures.

The impact of the length of the hydrophobic lipid tail on the self-assembly behavior and cellular-uptake properties of CPP-lipid tail amphiphiles has been evaluated by Burlina's group (Bode et al. 2012). They designed a small library of CPP amphiphiles by conjugating the Arg_4 CPP to alkyl tails with different lengths, varying from 0 to 18 carbons. Their results showed that the cellular uptake efficiency of Arg_4 conjugated with tails made of 12, 14, 16 and 18 carbons was significantly higher when compared to conjugates with shorter tails (less than 7 carbons). Further results revealed that Arg_4 conjugates with longer tails, containing more than 12 carbons, could promote better interactions with GAGs on the cell surface and trigger the GAG-dependent cellular uptake route. Furthermore, Arg_4 conjugates with longer tails (>12 carbons) could form micelle structures presenting a high density of cationic arginine residues at their surface. In contrast, Arg_4 conjugated with the tails with less than 7 carbons did not form any stable structures.

The effect of number of the hydrophobic lipid tails on the self-assembly behavior and cellular-uptake properties of CPP-lipid tail amphiphiles has also been studied. Lee's group has attached a stearic acid (C_{18}) to the N-terminus of Tat CPP and then increased the number of C_{18} tails to two and four to investigate the morphology and stability of their assemblies (Lim et al. 2007b). The critical aggregation concentration (CAC) of Tat conjugates with one C_{18} tail was not detected in the range of concentration up to 1 mM, while a CAC of 208 μM and 21 μM were determined for the Tat conjugates with two and four C_{18} tails respectively. Further examinations revealed that Tat conjugates with two C_{18} tails self-assembled into spherical micelles with a 6.5 nm diameter, while Tat conjugates with four C_{18} tails formed more stable cylindrical micelles with 6.8 nm diameter and 43 nm length. Studies from Cui's group investigated the effect of using a shorter octanoic acid tail (C_8) instead of stearic acid tail, conjugated to the Tat CPP (Zhang et al. 2013). In their design, only Tat conjugates with four C_8 tails were shown with the ability to form nanofibers with a high drug loading efficiency (89.7 ± 5.0%).

The use of non-linear hydrophobic lipid molecules, such as cholesterol, has also been investigated as building block for self-assembling CPP conjugates. An amphiphilic CPP-cholesterol conjugate was designed and synthesized containing both Arg_6 and Tat CPPs in the hydrophilic peptide headgroup (Liu et al. 2009). The designed CPP-cholesterol conjugates were shown to form core-shell structured nanoparticles with enhanced antimicrobial activity against a range of bacteria, yeasts and fungi. Moreover, the nanoparticles could cross the blood-brain barrier and suppress bacterial growth in the infected brains without causing significant toxicity to the major organs, which showed their potential to treat infections in the brain.

12.2.2 CPP-Drug Conjugates

Based on the fact that the majority of CPPs are hydrophilic while most chemotherapeutic drugs are hydrophobic, CPP-drug conjugates with self-assembly ability could be obtained following the first strategy, by conjugating these two segments together via specific linker. The linker between the CPPs and the chemotherapeutic drugs is typically used to guarantee the release of drugs upon

Table 12.2 Examples of commonly used linkers between CPPs and chemotherapeutic drugs and their cleavage susceptibility (when applicable) to trigger the drug release

Type of bond	Chemical structure	Trigger
Peptide (amide)	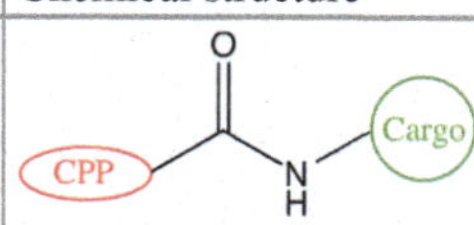	Enzyme (Cathepsin B (CatB), matrix metalloproteases (MMPs), other proteases)
Disulphide	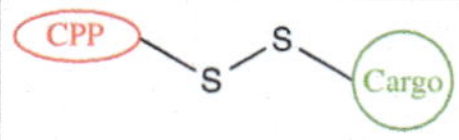	Reducing agent (glutathione), enzyme (glutathione-s-transferases, GSTs)
Maleimide	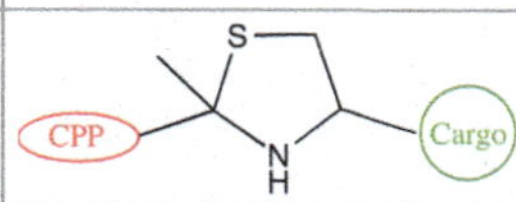	–
Hydrazone	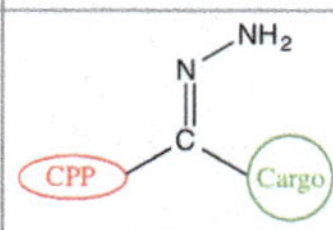	pH
Ether	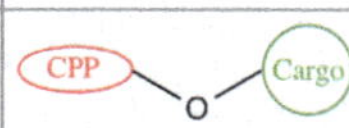	–
Sulphanyl (bis-functional)	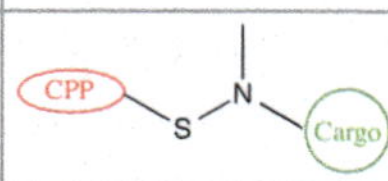	–
Phosphatydylethanolamine	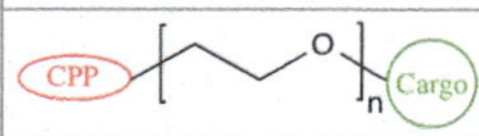	–

cleavage. In particular, some linkers are designed to be cleaved by specific triggers found in the local environment, such as altered pH or temperature, presence of hydrolytic or oxidative enzymes, to favor the controlled drug release at the target site (Table 12.2).

A multifunctional peptide pro-drug was designed by conjugating the antitumor drug doxorubicin to the hydrophobic tail of a CPP amphiphile, in which the hydrophilic peptide headgroup comprises a targeting peptide sequence (GRGDS) and an Arg_8 CPP (Chen et al. 2014). Owing to its amphiphilic nature, the designed peptide pro-drug could spontaneously self-assemble into spherical envelope-type nanoparticles with doxorubicin encapsulated in the hydrophobic inner core and the functional peptide sequences gathered on the outer shell. These spherical envelope-type nanoparticles were shown to transport and release doxorubicin into tumor cells specifically, presenting improved antitumor activity with minimal side effects.

A CPP-paclitaxel conjugate with the ability to self-assemble into supramolecular nanospheres with high drug loading efficiency (26.4%) was recently reported (Tian et al. 2015). The self-assembled nanospheres also served for the co-delivery of another anticancer drug, doxorubicin, to cancer cells without compromising the potency of paclitaxel.

12.2.3 CPP-Peptide Conjugates

Most of the CPP-peptide conjugates with self-assembly ability reported in the literature were designed based on the second strategy, by coupling a self-assembling (β-sheet or α–helix forming) peptide segment to the hydrophilic CPP. Substantiations for this design are summarized below.

A block peptide, TβP, consisting of a β-sheet forming peptide $(FKFE)_3$ and a Tat CPP was designed by Lee's group (Lim et al. 2007a). The peptide $(FKFE)_n$ is an artificial β-sheet forming peptide designed by the alternative placement of charged amino acids (K and E) and hydrophobic amino acids (F). This residues distribution promotes a proper β-sheet hydrogen bonding arrangement between amide hydrogen and carbonyl oxygen. Conjugation of the hydrophilic Tat CPP to the N-terminus of this β-sheet forming peptide through a flexible peptide linker (GSGG) led to the self-assembly into β-ribbon structures in which β-sheet interaction was the main driving force, while the hydrophobic and π-π-stacking interactions from the phenylalanine residues on one face of the β-tape also contributed to the stabilization of the bilayered β-ribbon structures. The hydrophobic interface inside the β-sheet ribbon structures allowed the encapsulation of hydrophobic molecules, suggesting the potential of TβP assemblies for drug delivery applications.

Later, a self-assembling rod-coil amphiphilic molecule composed of a polyproline rod and a Tat CPP coil was reported by the same group (Yoon et al. 2008). Among all the 20 naturally occurring amino acids, proline is the only amino acid with the side chain atoms forming a pyrrolidine ring with the backbone atoms. In aqueous solution, the proline-rich sequences tend to form a stiff helical rod structure, called polyproline type II (PPII) helix, since the cyclic structure of proline induces conformational constrains among the atoms in the pyrrolidine ring. The proline-rich sequences also present an increased hydrophobicity, when compared with proline itself as an isolated amino acid, because of the alignment of the three non-polar methylene groups at the outer part of the rod with the PPII helix formation. When the hydrophilic Tat CPP was conjugated to the relatively hydrophobic polyproline, the stiff rod character of the PPII helix enabled microphase separation from the hydrophilic Tat CPP, leading to the formation of self-assembled nanocapsules. These nanocapsules were shown to be stable enough to cross the cytoplasmic membrane of cells with efficient intracellular delivery of water-soluble drugs.

In addition to the PPII helix from polyproline, α–helices from polyleucine have also been exploited to develop self-assembling CPP-peptide conjugates. The formation of vesicular assemblies in aqueous solution was demonstrated using a block co-polypeptide composed of polyarginine and polyleucine ($R_{60}L_{20}$) owing to the combined interactions from both the α–helical hydrophobic polyleucine, that favored the formation of flat membranes, and the highly charged hydrophilic polyarginine segment, that imparted solubility and fluidity to these membranes (Holowka et al. 2007). The resulting vesicles were found to be transported into cells effectively with entrapment of water-soluble drugs.

Despite the effective cell internalization ability, most known CPPs lack tissue or cell specificity. Therefore, CPP-peptide conjugates with targeting capacity were also developed. Prohibitin is a protein that is highly expressed in adipose-tissue vasculature and shifts its position to the cell membranes and nuclei of differentiated adipocyte cells (Kolonin et al. 2004). A short peptide sequence (CKGGRAKDC), termed adipocyte-targeting peptide sequence (ATS) and able to specifically bind to prohibitin, was conjugated to polyarginine CPP (Arg_9) for developing an adipocyte-targeted gene carrier (Won et al. 2014). When mixed with short-hairpin RNAs for silencing fatty-acid-binding protein 4 (shTABP4), ATS-Arg_9/shFABP4 complexes self-assembled as a result of electrostatic interactions between the positively charged polyarginine CPP and the negatively charged shFABP4. The adipocyte-targeting fusion-oligopeptide gene carriers were shown to selectively transfect mature adipocytes by binding to prohibitin. Treatment of obese mice with ATS-9R/shFABP4 oligopeptide complex led to metabolic recovery and body-weight reduction.

To fully harness the power of CPPs for targeted intracellular delivery, MacEwan and Chikouti developed a smart approach by exploiting a temperature-triggered self-assembly process (Macewan and Chilkoti 2012, 2014). This approach was built upon earlier observation that cationic CPPs have a strong cut-off effect in their cell-penetrating ability and a threshold of eight consecutive arginines is needed for the effective cell internalization process (Futaki et al. 2001). Thus, Arg_5 was conjugated to an elastin-like polypeptide (ELP) that is composed of both hydrophobic and hydrophilic segments and is capable of temperature-triggered micelle assembly owing to its lower critical solution temperature (LCST) phase transition behavior. Below its LSCT, it exists as soluble monomer, while above its LSCT, it aggregates into insoluble micro-sized micelles. The self-assembly behavior of the designed Arg_5-ELP conjugates could be thermally controlled and, as a consequence, the cellular uptake and efficiency of anticancer activity could be tuned based on the large difference in the local density of arginine residues between the monomer and the micelle. A greater than eight-fold increase in cellular uptake was achieved when arginine residues are presented on the corona of the Arg_5-ELP micelles, as compared to the same Arg_5-ELP at the temperature in which it is a soluble monomer.

12.2.4 CPP-Polymer Conjugates

In most of the CPP-polymer conjugates, the CPP segment mainly serves to impart functionality rather than driving the self-assembly process due to the inherited amphiphilic property of the conjugated polymers. For example, an amphiphilic block copolymer, methoxyl poly(ethylene glycol)/ poly(ε-caprolactone) (MPEG-PCL), was used for conjugating Tat CPP via an ester bond (Kanazawa et al. 2011). The Tat-MPEG-PCL conjugates were shown to self-assemble into nano-sized micelles and facilitate targeted intranasal brain delivery.

Recently, a simple but multifunctional polymer-CPP-drug conjugate was developed, in which polyethylene glycol (PEG) was conjugated to the N-terminus of the Tat CPP through a MMP-2 sensitive peptide sequence, while doxorubicin was conjugated to the C-terminus of the Tat CPP via a maleimide bond (PEG-Tat-Doxorubicin) (Tu and Zhu 2015). Stable micelles, with a size around 70 nm, were formed from these multifunctional conjugates. Furthermore, improved cell internalization and intracellular distribution via Tat was achieved upon MMP2-triggered PEG deshielding. Inhibition of P-glycoprotein-mediated efflux was also observed, while showing anticancer efficacy in both drug sensitive and resistant cancer cells.

12.2.5 Other CPP Conjugates

In addition to the commonly used building blocks mentioned above, peptide nucleic acids, oligosaccharides, and fluorophores have also been used for the conjugation with CPPs to achieve functional structures via self-assembly.

Peptide nucleic acid (PNA) is a class of DNA mimics with the pseudopeptide backbone and antisense-antigene capacity. When conjugated with CPP, the CPP-PNA conjugates showed self-assembly behavior on an oligonucleotide scaffold, which significantly promoted the endocytosis of PNAs by at least 10-times in cell culture, demonstrating the potential of PNAs as antiviral therapeutics (Zhao et al. 2015).

Bis-β-cyclodextrin (bis-CD) is a cyclic oligosaccharide with a hydrophobic cavity that has been reported to interact with specific surface residues of insulin via hydrophobic interactions. After being site-specifically conjugated with penetratin, CPP-bis-CD was shown to co-assemble with insulin to generate insulin-loaded nanocomplexes (Zhu et al. 2014). Since insulin was hidden inside the hydrophobic cavity of the bis-CD, the obtained nanocomplexes showed better stability against high salt and enzyme concentration and greatly improved cellular uptake efficiency compared to the nanocomplexes co-assembled from CPP and insulin.

An amphiphilic CPP conjugate was also obtained by connecting a fluorophore and a quencher to a hydrophilic CPP as a proof-of-concept to design nanoprobes for cancer cell imaging (Lock et al. 2013). These conjugates formed supramolecular nanobeacons that were shown to be resistant to degradation by nonspecific enzymes.

12.3 Applications of CPP-Based Nanoassemblies for Intracellular Delivery

As demonstrated in previous sections, the use of self-assembling CPPs offer an effective method to transport a wide range of cargos, ranging from low molecular weight drugs to nanosized particles, across cell membranes. Here, we highlight few applications describing the use of self-assembling CPP conjugate nanostructures for intracellular delivery, which are summarized in Table 12.3.

The delivery of small-molecule anticancer drugs, which are usually hydrophobic, into tumor cells with the assistance of CPPs has been reported through both physical encapsulation and chemical conjugation approaches. When hydrophilic CPPs were conjugated with hydrophobic lipid tails, β-sheet peptides and polypeptides, different nanostructures, ranging from spherical micelles to nanofibers and nanoribbons, were formed by self-assembly. The hydrophobic core of these self-assembled nanostructures is able to host hydrophobic drugs, being promising as intracellular delivery vectors of anticancer drugs. In addition, CPP-drug conjugates have also been reported to self-assemble into nano-sized spherical micelles that allow higher drug loading content, as compared to drug-encapsulated nanomaterials (Chen et al. 2014). Moreover, spherical micelles assembled from CPP-Taxol conjugates were shown to serve as an efficient intracellular delivery vector for the co-delivery of other anticancer drug, doxorubicin, to cancer cells (Tian et al. 2015).

Hydrophilic nucleic acids are too large to pass through the cell membrane effectively, but their intracellular delivery has been achieved through physical adsorption or chemical conjugation to CPP-based nanostructures. Nucleic acids, which are negatively charged, could be physically adsorbed onto CPP-based nanostructures bearing a positive zeta-potential. Previous studies showed that coupling fatty acid chains to CPPs, such as stearylation of the CPP N-terminus, could increase the condensable ability of DNA and also the transfection efficiency by enhanced cytoplasmic delivery upon the formation of nanostructures (Wang et al. 2011). When conjugated with CPP, the CPP-PNA conjugates exhibited self-assembly behaviour on an oligonucleotide scaffold and its potential as antiviral therapeutics was further demonstrated (Zhao et al. 2015).

Like nucleic acids, proteins are also large molecules and their cell internalization has been enhanced with assistance of CPP-based nanostructures. Insulin, used for the treatment of diabetes, was encapsulated inside the hydrophobic cavity of bis-β-cyclodextrin (bis-β-CD), which was site-specifically conjugated with penetratin (Zhu et al. 2014). Improved cellular uptake efficiency was obtained owing to the cell internalization ability of penetratin and better stability of insulin against high salt and enzyme concentration was observed, since insulin was hidden and protected inside the hydrophobic cavity of bis-β-CD.

Enzyme sensors for the imaging of cancer cells have been developed based on CPP-fluorophore conjugates (Lock et al. 2013). These conjugates could self-assemble into spherical micelles and could be further internalized by cancer cells to detect the activity of the enzyme cathepsin B.

Some CPPs have also shown antibacterial activity in addition to the cell internalization ability. When conjugated to cholesterol, the antimicrobial activities of CPPs were strongly enhanced through the formation of spherical micelles, which could be further developed for treating infections in the brain (Liu et al. 2009).

12.4 Conclusions and Perspective

CPPs with the ability for effective cell membrane translocation have gained increased attention over the past 30 years. However, regardless of the promising *in vitro* cell-uptake results, the *in vivo* application of CPPs is much more complex. CPPs, like other pharmaceutical peptide products, suffer from short half-time in circulation caused by proteolysis and rapid renal clearance. The lack of cell specificity of CPPs also raises concerns when being considered for *in vivo* applications. To tackle these problems, attempts have been made to develop stable multifunctional and stimuli-responsive nanostructures via self-assembly. For example, the formation of an assembled structure, where the CPP functionality is buried inside the nanostructure could protect the CPP from proteolysis. In fact, previous studies showed that peptide linkers, when embedded within the core of the nanostructures, remain

Table 12.3. Building blocks used for the design of self-assembling CPP-conjugates into diverse nanostructures and for selected applications

Self-assembling CPP conjugates				
Building blocks		CPP	Supramolecular assemblies	Applications
Lipid tail	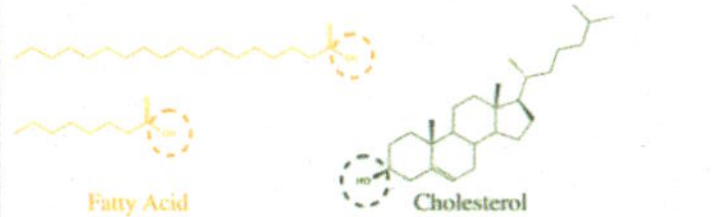	Tat and Arg_8	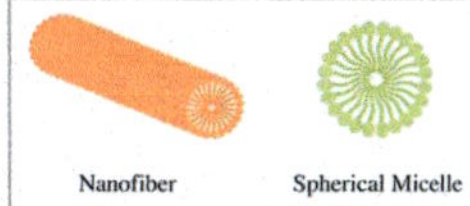	Intracellular delivery of paclitaxel for cancer chemotherapy (Zhang et al. 2013); targeted delivery of DNA into cell nuclei (Wang et al. 2011); antimicrobial agent for the treatment of brain infections (Liu et al. 2009).
Drug	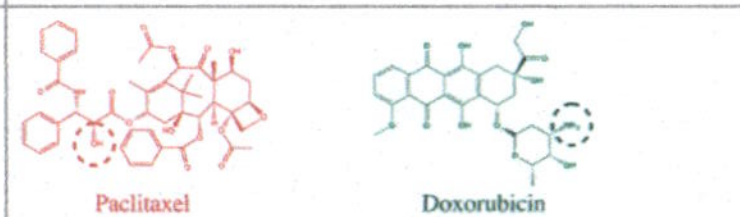	Arg_8	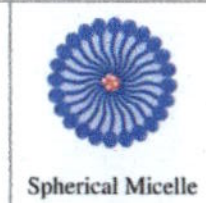	Intracellular delivery of doxorubicin (Chen et al. 2014) and paclitaxel (Tian et al. 2015) for cancer chemotherapies.
Peptide	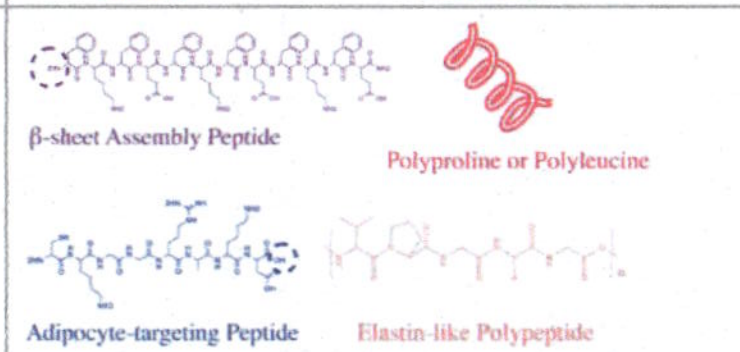	Tat and Arg_8	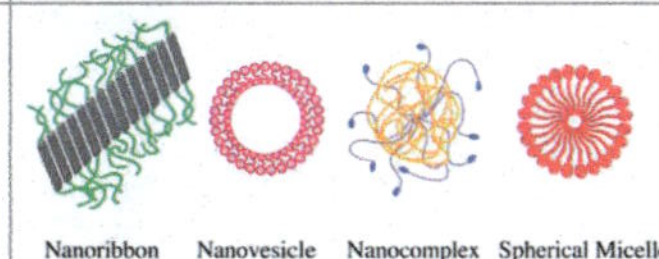	Intracellular delivery nanocarriers (Lim et al. 2007a; Yoon et al. 2008; Holowka et al. 2007); targeted gene delivery for anti-obesity therapy (Won et al. 2014); tumor targeted intracellular delivery by extrinsic thermal trigger (Macewan and Chilkoti 2012, 2014)
Polymer	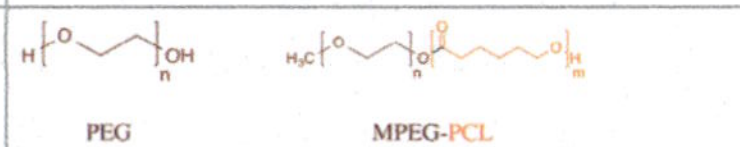	Tat	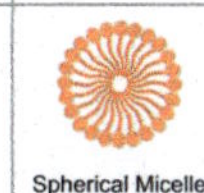	Targeted intracellular delivery of doxorubicin for enhanced cancer chemotherapies in a stimulus-sensitive manner (Tu and Zhu 2015); direct brain delivery via intranasal administration for glioblastoma treatment (Kanazawa et al. 2011)
Others	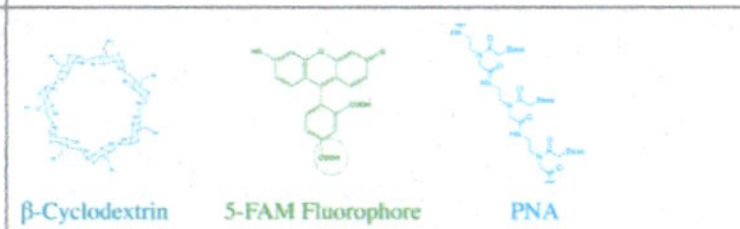	Penetratin, Tat and Arg_9	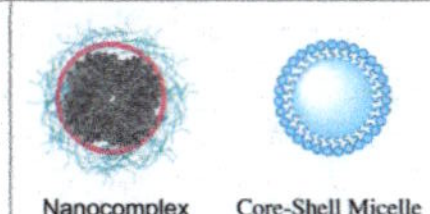	Intracellular delivery of insulin for the treatment of diabetes (Zhu et al. 2014); enzyme sensor for the imaging and diagnosis of cancer cells (Lock et al. 2013); intracellular delivery of PNAs for antiviral therapies (Zhao et al. 2015)

inaccessible to enzymes, minimizing their susceptibility to enzymatic degradation (Lock et al. 2013). In addition, the formation of stable nanostructures with dimensions greater than 20 nm could also contribute for reducing the amount of free peptide, especially containing positively charged residues, in the plasma. In this way, the interactions between positively charged CPPs and serum proteins could be minimized while avoiding clearance by the kidneys (Wang et al. 2014). It also allows the integration of targeting and stimuli-sensitive moieties at the nanostructure surface to sterically shield the CPP functionality and control its presentation at specific locations to enhance the targeted cellular internalization of the cargos (MacEwan and Chilkoti 2013).

In summary, employing supramolecular strategies to create self-assembling CPP-conjugates offers a simple approach to harness their potential and develop multifunctional and stimuli-sensitive intercellular delivery systems. However, optimizations of these self-assembling CPP-conjugates are still required. For example, the critical aggregation concentration of the self-assembling CPP-conjugates should be reduced to the nanomolar range to obtain nanostructures that remain stable during their circulation in the blood stream (Lock et al. 2013). It is also necessary to find a balance between the satisfactory cell uptake and the minimal density of CPPs within the assembly to minimize both the cytotoxic effects and costs associated with their synthesis (Macewan and Chilkoti 2012, 2014). The future of this research area will greatly benefit from computer molecular simulation models to predict their assembly, stability and interactions with cell membranes, enabling the development of more predictable and personalized nanocarriers for targeted intracellular delivery.

Acknowledgments Y. Shi thanks the China Scholarship Council for her PhD scholarship (No.201307060020) and J. Conde acknowledges Marie Curie International Outgoing Fellowship (FP7-PEOPLE-2013-IOF, Project No. 626386).

References

Bode SA, Thevenin M, Bechara C, Sagan S, Bregant S, Lavielle S, Chassaing G, Burlina F (2012) Self-assembling mini cell-penetrating peptides enter by both direct translocation and glycosaminoglycan-dependent endocytosis. Chem Commun 48(57):7179–7181. https://doi.org/10.1039/c2cc33240j

Chaloin L, Vidal P, Lory P, Mery J, Lautredou N, Divita G, Heitz F (1998) Design of carrier peptide-oligonucleotide conjugates with rapid membrane translocation and nuclear localization properties. Biochem Biophys Res Commun 243(2):601–608. https://doi.org/10.1006/bbrc.1997.8050

Chen JX, Xu XD, Chen WH, Zhang XZ (2014) Multi-functional envelope-type nanoparticles assembled from amphiphilic peptidic prodrug with improved anti-tumor activity. ACS Appl Mater Interfaces 6(1):593–598. https://doi.org/10.1021/am404680n

Copolovici DM, Langel K, Eriste E, Langel U (2014) Cell-penetrating peptides: design, synthesis, and applications.ACS. Nano 8(3):1972–1994. https://doi.org/10.1021/nn4057269

Crombez L, Aldrian-Herrada G, Konate K, Nguyen QN, McMaster GK, Brasseur R, Heitz F, Divita G (2009) A new potent secondary amphipathic cell-penetrating peptide for siRNA delivery into mammalian cells. Mol Ther 17(1):95–103. https://doi.org/10.1038/mt.2008.215

Cui H, Webber MJ, Stupp SI (2010) Self-assembly of peptide amphiphiles: from molecules to nanostructures to biomaterials. Biopolymers 94(1):1–18. https://doi.org/10.1002/bip.21328

Derossi D, Joliot AH, Chassaing G, Prochiantz A (1994) The third helix of the Antennapedia homeodomain translocates through biological membranes. J Biol Chem 269(14):10444–10450

Farkhani SM, Valizadeh A, Karami H, Mohammadi S, Sohrabi N, Badrzadeh F (2014) Cell penetrating peptides: efficient vectors for delivery of nanoparticles, nanocarriers, therapeutic and diagnostic molecules. Peptides 57:78–94. https://doi.org/10.1016/j.peptides.2014.04.015

Fonseca SB, Pereira MP, Kelley SO (2009) Recent advances in the use of cell-penetrating peptides for medical and biological applications. Adv Drug Deliv Rev 61(11):953–964. https://doi.org/10.1016/j.addr.2009.06.001

Frankel AD, Pabo CO (1988) Cellular uptake of the tat protein from human immunodeficiency virus. Cell 55(6):1189–1193

Futaki S, Suzuki T, Ohashi W, Yagami T, Tanaka S, Ueda K, Sugiura Y (2001) Arginine-rich peptides. An abundant source of membrane-permeable peptides having potential as carriers for intracellular protein delivery. J Biol Chem 276(8):5836–5840. https://doi.org/10.1074/jbc.M007540200

Green M, Loewenstein PM (1988) Autonomous functional domains of chemically synthesized human immunodeficiency virus tat trans-activator protein. Cell 55(6):1179–1188

Habibi N, Kamaly N, Memic A, Shafiee H (2016) Self-assembled peptide-based nanostructures: smart nanomaterials toward targeted drug delivery. Nano Today 11(1):41–60. https://doi.org/10.1016/j.nantod.2016.02.004

Holowka EP, Sun VZ, Kamei DT, Deming TJ (2007) Polyarginine segments in block copolypeptides drive both vesicular assembly and intracellular delivery. Nat Mater 6(1):52–57. https://doi.org/10.1038/nmat1794

Juliano RL, Alam R, Dixit V, Kang HM (2009) Cell-targeting and cell-penetrating peptides for delivery of therapeutic and imaging agents. Wiley Interdiscip Rev Nanomed Nanobiotechnol 1(3):324–335. https://doi.org/10.1002/wnan.4

Kanazawa T, Taki H, Tanaka K, Takashima Y, Okada H (2011) Cell-penetrating peptide-modified block copolymer micelles promote direct brain delivery via intranasal administration. Pharm Res 28(9):2130–2139. https://doi.org/10.1007/s11095-011-0440-7

Kolonin MG, Saha PK, Chan L, Pasqualini R, Arap W (2004) Reversal of obesity by targeted ablation of adipose tissue. Nat Med 10(6):625–632

Koren E, Torchilin VP (2012) Cell-penetrating peptides: breaking through to the other side. Trends Mol Med 18(7):385–393. https://doi.org/10.1016/j.molmed.2012.04.012

Lim YB, Lee E, Lee M (2007a) Cell-penetrating-peptide-coated nanoribbons for intracellular nanocarriers. Angew Chem 46(19):3475–3478. https://doi.org/10.1002/anie.200604576

Lim YB, Lee E, Lee M (2007b) Controlled bioactive nanostructures from self-assembly of peptide building blocks. Angew Chem 46(47):9011–9014. https://doi.org/10.1002/anie.200702732

Lim YB, Moon KS, Lee M (2009) Recent advances in functional supramolecular nanostructures assembled from bioactive building blocks. Chem Soc Rev 38(4):925–934. https://doi.org/10.1039/b809741k

Liu L, Xu K, Wang H, Tan PK, Fan W, Venkatraman SS, Li L, Yang YY (2009) Self-assembled cationic peptide nanoparticles as an efficient antimicrobial agent. Nat Nanotechnol 4(7):457–463. https://doi.org/10.1038/nnano.2009.153

Lock LL, Cheetham AG, Zhang P, Cui H (2013) Design and construction of supramolecular nanobeacons for enzyme detection. ACS Nano 7(6):4924–4932. https://doi.org/10.1021/nn400218a

Macewan SR, Chilkoti A (2012) Digital switching of local arginine density in a genetically encoded self-assembled polypeptide nanoparticle controls cellular uptake. Nano Lett 12(6):3322–3328. https://doi.org/10.1021/nl301529p

MacEwan SR, Chilkoti A (2013) Harnessing the power of cell-penetrating peptides: activatable carriers for targeting systemic delivery of cancer therapeutics and imaging agents. Wiley Interdiscip Rev Nanomed Nanobiotechnol 5(1):31–48. https://doi.org/10.1002/wnan.1197

MacEwan SR, Chilkoti A (2014) Controlled apoptosis by a thermally toggled nanoscale amplifier of cellular uptake. Nano Lett 14(4):2058–2064. https://doi.org/10.1021/nl5002313

Mae M, Langel U (2006) Cell-penetrating peptides as vectors for peptide, protein and oligonucleotide delivery. Curr Opin Pharmacol 6(5):509–514. https://doi.org/10.1016/j.coph.2006.04.004

Mendes AC, Baran ET, Reis RL, Azevedo HS (2013) Self-assembly in nature: using the principles of nature to create complex nanobiomaterials. Wiley Interdiscip Rev Nanomed Nanobiotechnol 5(6):582–612. https://doi.org/10.1002/wnan.1238

Milletti F (2012) Cell-penetrating peptides: classes, origin, and current landscape. Drug Discov Today 17(15-16):850–860. https://doi.org/10.1016/j.drudis.2012.03.002

Nakase I, Akita H, Kogure K, Graslund A, Langel U, Harashima H, Futaki S (2012) Efficient intracellular delivery of nucleic acid pharmaceuticals using cell-penetrating peptides. Acc Chem Res 45(7):1132–1139. https://doi.org/10.1021/ar200256e

Oehlke J, Scheller A, Wiesner B, Krause E, Beyermann M, Klauschenz E, Melzig M, Bienert M (1998) Cellular uptake of an alpha-helical amphipathic model peptide with the potential to deliver polar compounds into the cell interior non-endocytically. Biochim Biophys Acta 1414(1-2):127–139

Pooga M, Hallbrink M, Zorko M, Langel U (1998) Cell penetration by transportan. FASEB J 12(1):67–77

Raucher D, Ryu JS (2015) Cell-penetrating peptides: strategies for anticancer treatment. Trends Mol Med 21(9):560–570. https://doi.org/10.1016/j.molmed.2015.06.005

Reissmann S (2014) Cell penetration: scope and limitations by the application of cell-penetrating peptides. J Pept Sci 20(10):760–784. https://doi.org/10.1002/psc.2672

Saalik P, Elmquist A, Hansen M, Padari K, Saar K, Viht K, Langel U, Pooga M (2004) Protein cargo delivery properties of cell-penetrating peptides. A comparative study. Bioconjug Chem 15(6):1246–1253. https://doi.org/10.1021/bc049938y

Said Hassane F, Saleh AF, Abes R, Gait MJ, Lebleu B (2010) Cell penetrating peptides: overview and applications to the delivery of oligonucleotides. Cell Mol Life Sci 67(5):715–726. https://doi.org/10.1007/s00018-009-0186-0

Tian R, Wang H, Niu R, Ding D (2015) Drug delivery with nanospherical supramolecular cell penetrating peptide-taxol conjugates containing a high drug loading. J Colloid Interface Sci 453:15–20. https://doi.org/10.1016/j.jcis.2015.04.028

Toksöz S, Guler MO (2009) Self-assembled peptidic nanostructures. Nano Today 4(6):458–469

Torchilin VP (2014) Multifunctional, stimuli-sensitive nanoparticulate systems for drug delivery. Nat Rev

Drug Discov 13(11):813–827. https://doi.org/10.1038/nrd4333

Tu Y, Zhu L (2015) Enhancing cancer targeting and anticancer activity by a stimulus-sensitive multifunctional polymer-drug conjugate. J Control Release 212:94–102. https://doi.org/10.1016/j.jconrel.2015.06.024

Tünnemann G, Ter-Avetisyan G, Martin RM, Stöckl M, Herrmann A, Cardoso MC (2008) Live-cell analysis of cell penetration ability and toxicity of oligo-arginines. J Pept Sci 14(4):469–476

Wang HY, Chen JX, Sun YX, Deng JZ, Li C, Zhang XZ, Zhuo RX (2011) Construction of cell penetrating peptide vectors with N-terminal stearylated nuclear localization signal for targeted delivery of DNA into the cell nuclei. J Control Release 155(1):26–33. https://doi.org/10.1016/j.jconrel.2010.12.009

Wang F, Wang Y, Zhang X, Zhang W, Guo S, Jin F (2014) Recent progress of cell-penetrating peptides as new carriers for intracellular cargo delivery. J Control Release 174:126–136. https://doi.org/10.1016/j.jconrel.2013.11.020

Won YW, Adhikary PP, Lim KS, Kim HJ, Kim JK, Kim YH (2014) Oligopeptide complex for targeted non-viral gene delivery to adipocytes. Nat Mat 13(12):1157–1164. https://doi.org/10.1038/nmat4092

Yoon YR, Lim YB, Lee E, Lee M (2008) Self-assembly of a peptide rod-coil: a polyproline rod and a cell-penetrating peptide tat coil. Chem Commun 16:1892–1894. https://doi.org/10.1039/b719868j

Zhang P, Cheetham AG, Lin YA, Cui H (2013) Self-assembled Tat nanofibers as effective drug carrier and transporter. ACS Nano 7(7):5965–5977. https://doi.org/10.1021/nn401667z

Zhang Q, Gao H, He Q (2015) Taming cell penetrating peptides: never too old to teach old dogs new tricks. Mol Pharm 12(9):3105–3118. https://doi.org/10.1021/acs.molpharmaceut.5b00428

Zhao XL, Chen BC, Han JC, Wei L, Pan XB (2015) Delivery of cell-penetrating peptide-peptide nucleic acid conjugates by assembly on an oligonucleotide scaffold. Sci Rep 5:17640. https://doi.org/10.1038/srep17640

Zhu X, Shan W, Zhang P, Jin Y, Guan S, Fan T, Yang Y, Zhou Z, Huang Y (2014) Penetratin derivative-based nanocomplexes for enhanced intestinal insulin delivery. Mol Pharm 11(1):317–328. https://doi.org/10.1021/mp400493b

The Current Role of Cell-Penetrating Peptides in Cancer Therapy

13

Lucia Feni and Ines Neundorf

Abstract

Cell-penetrating peptides (CPPs) are a heterogeneous class of peptides with the ability to translocate across the plasma membrane and to carry attached cargos inside the cell. Two main entry pathways are discussed, as direct translocation and endocytosis, whereas the latter is often favored when bulky cargos are added to the CPP. Attachment to the CPP can be achieved by means of covalent coupling or non-covalent complex formation, depending on the chemical nature of the cargo. Owing to their striking abilities the further development and application of CPP-based delivery strategies has steadily emerged during the past years. However, one main pitfall when using CPPs is their non-selective uptake in nearly all types of cells. Thus, one particular interest lies in the design of targeting strategies that help to circumvent this drawback but still benefit from the potent delivery abilities of CPPs. The following review aims to summarize some of these very recent concepts and to highlight the current role of CPPs in cancer therapy.

Keywords

Cell-penetrating peptides • Peptide-drug conjugates • Drug delivery • Cancer therapy • Cancer targeting

13.1 Introduction

Cancer is still one of the most threatening diseases with the most cases of death worldwide. Thus, not only early diagnosis of cancer is of high importance, but also an effective therapeutic strategy. In most cases this is realized by means of surgical excision of the affected tissue, often combined with radiation therapy and chemotherapy. Several drugs are commonly used

L. Feni • I. Neundorf (✉)
Department of Chemistry, Biochemistry,
University of Cologne, Zuelpicherstr. 47a,
D-50674 Cologne, Germany
e-mail: ines.neundorf@uni-koeln.de

© Springer International Publishing AG 2017

A. Sunna et al. (eds.), *Peptides and Peptide-based Biomaterials and their Biomedical Applications*,
Advances in Experimental Medicine and Biology 1030, https://doi.org/10.1007/978-3-319-66095-0_13

for this purpose; however, upcoming resistances and the need for a more personalized treatment strategy still let researchers develop novel anti-cancer compounds. Although several of such new compounds show excellent activity profiles, their further development is hampered by an only poor pharmacokinetic profile. Usually this is based on their occasionally poor bioavailability due to limited cellular uptake. In fact, overcoming the cell's plasma membrane and reaching the intracellular target site resemble the major hurdles for an efficacious therapeutic agent. The plasma membrane surrounds all living cells and acts as a protective barrier that controls the in- and outflow of compounds within the environmental media. Drug molecules usually must overcome this barrier to reach their site of action. Once inside the cell, they act via different mechanisms, as e.g. the selective interruption or activation of a signal transduction pathway, or by direct interaction with the DNA in the nucleus. Beside the use of small molecule drugs, the current trend involves the use of macromolecules as anti-cancer agents, such as proteins, monoclonal antibodies, nucleic acids and nanoparticles, and the combination thereof. Despite the numerous advantages, the biodistribution and translocation of these hydrophilic macromolecular drugs is still a big challenge, prevented by the low permeability due to the intrinsic characteristics of biological membranes. In the course of time multiple approaches have been developed, such as the use of liposomes, microinjection, electroporation or also viruses and bacteria that are used particularly for gene transfer. Although through all these systems it is possible to import macromolecules into living cells, each of them presents a series of limitations that preclude their use in *in vivo* studies, in particular for possible therapeutic applications in clinic. The greatest obstacles can be summarized in their low internalization efficiency, complexity of manipulation, demand for expensive equipment, difficulty of release into the cytosol, and sometimes in cell toxicity and immunogenicity.

Recently, the direct transfer of macromolecules into cells has been obtained by so-called "cell-penetrating peptides" (CPP), a class of short peptides rich in basic amino acids, characterized by exceptional translocation properties across cell membranes (Kristensen et al. 2016; Bolhassani et al. 2017). Usually, CPPs consist of less than 35 amino acids, hold a positive net charge, and possess the ability to translocate across the plasma membrane. Thereby, CPPs can carry with them associated ligands, from small chemical molecules to nano-sized particles and large fragments of DNA, into the cell interior. The most studied CPP is the domain of the protein TAT (transactivating regulatory protein) of human immunodeficiency virus type 1 (HIV-1). The first evidence emerged in two articles published in the same issue of *Cell* in 1988 (Frankel and Pabo 1988; Green and Loewenstein 1988), which underlined the possibility for the protein TAT to enter mammalian cells when simply added to the culture medium. A domain between the amino acids 47 and 57, having the sequence YGRKKRRQRRR, is the region of the protein responsible for translocation. Some other examples found in nature, in addition to the TAT peptide, are presented by penetratin, a transcription factor from Drosophila (Derossi et al. 1994; Prochiantz 1996), VP22 from virus Herpes simplex (Elliott and O'Hare 1997) and pVEC, a peptide of 18 amino acids derived from the cadherin of murine vascular endothelium (Elmquist et al. 2001). It is interesting to note that these peptides have very different amino acid sequences and secondary structures while the mechanism of transfer within the cells seems to be similar. Trying to change these residues it was understood that the arginine (R) plays a fundamental role. Indeed same results in cell internalization were obtained with synthetic oligopeptides consisting in homoarginine, highlighting in particular maximum efficiency with R8/9 (Mitchell et al. 2000; Rothbard et al. 2000). Since the discovery of CPPs, intensive research has been carried out on the underlying entry mechanism in order to be able to totally exploit their transport properties but till now, in the literature, conflicting data are shown (Madani et al. 2011; Kauffman et al. 2015). It cannot be excluded that different internalization mechanisms are used concomitantly; furthermore, the permeation properties can vary in relation to the type of the associated cargo molecule and in respect to different cell types used. Despite this process of translocation remains unresolved, the effectiveness of the method is unequivocal, which promises to open new frontiers for research, holding great potential as *in vitro* and *in vivo* delivery vehicles.

However, although this strategy is very elegant and works well from a theoretical point of view, it has its difficulties and pitfalls in practice. First of all, it is important to choose a suitable conjugation method for each molecule that has to be carried by the CPP. Furthermore, the CPP to cargo ratio and the employment of peculiar linker systems are of relevance. Moreover, CPPs are normally not specific and they are consequently taken up in a variety of certain cells and tissues, leading to increased toxicity and side effects. That is why new strategies are now being developed in order to enhance their selectivity, and it is essential to understand which method is the best in every distinct context. Finally, there are different classes of cell-penetrating peptides, each of which works better for certain types of cells and it is important to choose the optimal sequence for the envisaged goals (Raucher and Ryu 2015). In particular, different sequences (exhibiting cationic or amphipathic characteristics) imply also distinct properties regarding internalization efficiency but also toxicity (Zaro 2015).

The community seems to evolve really vast, but still many details have to be taken in consideration for planning an optimal strategy when using CPPs. In this review, we summarize some of the innovative approaches that have been studied in order to go beyond the limitations with CPP application. In particular we focus on the selective delivery of drugs in the field of anticancer therapy, with specific emphasis on very recently published papers. We will first describe the different strategies that have been followed for synthesizing CPP-cargo conjugates, focusing our attention on covalent binding and fusion techniques. The cargoes described herein include small molecule drugs, peptides and proteins. Also non-covalent complexes and some applications for intracellular delivery of nanoparticles will be shortly depicted. In addition, we will cover strategies to obtain a targeted delivery towards cancer tissues. The last part of this review will deal with recently obtained advances of CPPs, to show how this new technique is actually reaching an increasing success in the treatment or diagnosis of cancer. Also drawbacks connected with the use of CPPs will be considered and discussed.

13.2 Different Strategies for the Synthesis of CPP-Cargo Constructs

Depending on the type of molecule that has to be transported inside the cell, there are various methods used allowing conjugation between the carrier moiety and the cargo of the drug delivery system. These different conjugation procedures involve a distinct synthetic pathway and may have impact on the route of administration, cell entry mechanisms, distribution inside the cell, and on other different effects on the cellular level. In addition, based on the therapeutic question and nature of the target, which the drug must act on, the choice of conjugation way plays a very important role. (Durzynska et al. 2015) This paragraph will contemplate in particular the formation of conjugates by covalent binding between CPPs and small molecule drugs, proteins and peptides. But also fusion techniques are described, which allow the synthesis of constructs including proteins. Other methods comprehending non-covalent complexes and nanoparticles constructs are finally shortly mentioned.

13.2.1 Covalent CPP-Cargo Constructs

Most CPP-cargo conjugates synthesized so far, particularly that one including small molecule drugs, are characterized by covalent bonding, including stable as well as cleavable linkages. However, this kind of conjugation method can be achieved by chemical reactions with or without the employment of such linker molecules. Thus, disulphide bond formation, thioether formation

Table 13.1 Examples of covalently connected CPP-cargo constructs. For more information refer to main text

Name	Sequence	Cargo	Connection	References
Oligoarginine	RRRRRRRRRRR	BSH	Disulfide bond	Michiue et al. (2014)
	RRRRRRRR	Taxol	Disulfide bond	Wender et al. (2012)
	RRRRRRRR	Doxorubicin	Thioether bond	Nakase et al. (2012)
	RRRRRRRR	Doxorubicin	Disulfide bond/ oxime linkage	Lelle et al. (2014)
	RRRRRRRR	Inhibitors of CyclinE/A-CDK	Amide bond	Dai et al. (2013)
	RRRRRRRRR	APT_{STAT3}	Peptide bond	Kim et al. (2014b)
Tumor homing CPP	RLYMRYYSPTTRRYG	Taxol	Ester bond	Tian et al. (2015)
Maurocalcine	GDCLPHLKLCKENKDCCSKKCKRRGTNIEKRCR	Pt chelator	Amide bond	Aroui et al. (2015)
sC18	GLRKFRLRKFRNKIKEK	Cymantrene	Amide bond	Splith et al. (2010a)
Low molecular weight protamine	VSRRRRRRGGRRRR	L-asparaginase	Disulfide bond	He et al. (2014)
Penetratin	CRQIKIWFQNRRMKWKK	KLA	Disulfide bond	Alves et al. (2014)
		LP4	Peptide bond	Prezma et al. (2013)
FHV	RRRRNRTRRNRRRVR	p53 C-terminal domain	Peptide bond	Ueda et al. (2012)
TAT	YARVRRRGPRR	PLHSpT	Peptide bond	Kim et al. (2014a)
		Peptides inhibiting autophosphorylation of EGFR	Peptide bond	Kuroda et al. (2013)
Transportan 10 (TP10)	AGYLLGKINLKALAALAKKIL	$SRC1_{LXXLL}$	Peptide bond	Tints et al. (2014)

or amide bonding were often utilized for connecting small molecule drugs to CPPs, (Regberg et al. 2012) but also for coupling proteins and peptides, peptide nucleic acids (PNAs) and morpholino oligonucleotides to CPPs (for examples see Table 13.1, Reissmann 2014).

Disulfide linkage is one of the most widely used methods for linking small molecule drugs to CPPs. For example, in the context of boron neutron capture therapy, mercaptoundecahydrododecaborate (BSH) was fused to the CPP R11 by a disulfide bond in order to allow cell penetration. The resultant conjugate was localized in the nuclei of glioma cells and showed a higher biological effect compared with the group treated with pure BSH, which stayed outside the cell. On the other hand, the compound could not be detected in the normal brain area (Michiue et al. 2014). In another study by Wender et al. R8 was conjugated to the drug taxol using again a disulfide linkage that is cleaved in the reducing environment of the cytosol, releasing there the free drug. This conjugate, in the treatment of human ovarian carcinoma, possessed comparable cytotoxicity and a better activity than the active drug alone, avoiding the efflux pump resistance (Wender et al. 2012). Also peptides were coupled via disulphide linkage to CPPs. For example He et al., coupled the CPP *"low molecular weight protamine"* to L-asparaginase by using the bifunctional cross-linker 3-(2-pyridyldithio)propionic acid N-hydroxysuccinimide (SPDP) leading to the formation of a disulfide bridge. This compound was then encapsulated into red blood cells for the treatment of acute lymphoblastic leukemia, interrupting L-asparagine supply in malignant cells. (He et al. 2014) Another example, in which the conjugation was done by formation of a disulfide bridge between two cystein residues incorporated in two peptidic moieties, was presented by Alves et al. To facilitate cellular uptake of the apoptotic peptide KLA it was conjugated to the CPP penetratin. The construct had a cytotoxic effect against cancer cell lines, including multidrug resistant cells, but not towards healthy ones, showing a selective effect *in vitro*, probably owing to differences in membrane composition. The mechanism of action of this conjugate was directed against mitochondria, in particular damaging their metabolic activity, and it could bypass the apoptosis resistance becoming an optimal alternative of synergistic strategy together with traditional chemotherapy. (Alves et al. 2014)

Nakase et al. used the CPP R8 again for conjugation with doxorubicin in order to prove the accumulation of the system in tumors. In this case the authors first prepared a doxorubicin-maleimide compound that was then coupled with R8 [(D-Arg)$_8$-Gly-Cys-amide], leading to the formation of a thioether bond. (Nakase et al. 2012)

The use of heterobifunctional crosslinkers was investigated by Lelle et al., who exploited a novel linking method containing a thiol and an aminooxy group. The CPPs used had two distinct sequences (R8 and a proline-rich amphipathic peptide), while the final conjugate contained both a cleavable disulfide bond that can be reduced inside the cell, and a stable oxime linkage to bind the drug doxorubicin. (Lelle et al. 2014)

Some other strategies like amide and ester coupling have also been applied. An example is the conjugation of taxol to a tumor homing CPP by a succinic acid linker, as described by Tian et al. Taxol has excellent self-assembly properties that permit the formation of a nanospherical construct. The taxol molecules are released by ester bond hydrolysis and can then exert the activity inside the cell. This system could be used for the co-delivery of other therapeutic molecules to cancer cells, like for example doxorubicin, in order to exhibit a synergistic effect (Tian et al. 2015).

During the last years there emerged a steady increasing interest in the use of metal-containing drug molecules. One of the most prominent examples is cisplatin that is frequently used as anti-cancer drug in combination therapy, and one of the best therapeutics against glioblastoma. Notably, cisplatin and related compounds are often characterized by their high toxicity and several side effects. Moreover, problems with solubility and water stability of such metal-containing compounds limit the application of also novel developed compounds that show interesting new activity spectra. These drawbacks can be significantly reduced by the attachment to a CPP, as it has been shown in the work by Aroui et al. They synthesized a novel platinum-maurocalcine conjugate by using an amide bond linkage. The conjugate showed a higher activity in U87 cells than cisplatin itself by targeting the intracellular redox system at lower doses and inhibiting the activity of ERK and AKT cascades. These additional activities could be essential in the treatment of resistant cancer cells. (Aroui et al. 2015) Recently, we synthesized several cymantrene-peptide conjugates by coupling functionalized cymantrene complexes via amide bonds to the CPP. (Neundorf et al. 2008; Splith et al. 2012) Cymantrenes are cyclopentadienyl manganese tricarbonyl metal complexes that exert cytostatic effects when in combination to CPPs. (Splith et al. 2010a; Hu et al. 2012) The CPP used in this study was the sC18 peptide, previously developed in our group (Neundorf et al. 2009). In some cases, a cathepsin B sensitive cleavage site was also introduced between the peptide and the metal complex. (Splith et al. 2010b) Cathepsin B is known to be over-expressed in several cancer cells, and thus, the selective release of the drug inside the cells can be enhanced. In fact, our study demonstrated high activity of these CPP-conjugates even again drug-resistant cancer cells. (Splith et al. 2010b) Amide bond formation was also used for the conjugation of the CPP R8 with novel inhibitors of cyclinE/A-CDK (cyclin-dependent kinases) in order to analyze the activity of the constructs against cancer cells. In fact, the conjugation via an amide bond of this potent inhibitor resulted in an accumulation of tumor suppressor p27, blockage of cell cycle progression and cell survival. Furthermore, the presence of the CPP allowed the inhibitor to easily permeate the cells reaching a promising activity in several tumor cell lines (Dai et al. 2013).

In another study presented by Ueda et al. in 2012, a multifunctional D-isomer peptide for the treatment of glioblastoma multiform (GBM) was designed. This was composed by a CPP (FHV, derived from flock house virus), a penetration accelerating sequence (Pas, derived from the retro sequence peptide of the cathepsin D cleavable sequence) and the C-terminus domain of p53, a biologically active tumor suppressor protein. The whole peptide was prepared by traditional solid phase peptide synthesis in line. As shown by *in vivo* studies it could inhibit the growth of GICs (glioma-initiating cells) and glioma cell lines with no effect on normal cells (Ueda et al. 2012). In the same year, p53 being frequently mutated in various human cancer types, was studied by Suhorutsenko et al., too. The research group synthesized short p53 protein analogues, starting from their C-terminal domain, and varying them by using CPP prediction algorithms. After modification with stearic acid to increase the transfection efficiency, they observed an increase in cellular uptake *in vitro* and certain selectivity in apoptotic activity against p53-mutant cells (Suhorutsenko et al. 2012). Not only had the protein p53 attracted the attention of scientists involved in the search for new possible appealing targets in tumor therapy. Plk 1 (Polo-like kinase 1), for example, plays key roles in regulating cell cycle events and is over-expressed in many cancer cell lines. For instance Plk 1 is essential during mitosis and in the maintenance of genomic stability. Its inhibition by specific phosphopeptide sequences has been proposed as an interesting strategy to inhibit tumor growth. In this framework, Kim et al. synthesized by solid phase peptide synthesis a new delivery system by conjugating PLHSpT, the minimal sequence necessary for the binding, to TAT peptide with the purpose of increasing the cell membrane penetration. *In vitro* studies showed inhibited cancer cell proliferation by blocking mitosis but also inducing apoptosis. (Kim et al. 2014) In addition to these examples, also the steroid receptor coactivator-1 (SRC-1) could be included in this group of proteins over-expressed in cancer, especially in breast cancer cells. This cofactor is characterized by the presence of a recognition motif LXXLL that is directly responsible for the binding to nuclear receptors. In this view, Tints et al. synthesized one of these sequences, able to bind estrogen receptors, and conjugated it to a cell-penetrating peptide transportan 10 (TP10), as an effective vehicle for the delivery of the active peptide to cellular targets. *In vitro* studies revealed high cytotoxicity in breast cancer cells with the induction of apoptosis. Importantly this effect was not affected by the estrogen receptors status, so that ER-negative breast cancer cells could be also treated by this strategy (Tints et al. 2014). With this in mind, many other representative peptides exhibiting selectivity to attractive tumor targets could be found; among all the cases, we mention STAT3-binding peptides (Kim et al. 2014b), the voltage-dependent anion channel 1 (VDAC1) – based peptides (Prezma et al. 2013) and oligopeptides inhibiting autophosphorylation of EGFR (Kuroda et al. 2013). All these active peptides against tumor disease have a hydrophilic sequence and cannot easily permeate the cell membrane. In these circumstances, a covalently conjugated cell-penetrating peptide represents an efficient carrier to enhance the cellular uptake and to obtain a potent antitumor activity. Within all of these studies dealing with cancer active peptides, the peptides were often simply attached to the CPP sequence by using solid phase peptide synthesis obtaining a final product without any inconvenient intermediate purification procedure.

One other important strategy for the delivery of proteins by the help of CPPs is the formation of fusion proteins generated by recombinant expression (see Table 13.2). As promising anticancer treatment, Yeh et al. reinvented the already known arginine depletion strategy trying to overcome the arginine deiminase (ADI) resistance in MDA-MB-231 cells. A pH-sensitive CPP-based fusion protein delivery system, which is able to carry ADI inside the cells, was constructed. The CPP HBHAc was incorporated with the pH-sensitive peptide HE and fused to ADI achieving tumor selective delivery in the mildly acidic tumor microenvironment of breast cancer cells (Yeh et al. 2016). In 2013, Lim et al. designed a new CPP starting from the sequence of the anticancer peptide, buforin IIb. This CPP, named BR2, can efficiently enter cancer cells by endocytosis thanks to the interaction with negatively charged gangliosides on the outer cell surface. Notably it shows no toxicity to normal cells. The ability of efficient drug delivery was proven by fusion to a single-chain variable fragment (scFv) antibody directed towards a mutated K-ras. The experiments were conducted with HCT116 cells causing a high level of apoptosis. This could be a useful and innovative drug delivery system with high selectivity toward cancer cells (Lim et al. 2013). Antibody targeting strategy and genetically engineered fusion technique were also employed by Shin et al., who proposed a new method to fight colorectal cancer by fusing to the sequence of TAT a molecule of so called gelonin, a very potent toxin that inhibits protein synthesis, but with an extremely poor cellular uptake. In order to obtain selectivity for this compound, a heparin conjugated anti-carcinoembryonic antigen (CEA) monoclonal antibody was associated *via* reversible electrostatic interaction. In this way, this CPP-fused chimeric protein was evaluated and showed a significant therapeutic efficacy against colorectal cancer therapy with a reduced toxicity to healthy tissues. (Shin et al. 2014a) Additionally, in a recent study by Orzechowska et al., cells were sensitized to cytotoxic drugs by delivery of the apoptotic protein BID (BH3-interacting domain death agonist) fused to the CPP TAT. This method gave good results in prostate and non-small human lung cancer cells providing a possible tool to improve the efficiency of therapeutic agents against this cancer cell types (Orzechowska et al. 2014).

Table 13.2 Examples of fusion proteins including a CPP. For more information refer to main text

CPP name	Sequence	Cargo	References
HBHAc	KKAAPAKKAAAKKAPAKKAAAKK	Arginine deiminase	Yeh et al. (2016)
BR2	RAGLQFPVGRLLRRLLR	scFv ab against mutated K-ras	Lim et al. (2013)
TAT	GRKKRRQRRRPQ	Gelonin toxin	Shin et al. (2014b)
		Apoptotic protein BID	Orzechowska et al. (2014)

13.2.2 Generation of Non-covalent CPP-Cargo Complexes

Covalent linking methods are sometimes limited by the concern that the synthetic covalent bond between CPP and the active moiety may alter the biological activity of the latter. This is the reason why many systems are often planned as non-covalent complexes, where all the entities are independent but at the same time connected to each other. Nowadays, this strategy is often performed with cell-penetrating peptides applied in gene therapy, e.g. for the delivery of genes, anti-sense oligodeoxynucleotides (ODNs), or small interfering RNA (siRNA) (Hoyer and Neundorf 2012; Margus et al. 2012). The negatively charged nucleic acids can in fact be easily complexed by electrostatic interaction with the often positively charged CPP, forming a stable complex. Frequently an excess of peptides is used that not only protects the nucleic acids from degradation but also helps to improve distribution, targeting and penetration of the nucleic acid in cells or tissues. Examples of recent works show how the systematically degradation of siRNA molecules can be avoided and their intracellular delivery promoted (Tanaka et al. 2013; Wang et al. 2014a; Golan et al. 2016). Non-covalent complexes have been also used for the delivery of small molecule drugs, even if the covalent conjugation is prevalently employed (see above). Li et al. described the formation of a complex between the active molecule doxorubicin and a particular CPP called CADY-1 that is a self-assembled peptide. This stable complex led to a longer blood residence time of the construct and better permeability of the drug with the subsequent improvement in therapeutic index (Li et al. 2012). Cyclic CPPs are known to be less susceptible to degradation and in a work by Mandal et al. cyclic cell-penetrating nuclear-targeting sequences were complexed with doxorubicin leading to efficient and targeted molecular transport (Mandal et al. 2011). Also we have recently investigated the impact of cyclization for the activity of CPPs. A shorter version of the CPP sC18 was cyclized using copper (I)-catalyzed alkyne-azide click reaction. The cyclized peptide exhibited increased proteolytic resistance and cytosolic cellular distribution. However, when complexed with plasmid DNA encoding for the enhanced green fluorescent protein (EGFP) the cyclized version demonstrated highly improved complexation and uptake of the plasmid in contrast to the linear CPP that was not able to transfect the used cancer cells at all (Horn et al. 2016). In another study, such cyclized CPPs containing a triazole were used to complex the drug daunorubicin in breast cancer MCF-7 cells. Also in this case the cyclization improved the transport efficiency of the herein used CPP (Reichart et al. 2016).

13.2.3 Generation of Multimodal Nanoparticles

Nanoparticles (NPs) are increasingly being studied as multimodal platforms for the grafting of bioactive molecules useful as diagnostic tools, or for therapeutic treatment. To get a deeper view into this emerging field and the use of NPs in anti-cancer therapy and diagnosis, the readers should refer to these excellent recent reviews (Sun et al. 2014; Bazak et al. 2015; Xu et al. 2015; Ma et al. 2016). However, the inability to pass through the lipid membranes of cells greatly limits their *in vitro* and *in vivo* use. To bypass this pitfall, CPPs can be set on their surface to facilitate and accelerate the cellular uptake, and to reduce possible cytotoxic effects. Moreover, owing to their size several other ligands or functionalities can be fixed on the same nanoparticle. Vehicles as polymeric nanoparticles or liposomes are typically used to develop a controlled release system. This approach can be used to improve the distribution, the absorption and the targeting of molecules, which otherwise would be quickly eliminated or would not be able to reach the target tissue (Koren and Torchilin 2012).

Among the simplest polymeric systems on the market, the classic example is the combination of liposomes with doxorubicin. Since the application of doxorubicin as anti-cancer drug may cause cardiac toxicity problems, researchers try to find solutions to circumvent these side effects.

By releasing the drug more slowly, a lower dose can be used and global toxicity can be reduced. These systems, however, after a certain time become ineffective because doxorubicin is a substrate for Pgp (glycoprotein P, an efflux pump) and the affected cells become resistant (Kopecka et al. 2014). Beside the addition of Pgp inhibitors, (Gao et al. 2014) the use of CPPs can be essential to overcome this resistance. In this context, mesoporous silica nanoparticles derivatized with an activatable CPP polyarginine and doxorubicin were synthesized by Liu et al. and the activity was efficiently demonstrated *in vivo* proving no side effects and tumor growth inhibition (Liu et al. 2015). In another study conducted by Wang et al. against multi drug resistance, *low molecular weight protamine* was used as CPP connected with poly(lactic-co-glycolic acid) (PLGA) nanoparticles additionally loaded with doxorubicin. The presented data suggest that this system could actually act against upcoming resistance by various mechanisms, like enhanced cellular uptake, accumulation in nuclei and diminished efflux (Wang et al. 2014c). *Low molecular weight protamine* with MMP2 cleavage site was also connected to paclitaxel-loaded PEG-co-PCL nanoparticles for targeted glioblastoma therapy inducing enhanced selectivity, cytotoxicity and cellular uptake in C6 glioma cells (Gu et al. 2013).

13.3 Different Strategies for Targeted CPP-Cargo Delivery

One of the main problems when using CPPs is their lacking target specificity. Avoiding unspecific uptake is mandatory to limit and exclude loading of healthy cells with CPP-drug conjugates. During the last years, different strategies have been described to obtain more selective CPPs in order to circumvent pathological changes in particular tissues provoked by the unspecific distribution of CPP-cargo conjugates. To circumvent such unspecific CPP uptake, also masking of the positive charges of the CPPs might be necessary and can be realized by the formation of so called activatable cell-penetrating peptides (ACPPs). Here, the CPPs are often fused to a polyanionic sequence, pH-sensitive polyethylene glycol (PEG) chains, or proteins.

13.3.1 Active Delivery Strategies

Changes in the local environment typically seen in cancer tissues can be used to actively deliver CPPs to the tumor tissue avoiding cellular uptake to normal cells. The following conditions may count to this, as reduced pH, presence of overexpressed metalloproteinases, and the accumulation of particular receptors on cell surfaces. However, also external triggers like heat, ultrasound and magnetic field can be used for a targeted drug uptake. Systems that combine both concepts of active and passive addressing of tumors are more and more popular, and in the following paragraphs some of these methods are taken into account.

A possible strategy that is often followed by researchers in order to obtain a selective CPP that can target tumor tissue without involving normal cells is the insertion of particular cleavage sites to the sequence of the CPP. These can be cleaved e.g. by metalloproteinases like MMP-2/-9, which play an important role in angiogenesis and metastasis of tumors, and are frequently over-expressed in cancer tissues. Another possibility is to make advantage of pH change in cancer tissue, which is normally characterized by mildly acidic conditions, differently from the neutral pH of the rest of the cells. For instance, an MMP-2 cleavage site was introduced by Li et al. between a CPP and a polyanionic peptide in order to block the penetration in normal tissue building an activatable pro-form. The ACPP was conjugated to protoporphyrin IX, a light-sensitive molecule, therefore utilized as therapy against different forms of cancer by photodynamic therapy. After cleavage and activation of the CPP in cancer tissue, this photosensitizer could be introduced inside the cells generating by irradiation reactive oxygen species. Tumor size was decreased without any systemic toxicity (Li et al. 2015a). Many other examples of similar ACPPs, containing a metalloproteinase cleavage site, describe the conjugation to different cargoes like methotrexate (Mae et al. 2012) and hTERT siRNA (Li et al. 2015b). Moreover, this method can also be used in order to control drug delivery and precisely track drug release in living cells. For example, Cheng et al. designed a novel drug delivery system made of three different components, in particular a fluorophor, a functionalized CPP with a cleavable site for metalloproteinase MMP-2 and the active drug doxorubicin. In the cancer tissue the structure is cleaved and the drug can easily pass through the cell membrane thanks to the activity of the CPP. Meanwhile the fluorophore will self-aggregate because of hydrophobic interactions, and turn on yellow fluorescence. By means of that, they could observe real time *in vivo* delivery of the drug (Cheng et al. 2016). Similarly, Savarian et al. projected an ACPP in order to evaluate the presence of metastases by means of Cy5 that is quenched by Cy7 till the linker between the two fluorophores is cut by MMP-2 and -9 in tumor tissues. The fluorescence emission is increased and the presence of the tumor and corresponding metastases can be easily detected. (Savariar et al. 2013) The same ratiometric activatable CPP system was also used by Hauff et al. in order to improve tumor identification (Hauff et al. 2014).

Not only metalloproteinases can be involved in the selective cleavage of CPPs in cancer tissue as was shown in the work by Liu et al., where TAT-liposomes loaded with doxorubicin were activated by the endoprotease legumain. This is a lysosomal cystein protein, whose expression directly corresponds to the malignancy of the tumor itself. Furthermore, the legumain, normally present in the cellular plasma, moves to the cell surface if conditions like starvation or hypoxia occur. The CPP TAT loses some of the permeation ability when conjugated to the legumain cleavage site, but this capacity is restored when in contact with this enzyme so that the CPP can enter efficiently and selectively in tumor cells but not in normal cells (Liu et al. 2014a).

As already mentioned, also the difference in environmental pH between normal and tumor tissue can be exploited to favor the selective therapeutic delivery. In fact, the tumor tissue is characterized by a slightly lower pH and many acid-labile systems have been designed in the last years (MacEwan and Chilkoti 2013). In this work by Fei et al., a $(HE)_{10}$ peptide was combined to the CPP MAP to mask the positive charges till reaching the tumor tissue. At this point, the lower pH would in fact protonate the histidine residues and allows the CPP to express its positive charges and the ability to pass over the cell membrane (Fei et al. 2014).

Another approach is to combine CPPs with receptor targeting moieties. Indeed, a great number of receptors are over-expressed in tumor cells, and can be targeted with extremely diverse ligands. Regarding their size as well as chemical structure these molecules are characterized by a very high heterogeneity. For instance, drug conjugates of the glycoprotein transferrin enable an efficient accumulation of drugs in cancer cells. Li et al. synthesized lipid nanoparticles for the delivery of siRNA loaded with the CPP R8 and the targeting ligand transferrin (Li et al. 2016), showing excellent gene silencing activity *in vitro* and *in vivo*. Transferrin can also be targeted by receptor-targeting sequences directly located within the CPP sequence, as shown by Youn et al. in a work about neuro-targeted siRNA delivery (Youn et al. 2014). Folic acid receptor is overexpressed in a variety of malignant cells. Vitamin folic acid can bind to this receptor with a high affinity and thus makes it an attractive target for the targeted drug delivery in tumors. Gao et al. used a combination of folate targeting and tumor microenvironment-sensitive polypeptides (with the presence of metalloproteinases cleavage sites) to deliver docetaxel loaded nanoparticles to tumor cells. The enhanced cellular uptake was caused by both the folate receptor and MMP-2 over-expression in tumor tissue (Gao et al. 2013).

On the other hand, another possible approach could be to address tumoral endothelial cells by targeting receptors, such as integrins. Integrins are one of the major families of transmembrane cell adhesion receptors; they are overexpressed on the surface of cancer cells and they are involved in tumor angiogenesis, progression and metastasis. In particular, they can be selectively targeted with cyclic and linear RGD peptide sequences. By inhibiting the endothelial cell proliferation, vital nutrients and oxygen will be no longer adequately provided from the tumor blood vessel system. In this way, the tumor growth and the formation of metastases would be suppressed. Many researchers are working in this field, and the interest in synthesizing different variables of RGD increased more and more since the development of *cilengitide* by Kessler et al. that failed in the phase III in the clinical development (Mas-Moruno et al. 2010). Chen et al. developed albumin-based nanoparticles composed of a CPP moiety, a targeting moiety (cRGD) and the active drug doxorubicin with pH dependent self-assembly behavior. After coming in contact with the endosomal environment the drug could be easily released and accumulated inside the nuclei (Chen et al. 2015). Liposomes were also loaded with paclitaxel and selectively targeted to tumor cells by a multifunctional CPP with a targeting moiety cyclic RGD showing selectivity to integrin receptors. This strategy was then applied in glioma cells inducing the strongest inhibition and apoptosis (Liu et al. 2014b). The targeting of integrin $\alpha_v\beta_3$ by the ligand cRGD was connected by Crisp et al. to the use of MMP-2 cleavage site in the CPP sequence to deliver the chemotherapeutic monomethylauristatin E (Crisp et al. 2014). Moreover, a dual targeting strategy was used to deliver paclitaxel loaded liposomes. The two targeting ligands were selective towards integrin and neuropilin I receptors, having a synergistic action and increasing the selectivity for glioma cells (Liu et al. 2016).

NGR was also used as a ligand for the delivery of doxorubicin by thermosensitive pegylated liposomes. In this work the drug delivery system was selectively targeted to tumor cells by the double action of NGR and thermosensitive liposomes that hinder the action of the CPPs till reaching the tumor tissue, where the temperature is a bit higher (Yang et al. 2014).

As another example of targeting ligands, a breast tumor homing cell-penetrating peptide was used for the selective delivery of the drug (-)-epigallocatechin-3-gallate. Silica nanoparticles were used as vectors in this case (Ding et al. 2015).

Since already used in prostate cancer detection, the two receptor prostate-specific antigen (PSA) and prostate specific membrane antigen (PSMA) can be utilized in the targeted delivery of anticancer drugs, in this case for the delivery of siRNA. Xiang et al. designed liposomes exposing an activatable CPP, which can be activated by PSA cleavage in the tumor tissue, and a folate moiety, selective to PSMA receptors. When the folate moiety attracts the liposome to the cancer cell surface, the CPP is activated and the system is taken up (Xiang et al. 2013).

During the last years, photosensitive approaches have gained increasing interest among researchers. In particular Yang et al. published some interesting work about this subject. siRNA molecules were delivered by cationic liposomes bearing an NGR peptide as targeting ligand and a CPP, shielded by photolabile groups able to neutralize its positive charges. In this study, they used NIR illumination, because of its characteristics of deep tissue penetration and being less harmful to cells. After light treatment, the CPP was exposed and its functionality restored allowing cellular penetration (Yang et al. 2015). The year after, the same group published another work in the same direction but this time adding a pH-responsive polypeptide. The deshielding of the CPP occurred in this case through the double action of intrinsic lowered pH in tumor tissue and external NIR illumination (Yang et al. 2016a). The same research group synthesized thermal and magnetic dual-responsive liposomes for siRNA delivery, too. In particular, magnetic fluid Fe_3O_4 was combined with thermosensitive lipids; the liposomes would accumulate at the tumor site by a magnetic force, replaced then by an alternating current magnetic field that induced the iron nanoparticles to produce heat. By this heat the thermosensitive lipids could undergo to a gel to liquid phase transition and the CPP-siRNA conjugate could pass inside the cell. Arriving in the cytosol the disulfide bond between CPP and nucleic acid would be reduced and the siRNA could silence the corresponding mRNA in the cytosol (Yang et al. 2016b). Hyperthermia was also employed by Ryu et al. in an experiment consisting in the synthesis of a construct based on an elastin-like polypeptide (ELP), a CPP named Bac and the C-terminal domain of the p21 peptide. Upon external application of localized mild hyperthermia, the ELP aggregates and accumulates in tumor tissue (Ryu and Raucher 2014).

13.3.2 Taking Advantage of the Enhanced Permeability and Retention Effect

Typically, tumors are characterized by strongly increased angiogenesis, which means that the new formation of blood vessels around the malignant tissue increases in order to satisfy the enhanced nutrient requirement of the cancer cells. However, these vascular systems significantly differ from the healthy one since the endothelium has a series of defects that make it permeable. Furthermore, the pressure of the interstitial tissue fluid of tumors is increased, with the result that the efficiency of small drug molecules is dramatically reduced, since these are easily eliminated. Nevertheless, precisely these two factors provide the solution approach for a possible targeted addressing of tumors. In fact, the principle of passive addressing of tumors with nanotherapeutics that are able to passively and selectively accumulate in the permeable tumor tissue has become the gold standard in our time and is often used with the term established by Matsumura and Maeda *"enhanced permeability and retention (EPR)" effect* (Matsumura and Maeda 1986).

For instance, PEG has the ability to shelter complexes that consist of CPPs and active molecules until they reach the tumor tissue, thereby prolonging their half-life in the blood circulation and avoiding unwanted metabolism. In addition, PEG induced steric hindrance may favor the accumulation in tumor tissue by the above-mentioned EPR effect. In a work by Veiman et al. the MMP cleavage site was introduced between a CPP molecule, complexed with a plasmid DNA, and a PEG molecule. As soon as the nanoparticles passively accumulated in the tumor environment through the EPR effect, the metalloproteases would cleave the substrate and the PEG would finally allow the CPP to come in contact with the cell surface. Then the CPP would penetrate inside, while transporting the gene inside the cell interior (Veiman et al. 2015). Recently, Wang et al. followed the same strategy for the delivery of siRNA targeting Plk1 (polo-like kinase 1) mRNA (Wang et al. 2014b). Also the work of Zhu et al. went in the same direction: here the nanoparticles were composed by self-assembling PEG and the active drug was paclitaxel, but the high tumor accumulation of the system was, here again, the result of the combined EPR effect with the up-regulated MMP-2 in the tumor (Zhu et al. 2013). Koren et al., on the contrary, investigated PEGylated liposomes containing doxorubicin, TAT and a targeting ligand mAb. The carried PEG molecules are characterized by different lengths and were conjugated by a pH-sensitive hydrazone bond. When liposomes accumulated in tumor tissue by the EPR effect and mAb active targeting, the mildly acidic environment led to the cleavage of the degradable bond and exposition of the CPP to the cell surface, enhancing cellular uptake of the small molecule drug (Koren et al. 2012).

In summary, all these examples show that particularly during the last years a multimodal approach is more and more used. Combination of a delivery unit, with a cytotoxic payload and a targeting sequence may be the right way to create a successful and efficient therapeutic strategy.

13.4 Concluding Remarks

One can observe that many recent publications are including *in vivo* experiments with CPPs, both for imaging or therapeutic applications. This fact highlights the increasing interest in developing peptide-based delivery vectors. Anyway, still many problems are connected with this kind of strategy. It is important to understand all the disadvantages and side effects in order to overcome them and synthesize a new effective drug delivery system that is active also *in vivo*. In fact, although many CPPs are being tested, only one CPP, called p28, has reached phase I in clinical trial in the context of cancer treatment, in particular against solid tumors expressing p53, as well as for CNS malignancies (Warso et al. 2013).

Problems connected with the use of CPPs include upcoming immunogenicity, since the sequences are novel to the organism to which they are being administered, as well as cytotoxicity, caused by the perturbation of plasma membrane dynamics (Dinca et al. 2016). Both side effects are deeply related to the particular sequences; for this reason, one cannot talk about a general problem of this class of peptides, since each CPP is defined by a distinct amino acid composition. Many studies have been recently done to establish a possible action on the immune system and a consequent immunological reaction in the organism, but no general immune response was observed. About toxicity one can state that in general cationic CPPs are less toxic than amphipathic CPPs, even if *in vivo* studies show positive results about the safety at the employed doses. Nevertheless, it is important to always analyze both immunogenicity and toxicity since, as already mentioned, every CPP is different from the others and their effect can also change in the presence of a cargo.

Another issue is the lack of selectivity, as already described in the previous paragraphs. Many are the strategies adopted, but sometimes the penetrating activity of the CPP is so strong that the targeting ability of the specific targeting moieties used can be completely hidden, and no positive result in selectivity to cancer tissue would be gained. (Shin et al. 2014a) Together with the use of targeting molecules, the ability of CPPs should be shielded by using pro-drug strategies based on electrostatic interactions or PEG systems that could block penetration by steric hindrance (Huang et al. 2013).

Blood stability is also a very important attribute that a drug should possess in order to reach the target without being destroyed by blood proteases before arriving to the tissue (Rizzuti et al. 2015). This obstacle can be circumvented applying different shielding strategies in order to protect the structure of the CPPs till reaching the desired tissue, utilizing for instance more stable D-amino acid configurations, (Jarver et al. 2010) backbone cyclization (Shirazi et al. 2013; Horn et al. 2016; Reichart et al. 2016) or, finally, backbone stabilization through β- and γ-peptoids inside the sequence. (Kristensen et al. 2016) In a recent study, Shen et al. tried to overcome all the problems connected with the systemic administration of CPPs, proposing a more favorable cell-based platform for local peptide or protein production within the target tissue. To improve the intercellular transport, they designed a new CPP based on a triple repeat of modified TAT and a secretory signal peptide, with improved transduction activity and secretion efficacy. This is still a developing method, but it can lead to a big improvement in cell-based delivery of CPPs precluding degradation by proteases in the blood, metabolism and too early excretion (Shen et al. 2014).

Endosomal escape is also a decisive concern, since for many CPPs the main entry pathways proceeds via endocytotic mechanisms. In fact, the CPP construct must be taken up by cells, but more importantly, cargoes have to be released and to reach their extra-endosomal targets in the cytosol or in the nucleus. This could be made by using endosomolytic sequences or fusogenic compounds (Farkhani et al. 2014; Reissmann 2014; Zaro and Shen 2015). It is also true that, even if the majority of the active molecules remain inside the lysosomes, the small quantity that succeeds in escaping and gets to the cytosol can be sufficient for the biological activity (Skotland et al. 2015).

Specific attention should be also paid on small molecule drug delivery with the aim to overcome the problem of multidrug resistance (MDR) that develops in tumor cells after repeated exposures to the same drugs. MDR can often be attributed to the up-regulation of efflux pumps, particularly active with lipophilic drugs inserted in the membrane. CPPs could help in this sense by changing the solubility properties of the drugs, favoring their entrance in the cytosol and releasing them by different mechanisms (Vargas et al. 2014; Zheng et al. 2010; Regberg et al. 2012). Such CPP-conjugates were also designed to promote some particular routes of administration. In the context of small molecule drug delivery, docetaxel cyclodextrin inclusion-loaded PLGA nanoparticles were administrated with CPPs to enhance its oral bioavailability. Bu et al. demonstrated how this system displayed the maximal cytotoxicity against breast cancer MCF-7 cells, enhancing absorption and bioavailability of the drug itself as a promising oral delivery carrier (Bu et al. 2015). Oral administration by new formulation approaches has been also studied in recent works even if this field is still growing and many more experiments have to be done (Khafagy el and Morishita 2012) (Kristensen and Nielsen 2016). Transdermal delivery capability could be also enhanced by cell-penetrating peptides. The pro-apoptotic peptide KLA, for example, was delivered by Gautam et al. to different cancer cells *in vitro* and into the skin *in vivo* by a new CPP IMT-P8. After internalization, the construct, containing a compartment-specific localization sequence, could localize to mitochondria causing cell death thanks to the peptide ability of disrupting the mitochondrial membrane. These results suggest that this could actually be used as topical delivery vehicle in dermal diseases (Gautam et al. 2016).

In general, cell-penetrating peptides offer a number of distinctive merits and can be involved in many strategies for the delivery of anticancer active drugs, the latter being easily conjugated in many different ways *via* either chemical or genetic engineering method without affecting their intrinsic activity. Furthermore, they can efficiently transport attached cargos into almost all types of cells and this property makes cell-penetrating peptides a very eclectic element in the design of new drug delivery systems. CPPs will be used to revolutionize drug design and development by providing better bioavailability of the traditional chemotherapeutics at a much earlier stage of drug development, facilitating effective transition from preclinical to clinical phase in drug development. The research is still increasing and many pitfalls are being overcome and solved by new strategies; so, very soon conjugation of old and new active drugs to cell-penetrating peptides will become a more widely established clinical modality for the treatment of those malignancies for which there currently are no good treatment options.

References

Alves ID, Carre M et al (2014) A proapoptotic peptide conjugated to penetratin selectively inhibits tumor cell growth. Biochim Biophys Acta 1838(8):2087–2098

Aroui S, Dardevet L et al (2015) A novel platinum-maurocalcine conjugate induces apoptosis of human glioblastoma cells by acting through the ROS-ERK/AKT-p53 pathway. Mol Pharm 12(12):4336–4348

Bazak R, Houri M et al (2015) Cancer active targeting by nanoparticles: a comprehensive review of literature. J Cancer Res Clin Oncol 141(5):769–784

Bolhassani A, Jafarzade BS et al (2017) *In vitro* and *in vivo* delivery of therapeutic proteins using cell penetrating peptides. Peptides 87:50–63

Bu X, Zhu T et al (2015) Co-administration with cell penetrating peptide enhances the oral bioavailability of docetaxel-loaded nanoparticles. Drug Dev Ind Pharm 41(5):764–771

Chen B, He XY et al (2015) Dual-peptide-functionalized albumin-based nanoparticles with pH-dependent self-assembly behavior for drug delivery. ACS Appl Mater Interfaces 7(28):15148–15153

Cheng Y, Huang F et al (2016) Protease-responsive prodrug with aggregation-induced emission probe for controlled drug delivery and drug release tracking in living cells. Anal Chem 88(17):8913–8919

Crisp JL, Savariar EN et al (2014) Dual targeting of integrin alpha(v)beta(3) and matrix metalloproteinase-2 for optical imaging of tumors and chemotherapeutic delivery. Mol Cancer Ther 13(6):1514–1525

Dai L, Liu Y et al (2013) A novel cyclinE/cyclinA-CDK inhibitor targets p27(Kip1) degradation, cell cycle progression and cell survival: implications in cancer therapy. Cancer Lett 333(1):103–112

Derossi D, Joliot AH et al (1994) The third helix of the Antennapedia homeodomain translocates through biological membranes. J Biol Chem 269(14):10444–10450

Dinca A, Chien WM et al (2016) Intracellular delivery of proteins with cell-penetrating peptides for therapeutic uses in human disease. Int J Mol Sci 17(2):263

Ding J, Yao J et al (2015) Tumor-homing cell-penetrating peptide linked to colloidal mesoporous silica encapsulated (−)-epigallocatechin-3-gallate as drug delivery system for breast cancer therapy *in vivo*. ACS Appl Mater Interfaces 7(32):18145–18155

Durzynska J, Przysiecka L et al (2015) Viral and other cell-penetrating peptides as vectors of therapeutic agents in medicine. J Pharmacol Exp Ther 354(1):32–42

Elliott G, O'Hare P (1997) Intercellular trafficking and protein delivery by a herpesvirus structural protein. Cell 88(2):223–233

Elmquist A, Lindgren M et al (2001) VE-cadherin-derived cell-penetrating peptide, pVEC, with carrier functions. Exp Cell Res 269(2):237–244

Farkhani SM, Valizadeh A et al (2014) Cell penetrating peptides: efficient vectors for delivery of nanoparticles, nanocarriers, therapeutic and diagnostic molecules. Peptides 57:78–94

Fei L, Yap LP et al (2014) Tumor targeting of a cell penetrating peptide by fusing with a pH-sensitive histidine-glutamate co-oligopeptide. Biomaterials 35(13):4082–4087

Frankel AD, Pabo CO (1988) Cellular uptake of the tat protein from human immunodeficiency virus. Cell 55(6):1189–1193

Gao W, Xiang B et al (2013) Chemotherapeutic drug delivery to cancer cells using a combination of folate targeting and tumor microenvironment-sensitive polypeptides. Biomaterials 34(16):4137–4149

Gao W, Lin Z et al (2014) The co-delivery of a low-dose P-glycoprotein inhibitor with doxorubicin sterically stabilized liposomes against breast cancer with low P-glycoprotein expression. Int J Nanomedicine 9:3425–3437

Gautam A, Nanda JS et al (2016) Topical delivery of protein and peptide using novel cell penetrating peptide IMT-P8. Sci Rep 6

Golan M, Feinshtein V et al. (2016) Conjugates of HA2 with octaarginine-grafted HPMA copolymer offer effective siRNA delivery and gene silencing in cancer cells. Eur J Pharm Biopharm

Green M, Loewenstein PM (1988) Autonomous functional domains of chemically synthesized human immunodeficiency virus tat trans-activator protein. Cell 55(6):1179–1188

Gu GZ, Xia HM et al (2013) PEG-co-PCL nanoparticles modified with MMP-2/9 activatable low molecular weight protamine for enhanced targeted glioblastoma therapy. Biomaterials 34(1):196–208

Hauff SJ, Raju SC et al (2014) Matrix-metalloproteinases in head and neck carcinoma-cancer genome atlas analysis and fluorescence imaging in mice. Otolaryngol Head Neck Surg 151(4):612–618

He H, Ye J et al (2014) Cell-penetrating peptides meditated encapsulation of protein therapeutics into intact red blood cells and its application. J Control Release 176:123–132

Horn M, Reichart F et al (2016) Tuning the properties of a novel short cell-penetrating peptide by intramolecular cyclization with a triazole bridge. Chem Commun 52(11):2261–2264

Hoyer J, Neundorf I (2012) Peptide vectors for the nonviral delivery of nucleic acids. Acc Chem Res 45(7):1048–1056

Hu W, Splith K et al (2012) Influence of the metal center and linker on the intracellular distribution and biological activity of organometal-peptide conjugates. J Biol Inorg Chem 17(2):175–185

Huang Y, Jiang Y et al (2013) Curb challenges of the "Trojan Horse" approach: smart strategies in achieving effective yet safe cell-penetrating peptide-based drug delivery. Adv Drug Deliv Rev 65(10):1299–1315

Jarver P, Mager I et al (2010) In vivo biodistribution and efficacy of peptide mediated delivery. Trends Pharmacol Sci 31(11):528–535

Kauffman WB, Fuselier T et al (2015) Mechanism matters: a taxonomy of cell penetrating peptides. Trends Biochem Sci 40(12):749–764

Khafagy el S, Morishita M (2012) Oral biodrug delivery using cell-penetrating peptide. Adv Drug Deliv Rev 64(6):531–539

Kim SM, Chae MK et al (2014a) Enhanced cellular uptake of a TAT-conjugated peptide inhibitor targeting the polo-box domain of polo-like kinase 1. Amino Acids 46(11):2595–2603

Kim D, Lee IH et al (2014b) A specific STAT3-binding peptide exerts antiproliferative effects and antitumor activity by inhibiting STAT3 phosphorylation and signaling. Cancer Res 74(8):2144–2151

Kopecka J, Salzano G et al (2014) Insights in the chemical components of liposomes responsible for P-glycoprotein inhibition. Nanomedicine 10(1):77–87

Koren E, Torchilin VP (2012) Cell-penetrating peptides: breaking through to the other side. Trends Mol Med 18(7):385–393

Koren E, Apte A et al (2012) Multifunctional PEGylated 2C5-immunoliposomes containing pH-sensitive bonds and TAT peptide for enhanced tumor cell internalization and cytotoxicity. J Control Release 160(2):264–273

Kristensen M, Nielsen HM (2016) Cell-penetrating peptides as carriers for oral delivery of biopharmaceuticals. Basic Clin Pharmacol Toxicol 118(2):99–106

Kristensen M, Birch D et al (2016) Applications and challenges for use of cell-penetrating peptides as delivery vectors for peptide and protein cargos. Int J Mol Sci 17(2):185

Kuroda Y, Kato-Kogoe N et al (2013) Oligopeptides derived from autophosphorylation sites of EGF receptor suppress EGF-stimulated responses in human lung carcinoma A549 cells. Eur J Pharmacol 698(1-3):87–94

Lelle M, Frick SU et al (2014) Novel cleavable cell-penetrating peptide-drug conjugates: synthesis and characterization. J Pept Sci 20(5):323–333

Li Y, Zheng X et al (2012) Self-assembled peptide (CADY-1) improved the clinical application of doxorubicin. Int J Pharm 434(1–2):209–214

Li SY, Cheng H et al (2015a) Protease-Activable cell-penetrating peptide-protoporphyrin conjugate for targeted photodynamic therapy in vivo. ACS Appl Mater Interfaces 7(51):28319–28329

Li H, He J et al. (2015b) siRNA suppression of hTERT using activatable cell-penetrating peptides in hepatoma cells. Biosci rep 35(2)

Li Y, Lee RJ et al (2016) Delivery of siRNA using lipid nanoparticles modified with cell penetrating peptide. ACS Appl Mater Interfaces 8(40):26613–26621

Lim KJ, Sung BH et al (2013) A cancer specific cell-penetrating peptide, BR2, for the efficient delivery of an scFv into cancer cells. PLoS One 8(6):e66084

Liu Z, Xiong M et al (2014a) Legumain protease-activated TAT-liposome cargo for targeting tumours and their microenvironment. Nat Commun 5:4280

Liu Y, Ran R et al (2014b) Paclitaxel loaded liposomes decorated with a multifunctional tandem peptide for glioma targeting. Biomaterials 35(17):4835–4847
Liu JJ, Zhang BL et al (2015) Enzyme responsive mesoporous silica nanoparticles for targeted tumor therapy in vitro and in vivo. Nanoscale 7(8):3614–3626
Liu Y, Mei L et al (2016) Dual receptor recognizing cell penetrating peptide for selective targeting, efficient Intratumoral diffusion and synthesized anti-glioma therapy. Theranostics 6(2):177–191
Ma Y, Huang J et al (2016) Cancer-targeted Nanotheranostics: recent advances and perspectives. Small 12(36):4936–4954
MacEwan SR, Chilkoti A (2013) Harnessing the power of cell-penetrating peptides: activatable carriers for targeting systemic delivery of cancer therapeutics and imaging agents. Wiley Interdiscip Rev Nanomed Nanobiotechnol 5(1):31–48
Madani F, Lindberg S et al (2011) Mechanisms of cellular uptake of cell-penetrating peptides. J Biophys 2011:414729
Mae M, Rautsi O et al (2012) Tumour targeting with rationally modified cell-penetrating peptides. Int J Pept Res Ther 18(4):361–371
Mandal D, Nasrolahi Shirazi A et al (2011) Cell-penetrating homochiral cyclic peptides as nuclear-targeting molecular transporters. Angew Chem 50(41):9633–9637
Margus H, Padari K et al (2012) Cell-penetrating peptides as versatile vehicles for oligonucleotide delivery. Mol Ther 20(3):525–533
Mas-Moruno C, Rechenmacher F et al (2010) Cilengitide: the first anti-angiogenic small molecule drug candidate design, synthesis and clinical evaluation. Anti Cancer Agents Med Chem 10(10):753–768
Matsumura Y, Maeda H (1986) A new concept for macromolecular therapeutics in cancer chemotherapy: mechanism of tumoritropic accumulation of proteins and the antitumor agent smancs. Cancer Res 46(12 Pt 1):6387–6392
Michiue H, Sakurai Y et al (2014) The acceleration of boron neutron capture therapy using multi-linked mercaptoundecahydrododecaborate (BSH) fused cell-penetrating peptide. Biomaterials 35(10):3396–3405
Mitchell DJ, Kim DT et al (2000) Polyarginine enters cells more efficiently than other polycationic homopolymers. J Pept Res 56(5):318–325
Nakase I, Konishi Y et al (2012) Accumulation of arginine-rich cell-penetrating peptides in tumors and the potential for anticancer drug delivery in vivo. J Control Release 159(2):181–188
Neundorf I, Hoyer J et al. (2008) Cymantrene conjugation modulates the intracellular distribution and induces high cytotoxicity of a cell-penetrating peptide. Chem Commun (Camb)(43): 5604-5606
Neundorf I, Rennert R et al (2009) Fusion of a short HA2-derived peptide sequence to cell-penetrating peptides improves cytosolic uptake, but enhances cytotoxic activity. Pharmaceuticals (Basel) 2(2):49–65
Orzechowska EJ, Kozlowska E et al (2014) Controlled delivery of BID protein fused with TAT peptide sensitizes cancer cells to apoptosis. BMC Cancer 14:771
Prezma T, Shteinfer A et al (2013) VDAC1-based peptides: novel pro-apoptotic agents and potential therapeutics for B-cell chronic lymphocytic leukemia. Cell Death Dis 4:e809
Prochiantz A (1996) Getting hydrophilic compounds into cells: lessons from homeopeptides. Curr Opin Neurobiol 6(5):629–634
Raucher D, Ryu JS (2015) Cell-penetrating peptides: strategies for anticancer treatment. Trends Mol Med 21(9):560–570
Regberg J, Srimanee A et al (2012) Applications of cell-penetrating peptides for tumor targeting and future cancer therapies. Pharmaceuticals (Basel) 5(12):991–1007
Reichart F, Horn M et al (2016) Cyclization of a cell-penetrating peptide via click-chemistry increases proteolytic resistance and improves drug delivery. J Pept Sci 22(6):421–426
Reissmann S (2014) Cell penetration: scope and limitations by the application of cell-penetrating peptides. J Pept Sci 20(10):760–784
Rizzuti M, Nizzardo M et al (2015) Therapeutic applications of the cell-penetrating HIV-1 tat peptide. Drug Discov Today 20(1):76–85
Rothbard JB, Garlington S et al (2000) Conjugation of arginine oligomers to cyclosporin A facilitates topical delivery and inhibition of inflammation. Nat Med 6(11):1253–1257
Ryu JS, Raucher D (2014) Anti-tumor efficacy of a therapeutic peptide based on thermo-responsive elastin-like polypeptide in combination with gemcitabine. Cancer Lett 348(1–2):177–184
Savariar EN, Felsen CN et al (2013) Real-time in vivo molecular detection of primary tumors and metastases with ratiometric activatable cell-penetrating peptides. Cancer Res 73(2):855–864
Shen Y, Nagpal P et al (2014) A novel cell-penetrating peptide to facilitate intercellular transport of fused proteins. J Control Release 188:44–52
Shin MC, Zhang J et al (2014a) Cell-penetrating peptides: achievements and challenges in application for cancer treatment. J Biomed Mater Res A 102(2):575–587
Shin MC, Zhang J et al (2014b) Combination of antibody targeting and PTD-mediated intracellular toxin delivery for colorectal cancer therapy. J Control Release 194:197–210
Shirazi AN, Tiwari R et al (2013) Design and biological evaluation of cell-penetrating peptide-doxorubicin conjugates as prodrugs. Mol Pharm 10(2):488–499
Skotland T, Iversen TG et al (2015) Cell-penetrating peptides: possibilities and challenges for drug delivery in vitro and in vivo. Molecules 20(7):13313–13323
Splith K, Neundorf I et al (2010a) Influence of the metal complex-to-peptide linker on the synthesis and properties of bioactive CpMn(CO)3 peptide conjugates. Dalton Trans 39(10):2536–2545

Splith K, Hu W et al (2010b) Protease-activatable organometal-peptide bioconjugates with enhanced cytotoxicity on cancer cells. Bioconjug Chem 21(7):1288–1296

Splith K, Bergmann R et al (2012) Specific targeting of hypoxic tumor tissue with nitroimidazole-peptide conjugates. ChemMedChem 7(1):57–61

Suhorutsenko J, Eriste E et al (2012) Human protein 53-derived cell-penetrating peptides. Int J Pept Res Ther 18(4):291–297

Sun T, Zhang YS et al (2014) Engineered nanoparticles for drug delivery in cancer therapy. Angew Chem 53(46):12320–12364

Tanaka K, Kanazawa T et al (2013) Cytoplasm-responsive nanocarriers conjugated with a functional cell-penetrating peptide for systemic siRNA delivery. Int J Pharm 455(1-2):40–47

Tian R, Wang H et al (2015) Drug delivery with nano-spherical supramolecular cell penetrating peptide-taxol conjugates containing a high drug loading. J Colloid Interface Sci 453:15–20

Tints K, Prink M et al (2014) LXXLL peptide converts transportan 10 to a potent inducer of apoptosis in breast cancer cells. Int J Mol Sci 15(4):5680–5698

Ueda Y, Wei FY et al (2012) Induction of autophagic cell death of glioma-initiating cells by cell-penetrating D-isomer peptides consisting of Pas and the p53 C-terminus. Biomaterials 33(35):9061–9069

Vargas JR, Stanzl EG et al (2014) Cell-penetrating, guanidinium-rich molecular transporters for overcoming efflux-mediated multidrug resistance. Mol Pharm 11(8):2553–2565

Veiman KL, Kunnapuu K et al (2015) PEG shielded MMP sensitive CPPs for efficient and tumor specific gene delivery in vivo. J Control Release 209:238–247

Wang H, Liang J et al (2014a) Cell-penetrating apoptotic peptide/p53 DNA nanocomplex as adjuvant therapy for drug-resistant breast cancer. Mol Pharm 11(10):3352–3360

Wang HX, Yang XZ et al (2014b) Matrix metalloproteinase 2-responsive micelle for siRNA delivery. Biomaterials 35(26):7622–7634

Wang H, Zhao Y et al (2014c) Low-molecular-weight protamine-modified PLGA nanoparticles for overcoming drug-resistant breast cancer. J Control Release 192:47–56

Warso MA, Richards JM et al (2013) A first-in-class, first-in-human, phase I trial of p28, a non-HDM2-mediated peptide inhibitor of p53 ubiquitination in patients with advanced solid tumours. Br J Cancer 108(5):1061–1070

Wender PA, Galliher WC et al (2012) Taxol-oligoarginine conjugates overcome drug resistance in-vitro in human ovarian carcinoma. Gynecol Oncol 126(1):118–123

Xiang B, Dong DW et al (2013) PSA-responsive and PSMA-mediated multifunctional liposomes for targeted therapy of prostate cancer. Biomaterials 34(28):6976–6991

Xu X, Ho W et al (2015) Cancer nanomedicine: from targeted delivery to combination therapy. Trends Mol Med 21(4):223–232

Yang Y, Xie X et al (2014) PEGylated liposomes with NGR ligand and heat-activable cell-penetrating peptide-doxorubicin conjugate for tumor-specific therapy. Biomaterials 35(14):4368–4381

Yang Y, Xie X et al (2015) Dual-modified liposomes with a two-photon-sensitive cell penetrating peptide and NGR ligand for siRNA targeting delivery. Biomaterials 48:84–96

Yang Y, Xie X et al (2016a) Polymer nanoparticles modified with photo- and pH-dual-responsive polypeptides for enhanced and targeted cancer therapy. Mol Pharm 13(5):1508–1519

Yang Y, Xie X et al (2016b) Thermal and magnetic dual-responsive liposomes with a cell-penetrating peptide-siRNA conjugate for enhanced and targeted cancer therapy. Colloids Surf B Biointerfaces 146:607–615

Yeh TH, Che YR et al (2016) Selective intracellular delivery of recombinant arginine deiminase (ADI) using pH-sensitive cell penetrating peptides to overcome ADI resistance in hypoxic breast cancer cells. Mol Pharm 13(1):262–271

Youn P, Chen Y et al (2014) A myristoylated cell-penetrating peptide bearing a transferrin receptor-targeting sequence for neuro-targeted siRNA delivery. Mol Pharm 11(2):486–495

Zaro JL (2015) Lipid-based drug carriers for prodrugs to enhance drug delivery. AAPS J 17(1):83–92

Zaro JL, Shen WC (2015) Cationic and amphipathic cell-penetrating peptides (CPPs): their structures and in vivo studies in drug delivery. Front Chem Sci Eng 9(4):407–427

Zheng Z, Aojula H et al (2010) Reduction of doxorubicin resistance in P-glycoprotein overexpressing cells by hybrid cell-penetrating and drug-binding peptide. J Drug Target 18(6):477–487

Zhu L, Wang T et al (2013) Enhanced anticancer activity of nanopreparation containing an MMP2-sensitive PEG-drug conjugate and cell-penetrating moiety. Proc Natl Acad Sci U S A 110(42):17047–17052

Index

© Springer International Publishing AG 2017

A. Sunna et al. (eds.), *Peptides and Peptide-based Biomaterials and their Biomedical Applications*, Advances in Experimental Medicine and Biology 1030, https://doi.org/10.1007/978-3-319-66095-0

Printed by Printforce, the Netherlands